The Solar Engine and Its Influence on Terrestrial Atmosphere and Climate

NATO ASI Series

Advanced Science Institutes Series

A series presenting the results of activities sponsored by the NATO Science Committee, which aims at the dissemination of advanced scientific and technological knowledge, with a view to strengthening links between scientific communities.

The Series is published by an international board of publishers in conjunction with the NATO Scientific Affairs Division

A Life Sciences B Physics	Plenum Publishing Corporation London and New York
C Mathematical and Physical Sciences D Behavioural and Social Sciences E Applied Sciences	Kluwer Academic Publishers Dordrecht, Boston and London
F Computer and Systems Sciences G Ecological Sciences H Cell Biology I Global Environmental Change	Springer-Verlag Berlin Heidelberg New York London Paris Tokyo Hong Kong Barcelona Budapest

NATO-PCO DATABASE

The electronic index to the NATO ASI Series provides full bibliographical references (with keywords and/or abstracts) to more than 30000 contributions from international scientists published in all sections of the NATO ASI Series. Access to the NATO-PCO DATABASE compiled by the NATO Publication Coordination Office is possible in two ways:

- via online FILE 128 (NATO-PCO DATABASE) hosted by ESRIN, Via Galileo Galilei, I-00044 Frascati, Italy.
- via CD-ROM "NATO Science & Technology Disk" with user-friendly retrieval software in English, French and German (© WTV GmbH and DATAWARE Technologies Inc. 1992).

The CD-ROM can be ordered through any member of the Board of Publishers or through NATO-PCO, Overijse, Belgium.

Series I: Global Environmental Change, Vol. 25

The Solar Engine and Its Influence on Terrestrial Atmosphere and Climate

Edited by

Elizabeth Nesme-Ribes

CNRS, Observatoire de Paris
5, place Jules Janssen
F-92195 Meudon, France

Springer-Verlag
Berlin Heidelberg New York London Paris Tokyo
Hong Kong Barcelona Budapest
Published in cooperation with NATO Scientific Affairs Division

Proceedings of the NATO Advanced Research Workshop on the solar engine and its influence on terrestrial atmosphere and climate, held in Paris, France, October 25–29, 1993

ISBN-13: 978-3-642-79259-5 e-ISBN-13: 978-3-642-79257-1
DOI: 10.1007/978-3-642-79257-1

Library of Congress Cataloging-in-Publication Data. The solar engine and its influence on terrestrial atmosphere and climate / edited by Elizabeth Nesme-Ribes. p. cm. – (NATO ASI series. Series I, Global environmental change; vol. 25) "Proceedings of the NATO Advanced Research Workshop on the Solar Engine and its Influence on Terrestrial Atmosphere and Climate, held in Paris, France, October 25-29, 1993" – T. p. verso. Includes bibliographical references and index. ISBN 3-540-58417-X (Berlin: acid-free paper) 1. Solar activity–Congresses. 2. Solar radiation–Congresses. 3. Climatic changes–Congresses. I. Nesme-Ribes, Elizabeth, 1942- . II. NATO Advanced Research Workshop on the Solar Engine and its Influence on Terrestrial Atmosphere and Climate (1993: Paris, France) III. Series. QC883.2.S6S6 1994 551.5'271–dc20 94-34430

Softcover reprint of the hardcover 1st edition 1994

Typesetting: Camera ready by authors
SPIN 10100917 31/3130 - 5 4 3 2 1 0 - Printed on acid-free paper

PREFACE

The message of sunspots from the interior of the Sun to the Earth's climate

When Galileo was summoned before the Inquisition on April 12, 1633, the main accusations laid against him concerned the doubts he expressed about Aristotle's theory of the universe. Aristotle's idea was that the Earth was the centre of the cosmos and that all of the stars, including the Sun, turned around it. Moreover, for Aristotle and the world of the Inquisitors, the Sun was a perfect celestial body.

Now, Galileo had discovered spots on the Sun. These spots were seen as imperfections, and not just surface markings, but coming from within the Sun. Worse yet, they revolved around the Sun. All this supported the newfangled theory of Copernicus, and undermined a system of thought that had reigned supreme for centuries. Man of science that he was, and a prudent Catholic too, Galileo strived all his life to prove that Copernicus' astronomical concept was compatible with the word of the Bible. He proposed that there were not two truths but a single divine truth. It was just expressed in two different languages : there was the language of the common people, with its imprecision and inconsistencies, but intuitively understandable by everyone ; and then there was the precise language of science with its strict regard for observation, which only a chosen few can grasp [L. Geymonat. 1992]. As we know, Galileo was unable to convince the Catholic Church of this, and a bitter fight ensued that led to his trial in 1663, and the subsequent bannings of his works.

What are the consequences four centuries later ? These sunspots that were so unacceptable to Galileo's prosecutors contain a wealth of information that is as explosive today as it is precious, even beyond what Galileo suspected. Not only do they tell us how the Sun turns on its axis, but they continually vary in number, hinting at secrets yet untold. They turn out to be sites of intense magnetic fields, for example, and are the most direct indicators we have of variations inside the Sun. Solar rotation begets these magnetic fields, acting like a dynamo that converts kinetic energy into magnetic and thermal energy. What is more, sunspots and other manifestations of the magnetic field could very well be the reason why the thermal energy the Sun radiates toward Earth varies too. And if these energy variations have any long-term components and are large enough, this can result in climatic fluctuations here on Earth. So sunspots are still feeding grist to the scandal mill ; they are the most direct evidence we have of solar variability, and this variability may influence the Earth's climate.

If Galileo were to be indicted by a scientific court to-day, he would have more counts against him than ever : now there is the effect of the Sun's magnetism on the Earth's climate. Instead of questioning a time-honoured philosophy, he would be challenging entrenched scientific dogmas. Of course nobody doubts that the Sun is our source of life, and governs the climate as a whole. But the possibility that ''short" term solar variations might play a role in climate has been given short shrift. And yet, is there any other realistic way of predicting any changes Man may unwittingly make in the Earth's climate, if we neglect to take solar variability into account in the climatic model ?

To gain a better understanding of the Sun-climate interaction over periods of the order of a century, we must first take a look at what we know about the mechanisms governing the solar cycle. Since Galileo's time, a rich harvest of data has been garnered, most of it in the form of sunspot observations. This we can use to trace out the history of solar variability over all this time since.

What are the pertinent facts about the Sun ? The Sun does not rotate like a solid body. Solar magnetic activity derives its energy from the non-uniformity of solar rotation. Sunspots come and go periodically in 11-y and 80-y cycles, and possibly other periodicities too.

What would we like to know more about ? The cause for underlying non-uniform solar rotation and the solar dynamo ; the mechanism behind the Maunder minimum and its consequences in terms of solar energy output ; and, of course, a prediction scheme for these minima. Along this line of thought, solar-type stars could provide good analogies for our Sun.

If solar magnetic fields have any significant effects on the solar output, the question naturally arises as to how much solar forcing there is on the Earth's atmosphere and climate.
From the climatic end of the process, many questions have to be addressed. Among the most urgent are : the response of the Earth's atmosphere to solar perturbations, the response of the complex atmosphere-ocean-biosphere system, the reality of secular climatic fluctuations linked with solar cycles, and the capability of climatic models to correctly reproduce the observed climatic fluctuations.

The climate has always been a central human concern, and humanity has usually faced it with fatalism. As J. C. André (Météo-France) once pointed out, Man has always wanted to control the elements to command rain and fine weather. By changing his environment, even unintentionally, Man is somehow achieving this goal (Mouvement Universel de la Responsabilité Scientifique, special issue, 1992). But it is perhaps under the very impetus of his anthropocentric way of thinking that he tends to draw hasty conclusions about why our environment is getting warmer today. While it is true that much is at stake, and that the individual and society should be made aware of their responsibilities, should not we still soberly admit that there are two ways of talking about the question - that of the intuitive man in the street and that of the scientists - just like at the dawn of the Enlightenment ? Indeed, if we want to be truly aware of what is threatening our environment, we must take an objective look at all of the factors that might enter into the equation, with their uncertainties and their limitations.

From October 25 to 29, 1993, as part of the NATO environmental global change program, forty-nine scientists gathered at the Observatoire de Paris to make some headway in our understanding of climatic change.

If only we have heard and understood Galileo's plea, echoing to us from beyond the centuries, then surely this effort will bring us closer to some more unified idea of the truth.

E. Nesme-Ribes

References

L. Geymonat, La vie de Galilée, translation, 1992.
J. C. André, Les Cahiers du Mouvement Universel de la Responsibilité Scientifique, 1992

TABLE OF CONTENTS

THE SOLAR DYNAMO

Paul H. Roberts
Institute of Geophysics and Planetary Physics
University of California
Los Angeles, California 90024
USA

Summary

A summary of solar dynamo theory today is presented.

1. Background and general strategy

The main theme of this meeting is the relationship between solar activity and irradiance. It seems to be well established that the solar output at sunspot minimum is of order 0.1% smaller than at sunspot maximum. There is also evidence that the solar luminosity was approximately 0.5% less during the Maunder minimum and that this may be connected in some way with what is often loosely termed "the little ice age"; see Nesme–Ribes *et al* (1993, 1994) and Sadourny (1994). These observations provide a powerful motivation for understanding in detail the solar dynamo mechanism. Although there is no prospect of a completely satisfactory solar dynamo model in the immediate future, it is not beyond the bounds of possibility that the solar dynamo, and the dynamics of the solar convection zone (SCZ) where it works, will eventually be so well understood that variations in solar irradiance will be predictable. It is probable however that, like Earth's atmosphere, forward prediction will be limited to a short period, though hopefully one that is of longer duration than that of Earth's atmosphere! Whatever the truth may be, it is clear that the SCZ and the solar dynamo hold the key, and the primary purpose of this review is to summarize where solar dynamo theory stands today.

It is sometimes said that the equations governing the solar dynamo are well–known and that only their solution is difficult. This statement is, however, misleading and unhelpful. It is admittedly true that the solar dynamo is governed by the classical equations of fluid mechanics, electrodynamics and thermodynamics, all of which were well known in the nineteenth century. But these "primitive equations" govern physical processes that, in the context of the SCZ, have significant roles to play on scales that differ by very

NATO ASI Series, Vol. I 25
The Solar Engine and Its Influence
on Terrestrial Atmosphere and Climate
Edited by Elizabeth Nesme-Ribes

many orders of magnitude. It is necessary to parameterize the small scales in some way. Anyone who prefers instead to employ the primitive variables, the fluid velocity $\mathbf{V}$, the magnetic field $\mathbf{B}$, the temperature T, the specific entropy, $S, \ldots$, and the full equations governing these quantities is free to do so, but he must then deal with diffusivities ν of momentum, η of field and κ of heat, that are characteristically of order $2.10^{-5}\mathrm{m^2s^{-1}}$, $1\,\mathrm{m^2s^{-1}}$ and $10^{-3}\mathrm{m^2s^{-1}}$ respectively, and he must therefore contend with impossibly large Rayleigh numbers and other dimensionless parameters. The resulting turbulent flows would require him to strive for unattainably high numerical resolution. Sooner or later he would be forced to accept an engineering, or "mixing length", approximation. In short, the classical equations in their original form are suited for neither analytic nor numerical studies of the solar dynamo. They must be transformed in such a way that physical processes operating on widely different scales are explicitly separated. A set of reduced equations must be generated that, at one and the same time, are not only physically sound but also simple enough to be analytically and/or numerically solved.

The reduced, or "mean field", equations govern $\overline{\mathbf{V}}$, $\overline{\mathbf{B}}$, $\overline{T}$ and $\overline{S}$, which are the macroscale parts of $\mathbf{V}$, $\mathbf{B}$, T and S, ie the parts that vary on length and time scales comparable with, or larger than, some L_M and τ_M. They contain terms that represent the averaged influence of the "microscale fields" of length and time scales smaller than L_M and τ_M, and characteristically on scales $L_m \ll L_M$ and $\tau_m \ll \tau_M$. We denote these microscale parts of $\mathbf{V}$, $\mathbf{B}$, T and S by $\mathbf{v}$, $\mathbf{b}$, T' and S'. We may regard L_M and τ_M as "cut-offs", dictated by the available computing power and decreasing monotonically as that power increases. The more extensive the computer resources, the higher the numerical resolution attainable, the smaller the chosen L_M and τ_M can be, the greater the range of scales that can be included in the macroscale (and solved for), the fewer modes that need be included in the microscale, and the less influence the microscale motions, which we call "turbulence", has on the macroscale.

For example, with the computers available to him in the early 80s, Glatzmaier (1985a,b, 1987) could determine the state of the SCZ by a spectral transform method in which there were $M = 16$ vertical levels and all spherical harmonics up to $N = 31$. The horizontal spatial resolution was therefore $L_M = \pi R_\odot/N \approx 7.10^7$m, which is somewhat greater than the scale of the supergranulation (3.10^7m). With the greater computer resources available to him today, he can work with $M = 129$ and $N = 319$, so that

$L_M \approx 7.10^6$m, which is of the same order as the supergranulation scale but larger than the granulation scale (3.10^6m). In the extreme limit of the ultimate computer in the infinite future, he would be able to take $L_m = \tau_m = 0$, ie to work with the primitive equations themselves.

The crucial point to stress here is that, despite the ever-growing capability of computers, and the increasing availability of massively parallel machines, we are still today very far from the ultimate computer: there is much power in the microscale, and its effect on the macroscale is enormous. It is still not possible to make L_M and τ_M so small that the macroscale embraced the energy containing scales of the turbulence. In such circumstances, a common engineering device is to represent the effects of the microscale on the macroscale by turbulent diffusivities of order $L_m v_{\rm rms} \sim 6.10^9 \rm m^2 s^{-1}$, which is many orders of magnitude greater than the values of ν, η and κ quoted above. (In making this estimate, we have taken $L_m = 2.10^6$m, as for the granulation, and $v_{\rm rms} = 3.10^3 \rm m\ s^{-1}$ for the rms turbulent velocity, $v_{\rm rms}$.) The disparity between the molecular and turbulent diffusivities is so enormous that it is very hard to believe that the simple addition of turbulent diffusivities to the molecular diffusivities in the mean field equations can adequately represent the effects of the microscale on the macroscale; a satisfactory parameterization cannot be that simple. New terms arise because the turbulence is, for a number of reasons (compressibility, Coriolis force, Lorentz force, ...), neither isotropic nor homogeneous, and therefore cannot be adequately represented by isotropic transport coefficients. But even attempts to include anisotropy by replacing scalar turbulent diffusivities by tensor diffusivities are doomed to failure (see below).

To obtain the mean field equations, one substitutes

$$\mathbf{V} = \overline{\mathbf{V}} + \mathbf{v}, \quad \mathbf{B} = \overline{\mathbf{B}} + \mathbf{b}, \quad T = \overline{T} + T' \quad S = \overline{S} + S', \quad \ldots \qquad (1)$$

into the equations governing the magnetohydrodynamics (MHD) of the solar plasma, averages those equations, and attempts to find simple expressions for nonlinear combinations of $\mathbf{v}$, $\mathbf{b}$, T' and S', such as

$$\overleftrightarrow{\mathbf{R}} = -\overline{\mathbf{v}\mathbf{v}}, \qquad \overline{\mathbf{Q}} = \overline{T'\mathbf{v}}, \qquad \overline{\mathcal{E}} = \overline{\mathbf{v}\times\mathbf{b}}, \qquad (2)$$

through which the microscale controls transport processes on the macroscale; those shown in (2) are related to the Reynolds stresses ($\overline{\rho}\overleftrightarrow{\mathbf{R}}$), the turbulent heat flux ($\overline{\mathbf{F}} = \overline{\rho c_p}\overline{\mathbf{Q}}$) and the electromotive force (emf) created by the microscale. A number of methods have been devised to carry out this program. For example, the equations governing $\mathbf{v}$, $\mathbf{b}$, T' and S' are solved in some approximation and the necessary nonlinear combinations are formed and averaged. The approximation must have the effect of closing the turbulent hierarchy. The simplest analytic way to do this is first order smoothing (FOS), also known as "the second order correlation approximation"; see for example Krause and Rädler (1980) and Rüdiger (1989). Other approaches employ the direct interaction approximation (DIA) or other closure methods; for example, see Yoshizawa (1990) and Canuto *et al* (1994).

As already mentioned, many significant effects arise from the inhomogeneities and anisotropies of the turbulence. To quantify these by solving the equations governing $\mathbf{v}$, $\mathbf{b}$, T' and S' is not easy, even using FOS or DIA. Although some guidance is available from the parity of the fields involved (i.e. their behavior under coordinate reflection), which excludes certain terms from appearing in the required averaged nonlinear combinations, this guidance is of limited utility, since it gives no hint about the magnitude of the terms that are permitted in those combinations, and leaves very many possibilities open. Even using FOS or DIA, it is usually necessary to make further simplifying assumptions that may remove the theory from the realm of physical reality. For this reason, and also because of the ever increasing power of computers, it has become increasingly popular to carry out numerical experiments aimed at estimating averaged nonlinear combinations such as (2). The idea here is that, although it is impossible even with today's supercomputers to resolve all significant length and time scales of the SCZ, one could solve for the microscale fields in a smaller system, say of the dimensions of a supergranule, and by analyzing the results be able to determine what the correct parameterization of the microscale is. One would then be an immensely better position to solve the MHD of the macroscale fields and motions. For a recent example of this kind of approach, see Pulkkinen *et al* (1993).

Another method is to tune the theory to the observations; having decided on the form that the nonlinear combinations should take, one tries to remove uncertainties in their magnitudes by making the theory fit the solar data. This approach is fraught with danger. I don't know who first said, "Give me three parameters and I will fit a camel", but there is certainly a possibility that, by forcing a given form for the nonlinear combinations to lead to "predictions" consistent with the observations, one may be deluding himself: the correct form may be something quite different, an unfortunate fact that is revealed only much later when new data becomes available. Moreover, even if this approach is successful, the question arises of how much one has really explained; the problem has not been solved but has been transformed into something different, namely that of explaining why the tuned parameters have the values found to fit the solar data best. It was apparently to avoid such criticisms that Gilman and Miller (1981), Gilman (1983) and Glatzmaier (1985a,b, 1987) preferred to represent the effect of the microscale turbulence on the macroscale by scalar turbulent diffusivities determined, in Glatzmaier's case, by the local shearing rate.

2. Kinematic mean field theory

Fortunately, an easier sub–problem can be split off from the formidable difficulties raised by the full MHD of the solar dynamo, namely the so–called "kinematic problem", which attempts to find $\overline{\mathbf{B}}$ and the statistical moments of $\mathbf{b}$ when $\overline{\mathbf{V}}$ and the statistical moments of $\mathbf{v}$ are given. Some remarks concerning the full problem will be made in §§3 and 4. The first matter to discuss is that of *mean field electrodynamics*, which attempts to answer the key question, "What is Ohm's law for the mean electromagnetic field?" The primitive Ohm's law is

$$\mathbf{J} = \sigma(\mathbf{E} + \mathbf{V}\times\mathbf{B}), \tag{3}$$

the all important induction equation

$$\frac{\partial \mathbf{B}}{\partial t} = \nabla\times(\mathbf{V}\times\mathbf{B} - \eta\nabla\times\mathbf{B}), \qquad \nabla\cdot\mathbf{B} = 0, \tag{4, 5}$$

which, together with appropriate boundary conditions, determines dynamo action. Here $\mathbf{E}$ is electric field, $\mathbf{J}$ is the electric current density, σ is the electrical conductivity, $\eta = 1/\mu\sigma$ is the magnetic diffusivity and μ is the permeability, assumed constant and that of free space. (SI units are used.) When (2) – (5) are averaged, they become

$$\overline{\mathbf{J}} = \sigma(\overline{\mathbf{E}} + \overline{\mathbf{V}}\times\overline{\mathbf{B}} + \overline{\mathcal{E}}), \tag{6}$$

and

$$\frac{\partial \overline{\mathbf{B}}}{\partial t} = \nabla\times(\overline{\mathbf{V}}\times\overline{\mathbf{B}} + \overline{\mathcal{E}} - \eta\nabla\times\overline{\mathbf{B}}), \qquad \nabla\cdot\overline{\mathbf{B}} = 0. \tag{7, 8}$$

The key question is now seen to be, "What is $\overline{\mathcal{E}}$?" A little consideration of the equation satisfied by $\mathbf{b}$, which is obtained by subtracting (7) from (4), soon convinces one that $\overline{\mathcal{E}}$ is[1] a linear functional of $\overline{\mathbf{B}}$ where, by using the word 'functional' instead of 'function', we imply that its value, at a particular point $\mathbf{x}$ and time t, depends on $\overline{\mathbf{B}}$ at other $\mathbf{x}$ and earlier t. Nevertheless, the situation is so complicated that usually the functional is expanded in terms of its value and its space and time derivatives at $\mathbf{x}$ and t. The resulting model is then said to be "local". (Similarly, when a model of turbulence is described as 'local', it is implied that all nonlinear combinations such as $\overline{\mathcal{E}}$ at $\mathbf{x}$ and t are expressed in some way as a truncated Taylor series.)

At first it seemed that turbulence could only make dynamo action more difficult. If $\eta = 0$, (4) leads to Alfvén's theorem, which shows that $\mathbf{B}$ is swept passively around with the fluid just as $\mathbf{V}$ dictates. It is known that when a passive scalar field, such as temperature in a thermally insulating fluid, is swept passively around by a turbulent fluid motion, it diffuses rapidly and, in isotropic conditions, this may be represented by a turbulent thermal diffusivity. Even when the fluid is not thermally insulating, but has a finite molecular diffusivity, this is often much smaller than the turbulent diffusivity. It was pointed out by Sweet (1950) that the same process should diffuse the mean magnetic field in a turbulently moving conductor. The point is made mathematically like this: if for simplicity one sets $\mathbf{V} = 0$ in (7) and supposes η is constant, one has

$$\frac{\partial \overline{\mathbf{B}}}{\partial t} = \nabla \times \overline{\mathcal{E}} + \eta \nabla^2 \overline{\mathbf{B}}. \tag{9}$$

If turbulent diffusion is to have the same general effect as molecular diffusion, represented by the last term in (9), one must have

$$\overline{\mathcal{E}} = -\beta \nabla \times \overline{\mathbf{B}}, \tag{10}$$

and then (9) becomes

$$\frac{\partial \overline{\mathbf{B}}}{\partial t} = \eta_t \nabla^2 \overline{\mathbf{B}}, \tag{11}$$

[1] For simplicity, we have excluded here the very real possibility that the induction equation governing $\mathbf{b}$ can support indefinitely a purely microscale dynamo, ie one in which $\overline{\mathbf{B}} \equiv 0$; see Kazantsev (1968), Meneguzzi *et al* (1981), Meneguzzi and Pouquet (1989), Novikov *et al* (1983), Molchanov *et al* (1985) and Nordlund *et al* (1992). Such microscale dynamos are of considerable interest but, by definition, cannot explain the existence of large scale fields, $\overline{\mathbf{B}}$, on the Sun.

just as though η were replaced by a total diffusivity $\eta_t = \eta + \beta$. The argument is very similar to the one that Reynolds used to replace the Reynolds stresses by an enhanced viscous term.

Sweet's argument though the truth, and nothing but the truth, is not the whole truth. Vector fields that move passively do not behave like scalar fields that move passively. Even when the turbulence is isotropic, its statistical properties need not be mirror-symmetric; it may be cyclonic (Parker, 1955), ie possesses helicity, $H = \overline{\mathbf{v}\cdot\boldsymbol{\omega}}$ where $\boldsymbol{\omega}$ is the microscale vorticity, the primitive vorticity being $\boldsymbol{\Omega} = \nabla\times\mathbf{V} = \overline{\boldsymbol{\Omega}} + \boldsymbol{\omega}$. In this case (10) is replaced by

$$\overline{\mathcal{E}} = \alpha\overline{\mathbf{B}} - \beta\nabla\times\overline{\mathbf{B}}, \tag{12}$$

and so enters the so-called "alpha–effect". The form (12) correctly suggests that the functional $\overline{\mathcal{E}}$ has been Taylor expanded. The ratio of successive terms in this expansion are, on the face of it, of order $L_m/L_M \ll 1$, so that the series converges rapidly. Indeed, one may at first wonder why just the first term of (12) will not suffice by itself. But the first term depends on the degree to which the turbulence lacks mirror-symmetry, or more precisely on the relative helicity $h = H/(v_{\rm rms}\omega_{\rm rms})$, where $v_{\rm rms} = \sqrt{(\overline{v^2})}$ and $\omega_{\rm rms} = \sqrt{(\overline{\omega^2})}$. It may happen that h is small and the ratio of the two terms on the right–hand side of (12), which is of order L_m/hL_M, is then not necessarily small. A little consideration of the way that convection and rotation creates helicity in turbulent flows (see below) soon convinces one that α should take opposite signs in opposite hemispheres of the Sun. Moreover, because of the compressibility of the solar plasma, rising parcels of fluid will expand, sinking parcels will compress. Since both will tend to conserve their angular momentum as they do so, H should be negative in the Northern hemisphere (and positive in the Southern); see Steenbeck *et al* (1966). Elementary arguments correctly show that α and H are anticorrelated: α should be positive in the Northern hemisphere of the Sun (and negative in the Southern).

Yoshizawa (1990) has recently questioned whether even (12) suffices. He has argued that more generally

$$\overline{\mathcal{E}} = \alpha\overline{\mathbf{B}} - \beta\nabla\times\overline{\mathbf{B}} + \beta_K\nabla\times\overline{\mathbf{V}}. \tag{13}$$

The new coefficient β_K vanishes with $\mathbf{B}$ and depends on the $\mathbf{B}$; it is a matter more properly left to §4.

The axisymmetry of the Sun's macroscale structure implies that all coefficients, such as α and β, are independent of longitude, ϕ. Thus solutions in which $\overline{\mathbf{B}}$ is permanently axisymmetric with respect to the polar axis Oz exist. Because of the alpha term in (12) such fields can last forever, i.e. self-sustaining dynamo action is possible with an axisymmetric $\overline{\mathbf{B}}$. Supplemented by choices of the solar differential rotation, $\mathbf{V} = r\sin\theta\,\overline{\Omega}(r,\theta)\mathbf{1}_\phi$, (12) has been the basis of many dynamo models of $\alpha\Omega-$type which successfully reproduce the observed solar activity cycle. [Here (r,θ,ϕ) are polar coordinates; $\mathbf{1}_q$ is a unit vector in the direction of increasing coordinate q.] The activity correctly progresses from poles to equator provided $\alpha > 0$ in the Northern hemisphere of the Sun, and provided that Ω predominantly increases with depth [ie $d\overline{\Omega}(r)/dr < 0$]; see for example Steenbeck and Krause (1969), Roberts (1972), Stix (1976), Parker (1979).

The question naturally arises,"How are H and α created and maintained?" Unless they have a source they will surely decay. The preferred method is through Coriolis forces and buoyancy, although, as Steenbeck *et al* (1966) showed in their seminal paper, they can also be created by rotation and a gradient of turbulent intensity. It is therefore necessary that the statistics of the turbulence have one or more preferred directions (ie be anisotropic) and to depend on position (ie be inhomogeneous). With these additions, the floodgates open and the difficulties multiply, even within the framework of mean field electrodynamics alone, let alone the full MHD problem! Much of the book by Krause and Rädler (1980) is devoted to just these matters, and it cannot be said that they have exhausted their subject! Considering first the effect of inhomogeneities, it is plausible that (12) should be replaced by

$$\overline{\mathcal{E}}_i = a_{ij}\overline{B}_j + b_{ijk}\nabla_k\overline{B}_j, \tag{14}$$

and the question arises,"What values do the tensors a_{ij} and b_{ijk} take?" A partial answer is

$$a_{ij} = \gamma\epsilon_{ijk}g_k + \alpha_1(\mathbf{g}\cdot\overline{\mathbf{\Omega}})\delta_{ij} + \alpha_2\overline{\Omega}_i g_j + \alpha_3\overline{\Omega}_j g_i + \ldots, \tag{15}$$

$$b_{ijk} = \beta\epsilon_{ijk} + \beta_1\overline{\Omega}_k\delta_{ij} + \beta_2\overline{\Omega}_j\delta_{ik} + \hat{\beta}\epsilon_{ijl}\overline{D}_{lk} + \ldots, \tag{16}$$

where $\mathbf{g}$ is the acceleration due to gravity and $D_{ij} = \frac{1}{2}(\nabla_i V_j + \nabla_j V_i)$ is the rate of strain tensor. Only terms linear in $\overline{\mathbf{\Omega}}$ and $\mathbf{g}$ are retained in (15) and (16). Other terms, represented by ..., are not only possible but likely if the Coriolis forces and buoyancy forces are strong; see Rüdiger & Kichatinov[2]

[2] To avoid confusion, we are using in this review the original English transliteration of his name rather than the more recent 'Kitchatinov'.

(1993). [A possible contribution of $\beta_3\overline{\Omega}_i\delta_{jk}$ has been omitted from (16); by (8) and (14), it would make no contribution to $\overline{\mathcal{E}}$.]

The terms in (15) and (16) may be interpreted as follows. The γ term represents the pumping of mean flux by the turbulent motions; see Drobyshevskii & Yuferev (1974), Moffatt (1974), Krause & Rädler (1980). It makes a contribution of $-\gamma\mathbf{g}\times\overline{\mathbf{B}}$ to $\overline{\mathcal{E}}$; this is quite similar to the term $\overline{\mathbf{V}}\times\overline{\mathbf{B}}$ appearing in (6) and (7) but $-\gamma\mathbf{g}$ is, unlike $\overline{\mathbf{V}}$, not restricted by a continuity equation[3]. The α_1 term is the genuine alpha effect sought, with $\alpha = \alpha_1(\mathbf{g}\cdot\overline{\boldsymbol{\Omega}})$ which takes opposite signs in opposite hemispheres. This term is attached to two others, the α_2 and α_3 terms, which are of the same order of magnitude. In other words, the α–effect is necessarily anisotropic. The β term in (16) accounts for the enhanced turbulent diffusivity. It dominates the β_2 and β_3 terms which otherwise would be capable of sustaining a dynamo; see §16.5 of Krause & Rädler (1980). The final term in (16) is included to show that, not only the rotational part of the mean motion, represented by $\overline{\boldsymbol{\Omega}}$, but also the rate of strain in the mean flow, represented by $\overline{D}_{ij}$, contributes to the mean emf; see Eq.(191) of Krause (1967) and Roberts & Soward (1975). Although (see above) we are excluding nonlinear terms, it is perhaps worth noting that β is so dominant in (16) that such terms will first have significance in modifying β. These are due to stratification or rotation and appear as additions to b_{ijk} proportional to $\epsilon_{ilk}g_jg_l$ and $\epsilon_{ilk}\overline{\Omega}_j\overline{\Omega}_l$; see Kichatinov *et al* (1993).

The constants γ, α_i, β, β_j, $\hat{\beta}$ appearing in (15) and (16) can be evaluated by applying FOS, by other approximations such as two–time scale methods (Roberts & Soward, 1975), by numerical experiments, or by tuning to fit the observed facts of solar activity. For recent discussions about the form of a_{ij} and b_{ijk}, see Rüdiger and Kichatinov (1993) and Kichatinov *et al* (1993).

In all this discussion, we have focussed on purely zonal mean flow, ie the solar differential rotation: $\overline{\mathbf{V}} = s\overline{\Omega}(r,\theta)$. The question of whether, in addition, a meridional flow exists on the solar surface (and presumably also at every depth in the SCZ) has, at various times, been the subject of intense debate. Such flows are at most of the order of a few tens of meters per second, and are therefore at the edge of detectability. The matter should be settled for once and all when definitive results emerge from the GONG

---

[3] The first dynamo model to include magnetic buoyancy in parameterised form was that of Leighton (1969). See also Moss *et al* (1990a, b).

program. Meanwhile (Hathaway 1991; Komm *et al* 1993), it seems that such flows do exist of order 10m/s, but at most of order 30m/s. It is worth emphasizing that even quite small meridional mean flows are significant for the dynamo process (eg see Roberts, 1972). Their main effect is to advect toroidal and poloidal field lines; they therefore cannot substitute for the α−effect in the dynamo process, but they can carry flux to positions favorable for regenerating field: for example, toroidal flux generated by intense shear can be carried by meridional motion to locations where α is large and *vice versa.* This requires that the sense of meridional circulation is in right direction, and that its strength is sufficient, ie is such that the time of meridional overturning is of the same order as the magnetic diffusion timescale, $R_\odot^2/\eta$.

3. Objections to early mean field theory

Even in 1984, it was widely believed that the basics of the solar dynamo and its activity had been uncovered by the mean field models referenced in §2. To be sure, there were some difficulties. Steenbeck and Krause (1969) had derived

$$\alpha = -\tfrac{1}{3}H\tau_m, \qquad \beta = \tfrac{1}{3}v_{\rm rms}^2\tau_m, \tag{17}$$

essentially by using FOS, which for its validity requires that either the microscale Reynolds number, $R_m = v_{\rm rms}L_m/\eta$, or the microscale Strouhal number, $S_m = v_{\rm rms}\tau_m/L_m$, is small. In the case of (17), it is assumed that $R_m \gg 1$ so that FOS cannot give reliable results unless $S_m \ll 1$. But, almost by definition, $S_m = O(1)$ in a turbulent flow. Thus the values of α and β given by (17) are unreliable, and are at best estimates. There were indications that (17) led to values of α and η that were too large (Köhler, 1973). These indications came from tuning: it was necessary to reduce estimates (17) in order to replicate the time scales of the solar activity cycle. In the independence of (17) on η, Steenbeck and Krause (1969) provided a foretaste of fast dynamo theory, which has been the subject of intense study over the past decade. See Childress (1992), Roberts & Soward (1992), Gilbert *et al* (1993), Soward (1993), Zheligovsky (1993), where references to earlier papers are given.

A more serious difficulty arose as a result of the numerical integrations of Gilman & Miller (1981) and Gilman (1983) who solved for both **B** *and* **V**, ie they solved the full MHD equations albeit in Boussinesq form and with enhanced (turbulent) diffusion coefficients. They found that $\overline{\Omega}$ on the surface of the Sun agreed approximately with observations but that beneath the surface it was approximately constant on cylinders (at least outside the so–called "tangent cylinder" that touches the radiative core at its equator), ie $\overline{\Omega} = \overline{\Omega}(s)$ where s is distance from the rotation axis. Below the equator, it therefore decreased weakly with depth, and so, as Gilman (1983) confirmed, the activity cycle moved in the wrong direction, ie from equator to poles; see also Köhler (1973) and Yoshimura (1975). It at first seemed possible that the omission of compressibility had led to these troubles, but that loophole

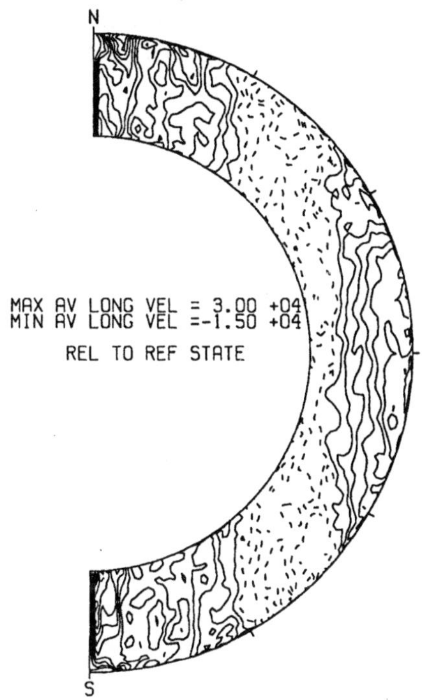

Fig 1. A simulation of the solar convection zone by Glatzmaier (unpublished). The numerical integrations were performed using a spectral transform method at high resolution (see text). The macroscale Rayleigh, Ekman and Prandtl numbers were 3.10^9, 5.10^{-6} and 0.05, respectively.

was effectively plugged by Glatzmaier (1985b) who found qualitatively that an anelastic model, which allows for the compressibility of the solar plasma, behaved in much the same way. The two–dimensionality of $\overline{\Omega}$ outside the tangent cylinder in this model was confirmed in recent integrations by Glatzmaier, one result of which is shown in Fig. 1. He obtained this at higher resolution ($M = 129$, $N = 319$; see §1) than he had been able to use in his earlier integrations ($M = 16$, $N = 31$). As a result, much more of the irregularity associated with the turbulence is evident; compare Fig. 1 with Fig. 2a of Glatzmaier (1985b). There remained a further possibility. Being quite unable (see above) to resolve even the supergranulation numerically, Gilman, Miller and Glatzmaier had had to include the effect on the mean fields of all scales from the supergranulation downwards by parameterizing the turbulent transport. They had done this in the simplest possible way: by enhancing scalar diffusivities such as ν, κ and η to large values appropriate to the turbulent shear. Perhaps this was too simplistic a way of dealing with the complications of solar turbulence, which is surely neither isotropic nor homogeneous? Possibly their integrations proved (see §1) that a more sophisticated parameterization of the turbulence was called for? There was therefore a tendency to continue to believe the kinematic $\alpha\omega$–models, despite the advances made by Gilman, Miller and Glatzmaier.

In 1984, clouds gathered on the horizon that could not be ignored, for they were new data from helioseismology. First, Duvall and Harvey (1984) and Duvall *et al* (1984) showed, from an analysis of the splitting of p–modes, that $\overline{\Omega}$ decreased weakly with increasing depth below the solar equatorial plane. In fact their findings were consistent with those of Gilman, Miller and Glatzmaier, but not with the mean field models. The old idea that $\partial\overline{\Omega}/\partial r < 0$ was no longer tenable. Worse news followed. Dziembowski *et al* (1989) demonstrated that, through most of the convection zone, a good approximation for $\overline{\Omega}$ would be $\overline{\Omega} = \overline{\Omega}(\theta)$, and this choice contradicts both the numerical work of Gilman, Miller and Glatzmaier and the simple $\alpha\Omega$–models that had so successfully reproduced the activity cycle. Nevertheless, the radiative core of the Sun appeared to be rotating uniformly, at least immediately below the so-called "overshoot layer", where convection from above penetrates into this otherwise stable region, and there therefore had to be strong radial gradients in $\overline{\Omega}$ within them. This makes the layer an attractive site for the Ω–source of an $\alpha\Omega$–dynamo, as was apparently first noted by Durney (1976). There has therefore been considerable interest in the dynamics of this region. See, for example, Pidatella and Stix (1986), where earlier references are given. Anyone who wonders how a large velocity shear can possibly be a permanent non-diffusing feature of the solar interior should consider the mechanism proposed by Spiegel and Zahn (1992). The reason why $\overline{\Omega}$ should be uniform at the top of the radiative core is also mysterious, although perhaps Spruit (1990) has the most promising explanation.

The uniform angular velocity, $\overline{\Omega}_c \approx 2.74\ 10^{-6}\mathrm{s}^{-1}$ of the core is intermediate between the surface angular velocity, $\overline{\Omega}_p = \overline{\Omega}(R_\odot, 0) \approx \overline{\Omega}(R_\odot, \pi) \approx 1.99\ 10^{-6}\mathrm{s}^{-1}$, at the poles and the greater surface angular velocity, $\overline{\Omega}_E = \overline{\Omega}(\frac{1}{2}\pi) \approx 2.78\ 10^{-6}\mathrm{s}^{-1}$, of the solar equator. This means that $\partial\overline{\Omega}/\partial r > 0$ in the overshoot layer near the equator of the radiative zone. It was argued above that, because of the expansion (contraction) of a rising (sinking) parcel of fluid in the bulk of the SCZ, $H < 0$ (> 0) in its northern (southern) hemisphere, suggesting that $\alpha > 0$ (< 0). If this were also true for the dynamo operating at the base of the SCZ, its magnetic activity would be towards mid-latitudes both from the poles and the equator. The integrations of the MHD dynamo equations indicate however that $\alpha < 0$ at great depth in the Northern SCZ. This is because downwelling (upwelling) fluid has to diverge (converge) as it approaches (leaves) the base of the SCZ, and the resulting Coriolis forces produce positive helicity; correspondingly, the helicity at depth in the Southern hemisphere is negative. The magnetic activity therefore moves from mid-latitudes towards poles and equator, in agreement with the observed polar motion of faculae at high latitudes and the equatorward moving sunspot activity in mid to low latitudes. It should be remarked that recent simulations of magnetoconvection in a rotating compressible fluid suggest that only downwelling plumes occur; see Vincent and Meneguzzi (1991), Nordlund *et al* (1992). Another reason for a change in sign of α is given by Rüdiger and Kichatinov (1993).

There are other theoretical attractions in placing the main seat of regeneration near the base of the SCZ. First, there has always been a difficulty in explaining how a toroidal field, strong enough to account for sunspots, could be held within the SCZ for a decade in the face of magnetic buoyancy (eg see Parker, 1979). The possibility of downward pumping was raised above, but it is still unclear how effective that mechanism would be; see Brandenburg *et al* (1990b). By manufacturing the flux tubes as deep as possible in the Sun, their inevitable rise and eruption in sunspots is delayed (Spiegel and Weiss, 1980; Van Ballegooien, 1982). Second, estimates of α have been consistently too large (see above). Because the density at the base of the SCZ is large in Glatzmaier's compressible models, momentum conservation implies smaller velocities and hence yields the smaller α-effect required. This was demonstrated in Glatzmaier's (1985b) paper which, by means of a drastic simplification, successfully simulated a dynamo operating at the base of the SCZ. The creation of α by magnetostrophic instabilities at the base of the SCZ has been investigated by Schmitt (1984, 1985, 1987) and Prautzsch (1993). A dynamo wave operating on the surface separating a region of nonzero $\overline{\Omega}$–gradients but zero α from a region of nonzero α but zero $\overline{\Omega}$–gradients has recently been investigated by Parker (1993). Rüdiger & Brandenburg (1993) have constructed an $\alpha\Omega$–model that incorporates an $\overline{\Omega}(r,\theta)$ that is consistent with the findings of helioseismology and which makes use of an anisotropic α as suggested by the mean field analysis of Rüdiger and Kichatinov (1993). The results are very promising.

4. Mean field hydrodynamics. Back reaction of B. Quenching. Mean field MHD

Kinematic theory is unsatisfactory from several points of view. It provides no mechanism by which the dynamo can equilibrate; either **B** dies out, or it grows without limit, or it balances precariously on the knife–edge of the "marginal state" between. It provides no insight into the form of **V**. Its only virtue lies in its linearity, and therefore comparative simplicity.

Some of the features of the full MHD problem — or "self–consistent problem" as it is often called — can be exhibited most simply for a uniform incompressible fluid. It is then convenient to measure **B** in velocity units, ie $\mathbf{B}/\sqrt{(\mu\rho)}$, where ρ is the fluid density, is re-written as **B**. The coupled equations governing the mean fluid velocity and mean magnetic field can then be written in the symmetric forms (Yoshizawa, 1990)

$$\frac{\partial\overline{\mathbf{V}}}{\partial t} + \nabla\cdot(\overline{\mathbf{V}}\,\overline{\mathbf{V}} - \overline{\mathbf{B}}\,\overline{\mathbf{B}}) = -\nabla P_M + \nabla\cdot\overleftrightarrow{\mathbf{R}}^K + \nu\nabla^2\overline{\mathbf{V}}, \tag{18}$$

$$\frac{\partial\overline{\mathbf{B}}}{\partial t} + \nabla\cdot(\overline{\mathbf{V}}\,\overline{\mathbf{B}} - \overline{\mathbf{B}}\,\overline{\mathbf{V}}) = -\nabla\cdot\overleftrightarrow{\mathbf{R}}^M + \eta\nabla^2\overline{\mathbf{B}}, \tag{19}$$

$$\nabla\cdot\overline{\mathbf{V}} = 0, \qquad \nabla\cdot\overline{\mathbf{B}} = 0, \tag{20, 21}$$

where ρP_M is the total pressure (including the magnetic pressure) and ν is the kinematic viscosity, while $\overleftrightarrow{\mathbf{R}}^K$ and $\overleftrightarrow{\mathbf{R}}^M$ are the kinetic (Reynolds) and magnetic correlation tensors:

$$\overleftrightarrow{\mathbf{R}}^K = -(\overline{\mathbf{vv} - \mathbf{bb}}), \qquad \overleftrightarrow{\mathbf{R}}^M = -(\overline{\mathbf{vb} - \mathbf{bv}}). \tag{22, 23}$$

The magnetic correlation tensor is simply related to $\overline{\boldsymbol{\mathcal{E}}}$ by

$$\overline{\mathcal{E}}_i = -\tfrac{1}{2}\epsilon_{ijk}R^M_{jk}, \quad \text{or} \quad R^M_{ij} = -\epsilon_{ijk}\overline{\mathcal{E}}_k. \tag{24, 25}$$

In a similar way to §2, we may solve the equations governing $\mathbf{b}$ and $\mathbf{v}$ in some approximation that "closes" the turbulence hierarchy and then evaluate $\overleftrightarrow{\mathbf{R}}^K$ and $\overleftrightarrow{\mathbf{R}}^M$. The resulting subject is sometimes called *mean field MHD*; when $\mathbf{B} \equiv 0$, it is called *mean field hydrodynamics* or *mean field fluid mechanics.* Then only (18) and (20) arise.

One of the main thrusts of mean field hydrodynamics is to determine the solar differential rotation, a central issue about which books have been written (Durney & Sofia 1987; Rüdiger 1989). It would serve no useful purpose to write at length about the subject here, the more especially since Rüdiger and Kichatinov are expounding it in their contribution to these Proceedings. We should observe however, because we need to refer to it later, that $\overleftrightarrow{\mathbf{R}}^K$ contains an analogue of the α−effect that appears in $\overleftrightarrow{\mathbf{R}}^M$, namely the Λ−effect which accounts for the first term on the right-hand side of the simple ansatz

$$R_{ij}^K = \Lambda_{ijk}\overline{\Omega}_k + 2\nu_{ijkl}^T\overline{D}_{kl}, \tag{26}$$

where the second term represents the turbulent viscous stress, which here has a tensor character because of the inhomogeneity of the turbulence. Rüdiger (1989) gives reasons why expressions such as

$$\Lambda_{ijk} = \Lambda_V(\epsilon_{ipk}g_j + \epsilon_{jpk}g_i)g_p + \Lambda_H(\mathbf{g}\cdot\overline{\mathbf{\Omega}})(\epsilon_{ipk}\overline{\Omega}_j + \epsilon_{jpk}\overline{\Omega}_i)g_p \tag{27}$$

are likely to arise and be significant. The Λ−effect is related to Anisotropic Kinetic Alpha (AKA) instability; see Frisch *et al* (1987), Sulem *et al* (1989). It also emerges from the recent analysis of stellar turbulence by Canuto *et al* (1994). The differential rotation in the SCZ excited by the Λ−effect has recently been analyzed by Kichatinov and Rüdiger (1993).

The differential rotation is really a thermal wind; it cannot be explained without including a buoyancy force in (18), introducing an equation of thermal balance, and interpreting the resulting set of equations as a Boussinesq approximation. This leads to a combination of mean field hydrodynamics and *mean field thermodynamics*; see Rüdiger (1989). The effect of the microscale convection, through term $\overline{\mathbf{F}} = \rho c_p\overline{T'\mathbf{v}} = \rho c_p\overline{\mathbf{Q}}$ dominates the macroscale heat flux and must be parameterized in some way. By analogy with the Reynolds stresses or with the mean emf (see §2), one might suppose that a law of Fourier form ($\overline{\mathbf{Q}} = -\kappa_t\nabla\overline{T}$) would suffice, but because of the anisotropy of the turbulence, $\overline{\mathbf{Q}}$ must be related to $\nabla\overline{T}$ tensorially:

$\overline{Q}_i = -\kappa_{ij}\nabla_j\overline{T}$; see Brandenburg **et al** (1992). In their article in these Proceedings, Rüdiger and Kichatinov suggest that this anisotropy is so significant that it resolves the disagreement between the computationally deduced $\overline{\Omega}(s)$ and the helioseismologically deduced $\overline{\Omega}(\theta)$.

To use the Boussinesq equations to describe the SCZ is an oversimplification. Strictly the compressibility of the SCZ should be recognized, most conveniently by replacing (18) – (21) with anelastic equations, including a buoyancy force depending on the excess temperature gradient $\nabla\overline{T} - \nabla\overline{T}_{\rm ad}$ where $\nabla\overline{T}_{\rm ad}$ is the adiabatic gradient. The ansatz just given for the Boussinesq fluid is then replaced by one in which the excess entropy gradient appears in place of the excess temperature gradient: $\overline{Q}_i = -\kappa_{ij}(\nabla_j\overline{T} - \nabla_j\overline{T}_{\rm ad})$.

A new analysis of mean field hydro- and thermo-dynamics has recently been presented by Canuto *et al* (1994) who consider that Gilman, Miller and Glatzmaier took a giant leap when they adopted a Boussinesq ansatz for the Reynolds stresses, an ansatz that originated from the instability of shear flows, and not from buoyancy–driven turbulence. Canuto *et al* (1994) seek to determine $\overleftrightarrow{\mathbf{R}}^K$ and $\overline{\mathbf{Q}}$ in terms of the mean rate of strain tensor $\overleftrightarrow{\mathbf{D}}$, the local angular velocity $\overline{\mathbf{\Omega}}$, the excess temperature gradient $\boldsymbol{\beta}$, and three parameters that describe the turbulence: the kinetic energy density per unit mass $\frac{1}{2}\overline{v^2}$, the energy dissipation rate per unit mass $\nu\overline{\omega^2}$, and the temperature variance $\overline{T'^2}$. In addition to recovering a Λ–effect, they claim to be able to explain *quantitatively* the observed correlation, $\overline{v_\theta v_\phi} = -R^K_{\theta\phi}$, on the surface of the Sun, as deduced from the solar data by Gilman & Howard (1984).

After this brief excursion into the non-magnetic problem, let us now suppose that $\overline{\mathbf{B}} \neq 0$. While $\overline{\mathbf{B}}$ remains "small", it is governed by the kinematic theory of §2 but, if that theory predicts its exponential growth, the time will come when terms in (18) and (19) that depend quadratically on the field will become important and prevent the further growth of field, ie the field amplitude equilibrates, although it may still be time dependent. This is sometimes described as "quenching" dynamo action, or as field "saturation". The most obvious way this can happen is through the $\overline{\mathbf{B}}\,\overline{\mathbf{B}}$ term on the left–hand side of (18), which reduces and modifies $\overline{\mathbf{V}}$, so that what was, according to the kinematic equation (19), a supercritical configuration becomes a marginal state. When saturation results from the changes in $\overline{\Omega}$ created by the Lorentz force $\overline{\mathbf{J}}\times\overline{\mathbf{B}}$, it is sometimes said that "the Malkus–Proctor mechanism" (Malkus and Proctor 1975) operates. An example that is particularly intriguing in the context of the present meeting is that of Brandenburg *et al* (1989b); see also Brandenburg *et al* (1988). This suggests that at times when the solar magnetic field has small energy, such as may be true during the Maunder minimum, its parity will be nearly quadrupolar; see Fig. 3 of Brandenburg *et al* (1989b).

Another way that field equilibration is brought about is through the effect of the fluctuating fields on the stresses $\overleftrightarrow{\mathbf{R}}^K$ and $\overleftrightarrow{\mathbf{R}}^M$ (or $\overline{\mathcal{E}}$). For example, we may imagine that

$$\alpha = \frac{\alpha_0}{1+\overline{B}^2/B_0^2}, \tag{28}$$

where α_0 defines the α-effect for very weak (zero) field and B_0 is a constant which, in the solar context, is of the order of the equipartition field, $B_{\rm eq} = \sqrt{(\overline{v^2})}$ in velocity units. The quenching law (28) is appealing on simple physical grounds: the presence of a field "stiffens" the medium, ie makes it less responsive to the generation of helicity by cyclonic motions, and so reduces H and α. Several simple axisymmetric mean field dynamo models have been integrated that make use of (28). See for example Jepps (1975), Rädler *et al* (1990), Moss *et al* (1991b), Brandenburg *et al* (1989a). As might be expected, the decrease in α with increasing $\overline{B}$ suffices to prevent exponential growth and, if the dynamo grows at all, (28) limits its amplitude to a finite value.

The quenching law (28) does not include an obvious physical effect: a field introduces a preferred direction even into otherwise isotropic turbulence. We must expect therefore that $\overline{\mathcal{E}}$ should depend on the direction of $\overline{\mathbf{B}}$ and not just on its magnitude. Assuming that the Lorentz force can be treated as a perturbation to isotropic turbulence, Roberts and Soward (1975) showed that (10) is modified to

$$\overline{\mathcal{E}} = -(\beta-\beta_B)\nabla\times\overline{\mathbf{B}} + \frac{1}{2}\nabla\beta_B\times\overline{\mathbf{B}} - \frac{\beta_B}{2\overline{B}^2}\overline{\mathbf{B}}\times(\overline{\mathbf{J}}\times\overline{\mathbf{B}}), \tag{29}$$

where $\beta_B > 0$ is a diffusion coefficient that they evaluated by FOS and $\mathbf{J} = \nabla\times\mathbf{B}$ in velocity units. It is clear from (29) that β is reduced by the mean field, which one might again expect from the stiffening of the medium by the field, which impedes the diffusive spreading otherwise created by turbulent mixing. The expansion of the last term in (29) gives rise to a term proportional to $(\overline{\mathbf{B}}\cdot\overline{\mathbf{J}})\overline{\mathbf{B}}$ whose existence was predicted by Roberts (1971) who gave a simple explanation of its origin. Equation (29) was derived for turbulence that, in the absence of $\overline{\mathbf{B}}$, is truly isotropic, ie mirror-symmetric. Corresponding results can be obtained when the turbulence is pseudo-isotropic, ie isotropic but not mirror symmetric. In particular, the modifications introduced into (16) and (17) by the Lorentz force of $\overline{\mathbf{B}}$ were computed (as a perturbation and in FOS) by Roberts and Soward (1975); the case of large $\overline{B}$ is treated by Rüdiger & Kichatinov (1993).

Yoshizawa (1990) has, in a sense, developed the analogue for mean field MHD of what Canuto *et al* (1994) have recently proposed for mean field hydrodynamics. Using a form of DIA, he evaluated $\overleftrightarrow{\mathbf{R}}^K$ and $\overleftrightarrow{\mathbf{R}}^M$ in terms of $\overleftrightarrow{\mathbf{D}}$, the corresponding tensor for $\overline{\mathbf{B}}$ [namely $\frac{1}{2}(\nabla_i\overline{B}_j+\nabla_j\overline{B}_i)$], the mean field $\overline{\mathbf{B}}$, the mean vorticity $\overline{\boldsymbol{\Omega}}$ and four quantities describing the turbulence: its energy densities per unit mass $\frac{1}{2}\overline{v^2}$ and $\frac{1}{2}\overline{b^2}$, the residual helicity $\overline{\mathbf{v}\cdot\boldsymbol{\omega}-\mathbf{b}\cdot\mathbf{j}}$, and the cross–helicity [4] $W=\overline{\mathbf{v}\cdot\mathbf{b}}$. The rate of energy dissipation, ϵ, is not in Yoshizawa's list and, as he did not include buoyancy effects in (18) – (21), he did not need to involve the temperature variance $\overline{T'^2}$ either, as Canuto *et al* (1994) had (see above). Yoshizawa (1990) was led to the unexpected final term in (13), which depends on W and vanishes with it.

As in §2, it makes sense to seek axisymmetric solutions of the governing equations, and to see what is required to make the theoretical predictions agree with the observed properties of the Sun. This activity, success in which should be well within the reach of anyone who possesses a modern workstation, is already under way. See, for example Brandenburg *et al* (1990a, 1991), who attempt to find a self–consistent $\alpha\Lambda$–model, ie one in which the differential rotation and meridional flow are obtained on the assumption that the Λ–effect is the most significant Reynolds stress in the momentum equation.

5. A recent dilemma

An interesting and clarifying debate was initiated by Vainshtein and Cattaneo (1992) the main conclusion of which, if correct, would make the α–effect an irrelevant part of the solar dynamo mechanism. The question centers on the form of the quenching law for α when R_m and $\overline{B}$ are large. Vainshtein and Cattaneo (1992) argued that a law such as (28) is unreasonable unless

[4] The present author finds this renewed interest in W fascinating – long ago, he and Tomomasa Tatsumi felt strongly that this quantity had a significant role to play; see Roberts & Tatsumi (1960).

$\overline{B}^2$ is replaced by $\overline{B^2} = \overline{B}^2 + \overline{b^2}$, so that

$$\alpha = \frac{\alpha_0}{1 + (\overline{B}^2 + \overline{b^2})/B_{\rm eq}^2}. \tag{30}$$

In the simplest picture, $\overline{b^2}$ is parameterized by a function of $\overline{\mathbf{B}}$, and a crucial question arises, "How large is

$$q = \frac{\overline{b^2}}{\overline{B}^2} \tag{31}$$

when R_m and $\overline{B}$ are large?" It is clear that, if (30) is correct and if $q\overline{B}^2 \gg B_{\rm eq}^2$, then $\alpha \ll \alpha_0$, ie the quenching of α is strong. For small $\overline{B}$, FOS gives $q = \mathrm{O}(R_m)$, and this implies a very small α. Taking $\rho = 200$kg m^{-3} and $v_{\rm rms} = 20 - 50$m s^{-1}, we have $B_{\rm eq} = 3 - 8$kG. If $R_m = 10^8$, the α–effect would be so weak that mean fields of the order of only 0.3 – 0.8G could be maintained by dynamo action, in agreement with the conclusion of Vainshtein and Cattaneo (1992). The $q = \mathrm{O}(R_m)$ estimate may, however, be unreasonable; for large $\overline{B}$, Eqns. (26) and (27) of Rüdiger (1974) suggest[5] that then $q \propto \overline{B}^{-2}$; see also §4 of Roberts and Soward (1975). Even if one does not accept this conclusion on the grounds that FOS is strictly inapplicable when $R_m \gg 1$ and $S_m = \mathrm{O}(1)$, not all calculations support the notion that $q = \mathrm{O}(R_m)$. The fast ABC dynamos of Gilbert *et al* (1993) indicate that $q = \mathrm{O}(R_m^n)$ with $n \leq 0.3$. Kleeorin *et al* (1990) found that $q = \mathrm{O}(\log R_m)$ and slowly decreases with increasing $\overline{B}$.

A detailed consideration of magnetic fields at large R_m soon convinces one that the expulsion of field by the violent microscale motions will shred any applied magnetic field into a web of intense flux tubes of small cross-section. The idea of the *filling factor*, f, naturally arises, ie the average fraction of the area occupied by flux tubes in an arbitrary section of the system. For large but steady microscale flows, it is known that $f = \mathrm{O}(R_m^{-1/2})$ for flux sheets and $f = \mathrm{O}(R_m^{-1})$ for flux ropes; for large and unsteady microscale flows, $f \ll R_m^{-1}$. Since $q = f^{-1} - 1$, it is again clear that $q \ll R_m$. This conclusion is reinforced by numerical simulations reported by Brandenburg *et al* (1994). This suggests that the Vainshtein and Cattaneo (1992) objection may not be as lethal as at first sight appears. In contrast, if one

[5] According to Rüdiger (1974), this conclusion is not valid when the Reynolds and magnetic Reynolds numbers are both infinite.

assumes that the kinetic Reynolds number, R_e is of the same order as R_m, the model of Jones and Galloway (1993), which they studied at such large R_m that intense flux concentration occurred, suggests that the $\alpha-$effect is quenched even more strongly (though only by a further logarithmic factor) than Vainshtein and Cattaneo (1992) suppose. It appears that the answer to the question raised by Vainshtein and Cattaneo (1992) is strongly model dependent. It is quite unclear which model, if any yet proposed, is representative of the SCZ.

6. Conclusions.

One need not be clairvoyant to see in the future an unending sequence of increasingly resolved numerical simulations of the solar dynamo, each making full use of the most powerful computational facilities of its time. A model that resolves the turbulence in the SCZ cannot be hoped for in the forseeable future, and questions of the parameterization of the microscale, which have been so central to the present review, will therefore remain of interest until all observations of the Sun are explained by a sufficiently sophisticated mean field MHD model. This may happen quite soon, since parameterizations seem already to be pointing in promising directions. One can hope that when this second plateau of solar dynamo theory has been scaled, it will not (as was the case for the first plateau of the subject) again crumble under the weight of future observational data! Assuming that the second plateau provides a sound foundation, further questions will arise, particularly those that engage the interests of the present meeting, namely long term variability of solar activity and its effects on the terrestrial climate.

Acknowledgments

I am grateful to Gary Glatzmaier for providing Fig.1 and for his permission to reproduce it here. I thank him, David Galloway, and Sasha Ruzmaikin for helpful discussions, and Axel Brandenburg for supplying the preprints of many papers prior to their publication, particularly Brandenburg *et al* (1994) on which §4 is based. Günter Rüdiger acted as referee, and made some helpful suggestions.

References

Brandenburg A, Krause F, Meinel R, Moss D, Tuominen I (1989a) The stability of nonlinear dynamos and the limited role of kinematic growth rates. Astron & Astrophys 213: 411 - 422.

Brandenburg A, Krause F, Tuominen I (1989b) Parity selection in nonlinear dynamos. In Turbulence and Nonlinear Dynamics in MHD Flows, M. Meneguzzi, A. Pouquet and P.L. Sulem eds., 35 – 40, Elsevier Dordrecht.

Brandenburg A, Krause F, Nordlund Å, Ruzmaikin A, Stein RF, Tuominen I (1994) On the magnetic fluctuations produced by a large scale magnetic field. To appear.

Brandenburg A, Moss D, Rieutord M, Rüdiger G, Tuominen I (1991) $\alpha\Lambda$-- dynamos. In The Sun and Cool Stars: Activity, Magnetism, Dynamos, I. Tuominen, D. Moss & G. Rüdiger eds., 147 – 150, Springer Berlin.

Brandenburg A, Moss D. Rüdiger G, Tuominen I (1990a) The nonlinear solar dynamo and differential rotation: a Taylor number puzzle? Solar Phys 128: 243 - 251.

Brandenburg A, Moss D, Tuominen I (1992) Stratification and thermodynamics in mean–field dynamos. Astron & Astrophys 265: 328 - 344.

Brandenburg A, Nordlund Å, Pulkkinen P, Stein RF, Tuominen I (1990b) 3D simulations of turbulent cyclonic magnetoconvection. Astron & Astrophys 232: 277 - 291.

Brandenburg A, Tuominen, I, Moss, D (1989) On the nonlinear stability of dynamo models. Geophys & Astrophys Fluid Dynam 49: 129 - 141.

Canuto VM, Minotti FO, Schilling O (1994) Differential rotation and turbulent convection: a new Reynolds stress model, Astrophys J, submitted.

Childress S (1992) Fast dynamo theory. In Topological Aspects of the Dynamics of Fluids and Plasmas, H.K. Moffatt, G.M. Zaslavsky, P. Comte & M. Tabor, eds., NATO ASI Series: Applied Sciences 218, 111 - 147, Kluwer, Dordrecht.

Drobyshevskii EM, Yuferev VS (1974) Topological pumping of magnetic flux by three–dimensional convection. J Fluid Mech 65: 33 - 41.

Durney BR (1976) On the constancy along cylinders of the angular velocity in the solar convection zone. Astrophys J 204: 596 - 598.

Durney BR, Sofia S, eds. (1987) The Internal Solar Angular Velocity. Theory, Observations and Relationship to Solar Magnetic Fields. Reidel, Dordrecht, xi + 374 pages.

Duvall TL Jnr, Harvey JW (1984) Rotational frequency splitting of solar observations. Nature 310: 19 - 22.

Duvall, TL Jnr, Dziembowski, WA, Goode PR, Gough DO, Harvey JW, Leibacher JW (1984) Internal rotation of the Sun. Nature 310: 22 – 25.
Dziembowski, WA, Goode PR, Libbrecht KG (1989) The radial gradient in the Sun's rotation. Astrophys J 337: L53 – L57.
Frisch U, She ZS, Sulem PL (1987) Large–scale flow driven by the anisotropic kinetic alpha effect. Physica 28D: 382 – 392.
Gilbert AD, Otani NF, Childress S (1993) Simple dynamical fast dynamos. In Solar and Stellar Dynamos, M.R.E. Proctor, P.C. Matthews and A.M. Rucklidge, eds., pp129 – 136. University Press, Cambridge.
Gilman PA (1983) Dynamically consistent nonlinear dynamos driven by convection in a rotating spherical shell. II Dynamos with cycles and strong feedbacks. Astrophys J Supp Ser 53: 243 – 268.
Gilman PA, Howard R (1984) On the correlation of longitudinal and latitudinal motions of sunspots. Solar Phys 93: 171 – 175.
Gilman PA, Miller J (1981) Dynamically consistent nonlinear dynamos driven by convection in a rotating spherical shell. Astrophys J Supp Ser 46: 211 – 238.
Glatzmaier GA (1985a) Numerical simulations of stellar convective dynamos. I. The model and the method. J Comp Phys 55: 461 – 484.
Glatzmaier GA (1985b) Numerical simulations of stellar convective dynamos. II. Field propagation in the convection zone. Astrophys J 291: 300 – 307.
Glatzmaier GA (1987) Numerical simulations of stellar convective dynamos. III. At the base of the convection zone. Geophys & Astrophys Fluid Dynam 31: 137 – 150.
Hathaway DH (1991) GONG Report #9, Tucson NOAO/NSO, p23.
Jepps SA (1975) Numerical models of hydromagnetic dynamos. J Fluid Mech 67: 625 – 646.
Jones CA, Galloway DJ (1993) Alpha–quenching in cylindrical magnetoconvection. In Solar and Stellar Dynamos, M.R.E. Proctor, P.C. Matthews and A.M. Rucklidge, eds., pp161 – 170. University Press, Cambridge.
Kazantsev AP (1968) Enhancement of a magnetic field by a conducting fluid, Soviet Phys JETP 26: 1031 – 1034.
Kichatinov LL, Rüdiger G (1993) $\Lambda-$effect and differential rotation in stellar convection zones. Astron & Astrophys 276: 96 – 102.
Kichatinov LL, Rüdiger G, Pipin, VV (1993) Nonlinear turbulent diffusion in rotating and magnetized fluids: viscosity, magnetic diffusivity & heat conductivity. To appear.

Kleeorin NI, Rogachevskii IV, Ruzmaikin AA (1990) Magnetic field reversal and instability in a plasma with advanced magnetohydrodynamic turbulence. Soviet Phys JETP 70: 878 – 883.

Köhler H (1973) The solar dynamo and estimates of the magnetic diffusivity and the α–effect. Astron & Astrophys 25: 467 – 476.

Komm RW, Howard RF, Harvey JM (1993) Meridional flow of small photospheric magnetic features. Solar Phys 147: 207 – 223.

Krause F (1967) A solution of the dynamo problem, based on the linear theory of magnetohydrodynamic turbulence. Habilitationsshrift, University of Jena. Partial English translation by Roberts and Stix (1971).

Krause F & Rädler K–H (1980) Mean–Field Magnetohydrodynamics and Dynamo Theory. Academy Press, Berlin; also Pergamon Press, Oxford.

Leighton RB (1969) A magneto–kinematic model of the solar cycle. Astrophys J 156: 1 – 26.

Malkus WVR, Proctor MRE (1975) The macrodynamics of α–effect dynamos in rotating fluids. J Fluid Mech 67: 417 – 444.

Meneguzzi M, Frisch U, Pouquet A (1981) Helical and nonhelical turbulent dynamos. Phys Rev Lett 47: 1060 – 1064.

Meneguzzi M, Pouquet A (1989) Turbulent dynamos driven by convection. J Fluid Mech 205: 297 – 312.

Moffatt HK (1974) Appendix to the paper of Drobyshevski and Yuferev, "Topological pumping of magnetic flux by three–dimensional convection". J Fluid Mech 65: 41 – 44.

Molchanov SA, Ruzmaikin AA, Sokoloff DD (1985) Kinematic dynamo in random flow. Soviet Phys Usp 28: 307 – 327.

Moss D, Tuominen I, Brandenburg A (1990a) Nonlinear dynamos with magnetic buoyancy in spherical geometry. Astron & Astrophys 228: 284 – 294.

Moss D, Tuominen I, Brandenburg A (1990b) Buoyancy–limited thin shell dynamos. Astron & Astrophys 240: 142 – 149.

Moss D, Brandenburg A, Tuominen I (1991a) Properties of mean field dynamos with nonaxisymmetric α–effect. Astron & Astrophys 247: 576 – 579.

Moss D, Tuominen I, Brandenburg A (1991b) Nonlinear nonaxisymmetric dynamo models for cool stars. Astron & Astrophys 245: 129 – 135.

Nesme–Ribes E, Ferreira EN, Sadourny R, Le Treut H, Li L (1993) Solar dynamics and its impact on solar irradiance and the terrestrial climate. J Geophys Res 98A: 18,923 – 18,935.

Nesme–Ribes E, Sokoloff D, Ribes J–C, Kremliovsky, M (1994) The Maunder minimum and the solar dynamo, these Proceedings.

Nordlund Å, Brandenburg A, Jennings RL, Rieutord M, Ruokolainen J, Stein RF, Tuominen I (1992) Dynamo action in stratified convection with overshoot. Astrophys J 392: 647 - 652.

Novikov VG, Ruzmaikin AA, Sokoloff DD (1983) Kinematic dynamo in reflection–invariant random field. Soviet Phys JETP 58: 527 – 532.

Parker EN (1955) Hydromagnetic dynamo models. Astrophys J 122: 293 – 314.

Parker EN (1979) Cosmical Magnetic Fields, their Origin and their Activity. Clarendon Press, Oxford.

Parker EN (1993) A solar dynamo surface wave at the interface between convection and nonuniform rotation. Astrophys J 408: 707 – 719.

Pidatella RM, Stix M (1986) Convective overshoot at the base of the Sun's convection zone. Astron & Astrophys 157: 338 – 340.

Prautzsch, T (1993) The dynamo mechanism in the deep convection zone of the Sun. In Solar and Stellar Dynamos, M.R.E. Proctor, P.C. Matthews and A.M. Rucklidge, eds., pp249 – 256. University Press, Cambridge.

Pulkkinen P, Tuominen I, Brandenburg A, Nordlund Å, Stein, RF (1993) Rotational effects on convection simulated at different latitudes. Astron & Astrophys 267: 265 – 274.

Rädler K–H, Wiedemann E, Brandenburg A, Meinel R, Tuominen I (1990) Astron & Astrophys 239: 413 – 423.

Roberts PH (1971) Dynamo theory in Lectures in Applied Mathematics, 14, ed. W.H. Reid. Am. Math. Soc., Providence RI.

Roberts PH (1972) Kinematic dynamo models. Phil Trans R Soc Lond A272: 411 – 451.

Roberts PH, Soward AM (1975) A unified approach to mean–field electrodynamics. Astron Nachr 269: 49 – 64.

Roberts PH, Soward AM (1992) Dynamo theory, Ann Rev Fluid Mech 24: 459 – 512.

Roberts PH, Stix M (1971) The turbulent dynamo. A translation of a series of papers by F. Krause, K.-H. Rädler and M. Steenbeck. NCAR Technical Note IA–60.

Roberts PH, Tatsumi T (1960) The decay of magnetohydrodynamic turbulence. J Math & Mech 9: 697 – 714.

Rüdiger, G (1974) The influence of a uniform magnetic field of arbitrary strength on turbulence. Astron. Nach. 295: 275 – 284.

Rüdiger G (1989) Differential Rotation and Stellar Convection, Sun and Solar-type Stars. Gordon & Breach, New York xii + 328 pages.
Rüdiger G, Brandenburg A (1993) A solar dynamo in the overshoot layer: cycle period and butterfly diagram. To appear.
Rüdiger G, Kichatinov LL (1993) Alpha–effect and alpha–quenching. Astron & Astrophys 269: 581 - 588.
Sadourny R. (1994) Climate sensitivity, Maunder minimum and the little ice age, these Proceedings.
Schmitt D (1984) Dynamo action of magnetostrophic waves. In The Hydromagnetics of the Sun, T.D. Guyenne & J.J. Hunt eds., 223 - 224, European Space Agency, Paris.
Schmitt D (1985) Dynamowirkung Magnetostrophischer Wellen, Doctoral Dissertation, University of Göttingen, 217 pages.
Schmitt D (1987) An $\alpha\omega-$dynamo with an $\alpha-$effect due to magnetostrophic waves. Astron & Astrophys 174: 281 - 287.
Soward AM (1993) Analytic fast dynamo solution in a two–dimensional pulsed flow. In Solar and Stellar Dynamos, M.R.E. Proctor, P.C. Matthews and A.M. Rucklidge, eds., pp275 - 286. University Press, Cambridge.
Spiegel EA, Weiss NO (1980) Magnetic activity and variations in solar luminosity. Nature 287: 616 - 617.
Spiegel EA, Zahn JP (1992) The solar tachocline. Astron & Astrophys. 265: 106 - 114.
Spruit HC (1990) Angular momentum transport and magnetic fields in the solar interior. In Inside the Sun, G. Berthomieu and M. Cribier, eds., 415 - 423, Kluwer, Dordrecht.
Steenbeck M, Krause F (1969) On the dynamo theory of stellar and planetary magnetic fields, I. A.C. dynamos of solar type. Astron Nach 291: 49 - 84 (1969). English translation by Roberts and Stix (1971).
Steenbeck M, Krause F, Rädler, K-H (1966) The generation of stellar and magnetic fields by turbulent dynamo action. Z Naturforsch 21a: 1285 - 1296. English translation by Roberts and Stix (1971).
Stix M (1976) Dynamo theory and the solar cycle. In Basic Mechanisms of solar activity, V. Bumba & J Kleczek eds., 367 - 388, Reidel, Dordrecht.
Sulem PL, She ZS, Scholl H, Frisch U (1989) Generation of large–scale structures in three–dimensional flow lacking parity–invariance. J Fluid Mech 205: 341 - 358.
Sweet PA (1950) The effect of turbulence on a magnetic field. Mon Not R Astr Soc 110: 69 - 83.

Tuominen I, Rüdiger, G (1989) Solar differential rotation as a multiparameter turbulence problem. Astron & Astrophys 217: 217 – 228.
Vainshtein SI, Cattaneo F (1992) Nonlinear restrictions on dynamo action. Astrophys J 393: 165 – 171.
Van Ballegooien AA (1982) The overshoot region at the base of the solar convection zone and the problem of magnetic flux storage. Astron & Astrophys 113: 99 – 112.
Vincent, A, Meneguzzi M (1991) The spatial structure of homogeneous turbulence at high Reynolds number. In The Sun and Cool Stars: Activity, Magnetism, Dynamos, I. Tuominen, D. Moss & G. Rüdiger eds., 75 – 79, Springer Berlin.
Yoshimura H (1975) Solar-cycle dynamo wave propagation. Astrophys J 201: 740 – 748.
Yoshizawa A (1990) Self-consistent turbulent dynamo modeling of reversed field pinches and planetary magnetic fields. Phys Fluids B7: 1589 – 1600.
Zheligovsky VA (1993) Fast kinematic dynamo sustained by Beltrami flow in a sphere. Geophys & Astrophys Fluid Dynam. To appear.

THE DIFFERENTIAL SOLAR ROTATION

Günther Rüdiger and Leonid L. Kitchatinov
Astrophysikalisches Institut Potsdam
An der Sternwarte 16
D-14482 Potsdam
Fed. Rep. Germany

Summary

Theory and observations of the solar differential rotation are reviewed, concentrating on recent developments. Reynolds stresses produced by stratified rotating turbulence have recently become available as fully nonlinear functions of the angular velocity. A model based on this theory has reproduced the solar rotation; almost no tuning of the theory to the observations was required. By taking into account the rotationally–produced anisotropy of the turbulent heat transport, it is even possible to resolve the "Taylor number puzzle". Temporal variations in the solar rotation during different periods in the past and their relation to magnetic activity are briefly discussed.

1. Introduction

The differential rotation is the most powerful and principal driver of the solar dynamo. An understanding of the solar magnetic activity without an understanding of the rotation law in the solar convection zone proves to be impossible.

We may state that definite progress has been made recently in understanding the solar differential rotation. Angular momentum transport by turbulent and laminar motions has been described from a common standpoint. Models based on this theory have few free parameters, but they are capable of reproducing the basic observational facts.

During these developments, new data has become available. The accuracy with which the horizontal Reynolds stresses have been determined by the tracer method has been considerably improved (Nesme-Ribes et al. 1993b).

NATO ASI Series, Vol. I 25
The Solar Engine and Its Influence
on Terrestrial Atmosphere and Climate
Edited by Elizabeth Nesme-Ribes

The same is true of the reliability of helioseismological inversions. New data on the differential rotation of other main sequence stars has made new tests of the theory possible.

2. Angular momentum conservation by laminar advection

The theory of non-uniform rotation is mainly the theory of one equation, i.e. the conservation law of angular momentum. Including temporal variations, meridional flow, Reynolds stress and Lorentz force it reads

$$\frac{\partial}{\partial t}(\rho\varpi^2\Omega)+\mathrm{div}\left\{\rho\varpi^2\Omega\overline{\mathbf{u}}^m + \rho\varpi < u'_\phi \mathbf{u}' > -\frac{\varpi}{\mu}\left(\overline{B}_\phi\overline{\mathbf{B}}+ < B'_\phi \mathbf{B}' >\right)\right\} = 0,$$

where Ω is the angular velocity; (ϖ, ϕ, z) are cylindrical coordinates with Oz parallel to the rotation axis; ρ is density, and μ is magnetic permeability; $\overline{\mathbf{B}}$ and $\overline{\mathbf{u}}$ are the ensemble averages of the magnetic field and fluid velocity, and $\mathbf{B}'$ and $\mathbf{u}'$ are their fluctuating parts. Except during the past two last decades, the interpretation of this equation has centred on analysing the effect of the meridional flow, $\overline{\mathbf{u}}^m$. To this end, the Lorentz force was ignored and the Reynolds stress was parameterised by

$$Q_{ij} \equiv < u'_i(\mathbf{x},t)u'_j(\mathbf{x},t) > = ... - \nu_T(\overline{u}_{i,j} + \overline{u}_{j,i}). \tag{2}$$

The coefficient ν_T in this relation is called the eddy viscosity. Its determination, for conditions in which the large-scale flow and a magnetic field are influential, is one of the basic problems of turbulence physics. The resulting equation describes angular momentum transport when the large-scale and small-scale flows are specified:

$$\mathrm{div}\left\{\rho\varpi^2\Omega\overline{\mathbf{u}}^m - \rho\varpi^2\nu_T\nabla\Omega\right\} = 0. \tag{3}$$

It is seems worth discussing this equation in its two limiting forms.

2.1 SLOW FLOW

By introducing the stream function A, we may write

$$\rho\overline{\mathbf{u}}^m = \mathrm{rot}\,\frac{A}{\varpi}\mathbf{l}_\phi, \tag{4}$$

where $\mathbf{l}_\phi$ is the unit vector in the $\phi-$direction. Since $\overline{\mathbf{u}}^m \cdot \nabla A = 0$, the streamlines are $A =$ const. On substituting the Legendre function expansion

$$A = (A_2 P_2^1 + A_4 P_4^1 + ...) \sin\theta \tag{5}$$

into (3) where θ is colatitude, we obtain a simple differential equation for A_2 with the solution

$$\Omega_{eq} - \Omega_{pole} \sim \int_{x_i}^{1} \zeta^3 A_2(\zeta) d\zeta, \tag{6}$$

where $x_i = r_i/R$ and r_i is the inner radius of the convection zone; R is the radius of the Sun. The equator is accelerated if $A_2 > 0$, and is decelerated in the opposite case. It may easily be shown that positive A_2 describes a meridional flow pattern with clockwise streaming (as seen in the NW quadrant of the meridional cross-section of the convection zone (CZ). Eq. (6) gives the most basic and most elementary discovery of the early theory (starting with Zöllner 1881).

2.2 RAPID FLOW

More complex is the rotation law for very rapid flow. Then, when reasonable Ω-gradients are assumed, the viscosity is unable to balance the Coriolis force and the angular velocity must adjust itself to satisfy

$$\rho \overline{\mathbf{u}}^m \cdot \nabla(\varpi^2 \Omega) = 0. \tag{7}$$

This relation requires the angular momentum to be constant on streamlines:

$$\varpi^2 \Omega = f(A). \tag{8}$$

Characteristic of this limit are therefore large values of Ω near the polar axis. The appearance of such a polar vortex is independent of the sense of the meridional circulation (Köhler 1969).

2.3 TAYLOR–PROUDMAN STATE

There is another important property of the Reynolds equation,

$$\frac{D\overline{u}_i}{Dt} = -\frac{1}{\rho}\frac{\partial p}{\partial x_i} - \frac{1}{\rho}\frac{\partial}{\partial x_j}(\rho Q_{ij}) + \frac{\partial \psi}{\partial x_i}, \tag{9}$$

(p pressure, ψ gravitational potential). This gives

$$\mathcal{D}(\overline{\mathbf{u}}^m) = \varpi \frac{d\Omega^2}{dz} + \frac{1}{\rho^2} (\nabla p \times \nabla \rho)_\phi , \tag{10}$$

in which the left–hand side represents the damping of meridional flow by the eddy viscosities. Two sources of flow exist: deformation and differential rotation. The flow resulting from non-parallel grad p and grad ρ is called the "barocline" or the "thermal wind".

When (10) is written in dimensionless units, the second term on the RHS acquires a dimensionless "Taylor number",

$$\mathrm{Ta} = \left(\frac{2\Omega R^2}{\nu_T} \right)^2 . \tag{11}$$

If this number is very large, (7) and (10) are simultaneously satisfied by

$$\overline{\mathbf{u}}^m = 0, \qquad \Omega = \Omega(\varpi). \tag{12}$$

Cylindrical iso-rotation contours are thus the immediate consequence of large Coriolis and centrifugal forces. The Taylor number for the Sun, $\mathrm{Ta} \sim 10^7$, is large enough to predict the Ω–contours as cylindrical. Several simulations led to this result. The helioseismic "observations" revealed, however, the opposite: disk–like iso-rotation contours dominate. We call the problem of resolving this striking contradiction the "Taylor number puzzle".

3. The observations

There are two basic methods for measuring the differential rotation, Ω_s, on the solar surface. One of these utilises the Doppler shifts of the radiating plasma, the other relies on the motion of usable tracers. Unfortunately, tracers are always associated with magnetism. The true plasma rotation is seen only by the spectroscopic method. The resulting rotation law relates directly to solar hydrodynamics. In contrast, the rotation laws derived from tracers must be used with caution.

The most frequently used representations for the surface angular velocity profile are

$$\Omega_s = A + B\cos^2\theta + C\cos^4\theta, \qquad A > 0. \tag{13}$$

When temporal variations due to cyclic magnetic phenomena are included, the representation (13) no longer makes sense.

3.1 PLASMA ROTATION

Using their Doppler measurements, Howard & Harvey (1970) found, for the mean motion of the solar plasma, that $A = 13.76$, $B = -1.74$ and $C = -2.19$, all in units of deg/day, which corresponds to 0.202 μ rad/s or to 140 m/s (at the equator). The rotation law can therefore be written as

$$\Omega_s = 2.67(1 - 0.033P_3^1/\sin\theta - 0.0043P_5^1/\sin\theta)\mu \text{ rad/s}. \tag{14}$$

We see that, in the general representation,

$$\Omega = \Omega_\odot \sum_{n=1,3,\ldots} \omega_{n-1}(r)P_n^1/\sin\theta, \tag{15}$$

where $\Omega_\odot \approx 2.67\mu$rad/s is the mean angular velocity, the coefficients decrease progressively:

$$\omega_0(1) = 1, \quad \omega_2(1) = -0.033, \quad \omega_4(1) = -0.0043. \tag{16}$$

Howard et al. (1983) published results covering an observation period of 14 years which included data from 188 rotations. The mean rotation law

$$\Omega_s = 2.76(1 - 0.032P_3^1/\sin\theta - 0.0045P_5^1/\sin\theta)\mu \text{ rad/s} \tag{17}$$

resulted. The corresponding equatorial velocity is about 3% larger than that given in (14).

According to Howard & Harvey (1970) the fractional overall pole–equator difference in the angular velocities

$$\delta\Omega = (\Omega_{pole} - \Omega_{eq})/\Omega_{eq}, \tag{18}$$

is about 30%.

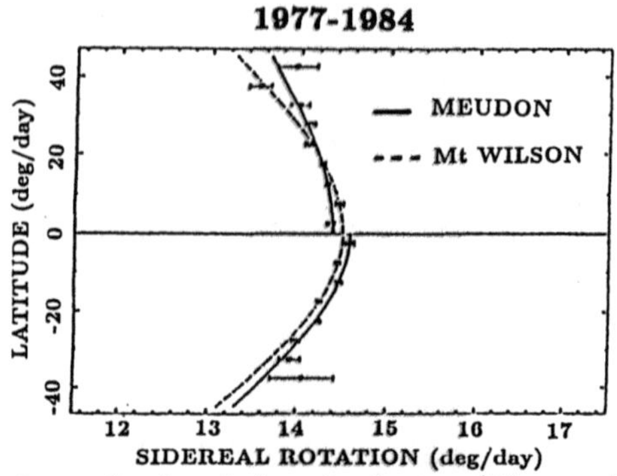

Fig. 1. Rotation laws for both hemispheres (Nesme-Ribes et al. 1993a)

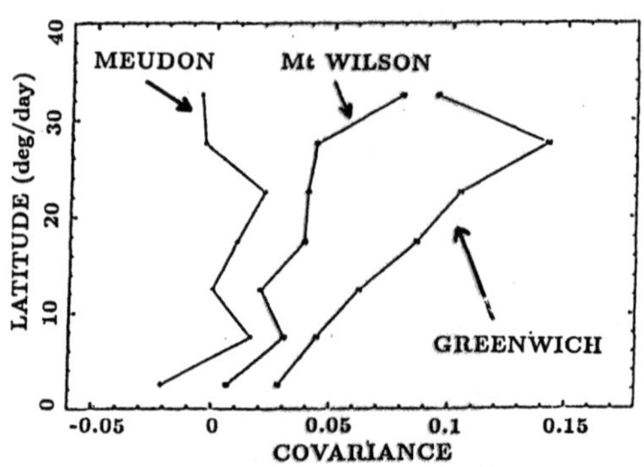

Fig. 2. Covariance $Q_{\theta\phi}$ (the "Ward profile") from sunspots (Nesme-Ribes et al. 1993b)

3.2 SPOT ROTATION

The result most often quoted for sunspots is due to Newton & Nunn (1951),

$$\Omega_s = 2.90(1 - 0.19 \cos^2 \theta)\mu \text{ rad/s}, \tag{19}$$

derived from individual long-lived sunspots. In comparison with the 1865 result of Faye, only the "gradient" has been reduced, from 0.22 to 0.19.

The most recent results from the tracer method are those of Nesme-Ribes et al. (1993a). An important example of parity breaking is given in Fig. 1. The angular velocity profile differs from hemisphere to hemisphere. NS-parity breaking is a highly nonlinear effect, and so far no satisfactory explanation has been offered. Of direct relevance is the result in Fig. 2 for the often discussed horizontal covariance $< u'v' >$ (the "Ward profile"). It is now believed to be generally rather small. The observations are relevant to the

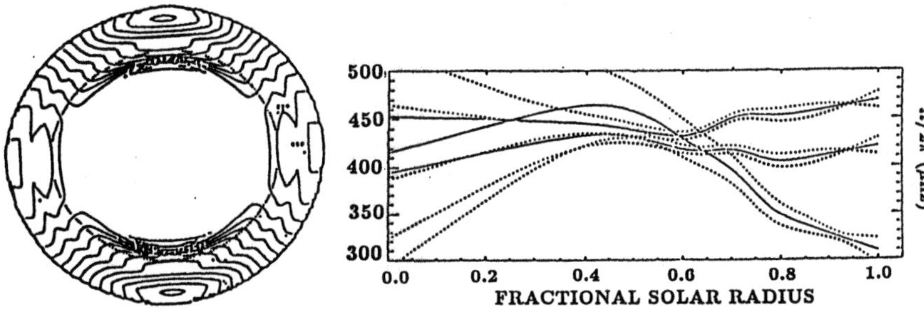

Fig. 3. Iso-contours of solar rotation. Left: Inversion by Libbrecht (1988). Right: Rotation laws by Schou (1993)

theory of differential rotation. The observed properties of the covariance are well explained by recent theoretical work by Canuto et al. (1993).

3.3 THE INTERNAL ROTATION

Helioseismology has taught us all we know about the internal rotation of the Sun. The oscillations observed are the surface manifestation of trapped standing sound waves. The solar oscillations are described by Legendre polynomials, $\xi_n(r)P_l^m(\theta)e^{i(\omega t-m\phi)}$, with ω the oscillation frequency. The n, l and m quantum numbers are the radial order, angular degree and angular order. If deformational forces, (rotation and/or magnetism), were absent in the Sun, then each (n,l)–modes would be $(2l{+}1)$-fold degenerate in m. It is best to view data in multiplets with a fine structure in m so that

$$\nu_{nlm} = \nu_{nl0} + L\sum_{i=1}^{N} a_i(n,l)P_i\left(\frac{m}{L}\right) \tag{20}$$

(Goode 1991, Schou 1991), where $L = \sqrt{l(l+1)}$, $N = 6$. The odd $a-$coefficients in (20) arise from the linear effects of rotation. The a_1, a_3 and a_5 are correlated with a rotation law of the form (15). The coefficient a_5 contains information about the depth-dependences of ω_4; a_3 and a_5 determine the depth-dependence of ω_2; and a_1, a_3 and a_5 yield integrals of $\omega_0(r)$.

The situation is very similar to that arising in dynamo theory for non-axisymmetric fields. The eigenfrequencies ω_{nlm} appear in the equations in the form

$$\omega_D = \omega_{nlm} + m\Omega(r,\theta), \tag{21}$$

where, however, ω_{nlm} is much smaller than the rotation rate. Fields with $m \neq 0$ are influenced by the internal rotation, their eigenfrequencies are functions of m and $\Omega(r,\theta)$. For uniform rotation it is simply

$$\Omega = \frac{1}{m}(\omega_{nlm} - \omega_{nl0}). \tag{22}$$

Various inversions procedures have revealed the following characteristics of the solar rotation law:

- a strong polar sub-rotation ($\partial\Omega/\partial r < 0$),
- a weak equatorial super-rotation ($\partial\Omega/\partial r > 0$),
- a reduced value of the equator-pole-difference in Ω at the bottom of the SCZ.

The overall result is a characteristically disk–like structure in the polar region and a characteristically Taylor–Proudman structure in the equatorial region (Fig. 3).

We do not yet know whether or not the equatorial Ω–profile possesses a maximum. Hill (1987), Rhodes et al. (1991) as well as Schou (1991) reported that the equatorial rotation rate is an increasing function of depth close to the surface. Recently, Schou (1993) shows profiles for the years 1987 and 1989 without any hint of local maxima.

4. Reynolds stress models

If only the contribution by turbulence to the angular momentum balance is considered, (1) becomes

$$\mathrm{div}\left(\rho\varpi < u'_\phi \mathbf{u}' >\right) \equiv \frac{\partial}{\partial x_i}(\rho\varpi Q_{i\phi}) = 0. \tag{23}$$

The correlation tensor (2) may be divided into several parts,

$$Q_{ij} = ... + Q^{\nu}_{ij} + Q^{\Lambda}_{ij}. \tag{24}$$

Here Q^{ν} is the diffusive part,

$$Q^{\nu}_{ij} = -\mathcal{N}_{ijkl}\overline{u}_{k,l}, \tag{25}$$

where the viscosity tensor, $\mathcal{N}$, provides the proportionality factors in the stress-strain relation. The remaining part, Q^{Λ}, does not include any spatial derivatives. It represents the non–diffusive transport of angular momentum by the turbulence (the "Λ–effect"). Eq. (23) implies that the differential rotation is maintained by a balance between the damping and driving by the turbulent flow.

The old treatments (Wasiutynski 1946) were restricted to the case of slow rotation. Only the radial Λ-effect survives in this limit, while $Q^{\Lambda}_{\theta\phi} = 0$. The Sun, is not, however, a slow rotator. The Coriolis number, $\Omega^* = 2\tau_{corr}\Omega$ is larger than unity for large-scale solar convection. The same is true for all single cool main-sequence stars.

When the rotation is not slow, a horizontal flux $Q^{\Lambda}_{\theta\phi}$ also exists (Rüdiger 1983). Hence, the equatorial acceleration may be produced by the Λ-effect. Pidatella et al. (1986) developed models to compare the efficiency of meridional flow and Reynolds stresses in generating the differential rotation. They found the Λ-effect as a more powerful generator. Many other indications confirm this (e.g. Durney & Spruit 1979, Kitchatinov 1986, Rüdiger & Tuominen 1990, Canuto et al. 1993). A non–trivial question remains however: can the required Reynolds stresses be derived from a reasonable turbulence model?

4.1 Λ-EFFECT AND EDDY VISCOSITIES

Only anisotropic and/or inhomogeneous turbulence under the influence of rotation can produce a Λ-effect. Although anisotropy can exist even in the absence of stratification, this is not the case for inhomogeneous turbulence. The homogeneous anisotropic case is most probably not relevant to the Sun. For $\Omega^* > 1$ its Λ-effect decreases with increasing Coriolis number, $Q^{\Lambda} \sim \Omega^{*-2}$, and is too small to reproduce the required solar equatorial acceleration for the solar $\Omega^* \simeq 6$. When the inhomogeneity is restored, the situation becomes more complicated but the resulting Λ–effect is more powerful. The quasi–isotropic model for inhomogeneous turbulence (Roberts & Soward 1975, Kitchatinov 1987) leads to a Λ–effect saturating at a constant value for large Ω^*.

Only two off-diagonal components of the Q^{Λ}_{ij}-tensor are relevant to angular momentum transport, namely

$$Q^{\Lambda}_{r\phi} = \nu_T \left(V^{(0)} + \sin^2\theta \; V^{(1)} \right) \Omega \sin\theta, \quad Q^{\Lambda}_{\theta\phi} = \nu_T H^{(1)} \Omega \sin^2\theta \cos\theta.$$

The dimensionless functions V and H for inhomogeneous quasi–isotropic turbulence were derived by Kitchatinov & Rüdiger (1993). When the main inhomogeneity is the density stratification and the mixing–length theory (MLT) is adopted, they are

$$V^{(0)} = \frac{\tau_{corr}^2 < u'^2 >}{H_\rho^2} \left[\mathcal{J}_0(\Omega^*) + \mathcal{J}_1(\Omega^*)\right],$$

$$V^{(1)} = H^{(1)} = -\frac{\tau_{corr}^2 < u'^2 >}{H_\rho^2} \mathcal{J}_1(\Omega^*), \tag{27}$$

where H_ρ is the density scale–height. The coefficient $\tau_{corr}^2 < u'^2 > /H_\rho^2$ is close to unity for the adiabatically stratified convection zone:

$$\frac{\tau_{corr}^2 < u'^2 >}{H_\rho^2} \simeq \left(\frac{l_{corr}}{H_\rho}\right)^2 \simeq \left(\frac{\alpha_{MLT}}{\gamma}\right)^2, \tag{28}$$

where α_{MLT} comes from $l_{corr} \simeq \alpha_{MLT} H_p$. Here l_{corr} is the mixing length, H_p is the pressure scale height, and γ is the ratio of specific heats. We assume that $\alpha_{MLT} = \gamma = 5/3$. The functions (27) for this case are shown in Fig. 4.

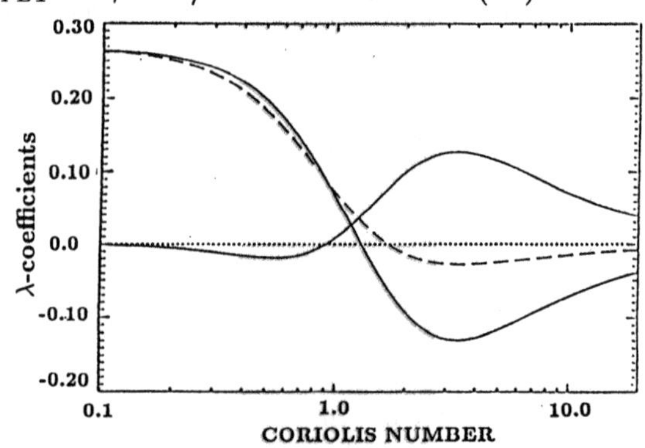

Fig. 4. $V^{(0)}$ and $V^{(1)} = H^{(1)}$ as functions of the Coriolis number Ω^*. Dashed line: $V^{(0)} + 0.8\ V^{(1)}$

There are marked differences between slow ($\Omega^* \ll 1$) and fast ($\Omega^* \gg 1$) rotation. For slow rotation $V^{(1)}$ is small while $V^{(0)}$ is finite. Therefore, only a radial non–uniformity of the rotation is generated in slow rotators. When $\Omega^* \gg 1$, $V^{(0)}$ becomes negative and simultaneously $V^{(1)}$ takes positive values of the same order. As a result, the quantity $V^{(0)} + \frac{4}{5}V^{(1)}$, which measures the effectivity of the Reynolds stresses in generating $\partial\Omega/\partial r$, becomes small (Rüdiger & Tuominen 1990). This is why the radial inhomogeneity in fast rotators is small. Note that the transition from the slow-rotation regime to the fast is rather sharp and is located near $\Omega^* \simeq 1$.

We must also consider the influence of rotation on the viscosity tensor,

$$\begin{aligned} \mathcal{N}_{ijkl} &= \nu_1 \left(\delta_{ik}\delta_{jl} + \delta_{jk}\delta_{il}\right) \\ &+\nu_2 \left(\delta_{il}\frac{\Omega_j\Omega_k}{\Omega^2} + \delta_{jl}\frac{\Omega_i\Omega_k}{\Omega^2} + \delta_{ik}\frac{\Omega_j\Omega_l}{\Omega^2} + \delta_{jk}\frac{\Omega_i\Omega_l}{\Omega^2} + \delta_{kl}\frac{\Omega_i\Omega_j}{\Omega^2}\right) \\ &+\nu_3\delta_{ij}\delta_{kl} - \nu_4\delta_{ij}\frac{\Omega_k\Omega_l}{\Omega^2} + \nu_5\frac{\Omega_i\Omega_j\Omega_k\Omega_l}{\Omega^4}, \end{aligned} \tag{29}$$

where

$$\nu_0 = \frac{4}{15}\,\tau_{corr} < u'^2 > , \qquad \nu_n = \nu_0\Phi_n(\Omega^*), \tag{30}$$

for $n = 1...5$. Here ν_0 is the isotropic shear viscosity for the non-rotating fluid. The viscosity quenching functions, $\Phi_n(\Omega^*)$, derived using the same approximations as those used to compute the Λ-effect (27), are known (Kitchatinov et al. 1994). Only $\Phi_1(0) = \Phi_3(0) = 1$; the remaining functions parameterise the viscosity anisotropy created by the rotation, and must be zero for $\Omega^* = 0$. All the functions decrease with increasing Ω^*.

Only two viscosity coefficients contribute to the diffusive flux of the angular momentum:

$$\begin{aligned} Q^{\nu}_{r\phi} &= -\nu_1 r \sin\theta\,\frac{\partial\Omega}{\partial r} - \nu_2 \sin\theta\cos\theta\left(r\cos\theta\,\frac{\partial\Omega}{\partial r} - \sin\theta\,\frac{\partial\Omega}{\partial\theta}\right), \\ Q^{\nu}_{\theta\phi} &= -\nu_1 \sin\theta\,\frac{\partial\Omega}{\partial\theta} - \nu_2 \sin^2\theta\left(\sin\theta\,\frac{\partial\Omega}{\partial\theta} - r\cos\theta\,\frac{\partial\Omega}{\partial r}\right). \end{aligned} \tag{31}$$

Note the presence in the radial flux of a latitudinal gradient in the angular velocity; in the latitudinal flux there is a radial gradient of Ω. This is the result of the anisotropy introduced by rotation (Kitchatinov 1986, Durney 1989, Rüdiger 1989).

4.2 THE MODEL

Eq. (23), with the Reynolds stresses defined by (27) and (31), has been solved numerically. The basic profiles (30) of the viscosity are taken from the solar convection zone model of Stix (1993). The usual stress–free boundary conditions were assumed. The model has only two adjustable parameters: the ratio $\alpha_{MLT} = l_{corr}/H_p$ and the basic angular velocity, Ω_0. Both are fixed for the Sun. There is no freedom to tune the model to the observations. Nevertheless the results agree very closely with the solar data (Küker et al. 1993, Fig. 5).

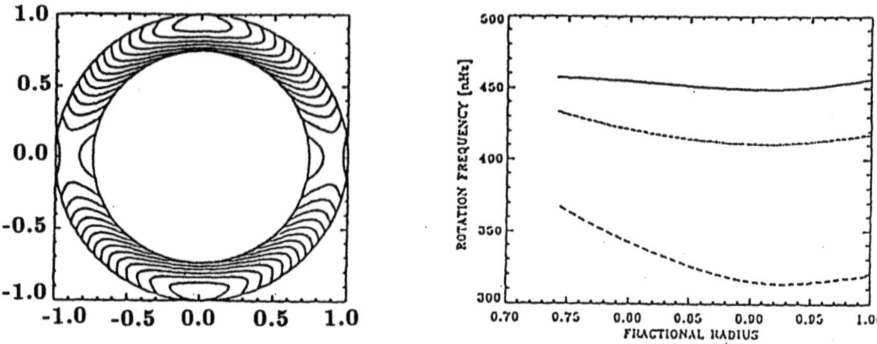

Fig. 5. The results for $\Omega_0 = 2.7 \cdot 10^{-6}\ s^{-1}$. Left: Iso-rotation contours. Right: Angular velocity for the equatorial plane (solid), the polar axis (dashed), and at 30°-latitude (dashed-dotted)

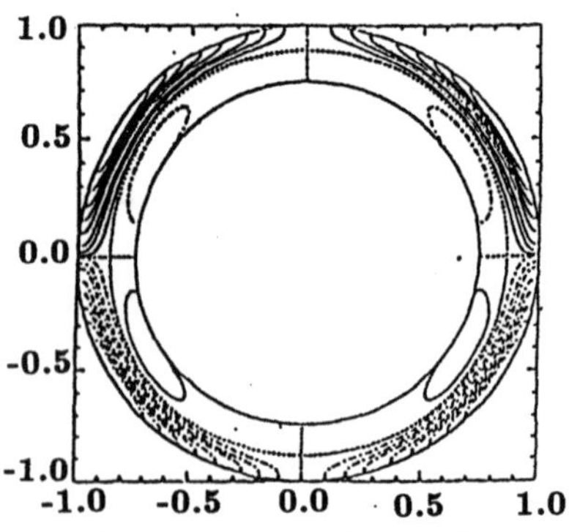

Fig. 6. Isocontours of covariance, $Q_{\theta\phi}$. Solid lines: negative values

Fig. 6 shows the isocontours of the velocity covariance $Q_{\theta\phi}$. Both positive and negative values are present in each hemisphere at different depths. It is often claimed that a positive covariance (in the northern hemisphere) is necessary to maintain the solar equatorial acceleration. This is generally not the case: Figs. 5 and 6 are consequences of the same model. The covariance $Q_{\theta\phi}$ is the sum (24) of the contributions from diffusion and from the Λ-effect. It is true that $Q^{\Lambda}_{\theta\phi}$ must indeed be positive, but the sum (24) can have either sign. Observations, of course, supply the complete covariance $Q_{\theta\phi}$. The recent measurements by Nesme-Ribes et al. (1993b) suggest relatively small values of $\mid Q_{\theta\phi} \mid$ of the order of 0.01 (deg/day)^2, where the sign depends on the nature of the tracers used. Possibly this is the result of different anchoring depths for different tracers.

5. Taylor number puzzle

5.1 MEAN-FIELD THERMODYNAMICS

Rotation laws such as those in 4.2 can never be exactly circulation–free. Their centrifugal force is no longer conservative so that, by (9), a meridional flow must develop. This flow, in turn, modifies the rotation law:

$$\frac{\sin\theta}{r^2}\,\frac{\partial}{\partial r}\left(\rho r^3 Q_{r\phi}\right) + \frac{\rho}{\sin\theta}\,\frac{\partial}{\partial\theta}\left(\sin^2\theta Q_{\theta\phi}\right) + \mathrm{div}\left(\rho r^2 \sin^2\theta\ \Omega\overline{\mathbf{u}}^m\right) = 0.$$

From the early model by Köhler (1970) to more recent formulations (Glatzmaier 1985, Gilman & Miller 1986, Brandenburg et al. 1990) the inclusion of a meridional flow always led to the "Taylor number puzzle": cylindrical iso-rotation contours are found, independently of the Λ-effect applied.

The only way to solve the paradox seems to be through the thermodynamics. This involves the second term on the RHS of (10) which contains a large dimensionless factor:

$$\mathcal{L} = \frac{L_\odot R}{4\pi\rho\nu_T^3}, \tag{33}$$

where $L_\odot$ is the solar luminosity. It is the working hypothesis (Gilman 1986, Durney 1987) that this term cancels the Taylor–Proudman state. This is not the case, however, when the turbulent viscosity and thermal conductivity are parameterised by scalars (Glatzmaier 1985, Gilman & Miller 1986). The anisotropy of the thermal conductivity, however, produces a promising effect.

Using the ideal gas relation, we find

$$\frac{1}{\rho^2}\left(\nabla\rho\times\nabla p\right) \;=\; -\frac{1}{\rho T}\left(\nabla T\times\nabla p\right) = -\frac{1}{\rho T}\nabla\Delta T\times\nabla p, \tag{34}$$

where $\nabla\Delta T$ as the superadiabatic temperature gradient. The thermodynamic term (34) is proportional to the deviation of the stratification from adiabaticity. The relative magnitude of this deviation is only 10^{-4}. The superadiabatic gradient can therefore be simplified by linearization, i.e. the superadiabatic part of the pressure gradient in (34) can be neglected. With $\nabla p = \rho \mathbf{g}$ we then find that

$$\mathcal{D}(\overline{\mathbf{u}}^m) = r \sin\theta \frac{\partial \Omega^2}{\partial z} - \frac{g}{T}\frac{\partial \Delta T}{\partial \theta}, \tag{35}$$

where

$$\mathcal{D}(\overline{\mathbf{u}}^m) = e_i \varepsilon_{ijk} \frac{\partial}{\partial x_j} \frac{1}{\rho} \frac{\partial}{\partial x_l} \rho \mathcal{N}_{klfn} \frac{\partial \overline{u}_f^m}{\partial x_n} \tag{36}$$

represents the diffusion of the meridional flow due to the eddy viscosity, ($\mathbf{e}$ eastward unit vector). Equatorial acceleration implies negative $\cos\theta \partial \Omega^2/\partial z$. The two terms in (35) can balance each other if $\cos\theta \ \partial \Delta T/\partial \theta$ negative. One needs a "hot pole" to solve the Taylor number puzzle (cf. Stix 1989).

Neglecting the viscous heating terms in the energy transport equation, we obtain

$$\mathrm{div}\left(\mathbf{F}^{conv} + \mathbf{F}^{rad}\right) + c_p \rho \overline{\mathbf{u}}^m \cdot \nabla \Delta T = 0, \tag{37}$$

where $\mathbf{F}^{conv}$ and $\mathbf{F}^{rad}$ are the convective and radiative heat fluxes,

$$F_i^{conv} = -c_p \rho \chi_{ij} \nabla_j \Delta T, \qquad \mathbf{F}^{rad} = -c_p \rho \chi_D \nabla T, \tag{38}$$

and $\chi_D = 16\sigma T^3/3\kappa\rho^2 c_p$. Because of the meridional circulation, (37) predicts a "differential temperature". It might be anticipated that the meridional flow due to the centrifugal term, $\partial\Omega^2/\partial z$, in (35) yields the desired hot pole. This is, however, not the case when the stratification is superadiabatic. Possibly, the penetration of the circulation into the sub-adiabatically stratified radiative interior can produce the desired suppression of the circulation. This possibility is not yet been investigated.

The only hope to get a hot pole is through the anisotropy of the thermal conductivity tensor χ_{ij}. The anisotropy can be produced by the influence of rotation on turbulent convection. The mixing–length representation for the turbulent thermal conductivity tensor, with rotational effects included, is

$$\chi_{ij} = \chi_T \delta_{ij} + \chi_{\|} \frac{\Omega_i \Omega_j}{\Omega^2}, \qquad \chi_T = \chi_0 \Phi(\Omega^*), \quad \chi_{\|} = \chi_0 \Phi_{\|}(\Omega^*), \tag{39}$$

where $\chi_0 = \tau_{corr} < u'^2 > /3$ is the isotropic thermal conductivity for a non–rotating fluid. The functions Φ and $\Phi_{||}$ are positive and represent the effects of rotation; $\Phi(0) = 1$ and $\Phi_{||}(0) = 0$. Explicit expressions are given in Kitchatinov et al. (1994). As a result of the $\chi_{||}$-term in (39), the radial thermal conductivity is larger at the poles than at the equator; a pole-ward latitudinal component in the convective heat flux (38) results. A hot pole can therefore indeed be expected.

5.2 THE SOLUTION

A new model for differential rotation was constructed by solving (35) and (37) for a spherical annulus. Kramers' opacity law was used in (38). The outer boundary was taken at $x_e = 0.95$. Eqs. (27) for the Λ–effect were adopted. The Ω^*–profile suggested by mixing–length theory was used. The coefficient ν_1 of the isotropic part of (29) was used in the Taylor number. The radial Ω^*–profile implies that this number is depth-dependent. Applying the mixing-length relation, $\tau_{corr} u' \cong \alpha_{MLT} H_p$ and with (30) and (11), we obtain

$$\mathrm{Ta} = \left(\frac{15}{4} \frac{R^2}{\alpha_{MLT}^2 H_p^2} \frac{\Omega^*}{\Phi_1(\Omega^*)} \right)^2 . \tag{40}$$

The constant value $\tau_{corr} < u'^2 >= 8 \cdot 10^{13}$ cm^2 s^{-1} was applied, corresponding to $\nu_0 = 2 \cdot 10^{13}$ cm^2 s^{-1}. The radial heat flux was assumed to be constant and latitude–independent at the bottom, while the top boundary radiated as a black–body.

The results for small Ω^* were the same as for the circulation–free model of 4.2. This is because the Taylor number (40) is relatively small. The two models remained close to each other until $\Omega^* \simeq 3$ when Ta $\simeq 10^7$. The results for $\Omega^* = 2.3$ (Ta $= 2.1 \cdot 10^7$) are displayed in Fig. 7. The general agreement with the solar data is quite obvious; only the surface meridional flow raises potential difficulties. The circulation is equatorwards, and the poles are warm.

The remaining open questions are probably due to our ignorance of the coupling between turbulence and super-adiabaticity. The real solar value of Ω_0^* is approximately 6. Viscosity quenching results, however, in Taylor number increased for this value to Ta $= 5.2 \cdot 10^9$. The corresponding rotation law (Fig. 8) is again close to the Taylor-Proudman state.

If we assume that the general trend of the rotation law with Ω^* is correct,

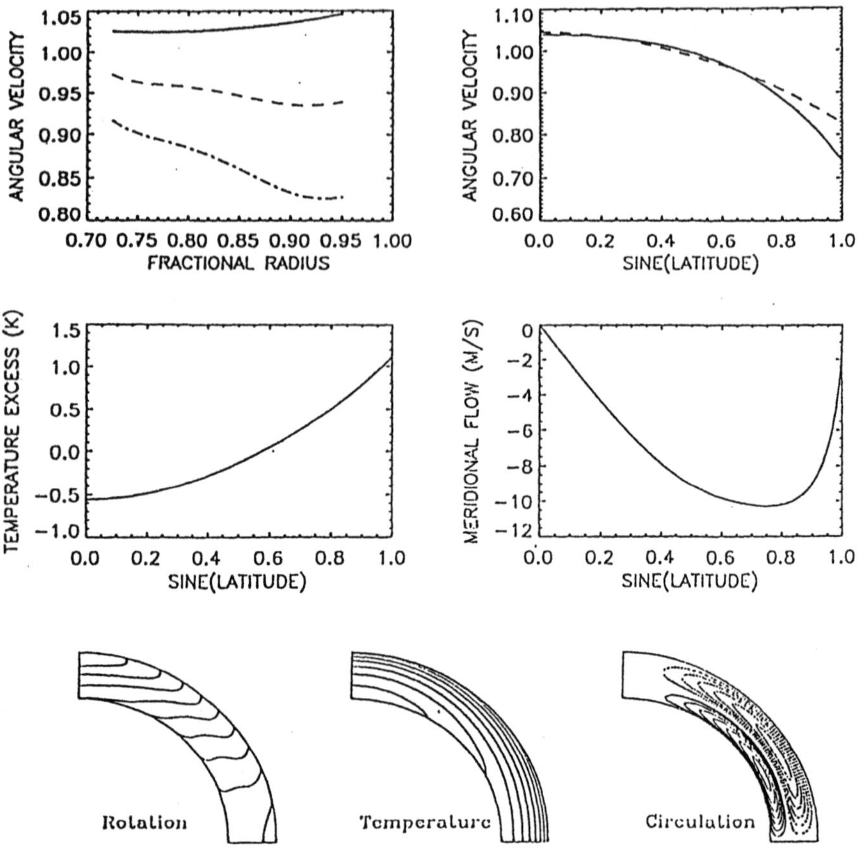

Fig. 7. Results for $\Omega_0^* = 2.3$ Upper row, Left: depth-dependence of Ω for equator (solid), for 45° (dashed), and pole (dashed-dotted); Right: the simulated (dashed) and observed (solid) Ω_s Middle row, Left: latitude-dependence of $T_s - T_{eff}$; Right: Surface meridional flow (negative value: equator-ward circulation). Bottom row: isolines of Ω, superadiabatic temperature, and the streamlines of meridional flow

stars rotating faster than the Sun should possess rotation laws such as in Fig. 8. For more rapid rotators, Taylor-Proudman solutions with equatorial deceleration are predicted. The observations by Donahue & Baliunas (1993) indicate that this may indeed be the case.

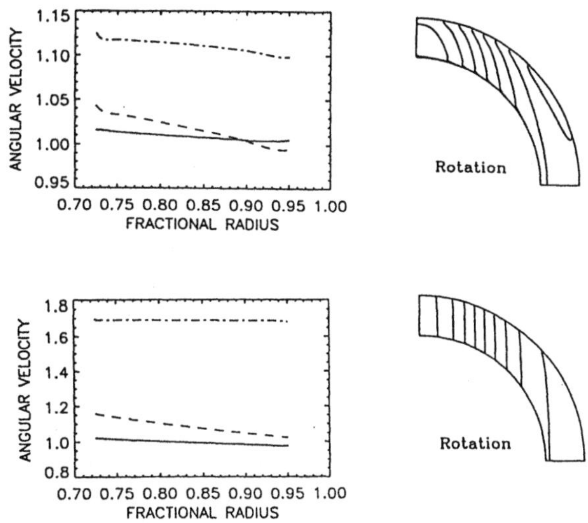

Fig. 8. Rotation laws for rapid rotators. Top: $\Omega_0^* = 7$, Ta $= 5.2 \cdot 10^9$. Bottom: $\Omega_0^* = 15$, Ta $= 1.5 \cdot 10^{11}$

6. Rotation in time

6.1 TORSIONAL OSCILLATIONS

There must also exist a feedback of the dynamo–created magnetic field on the rotation law. In the high-viscosity limit, we see from (1) that

$$\frac{\Delta\Omega}{\Omega} \sim \frac{v_A^2}{V_{rot}u'} \frac{H_\Omega}{l_{corr}} \varepsilon \sim \frac{H_\Omega}{10 l_{corr}} \varepsilon. \qquad (41)$$

v_A is the Alfvén speed, ε denotes the ratio between poloidal field and toroidal field, H_Ω is the characteristic scale of the angular velocity. Obviously, the Ω-gradients created by the Lorentz force are of order ε, i.e. of order 10^{-3}. The observations have the expected trend. Snodgrass & Howard (1985) found a variation of ω_2 and ω_4 during the solar cycle of

$$\delta\omega_2 \simeq 0.0024, \qquad \delta\omega_4 \simeq 0.0010 \qquad (42)$$

appeared for the difference in ω between maximum and minimum activity; ω_2, ω_4 and the magnetic activity cycle are in phase. At minimum, the angular velocity profile is slightly flattened at the equator but considerably ($\sim 5\%$) steepened at the poles. From spot measurements (Nesme-Ribes et al. 1993a) as well as from magnetic field data (Snodgrass 1991, Komm & Harvey 1993) one can show that the rotation is more "rigid" during activity maxima and more "differential" during minima, the difference being of the order of 1–2 % . A similar difference exists between the two solar hemispheres (cf. Fig. 1, Nesme-Ribes et al. 1993a). The asymmetry is stronger just after maximum activity.

The solar torsional oscillations are of smaller magnitude (Howard & LaBonte 1980). After subtracting the mean differential rotation from the observed angular velocity, a systematic flow structure, Ω_{cyc}, remains. The $\Omega_{cyc}(\theta)$ consists of alternating bands of fast and slow rotation drifting towards the equator; the cycle repeats every 22 years. The centre of magnetic activity is located between a pair of bands, the more rapidly moving of which is the closer to the equator. In other words, close to an activity belt $\cos\theta\partial\Omega_{cyc}/\partial\theta > 0$. The linear velocity corresponding to $\Omega_{cyc}(\theta)$ is only about 5 m/s.

6.2 LONG-TERM VARIATIONS

The series of presently available sunspot data is long enough to look for the long-term ($\sim$ 100 yrs) variability of the tracer rotation (Balthasar et al. 1986; Yoshimura & Kambry 1993a). The variation period of about 80 yrs was found. The equatorial acceleration has a minimum when the period of rotation is a maximum and vice versa. The variations in rotation rate are correlated with the magnetic activity. The mean rotation rate possesses a maximum close to the maximum of the 100-yr cycle of sunspot activity. There is, however, a phase-shift between them. The rotation follows the 100-yr magnetic cycle with a delay of about 23 yrs (Yoshimura & Kambry 1993b).

Solar magnetic activity experiences non-periodic long–term variations like the Maunder minimum, and one expects similar variations in rotation. Indeed, Ribes et al. (1987) and Ribes & Nesme-Ribes (1993) found a rotation rate slower by about 2% at the equator, and by about 6% at 18°–latitude, than at the present time. That rotation was much more differential than it is now. The more magnetic the Sun is, the faster and more rigidly its surface rotates.

7. Time-dependent models

The Lorentz force provides a natural explanation for the variations in rotation. Two possibilities exist: large-scale and small-scale feedbacks (Rüdiger & Kitchatinov 1990). Only the first possibility has been thoroughly studied so far. Schüssler (1981) and Yoshimura (1981) demonstrated that solar torsional oscillations might be explained by the mean-field Lorentz force. Rüdiger et al. (1986) appealed to a dynamo model that reproduced the phase relation between the torsional and dynamo waves. The oscillation amplitude was too small, however. The reason is the small poloidal field characteristic of solar $\alpha\Omega$-dynamos.

The micro–feedback can be driven by the toroidal field alone. The idea is that the field modifies the turbulence and the resulting Reynolds stresses; the small–scale Maxwell stresses may also be important. Estimates of the angular momentum fluxes created by the mean magnetic field are promising (Rüdiger & Kitchatinov 1990, Kitchatinov 1990). The efficiency of the micro-feedback has not, however, yet been demonstrated in any torsional oscillation model.

Acknowledgement. We are grateful to Professor Paul Roberts for critically reading this paper when in manuscript form, and for suggesting improvements.

References

Balthasar, H., Vaźquez, M., and Wöhl, H. (1986) Astron. Astrophys. **155**, 87

Brandenburg, A., Moss, D., Rüdiger, G., and Tuominen, I. (1990) Solar Phys. **128**, 243

Brandenburg, A., Moss, D., and Tuominen, I. (1992) Astron. Astrophys. **265**, 328

Canuto, V. M., Minotti, F. O., and Schilling, O. (1993) Astrophys. J. (submitted)

Donahue, R. A. and Baliunas, S. L. (1993) Astrophys. J. **393**, L63

Durney, B. R. (1987) In: The Internal Solar Angular Velocity, B. R. Durney & S. Sofia (eds.), D. Reidel, 235

Durney, B.R. (1989) Astrophys. J. **338**, 509

Durney, B. R. and Spruit, H. C. (1979) Astrophys. J. **234**, 1067

Gilman, P. A. (1986) In: Physics of the Sun, P. A. Sturrock (ed.), D. Reidel, **1**, 95

Gilman, P. A. and Miller, J. (1986) ApJ Suppl. **61**, 585

Glatzmaier, G. A. (1985) Astrophys. J. **291**, 300

Goode, P.R. (1991) In: The Sun and cool stars, I. Tuominen, D. Moss & G. Rüdiger (eds.), 157

Hill, F. (1987) In: The Internal Solar Angular Velocity, B. R. Durney & S. Sofia (eds), D. Reidel, Dordrecht, 45

Howard, R. and Harvey, J. (1970) Solar Phys. **12**, 23

Howard, R. and LaBonte, B. J. (1980) Astrophys. J. **239**, L33

Howard, R., Adkins, J.M., Boyden, F., Cragg, T.A., Gregory, T.S., LaBonte, B.J., and Webster, L. (1983) Solar Phys. **83**, 321

Kitchatinov, L. L. (1986) Geophys. Astrophys. Fluid Dyn. **35**, 93

Kitchatinov, L. L. (1987) Geophys. Astrophys. Fluid Dyn. **38**, 273

Kitchatinov, L. L. (1990) Geophys. Astrophys. Fluid Dyn. **54**, 145

Kitchatinov, L. L. and Rüdiger, G. (1993) Astron. Astrophys. **276**, 96

Kitchatinov, L. L., Pipin, V. V., and Rüdiger, G. (1994) Astron. Nachr. **315**, 97

Köhler, H. (1969) Thesis, Göttingen

Köhler, H. (1970) Solar Phys. **13**, 3

Komm, R. and Harvey, R. (1993) Solar Phys. (in press)

Küker, M., Rüdiger, G., and Kitchatinov, L. L. (1993) Astron. Astrophys. **279**, L1

Libbrecht, K. G. (1988) In: Seismology of the Sun and Sun-Like Stars, E. J. Rolfe (ed.), ESA SP-286, 131

Nesme-Ribes, E., Ferreira, E. N., and Mein, P. (1993a) Astron. Astrophys. **274**, 563

Nesme-Ribes, E., Ferreira, E. N., and Vince, I. (1993b) Astron. Astrophys.

276, 211

Newton, H. W. and Nunn, M. L. (1951) M.N.R.A.S. **111**, 413

Pidatella R. M., Stix M., Belvedere G., and Paterno, L. (1986) Astron. Astrophys. **156**, 22

Rhodes, E. J., Cacciani, A., Korzennik, S. G., and Ulrich, R. K. (1991) In: Challenges to Theories of the Structure of Moderate-Mass Stars, D. Gough & J. Toomre (eds), 285

Ribes, E., Ribes, J. C., and Barthalot, R. (1987) Nature **326**, 52

Ribes J. C. and Nesme-Ribes E. (1993) Astron. Astrophys. **276**, 549

Roberts, P. H. and Soward, A. (1975) Astron. Nachr. **296**, 49

Rüdiger, G. (1983) Geophys. Astrophys. Fluid Dyn. **25**, 213

Rüdiger, G. (1989) Differential rotation and stellar convection: Sun and solar-type stars, Gordon & Breach, New York

Rüdiger, G. and Kitchatinov, L. L. (1990) Astron. Astrophys. **263**, 503

Rüdiger, G. and Tuominen, I. (1990) In: Solar Photosphere: Structure, Convection, and Magnetic Fields, J. O. Stenflo (ed.), Kluwer, 315

Rüdiger, G., Tuominen, I., Krause, F., and Virtanen, H (1986) Astron. Astrophys. **166**, 306

Schou, J. (1991) In: Challenges to Theories of the Structure of Moderate-Mass Stars, D. Gough, J. Toomre (eds.), 93

Schou, J. (1993) PhD thesis, University of Arhus

Snodgrass, H. B. (1991) In: The Solar Cycle, Astr. Soc. Pacific conf. series No. 27, K. Harvey (ed.), 71 & 205

Snodgrass, H. B. and Howard, R. (1985) Solar Phys. **95**, 221

Stix, M. (1989) Rev. Mod. Astron. **2**, 248

Stix, M. (1993) (private communication)

Wasiutinski, J. (1946) Astrophysica Norvegica **4**, 1

Yoshimura. H. and Kambry. M. A. (1993a) Astron. Nachr. **314**, 9

Yoshimura, H. and Kambry. M. A. (1993b) Astron. Nachr. **314**, 21

Zöllner, F. (1881) In Wissenschaftliche Abhandlungen, Bd. 4 Commissionsverlag von L. Staackmann, Leipzig

OSCILLATORY DYNAMOS SHOWING CHANGE OF PARITY ON A LARGE TIME SCALE

F. Krause
Astrophysikalisches Institute Potsdam
An der Sternwarte 16
D-14482 Potsdam
Fed. Rep. Germany

Summary

More observational facts have become known in connection with the Maunder minimum, i.e. the time of weak solar activity dating between 1645 and 1720. It is thus worth noting that, within the solar-type dynamos dynamos developed so far, there exist models which reveal properties that help in our understanding of the weak intensity, such as the related change of the pattern of the butterfly diagram and the increase of the cosmic ray intensity.

1. Introduction

While much is being said today about the Maunder minimum, it should be noticed that this period was not only characterized by (i) the reduced appearance or even absence of sunspots and (ii) drastically reduced solar activity, but also by (iii) a significant asymmetry in the distribution of sunspots with respect to the equator (asymmetric butterfly diagram) (Ribes & Nesme-Ribes 1993) and (iv) an increase in the cosmic radiation in the Earth's environment (Beer *et al.*, 1991) .

Since sunspots are supposed to be an indication of a strong toroidal magnetic field in the solar interior it must be concluded from (i) that the magnetic field generated at the time of the Maunder minimum was of low intensity

NATO ASI Series, Vol. I 25
The Solar Engine and Its Influence
on Terrestrial Atmosphere and Climate
Edited by Elizabeth Nesme-Ribes

compared with normal cycles. However, there were some sunspots, and this suggests that the 11-year cycle persisted throughout the entire period (Ribes & Nesme-Ribes 1993). This conclusion was confirmed by the only other indicator of solar activity available at that time, the observations of aurorae (Schröder 1992).

Solar activity phenomena at the present time indicate that the magnetic field of the Sun is antisymmetric with respect to the equatorial plane. That is to say that the poloidal field part is dipolar and the toroidal field part consists of an even number of ring fields, arranged symmetrically above and below the equatorial plane but with opposite directions. This is inferred from the arrangement of sunspots with respect to the Sun's equator (evident from the butterfly diagram), and the observed distribution of the magnetic polarities not only between preceding and following spots in a given group but also between spot groups in both hemispheres. According to the observations, this symmetry was distorted during the Maunder minimum (Ribes & Nesme-Ribes 1993).

From the observed antisymmetry of the solar magnetic field during normal cycles, we infer that the present day field far into interplanetary space is dominated by its dipolar component. It is this component that affects the charged particles of the cosmic radiation most. A variation in the intensity of the cosmic radiation observed at the Earth's surface may be explained by a change in the large-scale structure of the magnetic field in the solar environment. We may thus infer that not only the toroidal field (as indicated by the different nature of the sunspot record) but also the poloidal field changed its large-scale structure during the Maunder minimum.
On the furthest reaching part of the Sun's magnetic field in interplanetary space, namely the dipole component, a significant quadrupolar component may have been superimposed (Sokoloff & Nesme-Ribes 1994, Nesme-Ribes *et al.*, 1994).

2. Solar-Type Dynamos

The question to be considered is whether these different observational findings and their physical interpretations may be explained in terms of the current dynamo theories of the solar magnetic field.

2.1 Symmetry properties of generated magnetic fields

Dynamo models close to the real Sun generally show that the most easily excited magnetic field mode is of odd parity (Steenbeck & Krause 1969, cf. Krause & Rädler 1980, Roberts 1994). This mode shows antisymmetry with respect to the equatorial plane. The poloidal field part is of dipolar-type and the toroidal field part consists of an even number of ring fields which are arranged symmetrical to the equatorial plane but with opposite orientation. This corresponds well with the observed field pattern of normal cycles, as described above. It is worth noting that the toroidal field vanishes at the equator.

These investigations further show that magnetic fields of even parity can also be excited by solar-type dynamos, although they require stronger inductive action (Roberts 1972). For magnetic fields of even parity, the poloidal field component is of quadrupolar type, and the toroidal part consists of an odd number of ring fields. One ring spans the equatorial plane and the others lie symmetrically above and below this plane. The magnetic field orientation is symmetrical with respect to the equatorial plane. In this case the toroidal field generally has at least local a maximum on the equatorial plane. The differences in the structure of the fields of the two parities are best visualized through the representations shown in a paper by Roberts (1972).

Real objects like the Sun will not generally generate a magnetic field of purely one parity. Investigations of nonlinear models reveal that only objects with excitation conditions just beyond the critical value will excite such single-parity fields (Krause & Meinel 1988) ; objects operating in more strongly nonlinear states excite fields of more complicated structure

(Brandenburg *et al.*, 1989, Rädler *et al.*, 1990, Jennings 1991, Jennings & Weiss 1991). For example, dynamos have been found that, in the saturated state, generate magnetic fields that oscillate between the two parities over a period long compared with that of the solar activity cycle.

2.2 Dynamo model showing " grand minima"

In the context of interpretations of the Maunder minimum and the related phenomena mentioned above, it is of interest that models of solar-type dynamos have been found (Brandenburg, Krause & Tuominen 1989, Brandenburg *et al.*, 1990) that generate an oscillating, nearly antisymmetric magnetic field similar to that observed on the Sun during normal cycles. However, on a time scale, which is large compared with the cycle period, it changes its parity, and a field with a significant contribution of a symmetric (even parity) part is formed over a period of a few cycles. It is worth noting that, at the time of the parity change, the total energy is small compared with its value during periods of normal parity.

Fig.**1** gives some characteristic parameters of the excited magnetic field. E denotes the total energy, P the parity (+1 corresponds to an even (quadrupolar-type) field and -1 to an odd (dipolar-type) field) and Q is a parameter describing the degree of asymmetry (the value Q=0 corresponds to a symmetric (even) as well as to an antisymmetric (odd) field).

The main point is that the model shows the simultaneous appearance of a minimal total energy over a period of several cycles. This is shown by the parameter E ; the maximal asymmetry is indicated by the parameters P and Q. The first is in agreement with the observed low sunspots count and weak solar activity, and the second with an asymmetric butterfly diagram (Fig.2). The second also indicates a change in the structure of the magnetic field in the Sun's environment : the dipole is weak and, remarkably, a quadrupole is superposed on it. The conditions determining the influx of cosmic ray particles on the Earth are clearly greatly changed.

3. Final Remarks

In the respects so far described, this model successfully simulates the behaviour of the Sun during the Maunder minimum. In addition, it predicts periodic reappearance of such "grand minima". Variations in the cosmic ray intensity indeed indicate that there have been several Maunder-minimum-type periods during the last few millennia (Beer *et al.*, 1991), although a periodic pattern of behaviour has not been inferred.

It should, however, be noted that, with respect to other properties (e.g. the assumed differential rotation), this model does not approximate the real Sun

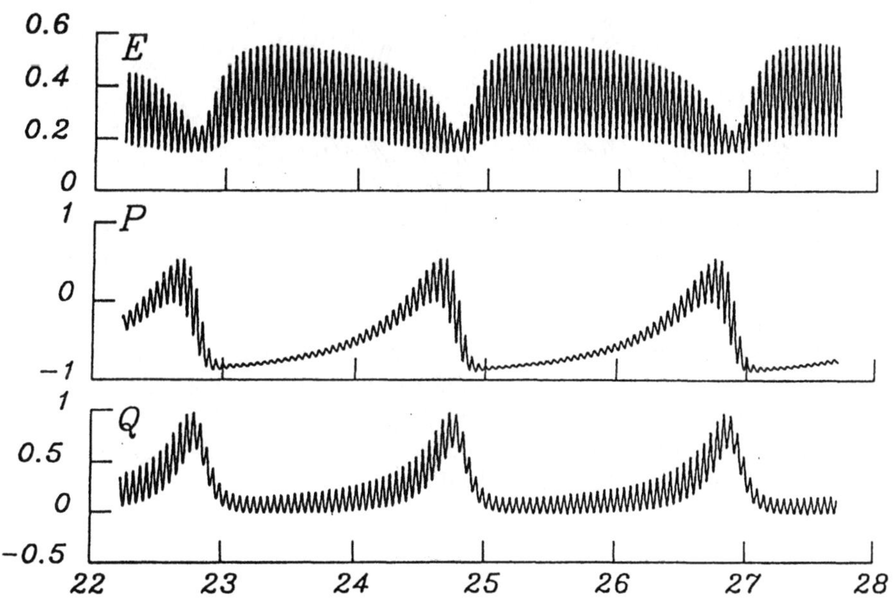

Fig. 1. The variation of the total magnetic energy E, the parity parameter P, and the degree of asymmetry Q for a solar-type dynamo model. Note the simultaneous appearance of an energy minimum, a change of parity, and a maximum of asymmetry (from Brandenburg, Krause & Tuominen 1989)

well. But we consider it of importance that, within this class of solar-type dynamos, some have been found that show such a strange phenomena like the "grand minima".

Future investigations should show whether models exist that approximate the Sun in nearly all essential respects. It would be desirable to find a class of models in which the "grand minima" are generic.

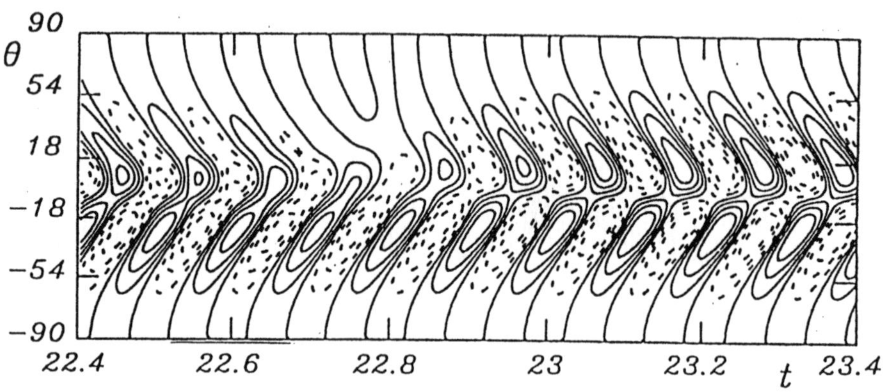

Fig. 2. Sunspots emerge at positions where the toroidal magnetic field is sufficiently strong. Thus lines of constant toroidal field strength derived from dynamo models lead to synthetic butterfly diagrams. At ≈22°5, an asymmetry can be seen : Sunspots emerge at, and a little south of, the equator (from Brandenburg, Krause and Tuominen 1989)

Acknowledgement

I am grateful to Professor Paul Roberts and to Dr. Elizabeth Nesme-Ribes for a critical reading of this paper when in manuscript form, and for suggesting improvements.

References

Beer, J., Raisbeck, G.M., Yiou, F. (1991). Time Variations of ^{10}Be and Solar Activity. In The Sun in Time, eds. C.P. Sonett, M.S. Giampapa, M.S. Matthews, The University of Arizona, Tucson: 343-359

Brandenburg, A., Krause, F., Tuominen, I. (1989). Parity Selection in Nonlinear Dynamos. In Turbulence and Nonlinear Dynamics in MHD Flows, eds. M., Meneguzzi, A., Pouquet and P.L., Sulem, Elsevier Science Publ. B.V. (North-Holland): 35-40

Brandenburg, A., Krause, F., Meinel, R., Moss, D., Tuominen, I. (1989). The Stability of Nonlinear Dynamos and the Limited Role of Kinematic Growth Rates. Astron. Astrophys., 213: 411-422

Brandenburg, A., Meinel, R., Moss, D., Tuominen, I. (1990). Variation of Even and Odd Parity in the Solar Dynamo. In Solar Photosphere : Structure, Convection and Magnetic Field, ed. J. O. Stenflo, IAU 138: 379-382

Jennings, R.L. (1991). Symmetry Breaking in a Nonlinear $\alpha\omega$-Dynamo. Geophys. Astrophys. Fluid Dyn., 57: 147-189

Jennings, R.L., Weiss, N.O. (1991). Symmetry Breaking in Stellar Dynamos. Mon. Not. R. Astr. Soc., 252: 249-260

Krause, F., Meinel, R. (1988). Stability of Simple Nonlinear α^2-dynamos. Geophys. Astrophys. Fluid Dynamics, 43: 95-117

Krause, F., Rädler, K.H. (1980). Mean-Field Magnetohydrodynamics and Dynamo Theory. Akademie Verlag Berlin/Pergamon Press Oxford

Nesme-Ribes, E., Sokoloff, D., Ribes, J.C., Kremliovski, M. (1994). The Maunder Minimum and the Solar Dynamo. This volume

Rädler, K.-H., Wiedemann, E., Brandenburg, A., Meinel, R., Tuominen, I. (1990). Nonlinear Mean-Field Dynamos : Stability and Evolution of Three-Dimensional Magnetic Field Configurations. Astron. Astrophys., 239: 413-423

Ribes, J.C., Nesme-Ribes, E. (1993). The Solar Sunspot Cycle in the Maunder Minimum AD 1645 to AD 1715. Astron. Astrophys., 276: 549-563

Roberts, P.H. (1972). Kinematic Dynamo Models. Phil. Trans. R. Soc. Lond., A., 272: 663-698

Roberts, P.H. (1994). The Solar Dynamo. This volume

Schröder, W. (1992). On the Existence of the 11-Year Cycle in Solar and Auroral Activity before and during the So-Called Maunder Minimum. J. Geomag. Geoelectr., 44: 119-128

Sokoloff, D., Nesme-Ribes, E. (1994). The Maunder Minimum, a Mixed-Parity Dynamo ?. Astron. Astrophys., in press

Steenbeck, M., and Krause, F. (1969). Zur Dynamotheorie stellar und planetarer Magnetfelder. I. Berechnungsonnenlicher Wechselfeldgeneratoren. Astrononische Nachrichten, 291: 49-84

A NEW RECONSTRUCTION OF SOLAR ACTIVITY, 1610-1993

Douglas V. Hoyt
Research and Data Systems, Corp.
7855 Walker Drive, Suite 460
Greenbelt, MD 20770

Kenneth H. Schatten
NASA/Goddard Space Flight Center
Code 914
Greenbelt, MD 20771

and

Elizabeth Nesmes-Ribes
Observatoire de Paris
Meudon, France

Summary

In the one hundred years since Wolf died, little effort has gone into research to see if improved reconstructions of solar activity can be made. We have gathered more than 348,000 observations of daily sunspot group counts from more than 350 observers active from 1610 to 1993. Based upon group counts alone, it is possible to make an objective and homogeneous reconstruction of solar activity. From our study, it appears that the Sun has steadily increased in activity since 1700 with the exception of a brief decrease in the Dalton Minimum (1795-1823). The significant results here are the greater depth of the Dalton Minimum, the generally lower activity throughout the 1700's, and the gradual rise in activity from the Maunder

NATO ASI Series, Vol. I 25
The Solar Engine and Its Influence
on Terrestrial Atmosphere and Climate
Edited by Elizabeth Nesme-Ribes

Minimum to the present day. This solar activity reconstruction is quite similar to those Wolf published before 1868 rather than the revised Wolf reconstructions after 1873 which used geomagnetic fluctuations.

Introduction

The eleven year or Schwabe cycle in solar activity was discovered by Heinrich Schwabe in 1843, based upon 17 years of observations. In the early 1850's Rudolf Wolf of Bern and later Zurich, Switzerland began a search of historic sunspot observations. Over the next 40 years, Wolf gathered thousands of observations from hundreds of observers and derived a record of sunspot numbers from the 1600's onwards. His index of solar activity is known as the Wolf Sunspot Number. It equals ten times the number of sunspot groups plus the number of individual sunspots, all multiplied by a correction factor (k) for each observer. The Wolf Sunspot Number is considered to be reliable from 1848 to the present, good from 1818 to 1847, questionable for 1749-1817, and poor for the earlier years. These subjective judgments do not give us a numerical measure of how confident we can be in Wolf's reconstruction of solar activity. Some measure of the random and systematic errors in his reconstructions are needed.

Wolf died on December 6, 1893. Since that time, little effort has been devoted to re-check Wolf's sunspot number reconstruction. These reconstructions are very important for studies of solar variability related to climate change. Did he get the correct answer? In this note, we will address this question, give a new reconstruction of sunspot numbers for 1610 to the present, and outline our plans for future work on this problem.

Data Collection

Wolf reconstructed solar activity using both the number of sunspot groups and the number of individual sunspots. We have shown previously (Hoyt and Schatten, 1992) that more than 90% of the variability in the Wolf Sunspot Numbers is attributable to changes in the number of sunspot groups. Schaefer (1993) similarly shows that the Wolf Sunspot Number may be set equal to a constant times the number of sunspot groups. An index based upon sunspot groups alone can simulate the Wolf Sunspot Number. This index will be called the Group Sunspot Number and will be discussed further below.

Based upon these considerations we started digitizing as many raw telescopic observations of daily counts of sunspot groups as we could find. Our objective is to relate all the observers to one standard observer so that all the sunspot group observations are self-consistent. Many of the observations are tabulated by Wolf in his journal, Astronomische Mittheilung (1852-1946). Other observations are scattered throughout the literature, such as the observations by Kunitomo in 1835-1836, which were tabulated and published by Yamamoto in 1935 in the Kyoto Bulletin. In all, our database contains 348,671 observations from more than 350 observers, several of whom were unavailable to Wolf. Although we continue to gather data, a sufficient number of observations now exist so that we make a significant re-appraisal of sunspot numbers for 1610-1993.

Data Analysis

The observations by the Royal Greenwich Observatory (RGO), 1874-1976, are among the most reliable sunspot observations made. Our approach is to use RGO as the standard observer and normalize earlier observers so that counts of sunspot groups would equal those of RGO. To do this, we sum up all observations made by RGO and the other observer on common days of measurement, and the ratio of the number of RGO groups to the number of groups by the other observers gives a correction factor (f) by which this observer's group counts must be multiplied to be on the same scale as RGO. For example, Schmidt of Athens had 3392 days of common measurements with RGO from 1874 to 1883. Schmidt observed 6221 groups and RGO observed 7017 groups. The correction for Schmidt therefore is 1.127. Schmidt also observed in the 1840's along with Schwabe who observed from 1826 to 1867. Schmidt and Schwabe had 1837 common observation days and observed 6260 and 6080 groups respectively. Because Schwabe ceased observing after 1867, it is not possible to directly compare his observations to RGO, but by taking the ratio Schwabe to Schmidt and the ratio Schmidt to RGO, one can find the ratio Schwabe to RGO. Other intermediate observers, such as Weber of Peckeloh and Wolf of Bern and Zurich, can be used to find the ratio of Schwabe to RGO. Our analysis indicates that Schwabe's sunspot group numbers need to be multiplied by 1.285 to equal those of RGO.

By using one or more intermediate observers, nearly all the observers from 1749 to the present can be put on the RGO scale. Our approach is to use the minimum number of intermediate observers, but to use maximum number of available paths of comparison. For example, for Christian Horrebow for whom we have daily observations for 1767 to 1776, we find a correction factor of 1.17 using four intermediate observers in each of 12 different paths or ratio determinations. Our analysis indicates that the monthly mean group number tabulation of Horrebow's sunspot groups by Thiele (1859) is less correct than the daily observations used by Wolf. Some isolated observers exist and the observers prior to 1749 are isolated from the modern day observers. The average modern observer needs to have his observations multiplied by 1.255 to bring them in agreement with RGO. For most of the earlier and isolated observers, we use this correction factor. It is important in these analyses to use primary sources or reliable secondary sources.

Given that all the observers are normalized to the RGO scale of observations, how do we convert the number of sunspot groups to Group Sunspot Numbers? For the period 1874 to 1991, we normalized the Group Sunspot Numbers to equal the average of the Wolf Sunspot Numbers for this period. This requires that the number of sunspot groups be multiplied by 11.93. The definition of the Group Sunspot Number (R_g) therefore is:

$$R_g = 11.93 G_{RGO}$$

where G_{RGO} is the number of sunspot groups which would be counted if RGO were observing.

Results

Our results can be most easily summarized by referring to the three figures accompanying this paper. Figure 1 shows the monthly mean Group Sunspot Numbers. For the present day period there is good overall agreement with the Wolf Sunspot Numbers as might be expected, although since about 1920 the Wolf Sunspot Numbers have risen faster than the Group Sunspot Numbers. Figure 2 shows this more clearly where the Group Sunspot Numbers minus the Wolf Sunspot Numbers are shown. Cycle number 11 peaking in 1870 is stronger in our reconstruction than in Wolf's reconstruction. The major differences occur for the years 1880 to 1848. Here the Wolf Sunspot Numbers are higher than the Group Sunspot Numbers. Curiously, our Group Sunspot Numbers are similar to the Wolf Sunspot Numbers published by Wolf in 1868 and earlier. In 1874, Wolf revised his original sunspot numbers by multiplying them by a factor of 1.25 for 1826 to 1848 and by about 1.2 to 1.5 for the earlier years. Wolf's correction was apparently determined using variations of the magnetic needle at Milan. Based upon our analysis, this correction is erroneous. The Maunder Minimum clearly shows up in our reconstruction. It appears that the Sun steadily increased in activity from 1700 to 1750 and then leveled off for 50 years before entering the Dalton Minimum (1795-1823). The significant result here is that there is a steady increase in activity following the Maunder Minimum, rather than a step-like increase to near modern values as is commonly assumed.

In Figure 3, we show the fraction of the time for which we have observations of the Sun. There is a paucity of observations in the 1740's and the 1780's, so the level of solar activity during these periods remain uncertain. The Sun was well observed during both the Maunder and Dalton Minima. For some of these earlier years, it appears the Sun was observed every day, based upon comments by some observers that no sunspots were observed during entire years. In fact, most of these years were observed on more than 50% of the days (Ribes et al., 1988) and future work will give us a better measure of the fraction of time that the Sun was observed. If more than 5% of the days in any one year are randomly observed throughout the year, a reasonable value for the yearly mean can be found. Most years meet this requirement. In 1877, Wolf reported he averaged 90 observation days/year for 1749-1783, 70 days/year for 1784-1818, and 260 days/year for 1819-1848. Our database has 139, 161, and 345 days/year respectively for these three groups of years. For cycle five, we find a peak in 1801 as opposed to Wolf's conclusion of a peak in 1805. For the years 1800 to 1810, we average about 192 days/year whereas Wolf reported that he had less than 10 days/year for many of these years. Thus, we are confident that for solar cycle five, we have a better reconstruction of the solar behavior.

Conclusions

The Group Sunspot Numbers here are preliminary and will undoubtedly improve as we acquire more data. It seems likely our understanding of long-term secular variations in sunspot numbers may require some revisions. The correct reconstruction of solar activity will eventually lead to a better understanding of the Sun and its influence upon climatic change.

Our work on reconstructing sunspot numbers is an on-going endeavor. More than 350 observers and 348,000 observations have been entered into computer readable files. Nonetheless we have identified more than 50 observers for which further research is required, involving location of their data or entering of their data into our database. Some observers, such as Thomas Hussey from 1826 to 1837, made drawings of the Sun, but these observations were never reduced to sunspot group counts, nor were they available to Wolf. Other observations not used by Wolf, which we have included in this study, are those by William and John Herschel, Shea, Greisbach, Lawson, Sharp, Wasse, Scheuchzer, Graham, Carbone, Willoughby, and Remoquietanus. Identified observers for which we plan future research are: 1) In the 1600's: Blancanus, Claramontius, Cysatus, Rheita, Fridanus, Grienbererus, Guldinus, Zucchius, Hortensius, Antoninus, Cornaeus, Marius, Frankenstein, Ehinger, Rosius, Vogelius, Weickmann, and Rentchius. 2) In the 1700's: Manfredi, Blanchini, Hoffmann, Alischer, Becker, Adelburner, Beyer, Boscovich, Bose, Frobés, Schubert, Horrebow, Muller, Dunn, Schulen, Wilson, Hornsby, Hahn, and the Kirch family. 3) In the 1800's: Soemmering, Henry, Kriel, Main, Chevallier, Borck, Schott, Schwabe, Wolf, Bernaerts, Weber, Tacchini, Braak, Koch, Jahn, Lichtenberger, Steinmann, Parpart, von Heneke,

Schurig, Charnoc, Carl, and Bogulawski. Many of these scientists made infrequent observations of the Sun so their future inclusion are expected to change only the details of our activity reconstruction.

If readers know of sunspot observations by these observers or others, particularly for 1848 and earlier years, we would be interested in adding them to our database. Once the database is complete, we plan to give it to the National Geophysical Data Center in Boulder, Colorado so it will be available to all researchers.

Acknowledgments

This work supported by NASA Contract NASW-4754. We wish to thank the Naval Observatory for the use of their library and particularly thank Gregory Shelton and Brenda Corbin for their hospitality there. In addition we appreciate the aid and hospitality of Peter Hingley at the Royal Astronomical Society Library, Allen Clark at the Royal Society Library, Adam Perkins at the Cambridge University Library, and Ruth Freitag of the Library of Congress.

References

Hoyt, D.V. and Schatten, K.H. (1992). New information on solar activity, 1779-1818, from Sir William Herschel's unpublished notebooks. Astrophys. J., 384: 361-384.

Ribes, E., Ribes, J.C., Vince, I. and Merlin, Ph. (1988). A survey of historical and recent solar diameter observations. Adv. Space Res., 8: 129-132.

Schaefer, B.E., 1993. Visibility of sunspots. Astrophys. J., 411: 909-919.

Thiele, T.N., 1859. De macularum solis antiquioribus quibusdam observationibus Hafniae Institutis. Astron. Nach., 50: 257-259.

Wolf, R., 1852-1865. In Mittheilungen uber die Sonnenflecken, all issues

Wolf, R., 1866-1893. In Astronomische Mittheilungen, all issues

Yamamoto, I., 1935. Kunitomo's sunspot observations during 1835-1836. Kyoto Bull., 4, No. 303.

Zurich Observatory, 1894-1946. In Astronomische Mittheilungen, all issues

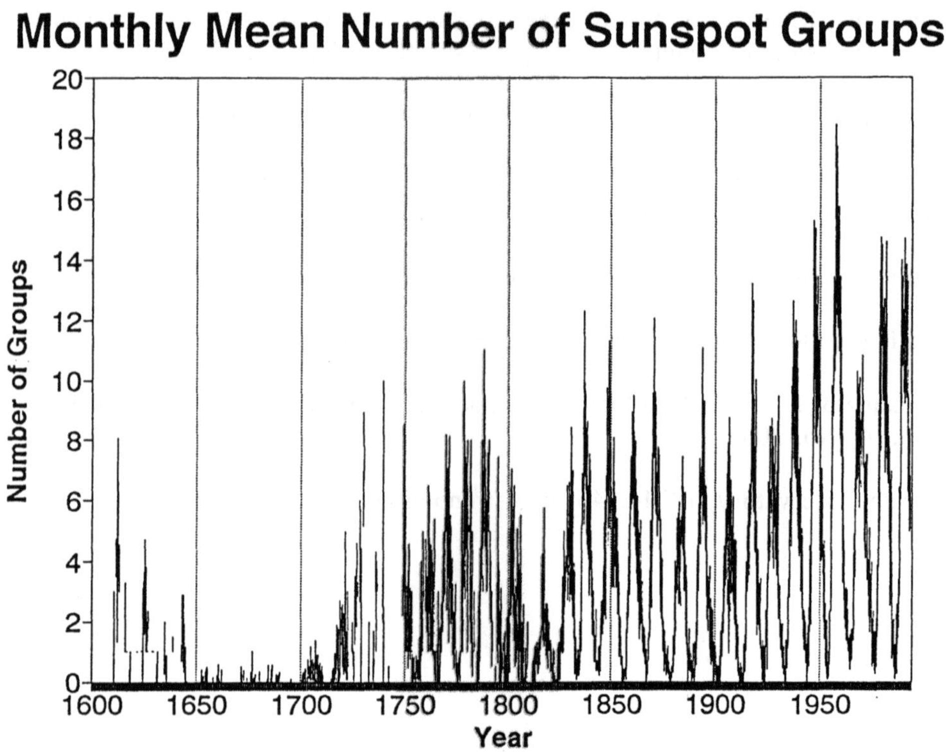

1. The reconstruction of sunspot numbers from 1610 to 1993 using only the counts of sunspot groups. More than 348,000 observations were used in the reconstruction.

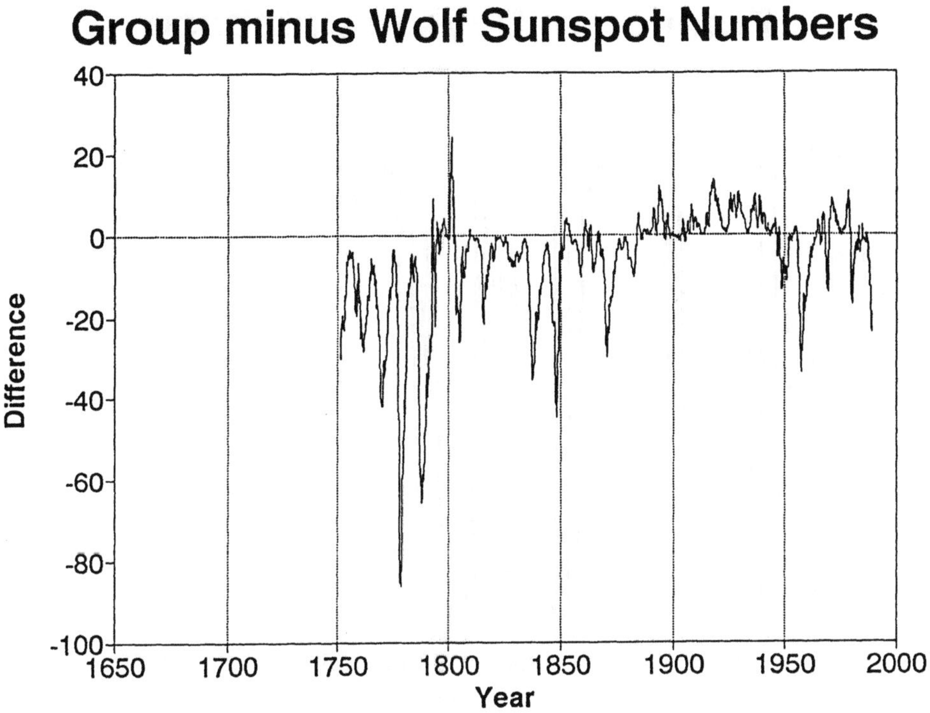

2. The differences of this reconstruction from the Zurich reconstruction. In general, we have lower numbers for sunspot numbers for 1800 to 1848, more consistent with Wolf's reconstructions prior to 1868.

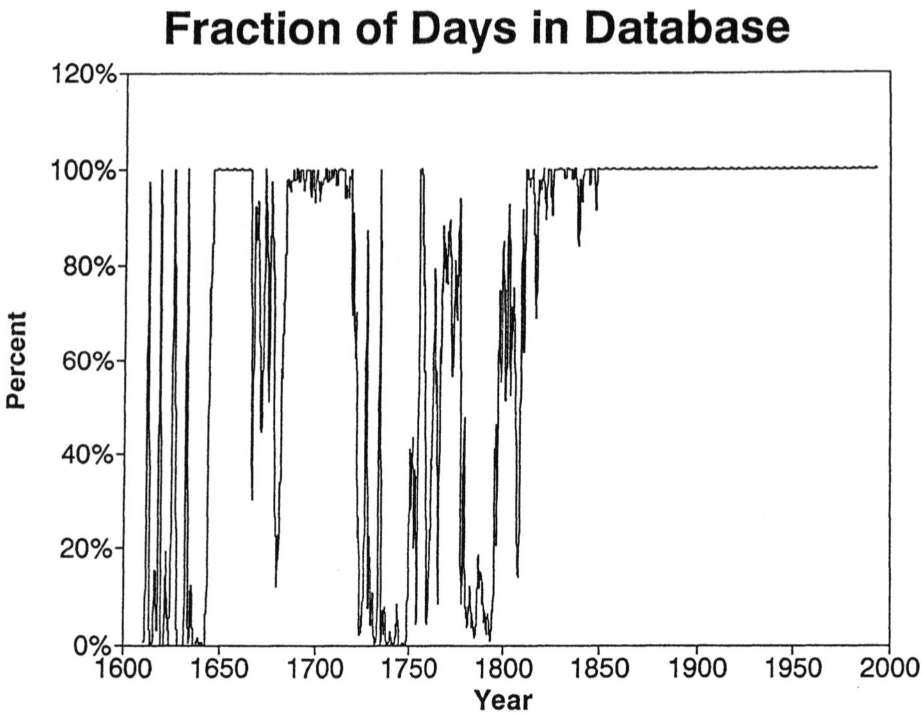

3. The fraction of the time for which we have observations. If more than 5% of the days are observed in a year, a good yearly mean can usually be found. Most years meet this criteria. Note that the Sun was well observed during the Maunder Minimum.

THE MAUNDER MINIMUM AND THE SOLAR DYNAMO

E. Nesme-Ribes[1], D. Sokoloff [1,2], J.C. Ribes[3] and M. Kremliovsky [4]

[1] Observatoire de Paris, 92195 Meudon, France
[2] Department of Physics, Moscow State University, 119899, Moscow, Russia
[3] Observatoire de Lyon, 69230 St Genis Laval, France
[4] Physics Department, University of California in San Diego, La Jolla, CA 92093, USA

Summary

On the basis of systematic observations of sunspots available since Galilei's time, we have been able to derive certain properties of normal and low-activity cycles, such as those of the Maunder minimum cycles. The characteristics used are the sunspot number, the north-south asymmetry of the sunspot distribution, the solar rotation, and other indicators of magnetic field strength. Furthermore, there is evidence that a long period of low-activity cycles began prior to the Maunder minimum that may have led to a decrease in solar output. A scenario is proposed for the long-term evolution of the magnetic field, with evident implications for the Earth's climate.

Introduction

Astronomers have often noticed and spoken about the prolonged absence of sunspots that marked the second half of the seventeenth century. They have also noticed the cold spell in Europe at the same time, referred to as the little Ice age. Let us report Cassini's comment to the French Royal Academy in 1694 : "The shift in the seasons that we have seen over the past years, and the frequent earthquakes at various locations, have hinted that some changes in the world economy has taken place" (see Cassini, 1729). Later, Cassini mentioned that astronomers were considering the possibility that astronomical

NATO ASI Series, Vol. I 25
The Solar Engine and Its Influence
on Terrestrial Atmosphere and Climate
Edited by Elizabeth Nesme-Ribes

effects were responsible for these climatic changes (in the form of irregularities in the orbital parameters) and concluded in the negative. It is clear from this that a relationship between solar variability and climate was being considered as early as the 17th century.

Two centuries later, Maunder again noticed the surprising coincidence between the solar anomaly and the coincident climatic change, and issued a series of papers published by the Royal society (Maunder, 1894). A century later, Eddy (1976) reopened the debate.

Perhaps the most fundamental objection to the causal solar variability-climate relationship was given by climatologists who argued that this "little ice age", which spanned more than two centuries (starting from the 16th and ending with the 18th century), was of much longer duration than just the Maunder minimum, usually defined as a 70-year lull (1645-1715).

The aim of this paper is to show that the Maunder minimum (1645 to 1715), (denoted hereafter to as Global Minimum or GM), might have been preceded by a number of low-activity cycles. This is based on certain characteristics of solar magnetic activity displayed in sunspot data. The weak sunspot activity of the early 17th and possibly 16th centuries might have led to a slow decrease in solar energy output, leading up to the Maunder minimum. This would have obvious implications on the Earth's climate.

Sources of Observations Used For Reconstructing Solar Activity

The first astronomical data on sunspot activity dates from the beginning of the telescopic era, i. e. 1609. During the first half of the 17th century, there were quite a few observers in a number of European countries. Solar observations consisted of sunspot drawings, until 1660. After that time, Picard developed a way of timing the edge passages of solar disk and the transit of sunspots across the disk, as well as their height (Ribes and Nesme-Ribes, 1993). So the Sun was in fact carefully observed throughout the 17th century.

For modern solar cycles, we have several homogeneous time-series of white-light images of the Sun, the most famous being those of Greenwich (1874-1976), which display sunspots and photospheric faculae. Spectroscopic time-series are also available and offer the advantage of displaying more magnetic tracers (sunspots, faculae, filaments, network and so forth).

Properties of Normal Cycles and GM Cycles

Sunspot Number

One way of measuring the strength of a cycle is by the mean monthly number of sunspot groups, as Hoyt *et al.*, did (for their method of reconstruction, see Hoyt *et al.*, this volume). Their results are shown in Figure 1, in which thcrc is clcar cvidcncc of a sunspot lull from 1645 to 1705. This period was thoroughly covered by the French school (a total of 10000 observing days regularly distributed over a 60 year-period). Furthermore, this lull seems to have been preceded by several decades of weak sunspot activity (the first half of the 17th century).

Butterfly Diagram and Sunspot Asymmetry

The sunspot butterfly diagram is reconstructed in Figure 2. Positions of sunspots on the solar disk can be reconstructed if we have sunspot drawings or timings. During the first part of the 17th century, sunspot drawings were not made systematically, rather than only when the sunspot sightings were frequent. This was the case for Scheiner, who observed at least from 1609 to 1630, but who published drawings only from 1624 to 1627. Similarly, Hevelius observed for many years but published drawings only from 1642 to 1644. Gassendi also observed the Sun in a fairly systematic way, but published sunspot drawings only from 1633 to 1635. To indicate that sunspots might have been present even in the absence of drawings, the periods without drawings are hatched in Figure 2a. So hatched areas would not correspond to a complete absence of sunspots but rather only to periods of low activity (see Hoyt *et al.*, this volume). This is true for the years 1615 to 1620, when sunspots were reported by Tarde (1622), and the years 1618 and 1623, for which Scheiner reported the presence of sunspots but did not make any drawings. We also examined the solar observations by Gassendi (drawings from 1633 to 1635). These are inaccurate, and the times of observation are often missing ; so it is difficult to extract any accurate information from them. However, we can assume that there was no multitude of sunspots so that what is displayed in Figure 2 is real and depicts the crest of solar cycles, as Eddy assumed it (1976).

During the second half of the seventeenth century, the French school performed the solar disk timings systematically (La Hire, 1683-1719). The sunspots were explicitly observed, so we know that they did not miss many. We have also plotted the butterfly diagram of the most recent solar cycles in Figure 2b, for comparison. The sunspot deficit (compared to modern 11-y cycles) during the 17th century, and the concentration of sunspots in the southern hemisphere, are both evident here. In conformity with the chronology cycles adopted by Wolf (see Schove, 1983), we extended the chronology beyond 1730, so that the telescope era (1609) begins with cycle -13 and the Maunder minimum (1645-1705) starts with cycle -9 and ends with cycle -4.

North-south asymmetry in the sunspot distribution seems to be a characteristic of GM and weak cycles. For comparison, strong solar cycles of the present epoch do not exhibit strong north-south asymmetry.

It is important to stress that the first cycle i. e. cycle -9 of the Maunder minimum is characterized by an absence of sunspots. Solar activity then remained very low, building up slowly until cycle -4. La Hire observed a resumption of activity from 1714 (cycle -3). Moreover, this cycle seems to be symmetric with respect to the equatorial plane. The number of sunspots in the northern hemisphere slightly exceeds the number in the southern. So we can consider the last cycle observed by La Hire as the first normal solar cycle after the Maunder minimum. This is of importance for understanding the time variation of the magnetic field.

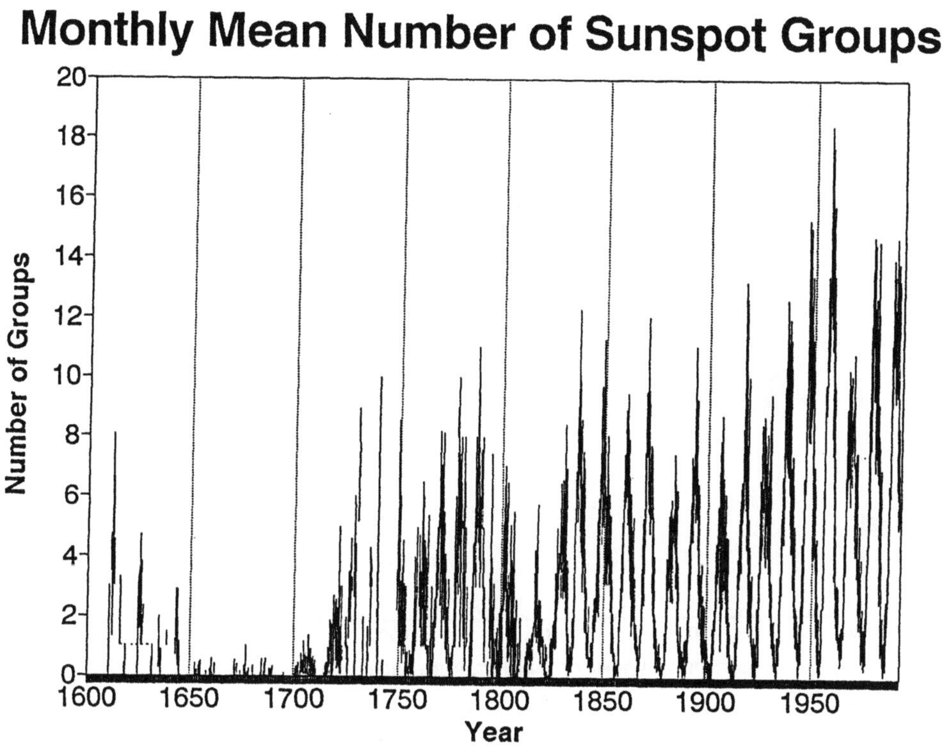

Figure 1 : Mean monthly sunspot group number reconstructed from 1609 onward (see Hoyt *et al.*, this volume)

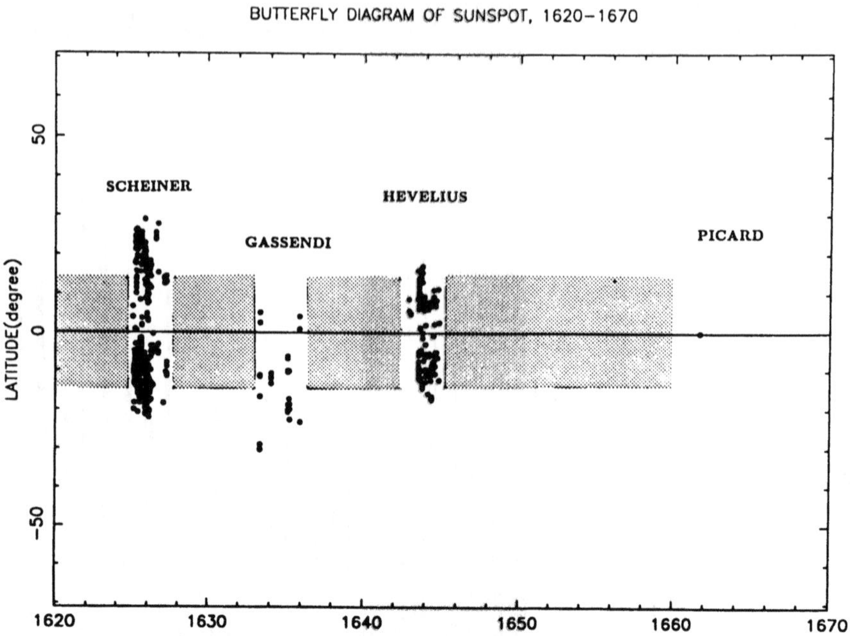

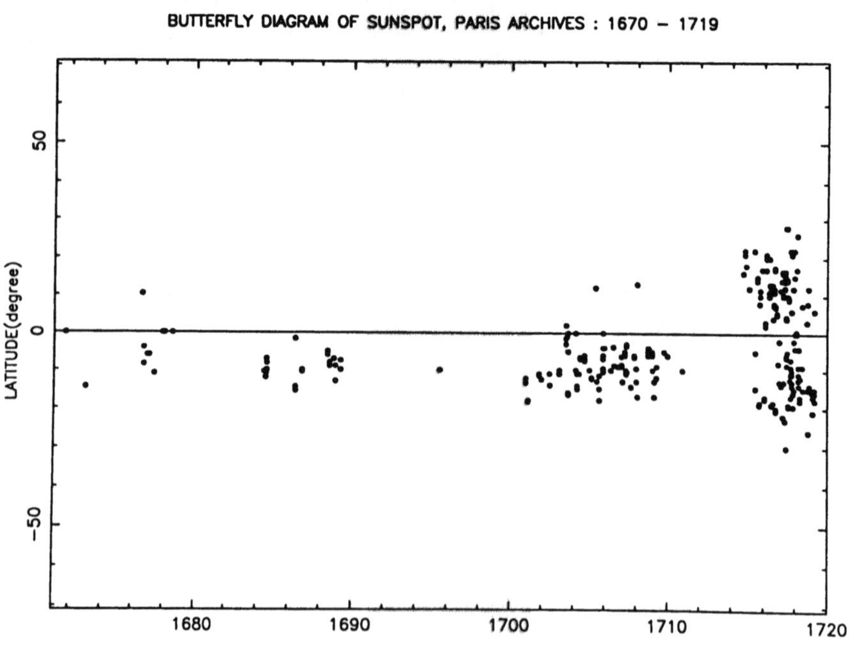

Figure 2a : Butterfly sunspot diagram (1620 to 1719) from historical observations. Hatched areas mean that some sunspots were sighted but no drawings are available.

SUNSPOT BUTTERFLY DIAGRAM: 1660-1720

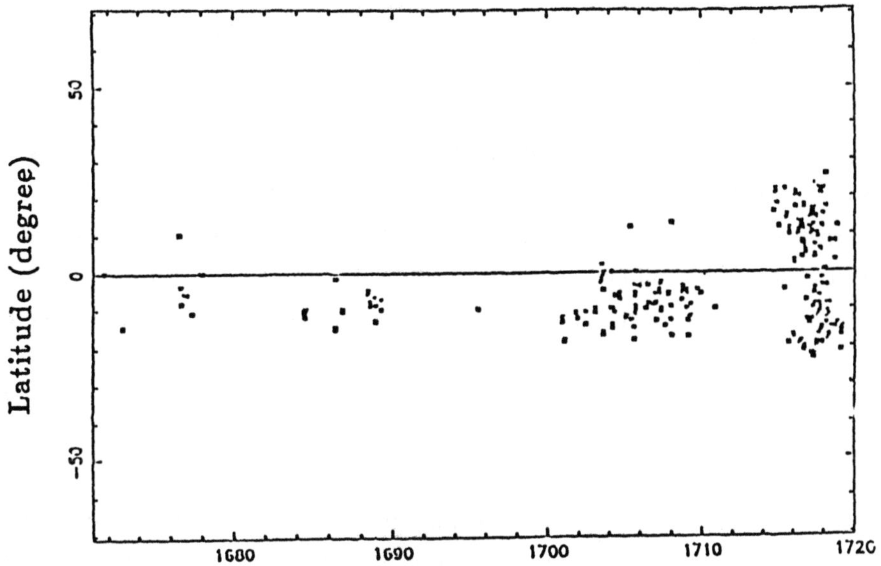

1945-1991

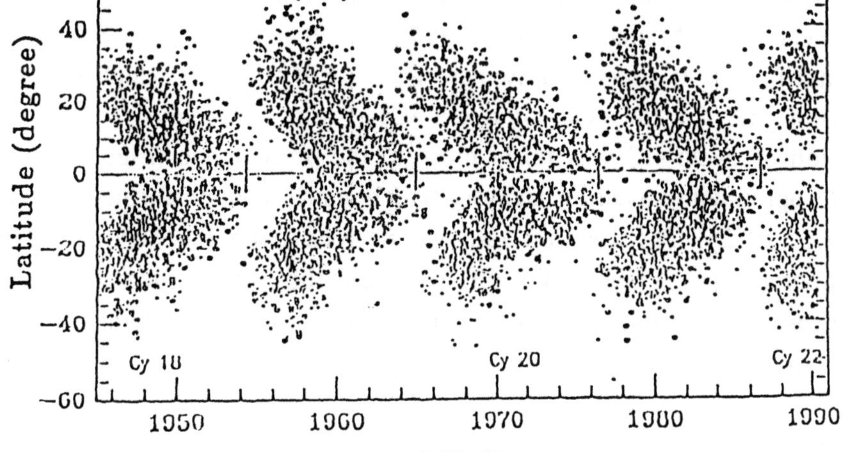

Figure 2b : The butterfly diagram for recent cycles is also plotted for comparison. The north-south asymmetry and the low sunspot number are a characteristics of the 17th century cycles.

Sunspot Rotation

It is a straightforward matter to obtain the sunspot rotation rate, as long as the sunspot positions on the solar disk are accurately known. This was the case as early as 1612, when Harriot left a series of sunspot drawings. Later, Scheiner published his solar observations, from 1624 to 1627, Hevelius from 1642 to 1644, and Picard and La Hire from 1666 to 1719. So the sunspot rotation rate could be reconstructed for the weak cycles of the first half of the seventeenth century, and also during the GM cycles. Results are shown in Figures 3. The rotation rate for cycle -13 has been investigated by Herr (1978) from Harriot's drawings ; it was found to be significantly more differential and slower than the modern rate (Figure 3a). A number of systematic errors could affect the determination of the solar rotation, leading to a certain underestimation of the rotation (Herr, 1978). However, the rotation in 1612 bears a close resemblance to that of sunspots timed by La Hire, during the Maunder minimum (Ribes and Nesme-Ribes, 1993). Furthermore, Galilei himself (1612) claimed that the sunspots he observed from 1609 to 1611 rotated at a synodic rate of approximately one lunar month, i. e. 29. 5 days. This value is in full agreement with Herr's analysis of Harriot data. The rotation rate decreased by 6 to 8 % in the 1610's and at the end of the Maunder minimum (1666-1719), i. e. cycle -3. Scheiner left a beautiful set of sunspot drawings, from 1624 to 1627, which provides a rotation rate for the peak of cycle -12. From our analysis, the sunspot rotation of cycle -12 was fairly close to the present one (Figure 3b). Our results agree with an earlier analysis of Scheiner's data performed by Yallop *et al.*, (1982). The rotation rate we derived from the Scheiner data is corroborated by Scheiner himself, who said that the sunspots he observed rotate with a synodic period of 27 to 28 days, which is in total conformity with our values (see Figure 3b).

For cycle -11, we have to rely on Gassendi's drawings. As we have said, these are not very accurate, and the times of observations are often missing. So, we can derive no information on the rotation rate for this cycle.

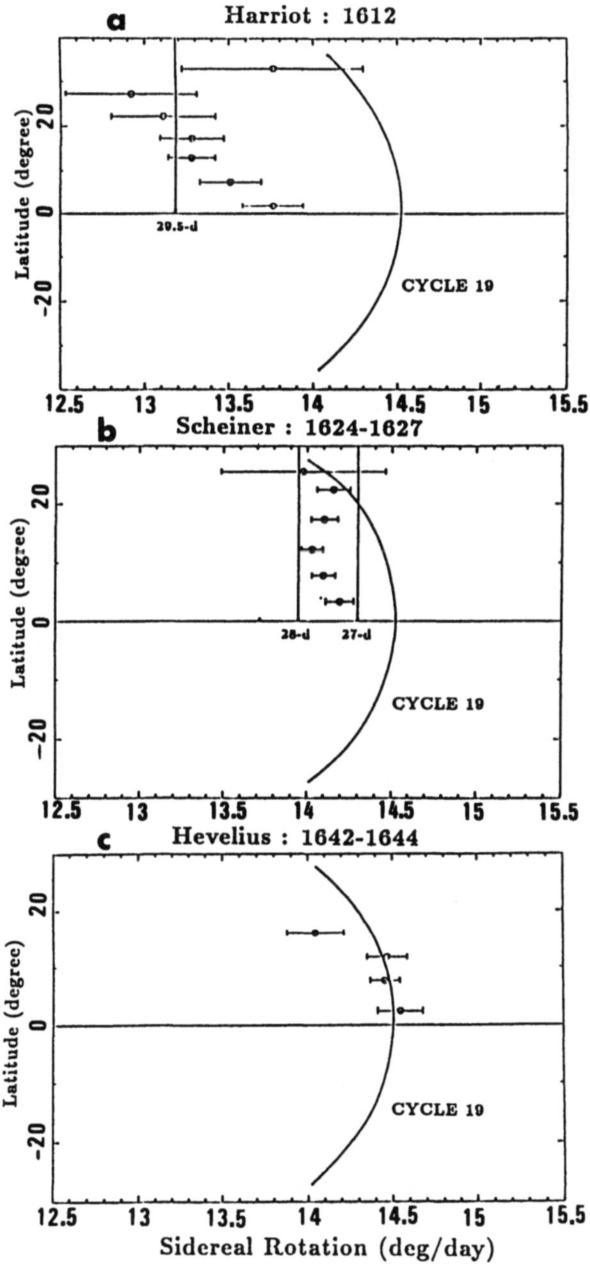

Figure 3 : Sunspot rotation rate for (a) cycle -13 (b) cycle -12 (c) cycle -10 ; modern rate (cycle 19) is plotted as well (solid lines). The synodic rotation of 27 days and 28 days is shown for Scheiner data (b). Sunspot number is given for each latitude range.

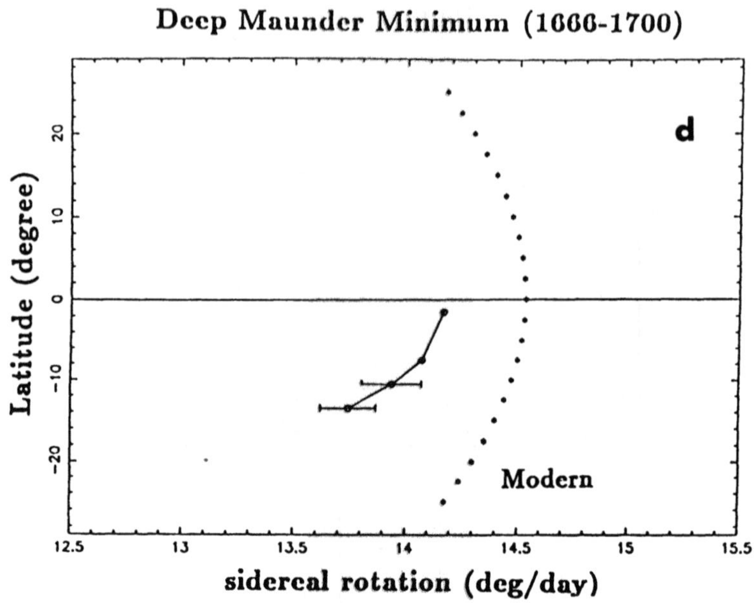

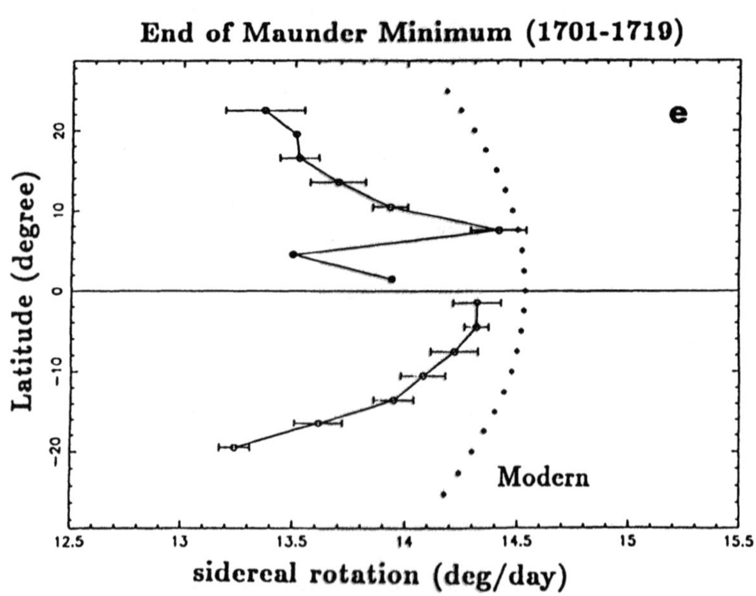

Figure 3 (continued) : Sunspot rotation rate for (d) cycle -5 (e) cycle -4 ; modern rate is plotted as well (dotted lines). During the 17th century, the rotation rate was slower and more differential than at present.

For cycle -10, we used Hevelius's sunspot drawings. The rotation rate is found to be very close to the present one (Figure 3c), with sunspots concentrated near the equator. Our analysis is in conformity with that of Abarbanell and Wöhl (1981) and by Yallop *et al.*, (1982). At this stage, we should mention that some conflicting results by Eddy *et al.*, (1976) have been presented for both the Scheiner and Hevelius sunspot rotations. We surmise that the difference lies in the data reduction procedure used by these authors. On the one hand, Abarbanell and Wöhl (1981), Yallop *et al.*, (1982) and Ribes and Nesme-Ribes (1993) measured the coordinates of individual sunspots on the drawings and calculated their heliographic coordinates, which is the most accurate method, and the result of all three are in complete agreement. On the other hand, Eddy *et al.*, (1976) used appropriate Stonyhurst Sun disk grids to position the sunspots on the solar disk. This is less accurate than the former method, and is subject to systematic biases which could be responsible for a rate which is systematically larger than our estimate. The final word is given by Scheiner himself, who analyzed his own data, concluding that the synodic rotation of sunspots is between 27 and 28 days (see Figure 3b). There is remarkable agreement between Scheiner's analysis and our findings as well as those of Yallop *et al.*, (1982), while Eddy *et al.*, (1976) give rotation rate falling outside the domain. A detailed study of the procedure and sources of errors will be presented in a forthcoming paper.

What is the rotation rate for modern solar cycles ? A detailed investigation is going on with the Meudon collection of spectroheliograms. It appears that the rotation is fairly rigid at the time of sunspot maximum and becomes more differential as sunspot activity decreases (Nesme-Ribes *et al.*, 1993 b ; see also Komm *et al.*, 1993). The amplitude of the change, within a modern solar cycle, is of the order of 1 % (from sunspot maximum to minimum). So there is some consistency between all our results, namely that the change of rotation is related to a change in the production of active regions. The conclusion is that the sunspot rotation rate exhibits both secular and century-type variability. The decrease ranges from 1 % for modern cycles to 8 % for GM cycles.

Aurora Activity

The aurora index, especially in the early days, is a very subjective one. What is responsible for the aurora ? There are essentially two sources of disturbances affecting the magnetosphere and the ionosphere (Legrand and Simon, 1989). One is a shock event caused by flaring sunspot activity, thereby directly related to the strength of the sunspot cycles. The other source of disturbances occurs when high speed solar wind is sweeping the Earth. High speed solar winds come mainly from the Sun's poles and depend on both the strength and orientation of the general solar dipole field. However, at the time of the dipole reversal i. e. near the sunspot maximum, the Earth's can also receive solar wind from the Sun's equator.

Can we estimate the level of sunspot activity from aurora frequency ? It should be noted that there are also preferred latitudes on Earth for observing aurorae (Legrand and Simon, 1989). Below latitudes of 50° on Earth, aurorae are mainly due to disturbances related to flaring sunspot activity, while those observed above 55° are due to solar wind. Until the last century, most aurorae were reported by amateurs living in mid-latitude, so aurora frequency should be more or less in phase with sunspot cycle. A compilation of historical aurora (Link, 1978 among others) clearly shows that aurora activity was low throughout the 16th and 17th centuries, though the figures given by Link (1978) should be taken with caution.

Apparent Solar Diameter

We know that the shape and size of the solar disk varies with the emergence of magnetic fields (Ribes *et al.*, 1992). However, the amplitude of the variation depends upon the physical processes and the time-scales involved (see Spruit, this volume). Measurements of the solar diameter were made even before telescopic era, but with an accuracy of 10 arcseconds. It is only since the advent of quantitative timings that it has become possible to know the Sun's size with an accuracy of a few arcseconds on each timing. One output from the systematic meridian timings performed by Picard and La

Hire from 1666 to 1719 was the size of the solar disc, after which it became possible to examine the variability of the apparent radius.

Two puzzling results were obtained. First, the solar radius was larger by 2 arcseconds during the deep Maunder minimum, and slowly returned to normal around 1705 when sunspot activity resumed (Ribes *et al.*, 1987). This increase, followed by a sharp decrease, seems to be real, since the measurements were made by the same observer using the same telescope. Another important fact was that the diameter exhibits a 10-y modulation, indicating that the solar cycle was still at work (Ribes *et al.*, 1989). A number of other periodicities were present in the Fourier analysis of historical radius observations, which are also found in modern radius observations (Ribes *et al.*, 1992 ; Delache *et al.*, 1993). Furthermore, similar variations are found in the abundancc of thc cosmogenic isotope deposits (see the contributions by Stuiver and Braziunas, 1993, and the two contributions by Stuiver and Beer *et al.*, this volume).

Three types of solar cycles can be defined from analysis of modern and historical cycles.

1. *"Normal" solar cycles*, in which
- the length is approximately 10 to 11 years
- the sunspot number is large, and roughly symmetrical about the equator. A slight time asymmetry is observed, but this is essentially due to the fact that each hemisphere is active at different times.
- sunspot rotation is fairly rigid, especially near the sunspot maximum
- auroral activity is observed, due to the presence of flaring active regions and also due to the dipole field structure.

2. *"GM" cycles*, in which
- the length is approximately 9-10 years
- the sunspot number is small, largely asymmetric
- sunspot rotation is slower, by several per cent
- auroral activity is reduced

3. *"Weak" cycles*, in which
- the length is possibly greater than 11-y ?
- the sunspot number is small (compared with normal cycles), but shows relatively little north-south asymmetry
- sunspot rotation is slowed by about 1 %
- auroral activity is intermediate between that of the normal and GM cycles.

From the four centuries of sunspot observations available, normal cycles may persist through six to eight cycles (the Gleissberg cycle ?), and GM cycles may persist over approximately the same time (several cycles). The transition time from one type to the other, however, is relatively short. Sunspot symmetry can be recovered after one or possibly two solar cycles (see the rapid transition between cycles -4 and -3). The solar cycles of the eighteenth century were much less intense than modern ones, but will be considered as "normal" cycles because the sunspot distribution exhibits north-south symmetry. The question of interest is whether or not the properties of the three types of solar cycles can be explained by an evolutionary scenario of the dynamo .

The Solar Dynamo During the Maunder Minimum, and Now

It is usually assumed that solar activity is generated by an α-ω dynamo (Parker, 1955). The depthwise non-uniformity of the Sun's rotation ω (the Sun's interior does not rotate at the same speed all the way down) distorts dipolar magnetic lines and generates an azimuthal field which is antisymmetrical with respect to the solar equator (toroidal ring fields of opposite directions on both sides of the equatorial plane). If the azimuthal field thus created emerges at the solar surface in the form of sunspots, the polarity of leading sunspots will be opposite in each hemisphere, as observed. These toroidal fields are first generated at relative higher latitudes, migrate towards the equatorial plane, and decay. The α-effect, due to the helicity of the convective motion, produces the regeneration of poloidal field from the azimuthal field and closes the chain of magnetic field self-excitation (Steenbeck and Krause, 1969).

If a weak quadrupole component field (symmetrical with respect to the equatorial plane) is superimposed on the dipole field, the resulting azimuthal field will be asymmetrical (Figure 4c). This is the case for modern 11-year cycles, where one hemisphere leads the other and vice versa, while the total sunspot number in each hemisphere (averaged over the whole cycle) remains roughly the same.

The excited magnetic field modes show different symmetry (quadrupole-type) or antisymmetry (dipole-type) with respect to the equator. The first antisymmetric dipole-type mode is the mode which is excited most easily in the α-ω dynamo. However, the first quadrupole-type mode is the next most easily excited .

It has been conjectured that the Maunder minimum type sunspot distribution could be explained if the magnetic field configuration was different from the configuration at a normal cycle. That is to say, a configuration with a quadrupole component of amplitude comparable to the dipole field (Sokoloff and Nesme-Ribes, 1994). The azimuthal field would be entirely suppressed in one hemisphere, yielding qualitatively the same sunspot butterfly diagram as the one obtained by Ribes and Nesme-Ribes (1993) during the second half of the seventeenth century (Figure 4d). Numerical simulations performed, for example by Brandenburg *et al.*, (1990) and Jennings and Weiss (1991), show that such episodes of Maunder minimum types could be obtained as a result of a negative feedback of the magnetic field on the solar plasma motion, i. e. on the helicity of turbulent motions. Let us also mention, that until now no clear example of magnetic field configuration of quadrupole-dominant stellar magnetic field has been reported.

One physical reason for getting sunspot concentrations close to the equatorial plane during the Maunder minimum has been suggested by F. Krause during this meeting.

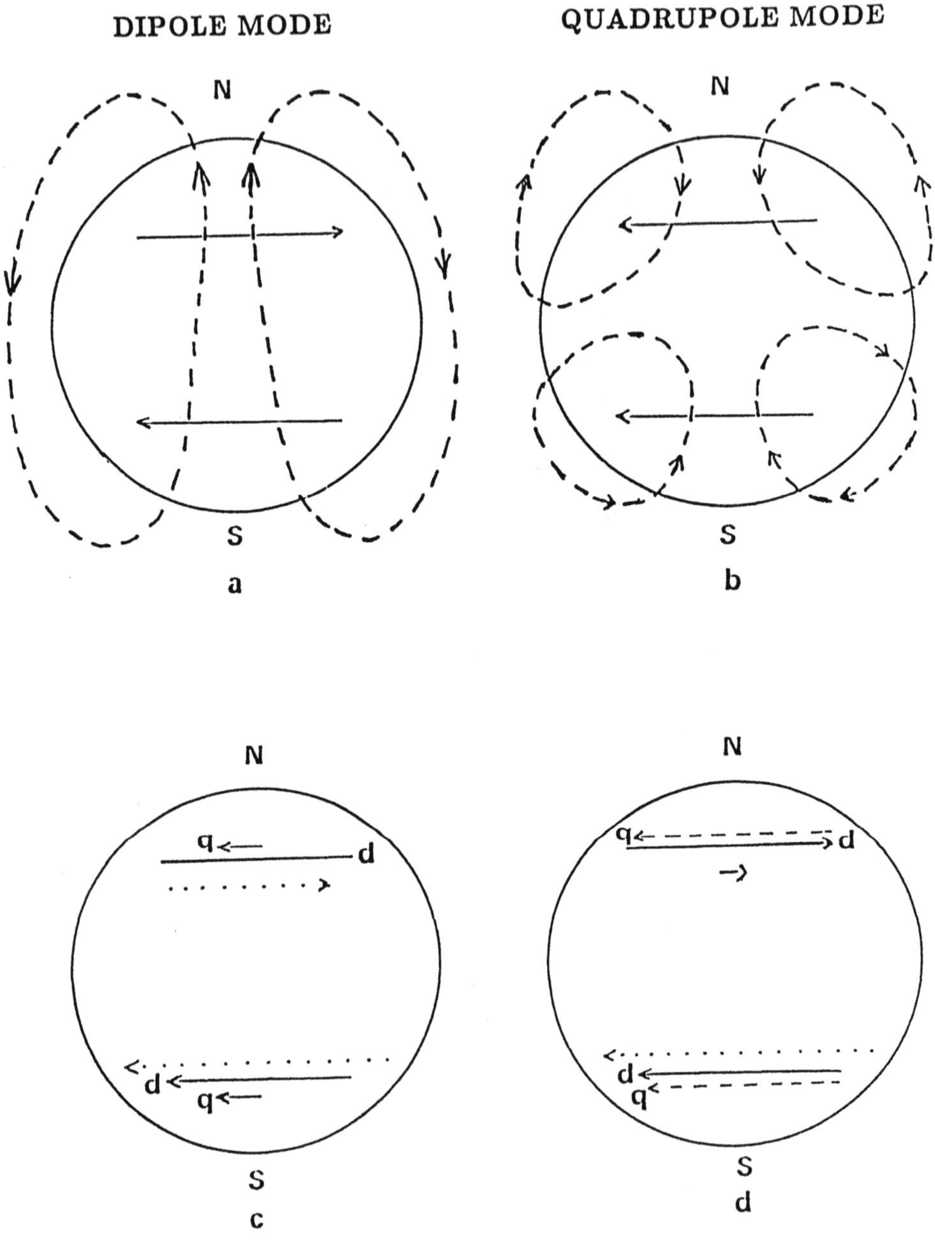

Figure 4 : Schematic representation of the magnetic field for normal and GM cycles.

The latitudinal distribution of the azimuthal field generated from a quadrupole mode should not vanish at the equator, but rather show a maximum. This means that the superimposition of dipole and quadrupole fields of approximately equal amplitude would shift the azimuthal field near the equator. Let us stress that most sunspots were located in the southern hemisphere during the Maunder minimum. However, a few sunspots were also observed close to the equator in the northern hemisphere. If the explanation of F. Krause is correct, these sunspots would have the same polarity as those in the southern hemisphere.

During the Maunder minimum, the general magnetic field of the Sun was still flipping on and off with a 10-y periodicity, as shown from the modulation of the solar radius and radio isotopes. This indicates that the period of the field oscillations did not vary when the parity was changing, otherwise we would see alternations of sunspots in each hemisphere rather than a suppression of them in the southern hemisphere only (corresponding solutions has been obtained by Brandenburg *et al.*, 1989). There are some modulations in solar activity data of the order of about a hundred years, e.g. variations in ^{10}Be, or variations in N-S asymmetry etc. One physical origin of these variations may have to do with a small difference between the frequencies of dipole and quadrupole modes of solar activity, $(\Omega_d - \Omega_q)^{-1}$. The duration of the Maunder minimum-type cycles may also be connected with the same splitting of frequencies.

Scenario For Solar Cycle Variation

We shall try to describe the time evolution of solar cycles using our description of historical and modern solar cycles (Kremlevskii *et al.*, 1992). The scenario proposed here is tentative, as observational data is still incomplete, and the modern dynamo theory is unable to reproduce all features of solar activity variation as yet.

We assume that the time evolution of the solar dynamo can be described in terms of the evolution of dipole and quadrupole modes. In this case, we are dealing with a low-dimensional nonlinear system. It is beyond the possibilities of the present paper to obtain this dynamical system from the equations of solar magnetohydrodynamics. However, we tentatively describe its properties from the observational data available. We consider the dipole and quadrupole mode intensities of the solar magnetic field as variables of a dynamical system describing the magnetic field variation. It is common to describe the sunspot number variation in terms of two variables : the intensity of the dipole mode, d, and its time derivative, d'. The solar cycle in such a diagram corresponds to rotations around the origin of coordinates (Figure 5a). The periods of rotation are approximately the same in normal and "GM" cycles, but their amplitude may vary considerably. It is more interesting to represent the behaviour of the dynamical system in a $\langle d \rangle$, $\langle q \rangle$ graph, where $\langle\ \rangle$ is the average over a cycle (Figure 5b). Here, we describe the trajectory for normal (strong) cycles as well as GM cycles. The trajectory corresponding to normal cycles is described as a random walk along the $\langle d \rangle$ axis (to the inside of the parabola of Figure 5b). The corresponding values of $\langle q \rangle$ are very small. Sometimes this trajectory wanders from the $\langle d \rangle$ axis. The attempt is usually unsuccessful and the trajectory rapidly comes back to the previous region. We have evidence of this from cycles showing a pronounced north-south asymmetry, such as cycles 12 and 13. Now if the magnetic field is sufficiently small (near the minimum of the parabola in Figure 5b), such deviations can be more and more successful. This was probably the case for cycles 5 and 6 (1798-1823), known as the Dalton minimum, or for the cycle -10 observed by Gassendi. Whenever the trajectory leaves the region of normal activity and stays very close to the line $\langle d \rangle = \langle q \rangle$, we obtain the Maunder minimum type behaviour, which corresponds to a long period of very low activity and large north-south asymmetry.

In the phase space, the point ($\langle d \rangle = 0$, $\langle q \rangle = 0$) is a saddle point of the dynamical system ; so as long as the trajectory continues in the same direction as the line $\langle d \rangle = \langle q \rangle$, it will go to the saddle point, so the Maunder minimum can persist. When this trajectory deviates from this line, we get only weak cycles, like the Dalton minimum cycles (5 and 6) or the last cycle observed by La Hire (cycle -3). The trajectories of solar cycles during the 17th century are schematically described in Figure 5b, showing how the magnetic field could evolve towards the Maunder minimum epoch before returning to a normal phase later in the 18th century. Later, several attempts to reach Maunder minimum-type cycles probably occurred, at the beginning of the 19th and 20th centuries.

Summarizing this discussion, we can conclude that the long-term evolution of solar activity spanning the Maunder minimum could be described by a low-dimensional chaotic dynamical system (say, Rössler attractor (Rössler, 1976)).
The unstable rest point of this system, responsible for chaotic behaviour, is presumably focus-saddle. Dimension of this system should be at least 4. (For some more about properties of this system, see Kremlevskii *et al.*, 1992).

How to Interpret Changes in 11-y Cycles in Terms of Total Solar Irradiance Variations ?

This is a relevant question at this meeting since climatologists need to have some estimate of the solar output for their climatic simulations.

The general problem consists of relating the physical parameters used as proxy data (magnetic activity, sunspot rotation, apparent diameter) with solar luminosity. In a general way, what is used by climatologists is not the solar luminosity but the Total Solar Irradiance (denoted hereafter as to T. S. Ir.), i. e. the energy radiated in the direction of the Earth. It is generally accepted that fluctuations in T. S. Ir., as measured by radiometers over the last solar cycle, are mostly due to sunspots (causing the T. S. Ir. to fall) and emissive features called "faculae" (causing it to rise). Some line intensities, such as the neutral Helium line or the Lyα total flux, are good surrogates for the T. S. Ir. , once the latter has been corrected for the sunspot blocking effect (Foukal and Lean, 1990). These lines are not available beyond 1978, so other indicators have to be used. This is the case for the 10.7 cm radio flux and the mean monthly sunspot number, R_z, which are related to the facular emission : the more active the Sun, the more sunspots there are, and thereby the more faculae.

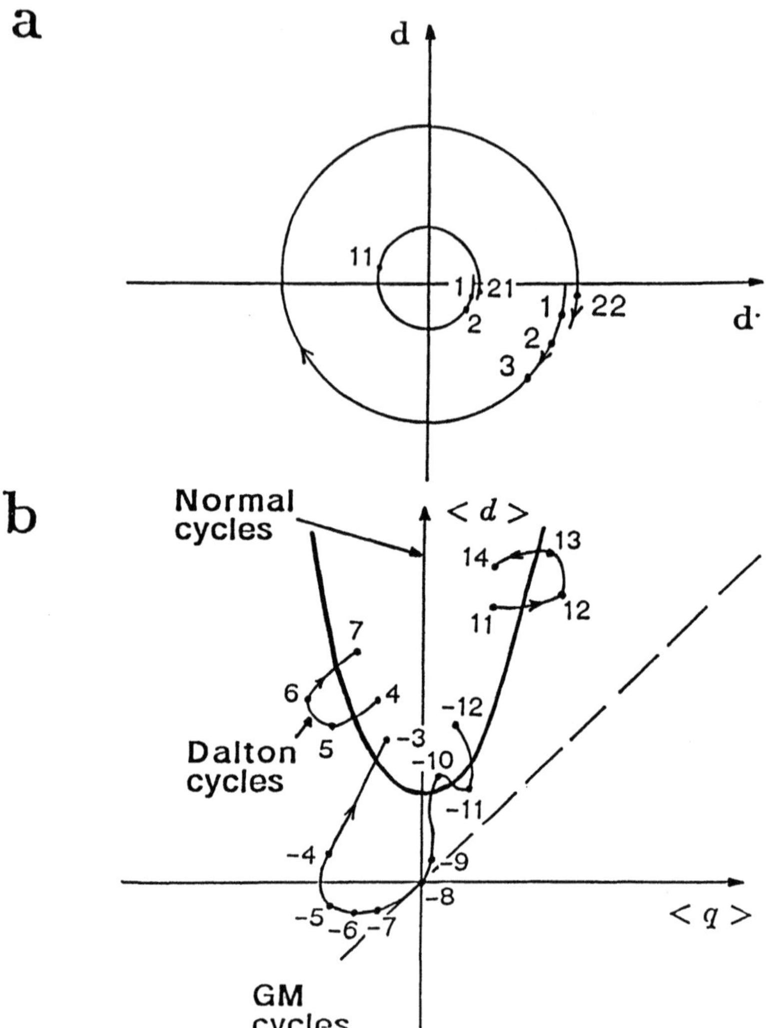

Figure 5 : (**a**) Representation of solar cycles in the (d, d') coordinate system, where d is the dipole mode and d', its time derivative. Solid lines describe the variation of the magnetic field through a complete normal 22-year solar cycle, while dashed lines describe the evolution of a GM cycle. Numbers 1, 2, 3,. . . 22 denote each year of a complete 22-y cycle. (**b**) Representation of the long-term evolution of the magnetic field in the (<d> <q>) coordinate system, where <d> and <q> represent the dipole and quadrupole components respectively. Normal cycles can be found in the region of large <d> and low <q> (delimited by a parabola). The dashed-line (<d> = <q>) going through the saddle point represents the variational track of GM cycles.

However, these are not one-to-one correlations, and they are not based on a theoretical modelling of the relationship between magnetic activity and solar luminosity (see the panel discussion by J. Pap). Furthermore, there is no reason to believe that the same correlations hold historically back to the Maunder minimum.

Similar criticisms apply to the use of the apparent solar diameter as surrogate for T. S. Ir. reconstruction (Nesme-Ribes *et al.*, 1993a.). This issue will be discussed by Ph. Delache in the panel discussion. Indeed, the apparent diameter should respond to perturbations in the convective zone or below : but we are still lacking a physical mechanism of radius change compatible with the observed variations of T. S. Ir. (see Spruit, this volume).

Analogies with solar-type stars can be criticized too. No mechanism driving brightness and magnetic cycles has been developed as yet ; so extrapolating the non-cycling star properties to the Sun in the Maunder minimum state (Baliunas and Jastrow, 1990 ; Lean *et al.*, 1992) is also questionable.

To address the problem of modelling the T. S. Ir. from surrogates, one has to solve the complete system of equations governing the magnetized plasma. An attempt has been made to relate the strength of the azimuthal field (sunspots) to their rotation rate (Nesme-Ribes and Mangeney, 1992). The physical idea is the following : at the sunspot minimum, the meridional circulation in the convective zone seems to be stronger than at the sunspot maximum (Ribes and Bonnefond, 1990), and causes a change in the rotation profile, from more rigid to more differential (Nesme-Ribes *et al.*, 1993a.). As a consequence, there will be excess kinetic energy at the sunspot minimum, and this should come from some energy reservoir. The authors surmise that it comes from the thermal reservoir. This model has been tested on modern 11-y cycles and could account for the 11-y trend present in the observed T. S. Ir.

Applying the model of Nesme-Ribes and Mangeney (1992) to the Maunder minimum during which sunspot rotation was weakened by about 6 %, a vigorous meridional circulation has been proposed to account for the observed rotation profile ; the associated excess kinetic energy caused a 0. 2 % decrease in the T. S. Ir. This estimate corresponds to the time of sunspot appearance (crest of the 10-y cycles). It is possible, however, that this estimate would be a lower limit as the decrease might have larger value at time of complete magnetic calm (for example, up to 1 % from 1645 to 1670 and from 1690 to 1700). We adopt a mean decrease of 0. 4 % as representative of the T. S. Ir. during the Maunder minimum, which corresponds to a solar flux decrease of 1 Wm^{-2} at the top of the Earth's atmosphere. This value is in broad agreement with various estimates deduced from other surrogates (Lean *et al.*, 1992 ; Soon *et al.*, this volume).

Prior to the Maunder minimum, the sunspot rotation seems to be slower by 1 % or more with respect to the modern rate. So, adopting the model of Nesme-Ribes and Mangeney (1992), the picture for the whole 17th century could be a decrease of the T. S. Ir. of a few per thousandths. Such a value is by no means negligible, and could lead to the cold spell observed during the 17th century (Sadourny, Royer *et al.*, this volume).

The model described by Nesme-Ribes and Mangeney (1992) still contains a number of simplifying assumptions in the solution of the momentum equations. Furthermore, to get the complete picture, the energy equations have to be solved self-consistently. This is far from being the case in the simple model considered by Nesme-Ribes and Mangeney (1992). However, their approach is a first step towards a better understanding of the physical mechanisms driving the magnetic cycles and corresponding brightness variations.

Incorporating solar forcing (and other natural effects) in climate simulations seems to be the only way to realistically evaluate the climatic trend induced by greenhouse gas forcing in the coming decades (Sadourny, 1993).

Prediction Scheme for GM-type Cycles

The next question to address concerns the prediction of the long-term evolution of solar activity, if it is proved that GM-type and weak cycles might be accompanied by a decrease in solar output.

There are several premonitory signs of the transition to GM cycles, namely sunspot deficiency, strong north-south asymmetry, a change in sunspot rotation, and an increase in the apparent size of the Sun. This can be done by monitoring sunspot activity from the ground. However, direct space observations of the solar spectrum output, especially in the UV range, are also necessary in order to estimate the consequences of these processes on the energy budget of the Sun.

Acknowledgements

This work is part of French-Russia collaborative effort undertaken under the initiative of the Délégation Générale de l'Armement. Our thanks go to the plasma physics group of the Direction des Recherches Etudes et Techniques (contract DRET n° 9220011/A) for their support. Scheiner, Hevelius and Gassendi data were found at the libraries of the Paris Observatory and of the Academy of Sciences (Mazarine library). We are very grateful to C. Brechet and J. Alexandre for their assistance in the data processing of these old manuscripts. D. Sokoloff would like to express his thanks to IZMIRAN, the Russian Foundation for Fundamental Research (grant N° 93-02-3638), the "Russian University" Foundation and Lunokhod Ltd for financial support.

References

Abarbanell, C., and Wöhl, H. (1981). Solar rotation velocity as determined from sunspots drawings of J., Hevelius, in the 17th century. Solar Phys., 70: 197-203

Baliunas, S., and Jastrow, J. (1990). Evidence for long-term brightness changes of solar-type stars. Nature, 348: 520-523

Brandenburg, A., Krause, F., and Tuominen, I. (1989). Parity selection in nonlinear dynamos, in Turbulence and Nonlinear Dynamics in MHD Flows, eds Meneguzzi, M., Pouquet, A., and Sulem, PL.. Elsevir Sci. Publ., North-Holland, 35-40

Brandenburg, A., Meinel, R., Moss, D., and Tuominen, I. (1990). Variation of even and odd parity in the solar dynamo. In Stenflo JO (ed) Solar Photosphere, Structure, Convection and Magnetic Fields. IAU, 379-382

Brandenburg, A., Krause, F., Meinel, R., Moss, D., and Tuominen, I. (1989). The stability of Nonlinear Dynamos and the limites Role of Kinematic growth Rates. Astron. Astrophys., 213: 411-422

Cassini, J.D. (1729). Mémoires Acad. Sc. Paris, Tome 8: 82

Delache, Ph., Laclare, F., Gavryusev, V., Gavryuseva, E., Regulo, C., and Roca Cortes, T. (1993). Time correlation between solar structural parameters : p-mode frequencies, radius, and neutrino flux. Astrophys. J., 407: 801-810

Eddy, J.A. (1976). The Maunder minimum. Science, 192: 1189-1202

Eddy, J.A., Gilman, P.A., and Trotter, D.E. (1976). Solar activity during the Maunder minimum. Solar Physics., 46: 3-14

Foukal, P., and Lean, J. (1990). An empirical model of total solar irradiance variation between 1874 and 1988. Science, 247: 556-558

Galilei, G. (1612). De maculis solaribus et stellis circa Iovem errantibus accuratior disquisitio. Academy of Lincei

Herr, R. (1978). Solar rotation determined from Thomas Harriot's sunspot observations of 1611 to 1623. Science, 202: 1079-1081

Hoyt, D.V., Schatten, K.H., and Nesme-Ribes, E. (1994). A new reconstruction of solar activity, 1610-1993. This volume

Jennings, R.L., and Weiss, N.O. (1991). Symmetry breaking in stellar dynamos. Mon. Not. Roy. Astron. Soc., 252: 249-260

Komm, R.W., Howard, R., and Harvey, J.W. (1993). In Brown, T.M. (ed.) Proceedings of the Gong meeting, Boulder, in press

Kremlevskii, M.N., Blinov, A.V., and Chervyakov, T.B. (1992). Reconstructing the phase-space trajectory of solar activity from sunspot numbers. Sov. Astron. Lett., 18: 423-426

Lean, J., Skumanich, A., and White, O.R. (1992). Estimating the Sun's radiative output during the Maunder minimum. Geophys. Res. Lett., 19: 1591-1594

Legrand, J.P., and Simon, P. (1989). Solar cycle and geomagnetic activity, a review for geophisicists. Part 2, The solar sources of geomagnetic activity and their links with sunspots cycle activity. Annales Geophysicae, 7: 579-593

La Hire, Ph. (1683-1719). Arch. Obs. Paris. Manuscripts, D 2: 1-10

Link, F. (1978). Solar cycles between 1540 and 1700. Solar Phys, 59: 175-178

Maunder, E.W. (1894). A prolonged sunspot minimum. Knowlegde, 17: 173 - 176

Nesme-Ribes, E., and Mangeney, A. (1992). On the plausible physical mechanism linking the Maunder minimum to the little ice age. Radiocarbon, 34: 263 - 270

Nesme-Ribes, E., Ferreira, EN., Sadourny, R., Le Treut, H., and Li, L. (1993a). Solar dynamics and its impact on Solar irradiance and the terrestial climate. J. Geophys. Res., 98, n° A11: 18923-18935

Nesme-Ribes, E., Ferreira, E.N., and Mein, P. (1993b). Solar dynamics over solar cycle 21 using sunspots as tracers. I. Sunspot rotation. Astron. Astrophys., 274: 563-570

Parker, E.N. (1955). Hydromagnetic dynamo models. Astrophys. J., 122: 293-314

Ribes, E., Beardsley, B., Brown, T.M., Delache, Ph., Laclare, F., Kuhn, J., and Leister, N.V. (1992). The variability of the solar diameter, in The Sun in Time, eds. C.P. Sonett, M. Giampapa and M. Matthews, Univ. Arizona, Tuscon: 59-97

Ribes, E., and Bonnefond, F. (1990). Magnetic tracers, a probe of the solar convective layers. Geophys. Astrophys. Fluid Dyn., 55: 241-245

Ribes, E., Ribes, J.C., and Barthalot, R. (1987). Evidence for a larger Sun with a slower rotation during the seventeenth century. Nature, 326: 52-54

Ribes, E., Ribes, J.C., Merlin, Ph., and Barthalot, R. (1989). Absolute periodicities in solar diameter, derived from historical and modern data. Annales Geophysicae, 7: 321-329

Ribes, J.C., and Nesme-Ribes, E. (1993). The solar sunspot cycle in the Maunder minimum AD 1645 to AD 1715. Astron. Astrophys., 276: 549-563

Rössler, O.E. (1976). An equation for continuous chaos. Phys. Lett., A57: 397-398

Sadourny, R. (1993). Climate sensitivity, Maunder minimum and the Little Ice Age. ESF/Bologna

Schove, D.J. (1983). Sunspot cycles. Hutchinson/Stroudsburg

Sokoloff, D., and Nesme-Ribes, E. (1994). The Maunder minimum, a mixed-parity dynamo ?. Astron. Astrophys., in press

Steenbeck, M., and Krause, F. (1969). Zur Dynamotheorie stellar und planetarer Magnetfelder. I. Berechnungsonnenlicher Wechselfeldgeneratoren. Astronomische Nachrichten, 291: 49-84

Stuiver, M., and Braziunas, T.F. (1993). Sun, ocean, climate and atmospheric $^{14}CO_2$. An evaluation of causal and spectral relationships. The Holocene, 3: 289-305

Tarde, G. (1622). Manuscript, Italy

Yallop, B.D., Hohenkerk, C., Murdin, L., and Clark, D.H. (1982). Solar rotation from 17th century records. Q.J.R.A.S, 23 : 213-219

Panel Discussion on the MODERN STATUS OF SOLAR DYNAMO

Leader : D. Sokoloff
Department of Physics, Moscow State University
119899, Moscow, Russia

Today, the solar dynamo theory is being thoroughly reviewed. On the one hand, it is the only mechanism which gives a consistent description of the solar cycle. On the other hand, such a theory is in need of improvement, to take into account new ideas and facts.

The initial concept of the solar dynamo activity was established by Larmor in 1919, namely that the motion of an electrically conductive medium in the Sun is what generates the solar magnetic field. However, it was only in the 1960's that the well-known mechanism of a α-ω dynamo was established. This concept identifies what kind of solar plasma motions are crucial for the dynamo process. In this framework, the large-scale solar magnetic field is assumed to be generated by the action of the mean helicity and the differential rotation, within the solar convective shell. The helicity of the turbulent fluid generates a poloidal field from a toroidal field, and the differential rotation produces a new toroidal field from this poloidal one, thereby connecting the chain of magnetic field self-excitation.

This mechanism does not claim to explain all types of solar magnetic field. For example, on a small-scale the solar magnetic field seems to be a superimposition of two components. One of these is an anisotropic field produced as a result of large-scale magnetic field disturbances by convective motions. This component is thought to be correlated with the solar cycle. However, there is another small-scale isotropic component which seems to be uncorrelated with the solar cycle, and which is presumably generated by an independent dynamo mechanism. Furthermore, the sunspot magnetic fields also should result from the magnetic field created by an α-ω dynamo. However, sunspot formation is not taken into account in the present model.

NATO ASI Series, Vol. I 25
The Solar Engine and Its Influence
on Terrestrial Atmosphere and Climate
Edited by Elizabeth Nesme-Ribes

The α-ω dynamo has not yet been fully explored. Originally, it was a kinematic theory describing the magnetic field generated by a given velocity field. But a theory can only hint at the complexities of solar activity. Many new nonkinematic ideas of the magnetic field evolution, including a negative feedback of the magnetic field on the velocity field, are being incorporated into this concept. In the 1990's, the dynamo theory could try to describe the long-term evolution of the solar magnetic cycle, including the Maunder minimum, normal cycles and intermediate processes.

However, it is clear that new observational and theoretical possibilities are the source of much of the disagreements between dynamo theory predictions and the real solar magnetic field. A revision of the present dynamo theory therefore seems to be unavoidable.

Among the various problems, one is well-known. Following the initial explanation by Parker of the solar activity cycle as a dynamo wave, the mean helicity of the flow is a key-ingredient of the dynamo. A standard explanation of the origin of helicity was given in 1966 by Steenbeck, Krause and Rädler (see the book by Krause and Rädler (1980)) : positive helicity, resulting from Coriolis forces and rotation, is predicted in the northern hemisphere, and negative helicity in the southern. The simplest model of differential rotation in a fluid (e. g. the keplerian distributions of accretion discs) leads to negative radial velocity shear. Of course, this argument has never been considered as a strong one, but it did lead to wave propagating equatorward, in agreement with the observed butterfly diagram of solar activity. However, the internal velocity gradient is now known from helioseismology : the radial velocity shear is positive, at least in the main part of the solar convective shell nearby equator. So it turns out that it provides a propagation of the dynamo wave in contradiction with observations. This is only one of a number of conflicts between the dynamo theory and certain observational data. We may mention other sources of discrepancy, such as the period of the solar cycle. The simple arguments mentioned above lead to a very short solar cycle, of about one year, while the real time scale of solar activity is one order of magnitude larger.

Recently Vainstein and Cattaneo (1992) drew attention to an additional difficulty in the dynamo theory. The generation of the large-scale magnetic field is accompanied by the generation of a small-scale magnetic field. The energy ratio between these small- and large-scale fields, according to an estimate of the two-dimensional flows in a kinematic regime by Zeldovich (1957), should be about $R_m^{\frac{1}{2}}$, where R_m is the magnetic Reynolds number. This dimensionless number is very large in the solar medium. So a pure kinematic description of the dynamo is not applicable, provided that the small-scale magnetic field is in equipartition with the kinetic energy of the solar convection. This leads to a large-scale magnetic field strength that is much less than the one observed. Although the complete sequence of argumentation is not completely clear at present, argument of Vainstein and Cattaneo would rule out the dynamo explanation in the solar convective shell, because it leads to a negligible intensity of magnetic field generation. As mentioned by P. Roberts, this opinion prompted an immediate negative reaction from the scientists involved in this field. More about the modern status of the dynamo theory can be found in this volume (Roberts, 1994).

A moderate viewpoint can be adopted on this problem. It is clear that the small-scale solar magnetic fields are really more intense than the large-scale. So the original estimate given by Vainstein and Cattaneo could be restricted to a kinematic stage of the dynamo process, or even to a two-dimensional magnetic field variation, which is not relevant to the real Sun.

In the three-dimensional nonkinematic picture, theory and computer simulations give some reason to believe that the strength of the small-scale magnetic field is stronger than the large-scale field, predicted by the Vainstein and Cattaneo estimation. However, a complete picture of the small-scale magnetic field generated together with a large-scale one remains to be developed.

There is now some concensus among dynamo specialists that the dynamo theory is the only way to describe the solar cycle, in spite of the existing problems. How to remove some of the above inconsistencies ?

L. Kitchatinov stressed that a dynamo theory should be based on a consistent description of solar convective motions in terms of equations for corresponding correlation functions, including the negative feedback of magnetic fields on the convective motions (see Rüdiger and Kichatinov, 1994). He drew attention to new computer simulations obtained by G. Rüdiger and A. Brandenburg (1993).

Considering the anisotropic nature of the mean helicity and the turbulent diffusion, these authors believe that it is possible to resolve the discrepancy between helioseismological data and dynamo theory, as well as the solar cycle period. The convention leading to positive helicity in the northern hemisphere are based on one component of the helicity tensor, in an anisotropic version of the α-ω dynamo. However, there is another component of this tensor that plays a role in the magnetic field self-excitation, and it is now possible to get a different sign for the helicity in realistic models of solar convection. This scenario should be considered as a confirmation of a suggestion by P. Gilman that helicity should be negative in the northern hemisphere.

It has been also mentioned that dynamo models could be based on a phenomenological description of mean properties of solar convection deduced from observations. Such descriptions are typical in another branch of macroscopic electrodynamics, e. g. the conductivity of silver is not deduced from physical principles, but from experiment.

A special feature of the dynamo theory is connected with the fact that helicity is a property that is very difficult to extract from observation or experimental measurements. Until now, there is no direct observation of helicity. This is a challenge. Helicity observation is in some senses more attractive than, say, problems of observating quarks or gravitational waves, because it is based on quite classical physics.

The problem of direct observations of helicity can hardly be solved for the case of the nontransparent solar convective shell. The transparent Milky Way would offer one such possibility, though an observational program of this kind would be very difficult. The possibility of direct measurements of helicity under laboratory conditions should be also considered.

Discussing this topic, P. Roberts stressed that one needs to be very careful when comparing these dynamo processes in solar, galactic and other physical conditions.

L. Kitchatinov drcw attcntion to a possiblc obscrvation of anothcr mcan valuc - the turbulent diffusivity - involved in the dynamo process.

This is because we need to understand the competition between the generation of magnetic field, due to helicity and differential rotation, and the destructive role of turbulent diffusivity. Until now, turbulent diffusivity has not been easy to measure. However, if sunspots are tracers of the toroidal magnetic field penetration from the convective shell up to the solar surface, it is reasonable to believe that sunspots or sunspot pair are subject to chaotic motion. The intensity of this motion is determined by the turbulent (or convective) diffusivity inside the shell. This means that sunspots can be used as tracers of turbulent diffusivity. The observational scheme for such measurements is basically the same as for the meridional circulation one. The effect of turbulent diffusivity can be extracted from the meridional circulation because the latitudinal distribution of sunspots is inhomogeneous.

Another basis for agreement between theory and observation could be found in a new understanding of all available observations. There is an obvious gap between "the observables" and the theoretical quantities. For example, in the framework of the dynamo theory, the solar dynamo wave is excited deep below the solar surface and is only tracked at the solar surface. Some magnetic tracers show poleward propagation, so one could in theory try to connect the kinematic dynamo activity in the equatorial region with these tracers, and interpret the main observational facts as a result of basic nonlinear processes.

One important idea of help in gaining a new understanding of observational data has been presented by Ph. Delache. Standard Fourier analysis of long term solar activity data is based on the concept of superimposition of periodical phenomena. This is a natural concept when analyzing, say, the brightness curve of a binary system ; but it is harder to accept for the case of solar activity. Of course, results corresponding to 11-year cycles are correct, but the real meaning of longer-term events like the Gleissberg cycle is not so evident. These should be interpreted possibly as a sequences of aperiodic chaotic phenomena. Dynamical system theory offers a rich choice of methods that are not based on the concept of long-term periodicity.

Let us now consider another problem we have in interpreting the observational data. According to the analysis of sunspot data obtained by the French Astronomical School during the XVII and early XVIII centuries, the Maunder minimum should not be considered as a period of total lull in solar activity. The 11-y cycles are present, though they are much less intense than modern cycles, as shown in the butterfly diagram (at least from 1700 to 1720). The periodicity of these cycles is remarkably stable, whereas their amplitude varies greatly. During the Maunder minimum, sunspots were located almost exclusively in the southern hemisphere. This fact has been interpreted by D. Sokoloff and E. Nesme-Ribes (1994) as a manifestation of a mixed parity dynamo solution. Mixed-parity solutions have been predicted by R. Jennings and N. Weiss (1991) from computer simulations. There is, however, another property of the sunspot distribution during the Maunder minimum which has to be explained in the dynamo theory. During Maunder minimum, sunspot latitudes were mainly below 20°, while during a normal cycle there are sunspots at around 40°. This might have to do with a change in differential rotation profiles which has been also observed (the rotation was then slower). Another alternative suggested by F. Krause is that this phenomenon is related to the different structures of dipole and quadrupole dynamo modes. This part of the discussion is discussed in greater detail by F. Krause (this volume).

Another problem of interest is the theoretical link between observations and numerical simulations. In the case of, say, the galactic dynamo, such a link is provided by asymptotic models. These semi-analytical models help sort out the basic physical properties, and indicate which observational quantities can be used for tests. Because the spherical dynamo model is a complex problem to handle, especially considering the numerous nonlinearities, such asymptotic models are needed in order to estimate the possible rates of magnetic field generation processes within the common α-ω dynamo.

W. Soon stressed that such analysis is even more important for the case of stellar activity. Much less information is available about stellar convection than solar, of course, but a variety of different situations is displayed by stars that are obviously an advantage for qualitative asymptotic methods.

REFERENCES

Jennings, R.L., and Weiss, N.O. (1991). Symmetry breaking in stellar dynamos. MNRAS, 252: 249-260

Krause, F. (1994). Oscillatory dynamos showing change of parity on a large time scale. This volume

Krause, F., and Rädler, K.H. (1980). Mean-Field Magnetohydrodynamics and Dynamo Theory. Akademie-Verlag, Berlin

Roberts, P.H. (1994). The solar dynamo. This volume

Rüdiger, G., and Brandenburg, A. (1993). A solar dynamo in the overshoot layer : cycle period and butterfly diagram. Astrophys. J., submitted

Rüdiger, G., and Kitchatinov, L.L. (1994). The differential solar rotation. This volume

Sokoloff, D., and Nesme-Ribes, E. (1994). The Maunder minimum, a mixed-parity dynamo mode ?. Astron. Astrophys., in press

Vainstein, S.I., and Cattaneo, F. (1992). Nonlinear restriction on dynamo action. Astrophys. J., 393: 165-172

Zeldovich, Ya.B. (1957). The magnetic field in the two-dimensional motion of a conductivity turbulent liquid. Sov. Phys., JETP, 4: 460-462

THEORIES OF RADIUS AND LUMINOSITY VARIATIONS

H.C. SPRUIT
Max Planck Institut für Physik und Astrophysik
Postfach 1523, D-85740 Garching, Germany

ABSTRACT. The main cause of variability of the Sun is its magnetic field. The theoretical work of the past decade has shown that the dominant effect of magnetic fields is a surface effect: a change of effective emissivity of the magnetic parts of the surface while the nonmagnetic part of the surface contributes very little to the irradiance variation on almost all time scales. No other processes have yet been found that would cause variations exceeding (at the current level of magnetic activity) the observed 0.1% irradiance fluctuation of the Sun. This implies that a knowledge of the surface magnetic fields [separated into its bright small scale (faculae, network) and dark large scale (spots) components] is sufficient for pre- or postdicting the solar irradiance. It is hypothesized that the discrepancy remaining between the measured irradiance variations and values reconstructed from proxies is due to the difficulty of finding a proxy that accurately correlates with the continuum contrast of a dispersed small scale magnetic field. Stellar structure theory predicts that the variations in the solar radius associated with magnetic activity are quite small, of the order of a few times 10^{-7}. Observations of radius variations much above this value would constitute a serious challenge for stellar structure theory.
keywords: Sun: magnetic fields – convection – radius – luminosity

1. Introduction

The solar luminosity varies on a wide range of time scales due to several processes. It increases as Hydrogen burns into Helium in the

NATO ASI Series, Vol. I 25
The Solar Engine and Its Influence
on Terrestrial Atmosphere and Climate
Edited by Elizabeth Nesme-Ribes

core. This increase is climatologically relevant ($\sim 30\%$ over the past $4.6\,10^9$ yr), but does not cause changes on human time scales (see review by Kasting and Grinspoon, 1991). The Sun oscillates in a large number of sound modes. The effect of these oscillations has been seen directly in the irradiance (see the reviews by Fröhlich, 1990, 1991). The effect is small ($\sim 10^{-5}$ level) and confined to short time scales (minutes). The random pattern of the granulation at the surface causes statistical noise at an amplitude of a few times 10^{-5} and similar time scales. Larger scale convective flows are known to exist and have longer time scales, but their effect on irradiance probably does not exceed that of the granulation. This leaves magnetically related variations as the main known cause of variability on all time scales from hours to millions of years.

In this lecture I summarize what is currently understood theoretically about radius and luminosity variations of stars like the sun, due to these magnetic effects, and which predictions can be made on the basis of this understanding. The reconstruction of irradiance variations from the observed properties of solar magnetic fields, and their distribution at the solar surface, is an important test of our theoretical understanding. In the conclusion I briefly discuss what appears to be needed, in terms of observations, to make this comparison more quantitatively accurate.

The approach in the following is a conservative one: the emphasis will be on known and independently tested theory, in particular the theory of stellar structure and of convection in stellar envelopes. In these theories quantitatively accurate models are possible, and specific predictions about radius and luminosity variations can be made. As an example of the maturity of these theories, note the remarkable success of the *ab initio* simulations of solar convection by Nordlund and coworkers (Nordlund and Stein 1991, for reviews see Spruit, Nordlund and Title 1990, Chan *et al.*1991) in reproducing the observed properties of the granulation.

The variations observable at the surface are connected with changes in the interior of the Sun. Independent tests of theoretical ideas are therefore possible through observations of p-mode frequencies and their variation with the solar cycle. This is discussed briefly in section 6.

1.1. LUMINOSITY MODULATION BY MAGNETIC FIELDS

To make a successful theory of luminosity variations by magnetic fields one needs two ingredients. A theory is needed that tells us how much magnetic field is present and where. Secondly, given the distribution of magnetic fields, a theory to compute the resulting luminosity variations during the Sun's magnetic cycle. The first of these questions is the *dynamo problem*. Dynamo theories have a long tradition and very extensive literature, but unfortunately have not progressed yet to the stage that the location and strength of the magnetic field in a star can be predicted. Hence one is forced to rely mostly on observations instead of theory. The second question turns out to be very much simpler. Once the magnetic fields are known, the problem of luminosity variations reduces to a calculation of thermal perturbations in which the sources are known. The theory of stellar structure is sufficiently well developed and tractable to produce reliable answers for this problem, at least at the 'up to factors of order unity' level. These answers, and the underlying physics are the subject of the following sections.

Conceptually, it is easiest to separate the thermal effects of magnetic fields into two classes:

1. 'Sources and sinks'. The generation of a magnetic field involves the conversion of kinetic energy of motion into magnetic energy. Since the motions in the solar envelope are thermally driven, this ultimately means conversion of thermal into magnetic energy: building up a magnetic field produces a thermal sink somewhere. The opposite happens when the field decays: magnetic energy is converted into heat. These thermal effects exist only during *changes* in the magnetic energy content of the envelope.

2. Changes in thermal transport coefficients. Magnetic fields interfere with convection, causing a reduction in the efficiency of

heat transport in the envelope. In contrast to (1), these changes last as long as the magnetic field itself is present.

Both these sources and sinks and changes in the transport coefficient cause thermal perturbations, varying with magnetic activity, which propagate through the envelope and cause variations in surface energy flux. Related to the second class of perturbations are the effects of magnetic fields at the surface of the star. Sunspots, being dark, radiate less than the surrounding photosphere, while the small elements that make up plage and the network have an excess emission.

2. Time scales

The response of the Sun to thermal perturbations is not governed by a single time scale, but by a wide range of time scales. The longest of these is the thermal time scale of the Sun as a whole, called the Kelvin-Helmholtz time scale, about 10^7 years. If the central heat source of the Sun were switched off, the internal structure and the luminosity would start to change on this time scale. More generally, we can define the thermal time τ_t scale as a function of depth:

$$\tau_t \equiv U(z)/L(z) \approx \frac{1}{L}\int_{R-z}^{R} 4\pi r^2 u dr, \tag{1}$$

where L is the luminosity at depth z, U the thermal energy of the envelope down to a depth z, and u the thermal energy per unit volume, approximately (for an ideal gas of constant γ) given by $u = P/(\gamma - 1)$. This is the time scale on which the structure of the envelope, and the observed luminosity, would start changing when the heat flux in the star were interrupted, by some magical means, at depth z. Some rough values for this quantity are $\tau_t \sim 10^5$yr at $z = 2\,10^5$km (depth of the convection zone), 10 yr at 20 000km (the size of a supergranule), 10 hrs at 2000km (size of a granule). This shows that the thermal time scale is a very sensitive function of depth in the Sun. As a result, the thermal response of the Sun also depends critically on the location of the disturbance. A fairly good approximation to the stratification of the convective envelope of the Sun is a polytrope of index $n = 2$ (this is better than the standard 'convective' value n=1.5 because $\gamma < 5/3$ due to partial ionization). Hence the gas pressure and thermal time scale vary roughly as $P \sim z^3$, $\tau_t \sim z^4$, where the depth z is counted from a level 3 scale heights above the surface $\tau = 1$.

A second kind of time scale involved in thermal readjustments is the *diffusive* time scale. In the mixing length approximation, transport processes in the convection zone can be computed with a turbulent diffision coefficient $\kappa_t \approx \frac{1}{3} l_c v_c$ where l_c and v_c are the convective length scale and velocity. In the solar convective envelope, this quantity varies only weakly, at a value of the order 10^{13} cm^2/s. Thermal inhomogeneities (more precisely: entropy inhomogeneities) of length scale d are smoothed by turbulent diffusion on a time scale

$$\tau_d = d^2 / \kappa_t. \tag{2}$$

For $d = 2\,10^5$km this is about 1 yr, for d = 2000km about 1 hr. Comparing τ_t and τ_d, we see that they are of similar magnitude close to the surface (to be precise: in the surface boundary layer where convection is not efficient enough to keep the stratification close to adiabatic). In deeper layers, the thermal time scale is much *longer* than the diffusive time scale, by a factor of up to 10^5.

The two time scales measure different types of thermal adjustment process. These same processes appear in the thermal behavior of, say, a spacecraft. Let me take the liberty of approximating a satellite as a chunk of aluminum with a heater in it. The thermal time scale is the time scale on which its temperature adjusts to a change in the heat input, such that the heat radiated from the surface into space balances the heat input again. It is determined by the heat capacity [U in eq. (1)] and the power level (L). The time scale on which different parts of the chunk equilibrate to the same temperature is governed by a different process, namely the thermal conduction (the equivalent of the turbulent diffusion in the Sun). The diffusion (conduction) time scale is much shorter than the thermal time scale, because of the large conductivity of Al. In the Sun, it is the very large turbulent diffusivity in the bulk of the convection zone that causes the very short diffusive time scale compared with the thermal time scale.

How do these different time scales come into play when the convection zone is thermally perturbed by, say, the storage of energy in a growing magnetic field? Such perturbations can be computed in detail, either by numerical methods (Endal *et al.*1985, Gilliland, 1988) or more analytically. We can, for example, consider the initial value problem in which a perturbation is allowed to evolve in time by thermal transport in the convection zone. In general this evolution has components on all the time scales of the problem, including the very long thermal time scale. Formal aspects of this problem are discussed briefly in the next subsection (for a more complete analysis, see Spruit, 1982ab, 1991, Arendt 1992). For quantitative results, skip to section (3).

2.1. Modes of thermal relaxation

The problem can be studied in detail within the diffusion approximation for convective energy transport:

$$\mathbf{F} = -\kappa \rho T \nabla S, \tag{3}$$

where $\mathbf{F}$ is the heat flux, κ the turbulent diffusivity, ρ the density, T the temperature and S the entropy. The energy equation, in the absence of sources, is given in terms of the entropy by

$$\rho T \frac{\mathrm{d}S}{\mathrm{d}t} = -\mathrm{div}\mathbf{F}. \tag{4}$$

As an example, consider the one dimensional problem in which all quantities depend on depth z only. In this case, combining (3) and (4), and neglecting flows we get

$$\frac{\partial S}{\partial t} = \frac{F_z}{\rho T H} + \kappa \frac{\partial S}{\partial z^2}, \tag{5}$$

where $H = \partial \ln \rho T/\partial z^{-1}$ is the pressure scale height. For perturbations in which the left hand side is balanced by the first term on the right, the time scale is of the order $F/(c_p\rho TH)$, which is just a thermal time scale (eq. 1). For modes in which the left hand side is balanced by the second term, we evidently have an ordinary diffusion problem, hence these modes evolve on diffusive time scales (eq. 2). The first term appeared as a consequence of the stratification of pressure, i.e. the inhomogeneity of the convection zone. In a homogeneous medium (as in a chunk of Al), the thermal time scale comes in only through the radiating surface boundary condition (which one could also consider as a form of inhomogeneity). If T_s is the surface temperature, this condition is

$$\sigma T_s^4 = F_z. \tag{6}$$

Since the perturbations in entropy outside spots and the magnetic elements making up the small scale field are small, the response of the convection zone can be computed by linearizing eqs. (3), (4), (6). An arbitrary initial perturbation may then be decomposed in the standard way in terms of the eigenmodes of the problem. These are found by setting $\delta S = e^{\eta t} f(\mathbf{r})$, and solving for $f(\mathbf{r})$ with the decay rate η as eigenvalue (the problem thus defined has only decaying solutions). Specializing to the one dimensional case ($f(z)$) one finds a series of modes ordered by the number of nodes n of f. The fundamental ($n = 0$) has a decay rate of the order $\eta_0 \sim \tau_t^{-1}$ that is, this mode decays on the thermal time scale. The higher modes decay with rates of the order $\eta_n \sim n^2\tau_d^{-1}$, they decay on diffusive time scales. This is illustrated in Fig. 1.

3. Effect of subsurface magnetic fields on L and R

In this section I consider the effect on luminosity of magnetic fields in the convection zone which vary on the time scale $t_c \approx 10$yr of the solar cycle (for a discussion of more general perturbations and for details of the calculations, see Spruit 1982a,b, 1991). The best

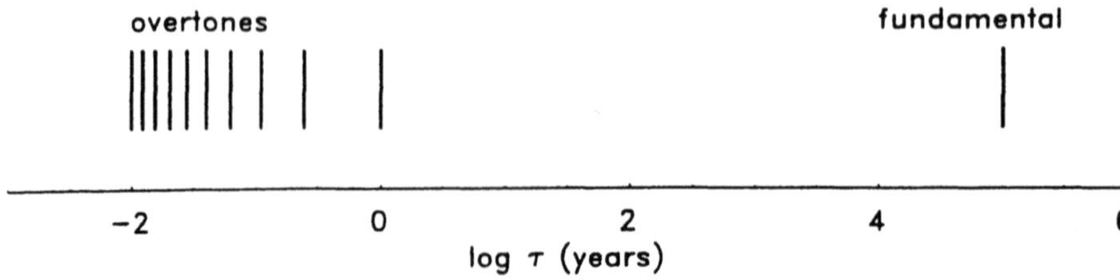

Figure 1: Spectrum of decay times of thermal perturbations in the solar convection zone (schematic)

current guess is that most of the flux, of the order $10^{23} - 10^{24}$ (Zwaan, 1978) Gcm^2 is located near the base of the convection zone, and has a strength of the order 50–100 kG (D'Silva and Choudhuri 1993, D'Silva and Howard 1993). These quantities are somewhat model dependent.

To reduce the dependence on these models, let me keep the depth z_b at which the field is located as a free parameter. To obtain upper limits to the luminosity effects, assume that the field at z_b covers the entire horizontal surface, and covers a depth range of one pressure scale height H (this is sufficiently general for the present purpose, since $H(z_b)$ is a significant fraction, of the order 1/3, of z_b itself). The convection zone, being nearly isentropically stratified, offers no restoring force to the motion of the magnetic field. If the field is in temperature equilibrium with its surroundings, the gas density in it is less, causing magnetic *buoyancy.* The speed at which the field starts rising under the influence of this force is of the order of the Alfvén speed. If the field is to remain inside the convection zone for a significant fraction of the cycle time, this yields the so-called buoyancy limit on the field of the order of 100G (Parker, 1984). Buoyancy can be avoided by creating the field with a slight temperature deficit, in such a way that it is neutrally buoyant. Such fields are found to be unstable however, on an Alfvén travel time scale (Parker 1979, 1984; Spruit and van Ballegooijen, 1982). This again limits the field to the buoyancy limit. This restrictive limit can be relaxed a bit by noting that convective downflows exist that are larger than the Alfvén speed at 100G. In principle, these flows can be used to prevent a field from rising through the convection zone. Making this assumption, the upper limit to the field becomes equal to the *equipartition* value, at which the energy density in the field equals the kinetic energy of the convective flows. This value increases with depth, to a maximum of about 10 000G near the base of the convection zone.

With these values for the strength, extent and location of the field, the predicted upper limits to fluctuations in luminosity can be estimated using the formalism of section (2.1). This is shown in Figure 2. The results depend on the nature of the perturbations. The result of conversion between thermal and magnetic energy is called a 'β-perturbation', that caused by a reduction of the convective energy transport efficiency an 'α-perturbation' (Endal *et al.*, 1985). In computing the α-perturbations, it is assumed that the reduction of convective efficiency by the field is of the order unity if the field strength is equal to the equipartiction value (Spiegel and Weiss, 1980). From Fig. 2 it is clear that α-perturbations are the more effective ones (Gilliland, 1988) except near the base of the convection zone. This has to do with the fact that the field can interfere with the convective flow as long as it exists, whereas the β-effect is limited by the magnetic energy content of the field. [Near the base the relative effectiveness of the two is reversed because convection there is so efficient that a reduction by the field has little effect]. The results in Fig. 2 agree qualitatively with results obtained from perturbed stellar structure models (e.g. Endal *et al.*1985, Gilliland, 1988).

Fig. 2 shows that luminosity variations on the order of the observed 0.1% irradiance variation can not be caused by subsurface magnetic fields unless they are located at a depth less than a few thousand km. This is so close to the surface that observations probably give a better estimate of the magnetic field at this depth than the theoretical limits used above. The surface field is nearly vertical, 1–3kG in strength, and generally has a filling factor of the order 1%

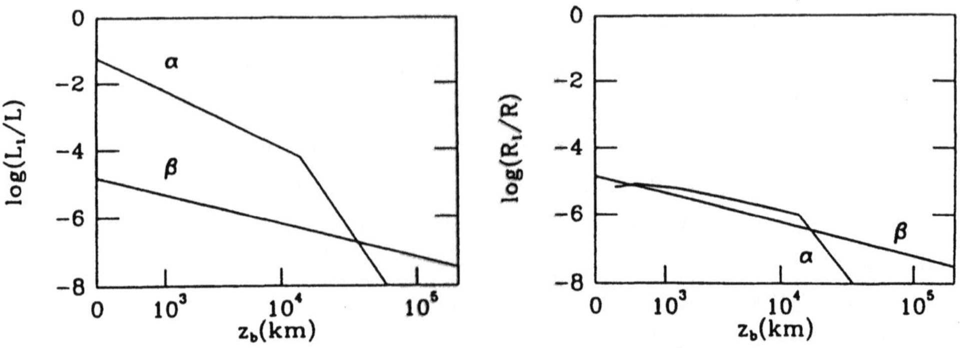

Figure 2: Upper limit on luminosity and radius variations on a solar cycle time scale caused by magnetic fields in the convection zone, as a function of their assumed depth z, for fields limited by equipartition with convection

or less. Because of the low filling factor (compared with the 100% filling factor assumed so far), such fields have much smaller effect than the fields assumed above. Their effect is discussed further in section 5.

At the base of the convection zone, the upper limit on the field strength based on the buoyancy argument is not valid, since the radiative core is quite stably stratified, and acts as a nearly solid surface towards motions in the convection zone. Fields close to this surface could plausibly be higher than the equipartion value. Ignoring the buoyancy limit, the next most plausible upper limit to the field strength is obtained by assuming that the entire luminosity of the Sun is spent in building up a magnetic field near the base of the convection zone during the ascending phase of the solar cycle, and this same amount released again during the descending phase. One finds that the luminosity effect in this case is $\delta L/L < 10^{-5}$ (for β-perturbations, which are the most effective ones in this case). This perhaps surprising result has to do with the very large heat capacity of the convection zone, related to its thermal time scale of 10^5yr.

3.1. 2-Dimensional effects, thermal shadows

In computing the effect of a reduction of the convective transport coefficient, we have assumed that this reduction takes place uniformly over a horizontal surface. If, instead, there are 'gaps' between the regions of reduced transport, the effects on the luminosity are much reduced. The heat flow blocked by the obstacles is effectively shunted around them. As in the case of thermal insulation of a home, it is not sufficient to have effective insulation around most of the building's surface, since the heat can still be lost effectively through the remaining holes. Examples of this shunting effect around obstacles buried inside the convection zone were given in Spruit (1977).

3.2. Effects of large scale flows

In the diffusive model for heat transport used in the above, possible irradiance variations due to large scale flows are not included. Their effect can be estimated simply from observations of the horizontal velocity amplitude of such flows. The supergranulation ($\sim$ 30 000 km), for example, has a horizontal velocity v_h of the order 300–500 m/s, upper limits for giant cells ($\sim$ 200 000 km) are on the order of a few tens of m/s. The vertical component of these flows carries extra energy to the surface, such that the change in heat flux δF due to such a cell is

$$\frac{\delta F}{F_0} \sim \frac{\delta v_z}{v_{z0}}, \tag{7}$$

where v_{z0} is the vertical velocity in the granulation (which carries most of the heat flux F_0 at the surface). For flows with a fixed aspect ratio (ratio of vertical to horizontal length scale), in which the horizontal scale L is large compared with the pressure scale height ($H_0 \sim$ 150km), the vertical velocity at the surface $z = 0$ scales as

$$v_z(z = 0) \sim v_h/L, \tag{8}$$

as a consequence of the continuity equation. The number n of cells of size L at the surface at any point in time is of the order $n =$

$\pi R^2/L^2$, where R is the solar radius. If the cells are uncorrelated, the irradiance variation due to large scale flows is then

$$\frac{\delta I}{I} = \frac{1}{\sqrt{n}}\frac{\delta F}{F_0} = \frac{\eta}{\sqrt{\pi}}\frac{v_h}{v_{h0}}\frac{L}{R}, \tag{9}$$

where v_{h0} is the horizontal velocity in the granulation, and η a numerical factor of order unity. By calibrating this relation with the observed contrast of the granulation, of the order 10%, we get $\eta \approx 0.1$. The predicted irradiance variation due to the granulation is then $2\,10^{-4}$ (time scales tens of minutes), for the supergranulation $2\,10^{-5}$ (days), for giant cells of 30 m/s $2\,10^{-7}$ (months). Except for granulation, these numbers are significantly smaller than the observed, magnetically related, variations, at the time scales indicated.

4. Radius variations

For the kind of thermal perturbations considered in the previous section, the variations in solar radius predicted are small. The maximum values in Fig. 2 may be observable, but if we take the current estimate (see above) that most of the field is located near the base of the convection zone, $z_b = 10^5$ km, a more plausible estimate is

$$\delta R/R \sim 10^{-7}. \tag{10}$$

The results in the previous section were derived for a sheet of horizontal fields in pressure equilibrium, but not at the same density as the surroundings. If magnetic fields are not at the same density, buoyancy forces will bring them to the surface. This was the reason for assuming the equipartition with convection as a plausible upper limit to the field strength. An exception to the buoyancy argument are *vertical* magnetic fields. Much of the magnetic fields we see at the surface may well continue down into the convection as more or less vertical fields. Are larger radius variations to be expected from such vertical fields? If the ratio of gas to magnetic pressure is β, the difference in density is roughly of the order $\delta\rho/\rho \sim 1/\beta$. The vertical magnetic fields, assumed to be in hydrostatic equilibrium, have displaced an amount of gas corresponding to this density difference. For a constant field strength B, covering the surface with a filling factor f, this amount is

$$\Delta M \sim fR^2B^2\ln(P_b/P_0)/g, \tag{11}$$

where g the surface gravity, and P_b, P_0 the gas pressures at the base and the surface of the convection zone. With $f \sim 10^{-2}$, $B \sim 10^3$, this yields $\Delta M/M_\odot \sim 2\,10^{-10}$. For the structure of the star, it is as if this amount of mass were added to its envelope. With the known adiabatic mass-radius relation for the Sun, this corresponds to a radius variation of the order

$$\Delta R/R \sim 10^{-10}. \tag{12}$$

The Sun is not strictly in hydrostatic equilibrium, it oscillates and contains convective flows. If the strength of these flows and oscillations varies with the solar cycle, they will cause a small radius effect. Such variation is known to exist; the oscillations as well as the speed of convective flows is reduced in magnetic regions (Title et al. 1992, Elsworth *et al.*1993). On average, the effect of flows is that of a turbulent pressure $P_t \sim \langle\rho v^2\rangle$. How much is the Sun expanded by this turbulent pressure? Both for convection and the oscillations, the effect on the solar radius peaks near the surface, and it yields a radius change of the order

$$\Delta R/R \sim 0.1\frac{H_0}{R}\left(\frac{v}{c_s}\right)^2, \tag{13}$$

where H_0 is the pressure scale height at the surface, and v/c_s the maximum Mach number of the flow. For convection, $(v/c_s)_{\max} \sim 0.3$, for the p-mode oscillations 0.1. This yields a radius change of the order $2\,10^{-6}$ due to convection and $2\,10^{-7}$ due to oscillations. Since not more than 20% of the solar surface is covered by regions in which the flows are reduced by magnetic fields, and the observed reduction in velocity amplitudes not more than a factor 2, the expected variation in radius during the cycle, due to changing strength of the flows, is not more than

$$\Delta R/R \sim 2\,10^{-7} \tag{14}$$

The effect is a *reduction* of R at maximum magnetic activity.

We conclude that for plausible assumptions about the location and strength of the magnetic fields, the known causes of radius change probably do not exceed a level of a few times 10^{-7}. Real measured variation much above this level would not be easily explainable by conventional mechanisms. Even changes ten times as large as this would be very hard to measure however. The main difficulty is that any experimental definition of the solar limb is affected by the presence of bright areas near the limb, whether they be due to identifiable faculae or to dispersed magnetic activity that is not easily recognized (cf. Brown 1982, and section IV in Ribes *et al.*1991).

The radius variations discussed above are defined as changes of the radius of the *nonmagnetic* part of the solar surface. In spots and small magnetic elements, the $\tau = 1$ surface is depressed (Wilson depression). If one were to compute a mean geometric radius including these depressions, radius changes up to 10^{-5} (a reduction at solar maximum) would result, due to the average depression in spots and the small scale fields. This mean radius is not a directly observably quantity however, and the radius of the nonmagnetic photosphere is, also observationally, the more meaningful quantitity.

5. Effect of surface magnetic fields

The strongest effects on luminosity are obtained from fields at the radiating surface itself (cf. Fig. 2). This is mostly a reflection of the steep depth dependence of the thermal inertia of the convection zone. The surface effects are caused by changes in the efficiency of energy transport associated with the field. A reduction of the convective energy flux due to the resistance of magnetic field lines against bending causes sunspots to have a (bolometric) surface emissivity of about 20-25% of that of the normal solar surface. In very small magnetic concentrations (less than about three pressure scale heights across) this is more than compensated by a net *excess* emissivity. Due to the lower internal pressure such concentrations are small 'pits' in the surface. Through the sides of these depressions, an extra energy flux is emitted, which is visible mainly when viewing them at an angle from the vertical i.e. when seen near the limb (Spruit, 1976). This effect led to a theoretical prediction of enhanced emissivity of the solar surface associated with the small scale magnetic field (Spruit 1977). Most of the magnetic field on the solar surface is in the form of these small concentrations, and it appears from the irradiance measurements of the last cycle that their effect dominates over the reduced emission of sunspots (this may not have been the case in all cycles, however, see Foukal 1993). Theoretical prediction is difficult, since the excess emission from a facular area is a sensitive function of the poorly known size distribution of the concentrations (see however Spruit and Zwaan, 1981). For a given size, the effect can be computed in principle through detailed numerical simulation of the energy transport in the concentration and its surroundings. Such calculations have already been undertaken by Knölker *et al.*(1991), and Schüssler (1991).

Changes in surface emissivity are much more effective than magnetic energy storage or release in the same structures. Consider for example a small concentration with a contrast ϕ (averaged over its surface cross section). The life time of such elements can be estimated from the decay time of an active region, on the order of a few months. During this time τ, the element emits, per unit of magnetic flux, an excess energy $E_e = \phi\tau F_\odot/B_s$, where B_s is the surface magnetic field strength. If the field lines of the element reach down to base $z = D$ of the convection zone, their magnetic energy content per unit magnetic flux is $E_B = DB/8\pi$. Putting in $B \sim$ 2kG, one finds that E_e is more than a factor 10 larger than E_B. This is only a simple energetic argument. In practice, the magnetic energy content of a magnetic structure is not so simply available to be emitted at the surface. In fact, there is no known way of transporting the energy in the deeper layer of the magnetic structure to the surface on the required time scale, even if all of it were available. This contrasts with the effect of the field on the surface emissivity. This effect develops as soon as the magnetic structure appears at the surface, because of the very short thermal time scale of the surface layers.

We conclude from the discussion so far that changes of the surface emissivity are predicted to be the main cause of irradiance variations. Since the brightness of sunspots and small scale fields can be measured, an irradiance variation can be calculated by just multiplying contrasts and areas and adding up. Is the situation as simple as this? What if some of the heat flux blocked by sunspots for example were reemitted elsewhere on the surface in diluted form? Calculations based on the formalism mentioned in section (2.1) predict that, in fact, this reemission 'at a distance' does not take place (Foukal *et al.*1983, Spruit, 1982a,b, Chiang and Foukal, 1984), on time scales of human interest. On time scales of the order $\tau_t \sim 10^5$ yr, the blocked flux *does* reappear. It does so in distributed form, nearly uniformly over the surface. Thus, the current nonmagnetic surface of the Sun radiates the blocked flux of all the spots on its surface avaraged over the past 10^5 yr (and similarly for the small scale fields). Due to this averaging effect, the nonmagnetic surface does not change on observable time scales.

Since the contrasts and numbers of magnetic elements of various sizes at the surface are in principle measurable at any point in time, the prediction that the brightness of the nonmagnetic surface does not take part in the cycle can in principle be checked. In practice this is rather difficult, since the largest effect comes from the small scale field. The predicted net emission of these small scale structures is a very sensitive function of their size (Spruit 1977, Spruit and Zwaan 1981), and the largest effect is predicted from structures that are near or below the current spatial resolution. To account for the small scale field meaningfully, an intensive study of the effective brightness of the small scale field, and its magnetic flux has to be made, as a function of its evolutionary state (since it is known that the typical size of magnetic elements is larger in young active regions than in old ones, for example). The current proxy indicators do not represent the relevant quantity (photospheric brightness) accurately enough. The prediction that the smallest elements give the largest contribution would mean that the contribution of the most dispersed small scale fields, which are hard to observe, may have been strongly underestimated. The results of Lean, Skumanich and White (1992) point in this direction.

5.1. Cancellation of spots and faculae

The contribution of spots and the small scale field on the irradiance roughly cancel in some average sense, with only a modest brightness excess remaining at maximum activity. On time scales less than a few months, the cancellation is only partial: the brightening due to the network and faculae has a longer time scale than the rather transient effects of spots (Foukal and Lean, 1988). There are indications that this cancellation may be a peculiarity of the Sun at its present level of activity. Stars more active than the Sun (larger filling factor of magnetic fields at the suface) are observed to be dimmer at the maximum of their activity cycle (Radick, Lockwood and Baliunas, 1991), the opposite of the Sun's behavior during the last cycle. The interpretation is that at greater activity than the Sun's, a larger fraction of the magnetic flux is in the form of spots, so that their dimming effect dominates. In stars with lower activity levels, the brightening due to small scale fields might dominate, as in the Sun. Since the level of the Sun's activity varies from cycle to cycle, it is possible that the ratio of spot dimming to weak field brightening varies as well. Such an effect has in fact been inferred by Foukal (1993) from an analysis of the Greenwich facula and spot data. He finds that in the strongest cycles the Greenwich facular areas are much below those expected from linear extrapolation of the weaker cycles. The near cancellation of spot dimming and facula-associated brightening may also be the reason (Foukal, 1993) why the irradiance variation in the Sun is so low compared with stars only slightly more active (Radick *et al.*1991).

5.2. THE 'BLOCKED HEAT FLUX PROBLEM' OF SUNSPOTS

Suppose at $t = 0$ a spot appears instantly on the surface, reducing the heat flux. At least a part of the blocked heat flux will find its way 'around' this obstacle, to show up as a 'bright ring'. How long does it take for this to happen, and what fraction of the blocked heat flux is reemitted elsewhere? To see this, consider a terrestrial analogy, consisting of an aluminum brick heated from below, and losing heat to its environments from the top. On this I put, again at $t = 0$, a thermally insulating patch. Does a 'hot ring' develop around the patch? Is all of the heat flux blocked by the patch radiated from some other part of the surface? The answer is no. The high thermal conductivity of aluminum does not allow strong inhomogeneities like a hot ring. By conduction an even temperature is maintained, and the blocked heat flux, instead of reappearing somewhere on the surface, appears in the energy balance as a slow heating up of the brick. A new equilibrium is reached on the thermal time scale of the brick, which is long on account of its high heat capacity. This situation also applies to a convective stellar envelope, with only a small additional complication. In a convective envelope both the thermal (turbulent) conductivity and the heat capacity are strong functions of depth. The comparison with aluminum applies only at depths where the convective stratification is close to adiabatic. Above this, convection is less efficient, and the heat capacity small. For this reason, it matters how deep into the envelope the blocking object extends. For the Sun, one finds (Spruit, 1982a, Foukal *et al.*1983) that less than 10% of the blocked flux reappears at the surface, on time scales of observational interest ($< 10^5$yr), if the blocking extends to a depth greater than about 1000km. The part that does return to the surface does so very close to the spot or facular element.

The prediction is thus that, for practical purposes, the solar surface outside magnetic regions *does not take part* in irradiance variations associated with the solar cycle. For dissenting views on the spot blocking problem and related matters, see Chapman *et al.*1984, Sofia *et al.*1982. Schatten and Mayr 1985, Fox, Sofia and Chan, 1991.

In numerical simulations of spot blocking (Foukal *et al.*1983, Fox *et al.*1991), the very large range of time scales causes problems. This is one of the reasons why the depth of the convection zone is reduced artificially in such calculations. As a result, the thermal time scale in such calculations is also much shorter than in the real convection zone. In the simulation by Fox *et al.*(1991) with a depth of only 2000km the thermal time scale is on the order of a few hours instead of 10^5yr for the real convection zone. The rapid return to a steady heat flux observed by Fox *et al.*is a consequence of this shallow depth. Secondly, the depth of their spot, 150 km, is much shallower than a real spot. The Wilson depression alone of a sunspot is about four times as deep as this. The strength of the sunspot field, 3000 G, is large enough to interfere strongly with the convective heat flow for several thousand km below the surface. Whereas a convective diffusion model predicts that less than 10% of the blocked heat flux returns to the surface if the blocking effect extends 1000 km or more (Spruit, 1982a), it predicts that 60% returns if the blocking is only 150 km deep. This is the consequence of the very steep increase of the turbulent thermal conductivity with depth (roughly proportional to ρ). Finally the upper boundary condition used by Fox *et al.*, which sets the temperature fluctuations to zero at the top surface instead of a proper radiative boundary condition, overestimates the fraction of the blocked heat flux that returns to the surface (Spruit 1976, section 4.2). In other words the results by Fox *et al.*are in qualitative agreement with the diffusion model, and are not directly relevant for sunspots.

For small magnetic concentrations, in which a lateral influx of heat enters into the tube rather close below the solar surface, a significant fraction of the influx of heat may derive from the immediate surroundings (forming a 'dark ring' around it). The extent to which this happens depends on the precise nature of the energy carrying flow near the tube, which is the subject of current numerical study. Since the concentrations that show a net excess are so small, the measured excess is usually an average over the concentration (see however Keller, 1992). This average can still be positive, because part of the lateral influx derives from the interior of the convection zone (this is a consequence of the rapid increase of the thermal conductivity with depth). In a simple diffusion model for the heat transport near the tube, an average excess emissivity of up to 100% was found (Spruit, 1977, Figures 16,17, see also Spruit and Zwaan, 1981).

6. Constraints from Helioseismology

Interpretation of observed irradiance and radius variations in terms of changes in the solar envelope are constrained not only by these observations, but also by the observed frequencies of p-mode oscillations. Variation of these frequencies with the solar cycle has now been observed in remarkable detail (Libbrecht and Woodard, 1990). From Libbrechts measurements it is found that: i) the changes in frequency originate in the active latitudes on the Sun, ii) the effect is due to a change in the outermost layers (down to about 200 km below the surface, Shibahashi 1991). The most significant finding is however that the frequencies *increase* with magnetic activity. If the outer convection zone were slightly hotter at solar maximum, as a naive interpretation of the increased irradiance at solar maximum would indicate, the p-mode frequencies would *decrease*. This is because the increase in frequency due to the higher sound speed is more than offset by the longer path length due to the expansion of the layers involved (for a detailed discussion of this issue, see Goldreich *et al.*1991). Instead, the frequency shifts are consistent with the increase in stiffness of the atmosphere due to the presence of a magnetic field (Goldreich *et al.*1991, Shibahashi, 1991).

The dependence of the frequency changes on mode frequency and horizontal wavenumbers contains detailed information that can be used to separate any non-magnetic effect, if present, from the magnetic increase in atmospheric stiffness. How large a contribution, not explainable by magnetic stiffness increase, may be present in the existing data is not clear at present. The potential of these measurements in putting tight limits is considerable however. Assume, as a computational example, that not more than 10% of the observed variation of 10^{-4} would be atributable to other causes than the magnetic stiffness increase, i.e. an upper limit $\Delta\nu/\nu \sim 10^{-5}$. Comparing this with the expected frequency change due to a uniform expansion of the solar convection zone (leaving the core unchanged), $\Delta\nu/\nu \sim \Delta R/D$, where D is the depth of the convection zone ($D = 0.3R$) we get an upper limit of $\Delta R/R \sim 3\,10^{-6}$ for radius changes of this kind. This is an order of magnitude smaller than the current sensitivity of direct radius measurements (cf. Brown 1988, Ribes *et al.*1991).

7. Conclusions, outlook for the future

The theory of structure and evolution of stellar envelopes gives a rather simple picture for the effect of magnetic fields on luminosity (irradiance). This is because it predicts that for irradiance variations on time scales short compared with 10^5 yr the non-magnetic surface[1] participates only at a very low level in the variations due to magnetic fields. Taking account of effects that are ignored in the calculations, such as systematic flows and uncertainties in the mixing length model used, may certainly change the contribution of the nonmagnetic surface by factors of order unity. The main conclusion is robust with respect to such changes because this contribution is so small in the first place.

Most of the uncertainty in making comparisons with observations comes from our poor knowledge of the intrinsic brightness and magnetic flux in the small scale magnetic field, especially in its most dispersed form, since this field contributes disproportionately to the

[1] More precisely: the surface outside about 1 granule diameter from a magnetic element, since the magnetic field has some influence on the convective flow in its neighborhood.

net irradiance. Its brightness for a given size and shape of the magnetic elements can in principle be obtained from numerical simulations (Knölker *et al.*1991). These simulations are now approaching such a level of realism that quantitative comparisons with observations become meaningful. One may hope that in the not-too-distant future simulations as good as the current ones for nonmagnetic granulation will be available. The spectrum of scales in the small scale field however depends on aspects of the solar dynamo process that are intrinsically difficult. Theories of the solar dynamo so far have hardly touched upon the question what determines this spectrum. For the time being, it seems more realistic to expect progress in this direction from observations. With new methods of image reconstruction, magnetic and intensity measurements on subarcsecond scale are becoming possible (e.g. Keller, 1992). Systematic investigation of the sizes and brightness contrasts of magnetic flux, as a function of global parameters like the stage of development of an active region, might be able to give the quantitative information with which proxy indicators could be calibrated more reliably.

Concerning radius variations, the implicit point of view in the above has been to regard current direct measurements as upper limits (in line with the views expressed by Brown, 1988). From the theoretical point of view, the solar radius is a rather robust quantity which, at the level of accuracy of current measurements, depends only on rather basic physics. In any case, if variations $\delta R/R$ much exceeding a few times 10^{-7} (section 4) were reliably detected, the implications would be much more serious than a mere contradiction between, say, a dynamo theory and observations of the solar cycle.

References

ARENDT, S. 1992, Large scale flows and solar luminosity variations *Astrophys. J* **389**, 421-427

BROWN, T.M. 1982, Seeing-independent definitions of the solar limb position *Astron. Astrophys.* **116**, 260-264

BROWN, T.M. 1988, in *Solar radiative output variation*, P. Foukal, ed.,Cambridge Research & Instrumentation Inc., 21 Erie st. Cambridge, MA 02139, p 174

CHAN, K.L., NORDLUND, Å., STEFFEN, M., STEIN, R.F., 1991, in *The Solar Interior and Atmosphere*, eds. A.N. Cox, W.C. Livingston and M.S. Matthews, University of Arizona Press, p223

CHAPMAN, G. A., HERZOG, A. D., LAWRENCE, J. K., SHELTON, J. C. 1984, Solar luminosity fluctuations and active region photometry *Astrophys.J.* **282**, L99-101

CHIANG, W. H. & FOUKAL, P. V. 1984, The influence of faculae on sunspot heat blocking *Solar Phys.* **97**, 9-20

D'SILVA, S. & CHOUDHURI, A. R., 1993 A theoretical model for tilts of bipolar magnetic regions *Astron. Astrophys.* **272**, 621-633

D'SILVA & HOWARD, R. A. 1993, *Solar Phys.* **148**, 1

Elsworth, Y., Howe, R., Isaak, G.R., McLeod, C.P., Miller, B.A., New, R., Speake, C.C., Wheeler, S.J., 1993, The variation in the strength of low-*l* solar p-modes: 1981-1992 *Mon. Not. R. astron. Soc.* **265**, 888-898

ENDAL, A. S., SOFIA, S. & TWIGG L. W. 1985, Changes in solar luminosity and radius following secular perturbations in the convective envelope *Astrophys. J.* **290**, 748-757

FOUKAL, P. V., FOWLER, P., & LIVSHITS, M. 1983, A thermal model of sunspot influence on solar luminosity *Astrophys. J.* **267**, 863-871

FOUKAL, P. V., & LEAN, J. 1988, Magnetic modulation of the solar luminosity by photospheric activity *Astrophys., J.* **328**, 347-357

FOUKAL, P. V. 1993, *Solar Phys.* **148**, 219

FOX, P., SOFIA, S., CHAN, K. L. 1991, *Solar Phys.* **135**, 15

FRÖHLICH, C. 1990, in *Climate Impact of Solar Variability*, K. Schatten, ed., NASA CP **3086**, p.269

FRÖHLICH, C., FOUKAL, P. V., HICKEY, J. R. & HUDSON, H. S., WILLSON, R. C. 1991, in *The Sun in Time*, eds. C. Sonett,

M. Giampapa & M. Matthews, University of Arizona, Tucson, p11

GILLILAND, R. L. 1988, in *Solar radiative output variation*, P. Foukal, ed., Cambridge Research & Instrumentation Inc., 21 Erie st. Cambridge, MA 02139, p239

GOLDREICH, P., MURRAY, N., WILLETTE, G., & KUMAR, P. 1991, Implications of solar p-mode frequency shifts *Astrophys., J.* **370**, 752-762

KASTING, J.F., & GRINSPOON, D.H. 1991, in *The Sun in Time*, University of Arizona, Tucson, C. Sonett, M. Giampapa & M. Matthews, eds., p447

KELLER, C.U. 1992, Resolution of magnetic flux tubes on the Sun *Nature*, **359**, 307-308

KNÖLKER, M. GROSSMANN-DOERTH, U., SCHÜSSLER, M., WEISSHAAR, E. 1991, *Adv. Sp. Res.* **11**, 285

LEAN, J., SKUMANICH, A. & WHITE, O.R. 1992, *Geophys. Res. Lett.* **19**, 1591

LIBBRECHT, K.G. & WOODARD, M.F. 1990, Solar-cycle effects on the solar oscillation frequencies *Nature* **345**, 779-782

NORDLUND, Å. & STEIN, R.F., 1991 in *Challenges to Theories of the Structure of Moderate-Mass Stars*, eds. D.O. Gough and J. Toomre, (Lecture Notes in Physics **388**), Springer, Berlin, p141

RADICK, R., LOCKWOOD, W., & BALIUNAS, S. 1991, *Science* **273**, 39

PARKER, E. N. 1979, *Cosmical Magnetic Fields* Oxford, Clarendon Press.

PARKER, E. N. 1984 Magnetic buoyancy and the escape of magnetic fields from stars *Astrophys. J.* **281**, 839-845

RIBES, E., BEARDSLEY, B., BROWN, T.M., ET AL. 1991 in *The Sun in Time*, eds. C. Sonett, M. Giampapa & M.S. Matthews, University of Arizona Press, Tucson, p59

SCHATTEN, K. H. & MAYR, H. G. 1985, On the maintenance of sunspots: an ion hurricane model *Astrophys. J.* **299**, 1051-1062

SCHÜSSLER, M. 1991, *Geophys. Astrophys. Fluid Dyn.* **62**, 271

SHIBAHASHI, H. 1991, in *Challenges to Theories of the Structure of*

Moderate-Mass Stars, eds. D.O. Gough and J. Toomre, (Lecture Notes in Physics **388**), Springer, Berlin, p101

SOFIA, S. OSTER, L., SCHATTEN, K. 1982, *Solar Phys.*, **80**, 87

SPIEGEL, E.A. & WEISS, N.O. 1980, Magnetic activity and variation in solar luminosity *Nature* **287**, 616-617

SPRUIT, H.C. 1976, Pressure equilibrium and energy balance of small photospheric flux tubes *Solar Phys.* **50**, 269-295

SPRUIT, H.C. 1977, Heat flow near obstacles in the solar convection zone *Solar Phys.* **55**, 3-34

SPRUIT, H.C. 1982a, Effect of spots on a star's radius and luminosity*Astron. Astrophys.* **108**, 348-355

SPRUIT, H.C. 1982b, The flow of heat near a starspot *Astron. Astrophys.* **108**, 356-360

SPRUIT, H.C. 1991, in *The Sun in Time*, eds. C. Sonett, M. Giampapa & M.S. Matthews, University of Arizona Press, Tucson, p118

SPRUIT, H.C. & ZWAAN, C. 1981, The size dependence of contrast and numbers of small flux tubes in an active region *Solar Phys.* **70**, 207-228

SPRUIT, H.C. & VAN BALLEGOOIJEN, A.A. 1982 Stability of toroidal flux tubes in stars *Astron. Astrophys.* **106**, 58-66, *Erratum* in **113**, 350

SPRUIT, H.C., NORDLUND, Å. & TITLE, A.M., 1990 Solar Convection *Ann. Rev. Astron. Astrophys.* **28**, 263-284

TITLE, A.M., TOPKA, K.P., TARBELL, T.D., SCHMIDT, W., BALKE, C., SCHARMER, G. 1992, On the difference between plage and quiet sun in the solar photosphere *Astrophys., J.* **393**, 782-794

ZWAAN, C. 1978, On the appearance of magnetic flux in the solar photosphere *Solar Phys.* **60**, 213-240

A technique for estimating long-term variations of solar total irradiance: Preliminary estimates based on observations of the Sun and solar-type stars

W. H. Soon, S. L. Baliunas, and Q. Zhang
Harvard-Smithsonian Center for Astrophysics
60 Garden Street, Cambridge, MA 02138

Summary

We analyze concurrent observations of Ca II H and K chromospheric emission fluxes (which are associated with the disk-averaged surface magnetic fluxes) and variability in the *b* and *y* bands of the Strömgren *uvby* photometric system (which are associated with the integrated total irradiance or brightness variation) in a group of stars, including the Sun. Those solar-type stars have a narrow range of mass and age, and similar observed activity variability. Their combined variability over several years reveals a range for the ratio of brightness change to chromospheric activity amplitude. We assume that the scatter in the ratios of brightness change to chromospheric activity change measured over a decade represents a distribution of variability that is *intrinsic* to the activity of a typical solar-type star (*i.e.*, the Sun) over centuries. Under that assumption, we infer an empirical bound of possible solar brightness variations over centuries (Zhang et al., 1994). The estimated increase in solar brightness accompanying the net increase in average activity from the Maunder Minimum interval (*ca.* A.D. 1660–1710) to the decade of the 1980's ranges from as low as 0.2 % to as high as 0.7 %; the indeterminacy of the brightness-chromospheric activity change ratio produces the large range of estimates. In addition, we estimate the timing of possible solar brightness change over the past three centuries. We also discuss the observed relation between the activity cycle amplitude and the activity cycle length, P_{cyc}, for the Sun and solar-type stars. The activity cycle amplitude is inversely related to the parameterized quantity, $(P_{cyc}/P_{rot})^2$, where P_{rot} is the rotation period.

Introduction

The question of the role of solar variability in long-term terrestrial climate change is a long-standing one. We assume the connection can be separated roughly into two components: (1) the external variability, namely, the latitudinal distribution of the solar radiative flux integrated over all wavelengths, *i.e.*, the solar total irradiance on Earth, and its long-term

NATO ASI Series, Vol. I 25
The Solar Engine and Its Influence
on Terrestrial Atmosphere and Climate
Edited by Elizabeth Nesme-Ribes

variability; (2) the internal dynamics of the Earth's climate. We will focus on the former component and will describe a technique that provides empirical estimates of the possible variation in the solar total irradiance on time scales of decades to centuries.

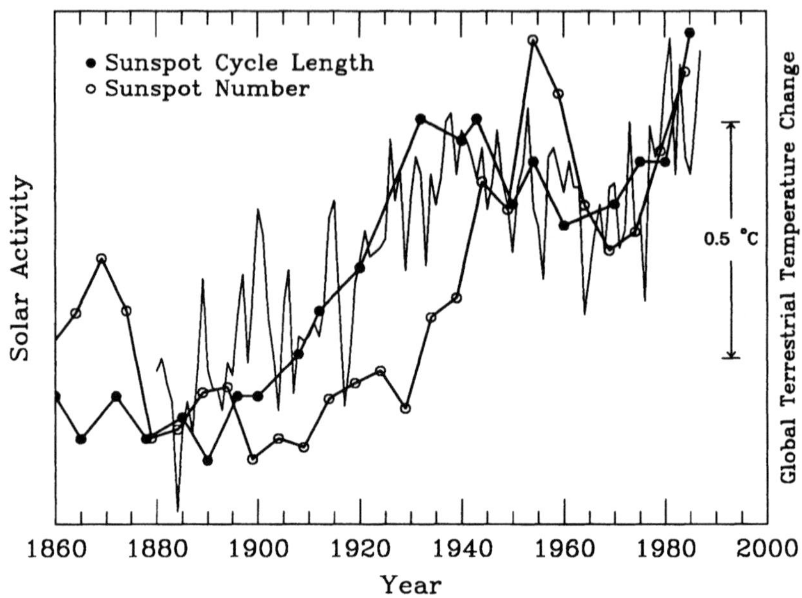

Figure 1: Timing of the variations in the sunspot cycle length (from Friis-Christensen and Lassen, 1991) and the sunspot number (11-year running-mean) versus the annual global terrestrial land air temperature change (relative to the mean of 1951-1980; Hansen and Lebedeff, 1987).

The record of the annual global land air temperature change (Hansen and Lebedeff, 1987) is shown in Figure 1. The net rise in the overall temperature profile of about 0.5–0.7 ^{0}C is non-monotonic. Also plotted in Figure 1 are two measures of long-term trends of solar activity: (a) the sunspot number (filtered by a 11-year running mean) and (b) the sunspot cycle length (smoothed with a running mean low-pass filter of weight 1-2-2-2-1, adopted from Friis-Christensen and Lassen, 1991). The terrestrial temperature (particularly the Northern Hemisphere land air temperature record) is strongly correlated with solar activity measured by the cycle length (Friis-Christensen and Lassen, 1991).
The rise of the terrestrial global temperature in the last 100-150 years in relation to the increase in anthropogenic greenhouse gas forcing and the potential solar forcing requires careful investigation. We estimate the potential variation of the solar brightness on time scales of decades to centuries from observations of the Sun and solar-type stars (see

Livingston, this meeting, for a discussion on available proxies of the solar total irradiance). The study of solar-type stars may expand the possible range of solar magnetic activity and brightness variability to time scales longer than the interval of satellite observations of solar total irradiance, about one solar cycle. The solar surface magnetic activity indicator is based on the Ca II K-line (K-index) measurements from the National Solar Observatory (NSO) (White and Livingston, 1981; White et al., 1992) and the solar irradiance variation is based on the Solar Maximum Mission (SMM) satellite measurements (Willson and Hudson, 1991). Information on the surface magnetic activity of solar-type stars is drawn from observations of the chromospheric Ca II H (396.8 nm) and K (393.4 nm) emission fluxes (S-index) made at Mount Wilson Observatory (MWO HK project) for the past 25 years (Baliunas et al., 1994; in preparation). The corresponding photometric (*i.e.*, brightness) measurements in the 15-20 nm wide passbands centered at 472 nm (b band) and 551 nm (y band) for some of those stars have been obtained at Lowell Observatory since 1984 (Lockwood et al., 1992).

The sample of solar-type stars is composed of all lower-main-sequence stars that have both activity and photometric measurements, whose color ranges from $0.55 \lesssim (B-V) \lesssim 1.4$ (corresponding roughly to masses from 1.1 to 0.5 $M_{\odot}$) and whose mean level of activity $\log\langle R'_{\rm HK}\rangle \lesssim -4.75$ and $P_{rot} \gtrsim 20$ days (corresponding roughly to ages $\gtrsim$ a few billion years; Soderblom et al., 1991). To compare stars of different masses, we converted the instrumental S-index to units of $R'_{\rm HK}$, where $R'_{\rm HK}$ is the ratio of the magnetic component of the chromospheric flux to bolometric luminosity (Noyes et al., 1984).

Results and Discussion

Precise measurements of the solar total irradiance from SMM and Nimbus-7 satellites reveal variations on time scales from minutes up to the current length of the observational window, about one solar cycle (Hudson, 1988). Irradiance variations observed on different time scales may be attributed to various varying components of surface magnetic inhomogeneities such as sunspots, faculae and the active network (Hudson, 1988). (Possible contributions to secular irradiance changes from a non-magnetic component cannot yet be ruled out; e.g., Kuhn et al., 1988 and Nesme-Ribes et al., 1993.) Those competing components, *i.e.*, sunspots and faculae, together with bright network elements, roughly account for the net irradiance change (~ 0.1 %) observed over the cycle (Foukal and Lean, 1990). However, observations near the activity cycle maxima of Cycles 21 and 22 indicate a substantial excess of solar irradiance compared to the models (see Willson and Hudson, 1991; Livingston, Pap and Willson, this meeting). The disagreement may be important as an indication of a 'missing' component of solar irradiance change existing within an activity cycle, but not present in the models (Willson and Hudson, 1991). Such a component may dominate at times in the Sun's history and produce a different ratio of irradiance change to magnetic activity change than currently observed. Although not definitive, a change in slope of the brightness-activity relation may be present in the observed upper limit of 0.4 % (peak-to-peak) variation of solar irradiance from 1967–1984 estimated from several rocket and balloon measurements, however, they are less accurate than the satellite measurements (Fröhlich, 1987; see also Reid, 1991, for additional discussion on the quality of the balloon data).

(a) Activity Cycle Amplitude versus Brightness Variation in Solar-type Stars

An alternate way of gathering information on the possible range of solar variations is to observe a large sample of solar-type stars over a relatively shorter interval of time. This assumes that stars of similar physical properties show similar magnetic variability. That hypothesis of solar-stellar magnetic equivalence is based on two main observational results. First, the time-averaged levels of magnetic activity are similar in stars with similar mass and age (Noyes et al., 1984; Baliunas, 1991). Second, the ensemble of records of activity variations in solar-type stars observed over decades give evidence for longer-term (centuries) behavior (Baliunas and Jastrow, 1990). Furthermore, irradiance (or brightness) and surface magnetism are positively correlated for the Sun (Foukal and Lean, 1990; Willson and Hudson, 1991) and for solar-type stars (Radick et al, 1990), although the correlations have different slopes.

We assume that in response to changes in magnetic activity, brightness variations in solar-type stars follow stellar equivalence. Based on this assumption, observations of a number of solar-type stars over a limited period may provide information on both the magnetic activity and brightness variability of one solar-type star at random phases in its history. If the sample is sufficiently large, the observations may yield information on the range of possible variations of both variables over an extended period of time (e.g., decades to centuries).

Figure 2 shows the distribution of variability between the rms brightness changes (in %) and rms surface magnetic activity changes observed from 1984 to 1991 for ten solar-type stars plus the Sun (from Zhang et al. 1994). The rms dispersion of the yearly averaged data was used in order to focus on long-term variability and to minimize short time scale (*i.e.*, < 1 year) variations due to rotational modulation (Foukal and Lean, 1990; Lockwood et al., 1992).

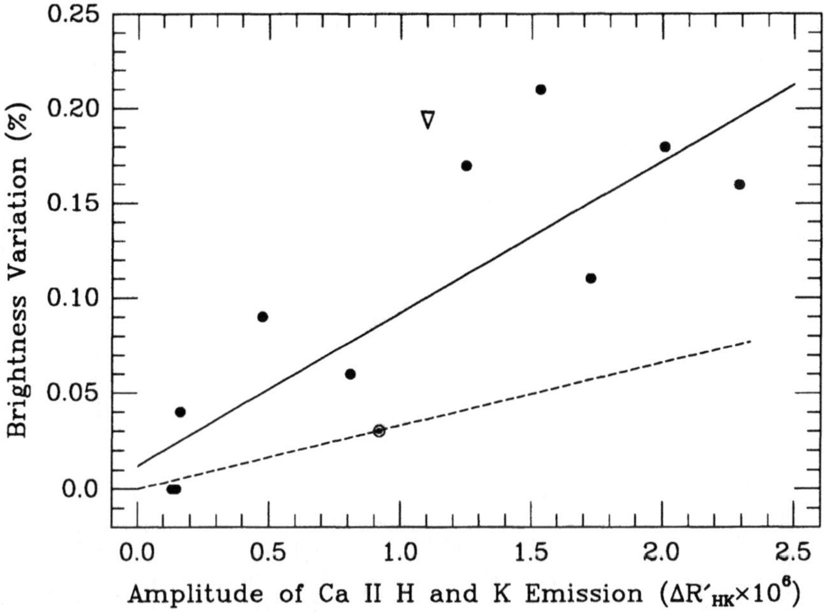

Figure 2: The rms variation in brightness (in %) versus the rms variation in surface chromospheric activity, $\Delta R'_{HK}$. The solar total irradiance variation (⊙) is from SMM satellite measurements and the solar $\Delta R'_{HK}$ is based on Ca II K-line measurements. The dashed line defines the solar brightness-chromospheric activity change ratio based on the yearly-averaged data of SMM and NSO from 1980 to 1988. The inverted triangle (∇) is the upper bound of the solar total irradiance variation from 1967-1984 taken from the rocket and balloon measurements (Fröhlich, 1987) and the corresponding value of $\Delta R'_{HK}$ is estimated from the combined solar measurements of MWO (1967-1978) and NSO (1976-1984). The solid line is the linear regression using all data [Eq. (1)], except the upper limit for the Sun (∇).

The linear regression fit for the data in Figure 2 gives

$$\Delta \text{ brightness } (\%) \approx 8.0 \times 10^4 \Delta R'_{HK} \quad , \tag{1}$$

with a 95 % confidence interval of 4.1×10^4 in the mean slope (the measurements error of the quantities in both axes are too small to explain all the scatter in Figure 2; see Zhang et al., 1994 for a discussion on possible causes of scatter in Figure 2). We assume that at least part of the scatter in Figure 2 can be associated with brightness variability in response to the appearance of different magnetic components at different times. That assumption means that the combination of slopes in Figure 2 would apply to time scales longer than a solar cycle, although the upper limit on the time scale would be difficult to specify. If results from solar-type stars are applicable to the long-term variations of the Sun, then two estimates can be made. First, the 95 % confidence interval of equation (1) can be used to define the range of possible brightness from the scatter in the slopes of brightness-activity change. Second, the relation in Figure 2 can be used to examine the timing of variations in the Sun's brightness over the past few centuries.

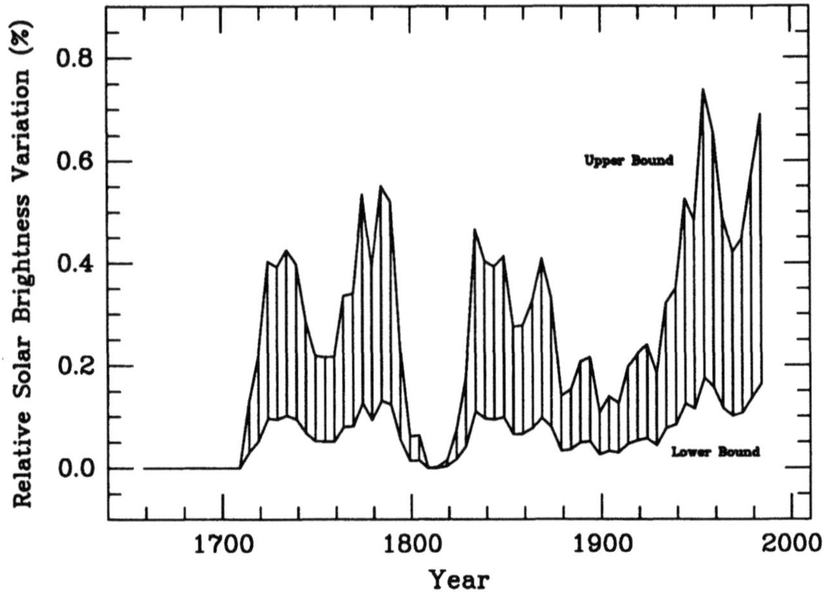

Figure 3: The relative change in solar brightness reconstructed using the 11-year running mean of sunspot numbers and the brightness variation versus the variation of surface chromospheric activity relation of Fig. 2. The indeterminacy of the retrodiction, indicated by the shaded area, includes both the effects of 95 % confidence interval in the slope of Eq. (1) and the uncertainty in the solar S-index (± 0.005) during the Maunder Minimum. The interval 1660-1710 (the Maunder Minimum) was chosen as the arbitrary reference point of the computed relative brightness change.

Sunspot and terrestrial ^{14}C (e.g., Stuiver, this meeting) and ^{10}Be (e.g., Beer, this meeting) isotope records show that solar activity was significantly depressed during the middle to late seventeenth century, *i.e.*, the Maunder Minimum (Eddy, 1976). The combination of the information on recent solar $\langle S \rangle$ values (transformed from the K-index of NSO; White et al., 1992) with an assumed mean S-value of 0.145 for the Sun in the Maunder Minimum (Baliunas and Jastrow, 1990; White et al., 1992), with equation (1) yields an estimate of the possible net change in the solar brightness between those two times. As stated above, the 95 % confidence interval of the fitted slope in Figure 2 envelopes the range of ratios of brightness-activity change. In addition, we include an uncertainty of ± 0.005 in the value of S during the Maunder Minimum. The net increase in magnetic activity from the Maunder Minimum to the decade of the 1980's should have been accompanied by a net increase in brightness of (0.2–0.7)% with a mean of 0.4 %. The lower limit of 0.2 % in solar brightness increase from the Maunder Minimum to the 1980's is similar to the result of a current solar irradiance model (Lean et al., 1992; Hoyt and Schatten, 1993). The upper limit of 0.7 % is in reasonable agreement with Reid's (1991) results which indicate a change in solar irradiance of about 1 % from the Maunder Minimum to the 1980's. (Reid's estimate used the slope of the brightness-activity change derived from the rocket and balloon data presented in Figure 2.)

Although beyond the scope of this report, our bounds are not inconsistent with the values provided by an independent estimate of 0.1 % to 1 % decrease in total irradiance during the roughly 50-year Maunder Minimum interval (*i.e.*, changes within the beginning [almost spot-less] and the ending of the 50-year Maunder Minimum interval) based on historical studies of solar diameter by Nesme-Ribes et al. (1993). The estimates from the 17th century solar diameter suggest a mean decrease of 0.4-0.5 % (averaged over the 50-year interval) of solar total irradiance during the Maunder Minimum when compared to the reference satellite radiometers value of the 1980's (see discussion by Nesme-Ribes, Sokoloff and Sadourny, this meeting). The range of our estimates of the solar brightness variation also includes those inferred from the terrestrial temperature record by Kelly and Wigley (1992) and Schlesinger and Ramankutty (1992).

To provide the detailed timing of the solar brightness variation, we used the record of the sunspot number as a proxy of solar magnetic activity (see Zhang et al., 1994 for details and uncertainties of the transformation). The reconstructed 330-year (1660-1990) history of possible solar brightness variations is given in Figure 3. The substantial range of indeterminacy is indicated by the shaded area. The available sample of solar-type stars must be expanded in order to improve the indeterminacy of the estimate provided.

(b) Activity Cycle Amplitude versus Activity Cycle Length in Solar-type Stars

Another measure of solar activity that may be connected with terrestrial climate change is the length of the activity cycle[1], P_{cyc} (Friis-Christensen and Lassen, 1991). Figure 4 shows the relation between peak-to-peak activity cycle amplitude, $\Delta R'_{HK}$, and the parameterized quantity $(P_{cyc}/P_{rot})^2$ for the solar-type stars. P_{cyc} and P_{rot} are the activity cycle periods and rotation periods detected by a periodogram analysis (Baliunas et al., 1994;

[1] An extension of the definition of 'period' is needed for behavior that is beyond the 'cyclic' phases of solar activity, such as that observed during the Maunder Minimum interval.

in preparation). The quantity, $(P_{cyc}/P_{rot})^2$, is a useful parameterization of stellar chromospheric activity and $(P_{cyc}/P_{rot})^2$ has been proposed as the observational equivalent of an effective dynamo number, N_D (Soon et al., 1993). That suggestion attempts to bridge the gap between observational results and existing physical theory (see, e.g., Jennings, 1993).

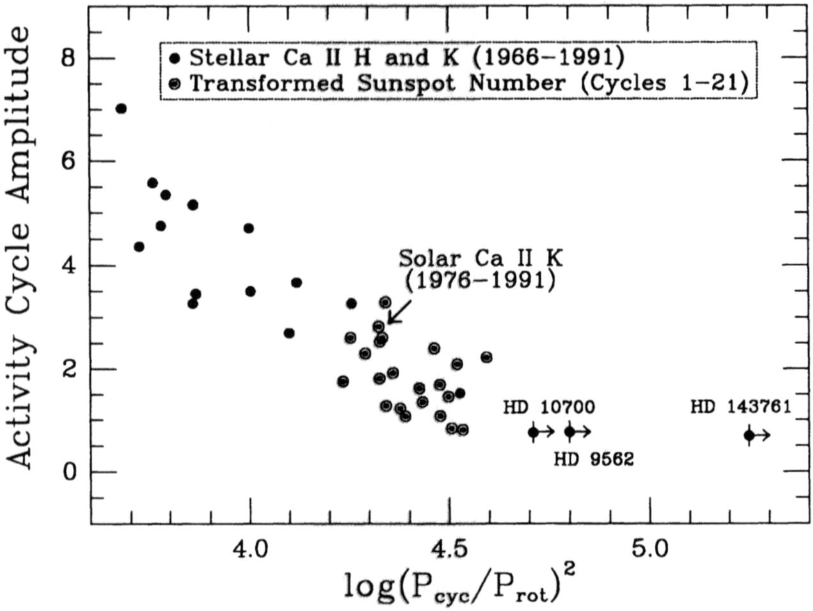

Figure 4: Activity cycle amplitude (peak-to-peak; in units of $\Delta R'_{HK} \times 10^6$) versus $(P_{cyc}/P_{rot})^2$ for the solar-type stars (derived from the 25-year records of S-index of the MWO HK project) and the Sun (transformed from sunspot record; Cycles 1 to 21). The value for the current Sun (1976-1991, marked by arrow) was derived from Ca II K record which is directly comparable to the stellar Ca II H and K records. In addition to the stars with cyclic behavior, we have also included three solar-type stars which exhibit low and quiescent Ca II H and K emission that resembles the solar activity behavior during the late 17th and early 18th centuries (*i.e.*, the Maunder Minimum interval). We have indicated the approximate activity amplitudes and lower bounds of $(P_{cyc}/P_{rot})^2$ (labeled with horizontal arrows).

A comparable relation between $\Delta R'_{HK}$–$(P_{cyc}/P_{rot})^2$ can be derived for the Sun (Figure 4). The current, solar $\Delta R'_{HK}$ value is based on the direct Ca II K-index measurements from the NSO. We converted the solar K index to the S index of MWO by using $S = 0.04 + 1.53 \times K$ (White et al., 1992). The historical series of transformed solar activity cycle amplitude, $\Delta R'_{HK}$, was derived from the sunspot record in the following way. From the calibration between the annual mean sunspot number and available solar S of 1976-1991, we first transformed the sunspot series into an equivalent of S series. Then, the record was transformed into the units of R'_{HK} by using the formula of Noyes et al. (1984) and the peak-to-peak activity cycle amplitude for each cycle was calculated. The values of the sunspot cycle lengths, P_{cyc}, are from Waldmeier (1961). The average rotation periods, P_{rot}, for Cycles 12 to 20 are from Balthasar et al. (1986) which were derived from the historical record of the positions of sunspot groups measured at the Royal Greenwich Observatory. The values of P_{rot} for other solar cycles are taken to be 25 days.

Figure 4 also includes those solar-type stars whose behavior resembles the Sun's during the Maunder Minimum interval. We considered the Maunder Minimum interval as a long period with very low amplitude. The lower limit of the cycle length is defined by the observing interval; the activity amplitudes and stellar rotations were derived from the 25-year chromospheric activity records.

The results in Figure 4 may help to elucidate the close correlation (*i.e.*, the timing) between the sunspot cycle length and the global (or northern hemisphere) terrestrial temperature change (Friis-Christensen and Lassen, 1991). Combined with the results from Figure 2, the inverse relation between activity cycle amplitude and $(P_{cyc}/P_{rot})^2$ (or dynamo number, N_D) suggests that on time scales of decades to centuries, the amplitude of solar brightness variability decreases as $(P_{cyc}/P_{rot})^2$ increases.

(c) A progress report

In section (a) we illustrated a method of estimating solar irradiance change on time scales longer than the current limit of direct solar observations through the introduction of solar-type stars. Providing quantitative estimates of brightness variations from the relation discussed in section (b) will have to await further study because the detailed physical mechanism of solar total irradiance variability is still poorly understood. Furthermore, our approach rests on the assumption that long-term solar variability (*i.e.*, in total irradiance and Ca II emission) can be statistically inferred from solar-type stars.

Understanding the solar relation to terrestrial climate on time scales of decades to centuries has been complicated by the recent addition of anthropogenic gases since the 19th century. Some progress is being made with direct decade-long observations of the Sun and solar-type stars. Further progress requires the understanding of the nonlinear interactions of the solar-terrestrial system.

Acknowledgments

We are grateful for the dedicated efforts of our colleagues at Mount Wilson Observatory. Special thanks are extended to H. Hudson, W. Livingston, W. Lockwood, E. Nesme-Ribes, R. Radick, B. Skiff, D. Sokoloff, R. White and R. Willson.
The work was supported by the Electric Power Research Institute, Mobil Foundation Inc., Scholarly Studies Program and Langley-Abbot fund of the Smithsonian Institution, American Petroleum Institute and Richard C. Lounsbery Foundation.
This research was made possible by a collaborative agreement between the Carnegie Institution of Washington and the Mount Wilson Institute.
W. H. S. acknowledges travel support provided by the NATO Advanced Research Workshop (ARW 929046).

References

BALIUNAS, S. L. (1991) The past, present and future of solar magnetism: Stellar magnetic activity. In The Sun in Time (eds. C. P. Sonett, M. S. Giampapa, and M. S. Matthews) The University of Arizona Press, Tucson: 809–831

BALIUNAS, S. L. and JASTROW, R. (1990) Evidence for long-term brightness changes of solar-type stars. Nature 348: 520–523

BALTHASAR, H., VÁZQUEZ, M., and WÖHL, H. (1986) Differential rotation of sunspot groups in the period from 1874 through 1976 and changes of the rotation velocity within the solar cycle. Astron. Astrophys. 155: 87–98

EDDY, J. A. (1976) The Maunder Minimum. Science 192: 1189–1190

FOUKAL, P. and LEAN, J. (1990) An empirical model of total solar irradiance variation between 1874 and 1988. Science 247: 556–558

FRIIS-CHRISTENSEN, E. and LASSEN, K. (1991) Length of the solar cycle: An indicator of solar activity closely associated with climate. Science 254: 698–700

FRÖHLICH, C. (1987) Variability of the solar constant on time scales of minutes to years. J. Geophys. Res. 92: 796–800

HANSEN, J. and LEBEDEFF, S. (1987) Global trends of measured surface air temperature. J. Geophys. Res. 92: 13345–13372

HOYT, D. V. and SCHATTEN, K. H. (1993) A discussion of plausible solar irradiance variations, 1700-1992. J. Geophys. Res. 98: 18895–18906

HUDSON, H. S. (1988) Observed variability of the solar luminosity. Ann. Rev. Astron. Astrophys. 26: 473–507

JENNINGS, R. L. (1993) A nonlinear model of the solar dynamo. Solar Phys. 143: 1–17

KELLY, P. M. and WIGLEY, T. M. L. (1992) Solar cycle length, greenhouse forcing and global climate. Nature 360: 328–330

KUHN, J. R., LIBBRECHT, K. G., and DICKE, R. H. (1988) The surface temperature of the sun and changes in the solar constant. Science 242: 908–911

LEAN, J., SKUMANICH, A., and WHITE, O. (1992) Estimating the sun's radiative output during the Maunder Minimum. Geophys. Research Let. 19: 1591–1594

LOCKWOOD, G. W., SKIFF, B. A., BALIUNAS, S. L., and RADICK, R. R. (1992) Long-term solar brightness changes estimated from a survey of Sun-like stars. Nature 360: 653–655

NESME-RIBES, E., FERREIRA, E. N., SADOURNY, R., TREUT, H. LE., and LI, Z. X. (1993) Solar dynamics and its impact on solar irradiance and the terrestrial climate. J. Geophys. Res. 98: 18923–18936

NOYES, R. W., HARTMANN, L. W., BALIUNAS, S. L., DUNCAN, D. K., and VAUGHAN, A. H. (1984) Rotation, convection and magnetic activity in the lower main-sequence stars. Astrophys. J. 279: 763–777

RADICK, R. R., LOCKWOOD, G. W., and BALIUNAS, S. L. (1990) Stellar activity and brightness variations: A glimpse at the Sun's history. Science 247: 39–44

REID, G. C. (1991) Solar total irradiance variations and the global sea surface temperature record. J. Geophys. Res. 96: 2835–2844

SCHLESINGER, M. E. and RAMANKUTTY, N. (1992) Implications for global warming of intercycle solar irradiance variation. Nature 360: 330–333

SODERBLOM, D. R., DUNCAN, D. K., and JOHNSON, D. R. (1991) The chromospheric emission-age relation for stars of the lower main sequence and its implications for the star formation rate. Astrophys. J. 375: 722–739

SOON, W. H., BALIUNAS, S. L., and ZHANG, Q. (1993) An interpretation of cycle periods of stellar chromospheric activity. Astrophys. J. 414: L33–L36

WALDMEIER, M. (1961) The Sunspot Activity in the Years 1610-1960, Schulthess, Zurich

WHITE, O. R. and LIVINGSTON, W. C. (1981) Solar luminosity variation. III. Calcium K variation from solar minimum to maximum in Cycle 21. Astrophys. J. 249: 798–816

WHITE, O. R., SKUMANICH, A., LEAN, J., LIVINGSTON, W. C., and KEIL, S. L. (1992) The Sun in noncycling state. Publ. Astron. Soc. Pacific 104: 1139–1143

WILLSON, R. C. and HUDSON, H. S. (1991) The Sun's luminosity over a complete cycle. Nature 351: 42–44

ZHANG, Q., SOON, W. H., BALIUNAS, S. L., LOCKWOOD, G. W., SKIFF, B. A., and RADICK, R. R. (1994) A method of determining possible brightness variations of the Sun in past centuries from observations of solar-type stars. Astrophys. J. 427: L111–L114

Surrogates for total solar irradiance

W. Livingston
National Solar Observatory
PO Box 26732, Tucson, AZ 85726, USA

Summary

Variations in the total solar irradiance as measured in space by ACRIM and ERB radiometers correlate well with a variety of ground-based activity indices such as sunspot number, 10-cm radio flux, and Ca K line intensity. These indices, in turn, correlate with the much longer geophysical records like that of radiocarbon in tree rings, and these time series can then serve as surrogates to extrapolate the solar output into the past and help predict the future. Surrogates also clarify the source of variability and how, for example, the Sun could change its color slightly, even while maintaining constant luminosity, so as to encrease or decrease the absorption of energy by the Earth's oceans.

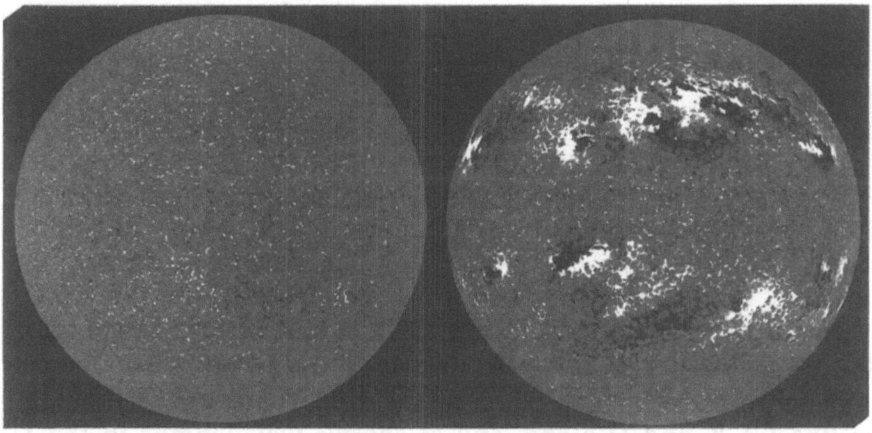

Fig 1. Left: full disk magnetograms taken on 27 Dec 1985 near solar minimum. Right: same for 12 Feb 1989 near maximum. Magnetic polarity is coded bright (N) and dark (S). Toward the limb magnetic features are co-spatial with faculae, although polarity independent (see Fig. 2).

NATO ASI Series, Vol. I 25
The Solar Engine and Its Influence
on Terrestrial Atmosphere and Climate
Edited by Elizabeth Nesme-Ribes

Introduction

The Sun is a magnetically variable star, Fig. 1. Magnetic flux surfaces in active regions, fragments, and is distributed over the disk by a hierarchy of horizontal flow patterns. Zeeman images of the disk at minimum and maximum phase of the 11-year cycle display a 'salt and pepper' background of small flux concentrations, while at maximum there are, in addition, sunspots, plage, and the active network. In the visible at 500 nm a white light photograph of the Sun shows granulation, dark sunspots, sometimes a subtle bright ring around spots, and faculae near the limb, Fig. 2.

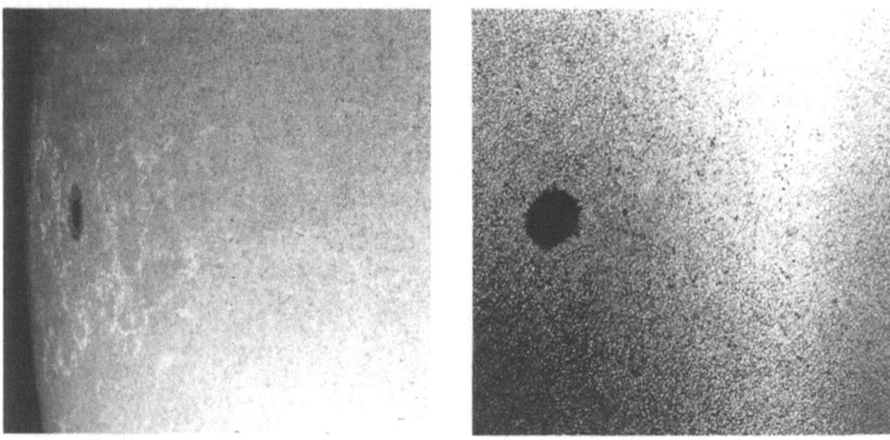

Fig 2. Left: a high contrast white-light photograph of the solar limb showing faculae, limb darkening, and a sunpot. Right: a center disk sunspot. Note the bright enhancement around the sunspot; this is due to 'abnormal granulation' in the spot's extended magnetism, and is not an indicator of excess radiation (see sec. on convection below and Spruit, this conference).

In monochromatic images taken at the centers of strong Fraunhofer lines like Ca K and H-alpha, and at shorter wavelengths in the UV, magnetic features show up as disk brightenings. The also happens in the visible continuum as white-light faculae. In this way surface magnetism modulates slightly the general energy output of the Sun, faculae adding, sunspots subtracting (Foukal and Lean 1986). This modulation, which may affect the Earth's climate in the long term, is sensed by many radiative indices (like the 10-cm flux) and in a more complicated way by geomagnetic indices (auroras), and via the influence of the heliosphere on galactic cosmic rays, by ^{14}C

deposition. We call these indices surrogates because their variability signatures represent, to a degree yet to be determined, how the Sun's radiation has varied in the past.

Implicit to this discussion is that we are dealing only with surface phenomena, that all is visible and nothing hidden. If a deep-seated dynamo process actually alters the energy output from the solar interior in some time-dependent way, this change would only be detected by sustained, carefully calibrated, radiometric observations from space, or from solar radius determinations past, present, and future.

Solar Activity Surrogates

In this section we catalog surrogates that serve some purpose; we illustrate their time series, and briefly comment on salient properties.

Radiometric indices:

ACRIM and **ERB**. This is the basic data on the Sun's output (Willson and Hudson 1988; Hoyt et al 1992). From Nimbus 7 (ERB) and SMM (ACRIM), we have good agreement: 1371.40 W/m^2 and 1367.10 W/m^2, the disk transit of large spots depresses the observational signal slightly, and we find that the activity cycle modulation is about 0.1% (Fig. 3).

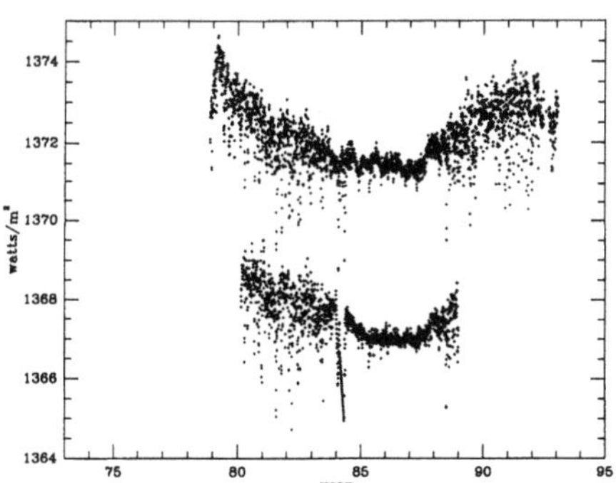

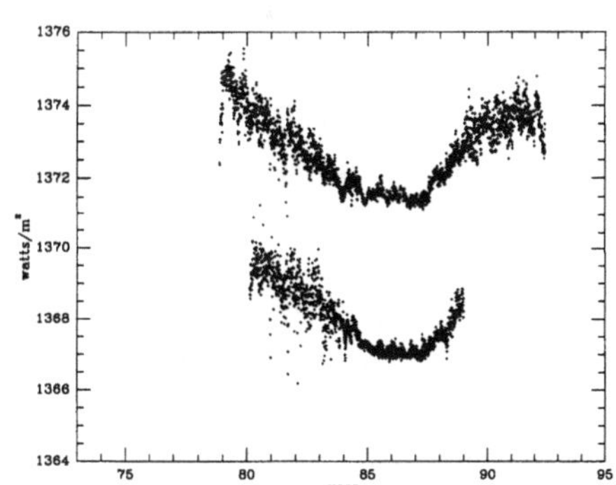

Fig 3. Left: raw total irradiance as measured by ERB (top) and ACRIM (bottom). Right: the same data corrected for sunspot blocking.

Magnetic indices:

1. **Absolute Field |B|**, Fig. 4. Regarding the Kitt Peak data, Jack Harvey comments as follows: daily measurements of the line-of-sight

field B are observed using the Zeeman Effect in Fe 868.8 nm with 1 arc-sec pixels over the entire resolved disk, scanned with a 512 arc-sec wide swath. The absolute value of B is then averaged over the disk in units of Gauss. Systematic error is a few tenths of a Gauss. Obvious bad data has been deleted through editing.

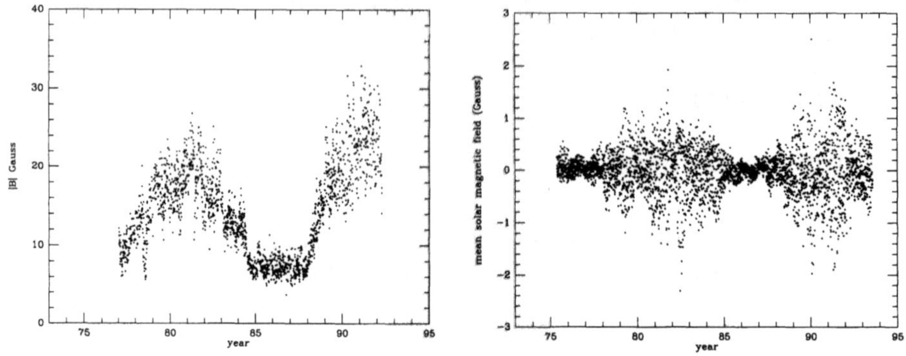

Fig 4. Left: the Kitt Peak record of |B|. Right, the Wilcox Observatory, Stanford, record of $\bar{B}$.

Peak-to-peak flux variation over the solar cycle was about a factor of 5 in cycle 22. This is less than one might guess from an inspection of the magnetogram extremes from minimum and maximum (Fig 1). Partly this has to do with spatial resolution and partly with how the pictures are made. The look-up table is non-linear with weak fields emphasized. Compared to irradiance data (integrated light), magnetograms are resolution sensitive; unresolved bipolar fields are lost.

2. **Solar mean field** $\bar{B}$ is measured daily at the Wilcox Observatory, Stanford, Fig. 4. Resolution is not a factor since the full disk is averaged optically and most bipolar aspects are lost. Note, however, that at solar minimum the annual tilt of Sun's heliographic pole induces a sinusoidal modulation indicative of a global bipole.

Radiative indices:

1. **Sunspot numbers**, Fig. 5. One day this past September (9/10/93), while at the telescope, I found no spots on the Sun, and Boulder also reported none. Yet the Ca K index for the day lay about mid-way between the minimum and maximum values in the cycle. Yes, sunspot number is an imperfect index, like all others. But it works remarkably well when monthly, yearly, or cycle means are taken. The great value of the sunspot number record is its simplicity

(although that aspect has been questioned, e.g. Schaefer [1993]) and its historical length (Eddy 1976; Foukal and Lean 1990; Hoyt and Schatten 1992; Ribes and Nesme-Ribes 1993).

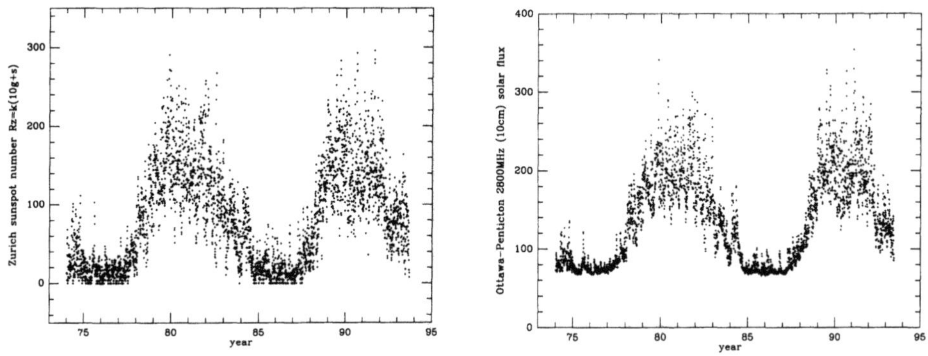

Fig 5. Left: sunspot numbers for cycles 21 and 22. Right: 10-cm radio flux for the same approximate interval.

2. **Sunspot areas**. The PSI (photometric sunspot index), or sunspot blocking functions are useful for finding the radiative excess in total irradiance. Lawrence and Chapman (1990) have been working on an objective measure of this useful parameter. This correction is applied to the right-hand data in Fig. 3.

3. **Plage area index** is difficult to determine, not only because it is ambiguous (plage vs. active network vs. network?) but because its measurement is seeing dependent. McMath-Hulbert, followed by Big Bear Solar Observatory, did report this index for many years, but observations appear to have stopped.

4. **10-cm radio flux** is one of our best and longest of modern indices, Fig. 5. It has been obtained rain or shine since 1947, thanks to the perseverance of Arthur Covington and Ken Tapping. A defect is an insensitivity at low levels of activity (Tapping and Charrois 1992), and its two component origin (Gelfrikh 1992).

5. **Irradiance spectrum line intensity**.

The central depths and equivalent widths of selected Franhofer lines are good surrogates of magnetic activity.

a) *Ca K 393.2 nm* is the strongest Fraunhofer line in the visible spectrum of the Sun and solar-type stars. Its cyclic variability (central depth varies 20%) is the greatest of any line in terms of excursions of the core profile (Figs. 6 and 7). An instrumental demand is that of high spectral resolution and low scattered light.

b) *CN 388.3 nm* was shown by Sheeley (1969) to be co-spatial with photospheric magnetism (as compared to higher level lines formed where the field has greatly expanded and geometrically distorted). As a surrogate it is particularly useful at solar minimum (Livingston et al 1988), Fig 8.

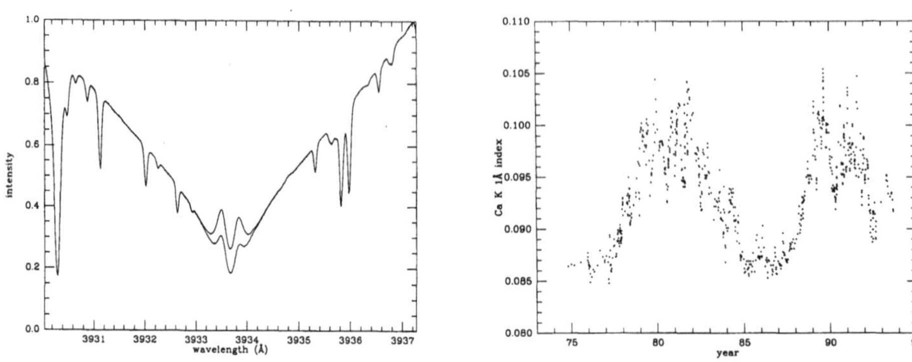

Fig 6. Left: profile of Ca K 393.2 nm in the irradiance (full disk) spectrum. Solar maximum is compared with solar minimum, lower. The Ca K index is an area under the line core of width 0.1 nm (1 Å). Right: time series record of the Ca K index as sampled at Kitt Peak 3-4 times a month since 1974 (White and Livingston 1978).

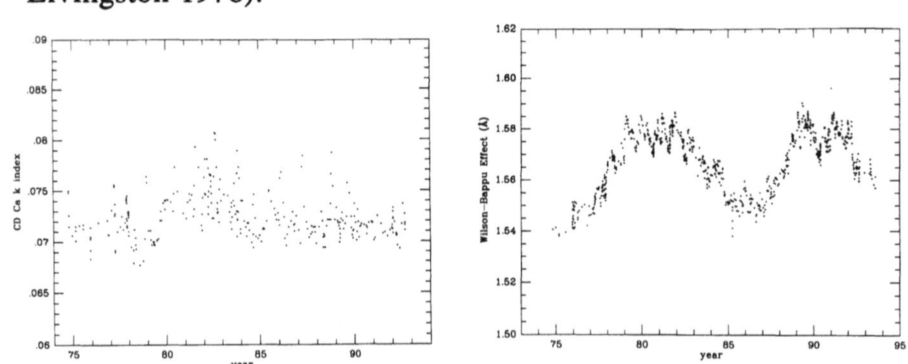

Fig 7. Left: time series of the Ca K index recorded near disk center in quiet regions, showing the absence of cyclic modulation. Conclusion: the observed variability is a localized, not global, phenomenon. Right: the Wilson-Bappu Effect for solar irradiance. Unlike other intensity measurements, this is simply the width of the Ca K emission feature. Applied to stars this

parameter correlates with stellar luminosity.

c) *Mn 539.4 nm.* Most Fraunhofer lines formed in the mid- or high-photosphere weaken in magnetic regions. For reasons apparently not understood, Mn 539.4 weakens a lot, Fig. 8.

d) 0 *777.4 nm* represents the so-called oxygen triplet. It shows cycle variability opposite to other lines, i.e. it strengthens in plage.

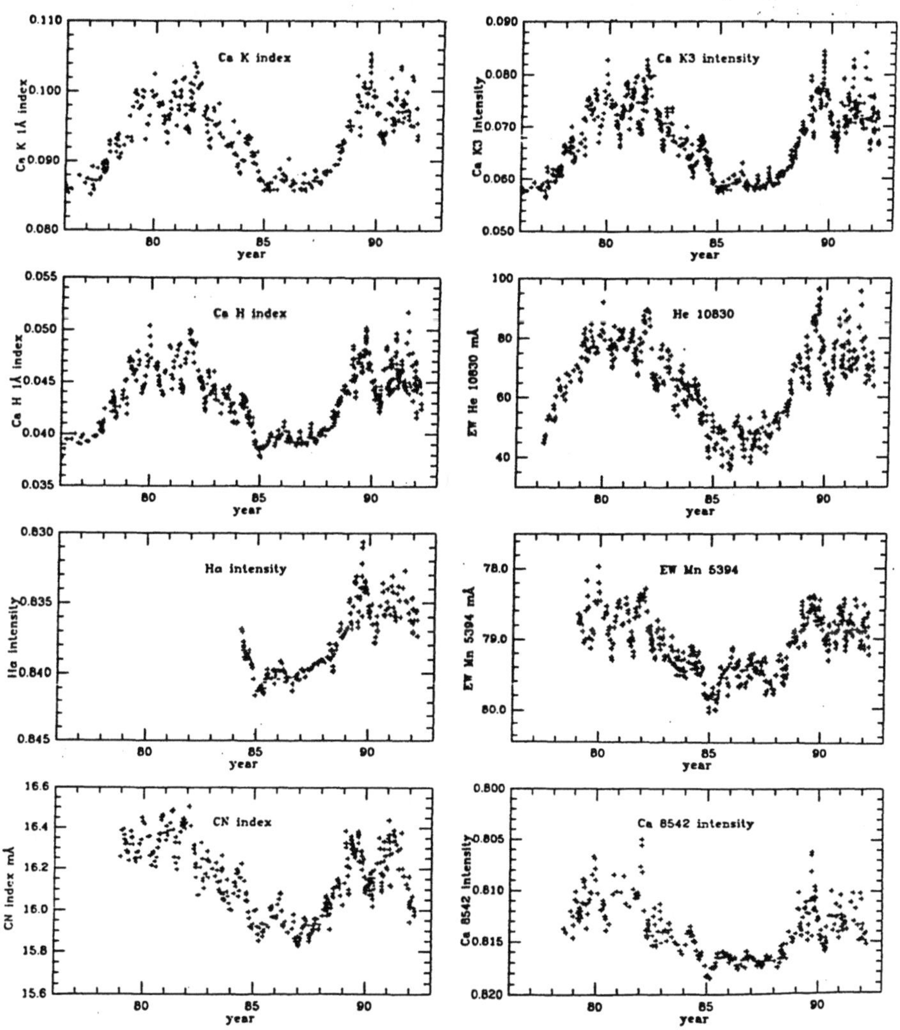

Fig 8. Fraunhofer lines which display a large cyclic modulation: Ca K, Ca H, He 1083.0 nm, H alpha 656.2 nm, Mn 539.4 nm, the CN index at 388.3 nm, and Ca 854.2 nm.

e) *H-alpha 656.2 nm* shows a clean cyclic signature comparable to popular Ca K. It is of interest because it can be observed in stars easily, i.e. it is a broad line so that spectral resolution is not a problem (Fig. 8).

f) *He 1083.0 nm* is observed daily at the Kitt Peak Vacuum Telescope, leading to one of the longer time series archives of line intensity (Fig. 8). This high-excitation line reflects conditions in the upper chromosphere and, through NLTE pumping, activity of coronal He II 304 nm (Harvey and Sheeley 1977).

g) *C 538.0 nm.* This weak, high-excitation line is formed almost coincident with the continuum. It displays almost no indication of the 11-year cycle (Fig. 9). Statistically, it is increasing in strength, but the available time series is too short (15 years) for confidence.

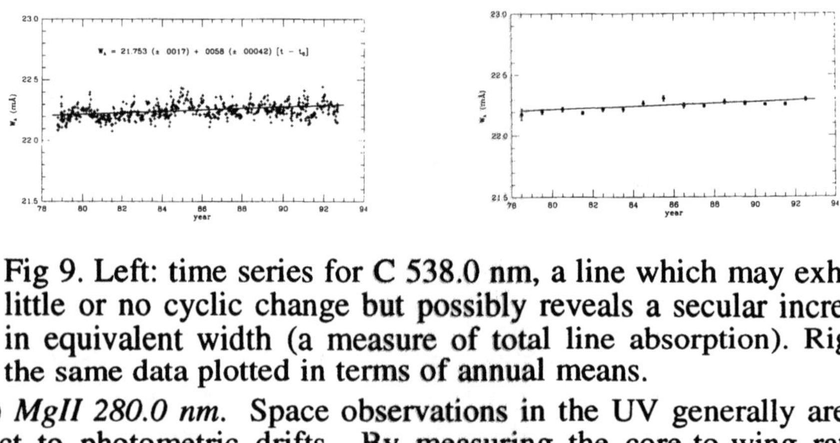

Fig 9. Left: time series for C 538.0 nm, a line which may exhibit little or no cyclic change but possibly reveals a secular increase in equivalent width (a measure of total line absorption). Right: the same data plotted in terms of annual means.

h) *MgII 280.0 nm.* Space observations in the UV generally are subject to photometric drifts. By measuring the core-to-wing ratio of MgII h and k, such drifts are compensated for to yield a unique surrogate for UV flux (Heath and Schlesinger 1986).

i) *Ly-alpha 121.6 nm* is another useful UV surrogate. The energy flux from this emission line itself is a significant fraction of the total at these wavelengths.

j) *X-rays (0.1 - 0.8 nm)* reports mostly on the maximum phase. X-ray emission is lost in the noise at minimum (Bouwer 1992).

Convective indices:

1) *Granulation size.* High-resolution photographs of solar granulation 1976-83 indicate a 25% increase of their the number density at minimum as compared to maximum (Macris and Rösch 1983; Muller and Roudier 1984). An explanation is found in the way granules

fragment or otherwise change when threaded by magnetic flux.

Abnormal granulation (Bray et al. 1984) occurs near sunspots where the surrounding photosphere is threaded by magnetism. The curvature of spectrum line bisectors (Fig. 10), a convection indicator, is reduced in plage (Livingston 1982). So increased surface magnetism inhibits convection and leads to smaller, or less contrasty, granules and their apparent number density decreases.

2) *Radial velocity of Sun as a star* depends on the average blue shifts from the ~2×10^6 rising granules on the visible disk (Deming and Plymate 1993; Jimenez et al. 1986). Absolute wavelengths at 2.3 μm, after ephemeris corrections, yield a 20-25 m sec^{-1} cyclic variation, in agreement with the smaller size of granules at maximum (greater red-shift with increasing field). However, McMillan et al (1993) find no such shift, so these results are controversial at present.

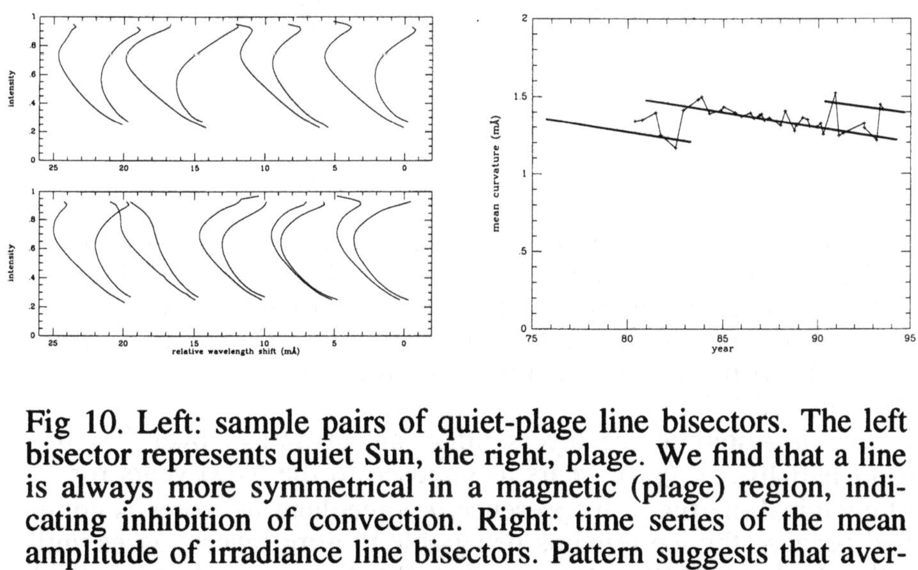

Fig 10. Left: sample pairs of quiet-plage line bisectors. The left bisector represents quiet Sun, the right, plage. We find that a line is always more symmetrical in a magnetic (plage) region, indicating inhibition of convection. Right: time series of the mean amplitude of irradiance line bisectors. Pattern suggests that average granular convection is linearly suppressed by expanding surface magnetism in a repetitious fashion.

3) *Line bisector curvature* in the irradiance spectrum has been observed since 1980 with the 1-m FTS. We know from resolved disk observations that the amount of curvature depends on granular convection: velocity amplitude of the cells weighted by their brightness contrast (Fig 10). The irradiance record is unclear but may be interpreted as a cycle variation, with phase delay, in support of above (Fig 16). Or, one can suppose there is a secular trend toward decreased convection in progress.

Geospheric/Heliospheric Indices:

1) *^{14}C from tree rings.* Galactic cosmic rays, modulated by the interplanetary magnetic field, produces radiocarbon in the atmosphere with a half-life of 5700 years (Damon 1988). Eventually residing in the calendar of tree rings, this isotope shows a remarkable correlation with the sunspot record. Sonett and Jirikowic (1993) find, from new data by Stuiver (1993), that the radiocarbon activity cycle record persisted through the Maunder Minimum (Fig. 11).

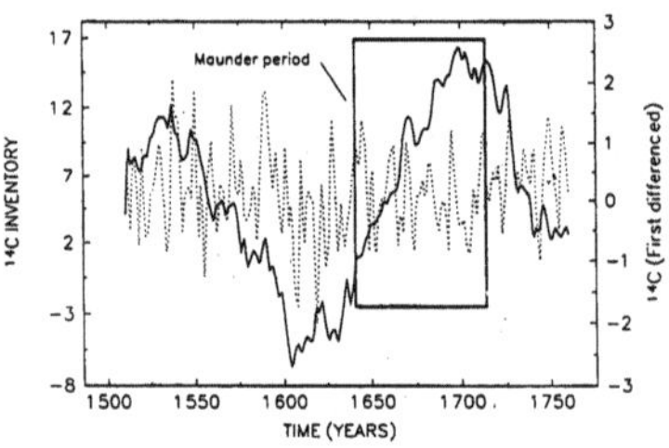

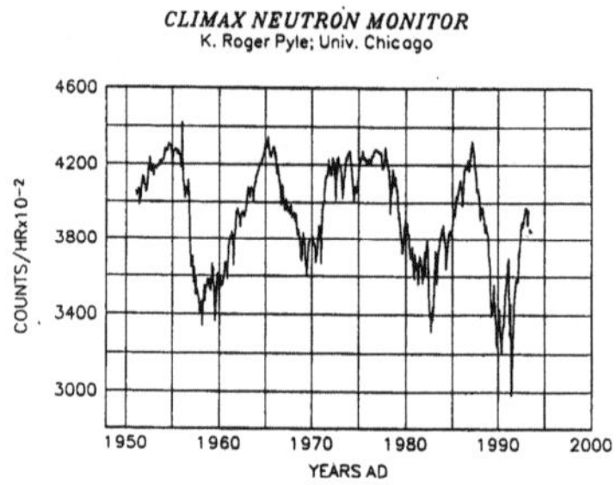

Fig 11. Left: by simply taking the first differences in the Stuiver radiocarbon data, Jirikowic and Sonett (1993) produced this ^{14}C time series which, strangely, displays a solar cycle which persists through the period of the Maunder Minimum (box). Right: time series from the Climax neutron monitor (Jokipii 1991; see also Beer et al 1991).

2) *^{10}Be* in Antarctic ice layers offers an advantage over ^{14}C in its longer half-life (1.5×10^6 years as opposed to the 5700 years for ^{14}C) and chemical inertness. Its presence in Greenland ice cores is complicated because the ice remains penetrable to atmospheric gases until a depth of about 300 meters so that the time record is somewhat smeared (see Beer this conference).

3) *Climax neutron monitor*, which has been in operation since 1951, clearly displays the 11-year and 22-year (Hale) cycles. This is the one surrogate that may reveal the Hale polarity cycle by its sharp vs flat top in alternate cycles (Fig. 11).

4) *Aurora* activity depends on the solar wind and has potential to serve as a sunspot proxy in past times before the telescope. The data record is of variable quality, unfortunately (Feynman 1988; Schröder 1988).

5) *Geologic varves.* Any varve must first be proven annular to be valid. Sonett and Williams (1985), believe the Skilak Lake, Alaska, laminae passes the this test and contains an 11-year modulation after an adjustment for summer thunderstorms (Jirikowic and Sonett 1993).

Indices reporting on the solar interior:

1) *Solar radius* measurements offer the possibility of sensing non-surface, deep-seated temperature changes. Because classical geometers attached great importance to the size of the Sun, and because where a total eclipse happens depends critically on the solar radius, historic reconstructions may be possible (Ribes et al 1987; Maier et al 1992).
2) *Neutrino flux* may contain an activity cycle signature (Davis, Jr, and Cox 1991; Bieber et al. 1990; Delache et al. 1993).

Different Surrogates often exhibit phase shifts when compared. Why?

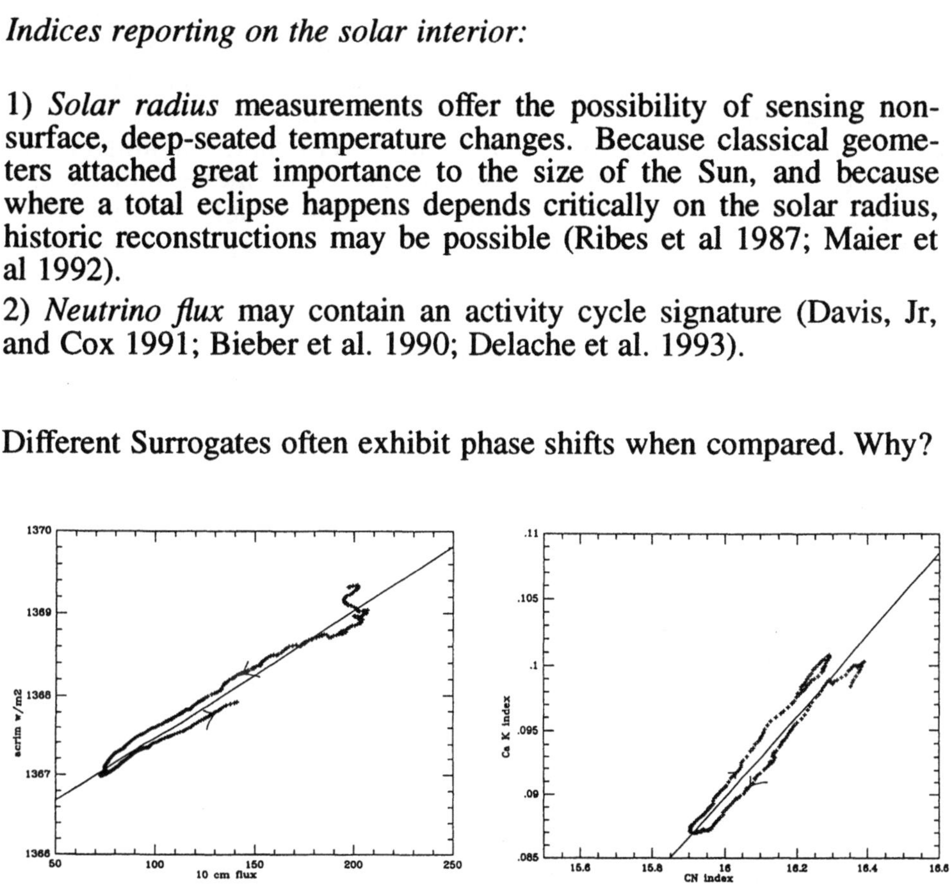

Fig 12. Left, correlation plot of ACRIM total irradiance vs 10-cm flux with a one year running mean, 1980 to 1988. Solar minimum occurs at the lower tip. Right, similar plot for Ca K vs CN index.

A time element or phase shift is introduced between indices depending on their place of origin: magnetic flux initially surfaces (spot umbrae), this flux expands locally (plage), spreads more extensively (by supergranular motions), spreads globally (by differential rotation), and finally expands with height (evolving through the photosphere-chromosphere-corona-heliosphere). See Fig. 12 for examples of hysteresis effects.

Donnelly (1991) has given a detailed description of the reasons for hysteresis in a Mg II core-wing (c/w) ratio vs. 10-cm correlation plot. The 10-cm radio flux originates with two mechanisms: coronal thermal plus bremsstrahlung, and coronal gyrocyclotron emission about the magnetic fields above active regions. While the coronal magnetic fields adjust quickly it takes a some time for an active region to develop its maximum UV output. As a result there is a delay of about 27 days between Mg II c/w and 10-cm flux. Bachmann and White (1993) find lag times between several radiative indices and sunspot number. Unfortunately, there are inexplicable exceptions to a simple picture of height dependence, and the topic remains under study.

On the goodness of fit for total irradiance surrogates.

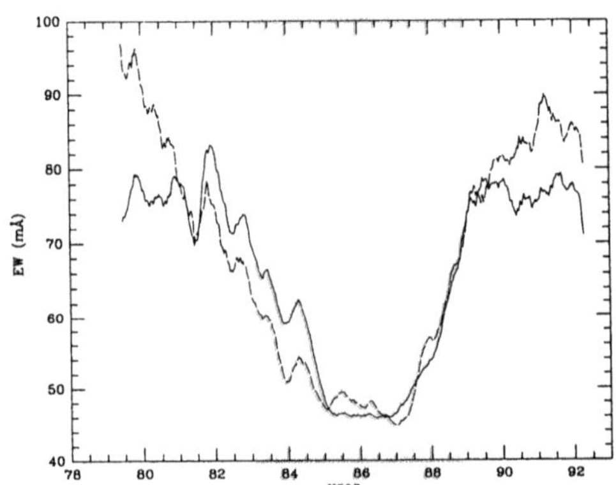

Fig 13. Over-plot of the ERB total irradiance (dashed) and He 1083.0 nm line intensity (solid). We cannot obtain a satisfactory fit both at solar maximum and minimum. Thus, for reasons unknown, the surrogate relationship fails.

On a day-to-day time scale most radiative surrogates show differences which are probably related to how the particular index responds to the presence of magnetic field elements. Donnelly has pointed out that such differences diminish when the time series is subjected to a running mean of 27 days (or about one solar synodic rotation period). Even more effective is a mean about three times longer or, say, 85 days, which tends to suppress the signature related to the appearance and disappearance of active regions (Fig 14).

Some surrogates track solar total irradiance better at minimum. The CN index, for example, is found to correlate well at minimum, while 10-cm flux does poorly there (Livingston et al 1988). At maximum several indices work well: 10-cm flux, Ca K index, and He 1083.0 nm equivalent width for example.

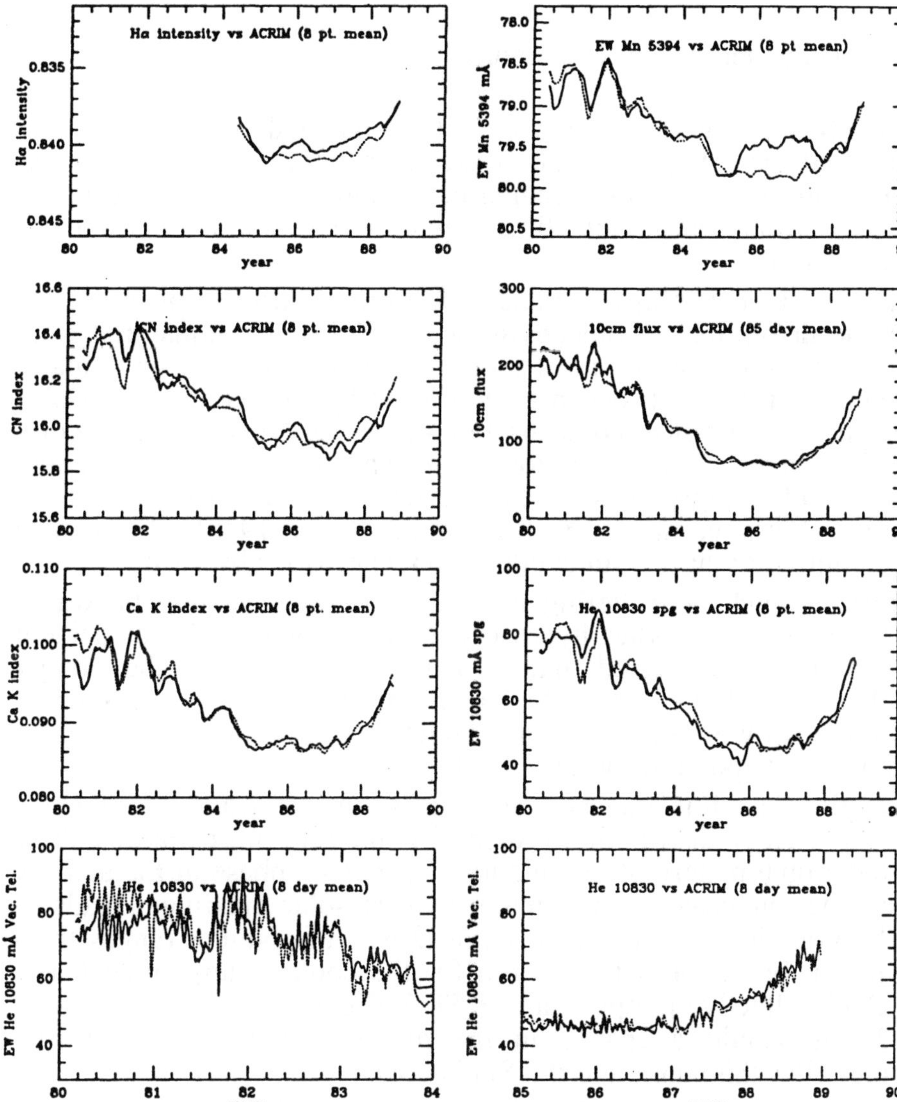

Fig 14. Comparison of various spectrum line intensities with ACRIM total irradiance.

Surrogates like radiocarbon which involve interactions with the heliosphere followed by extended deposition times in the biosphere naturally require long time bases to reveal a signal. For this reason ^{14}C is superior to ^{10}Be, because response time for the former is shorter than for the latter, which takes some time to be absorbed by ice, for example.

Color change and ocean temperature.

Can the Sun change its color while total solar irradiance remains constant? And what might be the consequences for ocean temperature. We end this review with speculation on how a subtle change in the Sun's radiation could alter climate even if the total solar irradiance remains fixed.

Ocean temperature plays a crucial role to the Earth's climate. (Consider the disruption accompanying a 2-3 C El Niño equatorial transient). Solar input occurs mostly in the visible and is not particularly wavelength dependent; ocean albedo is set by surface reflection plus atmospheric scattering and absorption processes (see conference comment by Shine). The absorption coefficient of water per unit length does depend strongly on wavelength, increasing 30% from 510 nm to 520 nm (Plass et al. 1978; Smith and Baker 1978). But albedo matters little at what ocean depth radiant energy is absorbed.

Suppose, though, a scattering medium is introduced into the water, say by plankton or other hydrosols. Now a fraction of the incident light will be redirected back to the surface, and the longer wavelengths will be more efficiently absorbed than to the shorter. Ocean albedo will become more a function of wavelength and ocean heating will respond to the Sun's color.

Convective and line intensity surrogates suggest ways to influence the solar energy distribution (color). Pic du Midi observations show that the global pattern of granulation depends on phase of the activity cycle. The enhanced surface magnetism at solar maximum leads to smaller granules. Normally, the color (λ_{max}) of granules ranges from 498 nm in bright ones to 525 nm for the cooler intergranule lanes. The Sun's effective temperature of 5777 K corresponds to a λ_{max} of 500.2 nm. Could a change in the mix of bright and less-bright granules alter this λ_{max} and the Sun's color?

Another way to induce color change is by line-blanketing (Mihalas 1978; Mitchell Jr and Livingston 1991). As spectrum line strength increases, continuum effective temperature must rise to maintain luminosity. This effect is well observed in certain stars. Since surface magnetism leads to line weakening and hence to lower temperatures, the result may be opposite to the above convective mechanism.

We conclude that if the oceans became fouled by scatterers and if the Sun changed its color for either convective or line intensity reasons, the sea surface temperature might respond. Reid (1991) has noted a remarkable correlation between sea surface temperature and sunspot number. From ACRIM and ERB we know the Sun's total irradiance but not its spectral distribution. The latter could be a driver of climate change.

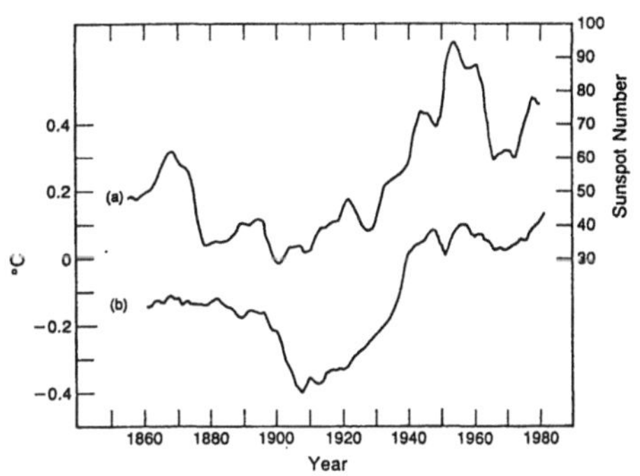

Fig 15. Sunspot number (curve a) vs sea surface temperature (curve b), with a one cycle running mean (Reid 1991).

References

Bachmann, K.T. and White, O.R. (1993) Observations of hysteresis in solar cycle variations of six radiative indices and sunspot number. Preprint.

Beer, J. Raisbeck, G.M., and Yiou, F. (1991) Time variations of ^{10}Be and solar activity. 351 in *The Sun in Time*, Sonett, C.P. et al (eds.), Univ. Arizona Press, Tucson.

Bieber, J.W., Seckel, D., Stanev, T., and Steigman, G. (1990) Variation of the solar neutrino flux with the Sun's activity. Nature 348: 407.

Bouwer, S.D. (1992) Periodicities of solar irradiance and solar activity indices, II. Solar Phys 142: 365-389.

Bray, R.J., Loughhead, R.E., and Durrant, C.J. (1984) p. 86 in *The Solar Granulation*, Cambridge University Press, Cambridge.

Damon, P.E. (1988) Production and decay of radiocarbon and its modulation by geomagnetic field-solar activity changes with possible implications for global environment. 267-285 in Secular Solar and geomagnetic Variations in the last 10,000 years , Stephenson, F.R. and Wolfendale, A.W. (eds), Kluwer Publ.

Davis, Jr., R. and Cox, A. N. (1991) Solar neutrino experiments. 51-85 in Solar Interior and Atmosphere, Cox, et. al. (eds), Univ. Arizona Press.

Delache, Ph., Gavryusev, V. Gavryuseva, E., Laclare, F., Régulo, C., and Cortés, T. Roca (1993) Time correlation between solar structure parameters: *p*-mode frequencies, radius, and neutrino flux. Ap. J. 407: 801-805.

Deming, D. and Plymate, C. (1993) On the apparent velocity of integrated sunlight II. 1983-1992, and comparisons with magnetograms. Ap. J. (submitted)

Donnelly, R.F. (1991) Solar UV spectral irradiance variations. J. Geomag. Geoelectr. 43: Suppl. 835-842.

Eddy, J.A. (1976) The Maunder Minimum. Science 192: 1189-1202.

Feynman, J. (1988) Solar, geomagnetic and auroral variations observed in historical data. 141-159 in *Secular Solar and Geomagnetic Variations in the last 10,000 years*, Stephenson, F.R. and Wolfendale, A.W. (eds), Kluwer Publ.

Foukal, P. and Lean, J. (1986) The influence of faculae on total solar irradiance and luminosity. Ap. J. 302: 826-835.

Foukal, P. and Lean, J. (1990) An empirical model of total solar irradiance variation between 1874 and 1988. Science 247: 505-604.

Gelfreikh, G.B. (1992) Solar radio emission at centimeter wavelengths. 196-226 in Proceedings of the Workshop on Solar Electromagnetic Radiation Study for Solar Cycle 22, Donnelly, R.F. (ed), NTIS publication.

Harvey, J.W. and Sheeley, Jr., N.R. (1977) A comparison of He II 304 Å and He I 10830 Å Spectroheliograms. Solar Phys. 54: 343-351.

Heath, D.F. and Schlesinger, B.M. (1986) Mg 280-nm doublet as a monitor of changes in solar ultraviolet irradiance. J.G.R. 91: 8672-8682.

Hoyt, D.V. and Schatten, K.H. (1992) New information on solar activity, 1779-1818, from Sir William Herschel's unpublished notebooks. Ap. J. 384: 361-384.

Hoyt, D.V., Kyle, H.L., Hickey, J.R. and Maschhoff, R.H. (1992) The Nimbus 7 solar total irradiance: a new algorithm for its derivation. J.G.R. 97: 51.

Jokipii, J.R. (1991) Variations of the cosmic-ray flux with time. in *The Sun in Time*, Sonett, C.P. et al (eds.), Univ. Arizona Press, Tucson.

Jimenez, P.L., Palle, C., Regulo, C., Roca Cortes, T., Elsworth, Y.P., Isaak, G.R., Jefferies, S.M., McLeod, C.P., New, R., van der Raay, H.B. (1986) Variations in the mean line of sight velocity of the sun: 1976 - 1985. 215-218 in Adv. in Helio- and Asteroseismology, J. Christensen-Dalsgaard and S. Fandsen (eds) IAU Symp. 123.

Jirikowic, J.L. and Sonett, C.P. (1993) 'Varve' counting vs. tephrochronology and ^{137}Cs and ^{210}Pb dating: a comparative test at Skilak Lake, Alaska: comment and reply. Geology p. 763, Aug. 1993.

Lawrence, J.K. and Chapman, G.A. (1990) Photometric observations of the energetics of small solar active regions. Ap. J. 361: 709-717.

Livingston, W.C., Wallace, L, and White, O.R. (1988) Spectrum line intensity as a surrogate for solar irradiance variations. Science 240: 1765-1767.

Livingston, W.C. (1982) Magnetic fields, convection and solar luminosity variability: new observations. Nature 297: 208-209.

Macris, C. and Rosch, J. (1983) C.R. 296: 268.

Maier, E., Twigg, L., and Sofia, S. (1992) Preliminary results of a balloon flight of the solar disk sextant. Ap. J. 389: 447-452.

McMillan, R.S., Moore, T.L., Perry, M.L., and Smith, P.H. (1993) Radial velocity observations of the sun at night. Ap. J. 403: 801.

Mihalas, D. (1978) p 167 in Stellar Atmospheres, Freeman, San Francisco.

Mitchell Jr., W.E. and Livingston, W.C. (1991) Line-blanketing variations in the irradiance spectrum of the sun from maximum to minimum of the solar cycle. Ap. J. 372: 336-348.

Muller, R. and Roudier, Th. (1984) Variability of the structure of the granulation over the solar activity cycle. 4th Eur. Solar Phys. Conf. p. 51.

Plass, G.N., Humphreys, T.J., Kattawar, G.W. (1978) Color of the ocean Appl. Optics 17: 1432.

Reid, G.C. (1991) Solar total irradiance variations and the global sea surface temperature record. J.G.R. 96: 2835.

Ribes, E., Ribes, J.C. and Barthalot, R. (1987) Evidence for a larger sun with a solar rotation during the seventeenth century. Nature 326: 52-55.

Ribes, J.C. and Nesme-Ribes, E. (1993) The solar sunspot cycle in the Maunder Minimum AD 1645 to AD 1715. Astron. Astrophys. 276: 549-563.

Schaefer, B.E. (1993) Visibility of sunspots. Ap.J. 411: 909.

Sheeley,Jr.,N.R. (1969) The evolution of the photospheric network. Solar Phys 9: 347-357.
Schröder, W. (1988) Aurorae during the Maunder Minimum. Meterol. Atmos. Phys. 38: 246-251.
Smith, R.C. and Baker, K.S. (1981) Optical properties of the clearest natural waters (200-800 nm). Appl. Optics 20: 177.
Sonett, C.P. and Jirikowic, J.L. (1993) Is the Maunder Minimum a faire tale? Evidence for decoupled solar dynamo modes. J.G.R. (submitted)
Sonett, C.P. and Williams, G.E. (1985) Solar periodicities expressed in varves from Skilak Lake, Alaska. J.G.R. 90: 12019.
Stuiver, M. (1993) A note on single-year calibration of the radiocarbon time scale, AD 1510-1954. Radiocarbon 35: 67-72.
Tapping, K.F. and Charrois, D.P. (1992) The 10.7 cm flux as an index of solar activity.
White, O.R. and Livingston, W. (1978) Solar luminosity variation. II. Behavior of Calcium H and K at solar minimum and the onset of cycle 21. Ap. J. 226: 679-686.
Willson, R.C. and Hudson, H.S. (1988) Solar luminosity variations in solar cycle 21. Nature 332: 810-812.

Comment

Shine: I do not agree that the wavelength dependence of the absorption coefficient of water can be used to indicate the wavelength dependence of the albedo, which is energetically the important quantity. Albedo shows a very small variation with wavelength except in certain cases. The spectral variation of the albedo/opacity of the atmosphere itself due to Rayleigh scattering and selective absorption is a far more important factor. Changes in the Sun's spectra are unlikely to have much effect on energy absorbed by the ocean.

Livingston: You are correct and I was in error in my oral presentation. This point is clarified in the revised text above. However, the possibility remains that the introduction of scattering material could alter the ocean's albedo to make it more sensitive to the Sun's color.

SOLAR FORCING OF GLOBAL CHANGE

Judith Lean
E. O. Hulburt Center for Space Research
Naval Research Laboratory
Washington, DC 20375, USA

Abstract

To reduce uncertainties in anthropogenically forced global change superimposed on natural climate variability, knowledge of changing solar radiative output, a primary natural forcing agent, is essential. Estimates of past solar forcing are derived from radiometric measurements and empirical models of contemporary solar irradiance variations. Implications for global change are discussed.

Introduction

Evidence for solar variability exists on contemporary and climatological time scales, as demonstrated in Figure 1. Solar magnetic surface phenomena such as sunspots indicate an approximate 11-year occurrence cycle that has persisted for ~300 years, although with neither constant amplitude nor phase. Cosmogenic isotopes in ice cores and tree rings provide indirect evidence for solar variability on even longer time scales (Beer et al., 1991; Stuiver and Reimer, 1993).

Significant progress has been made during the past decade in defining the amplitude and temporal structure of solar forcing of global change. This forcing arises from changes in the energy output of the Sun, especially the visible radiative energy that is deposited at the Earth's surface and the ultraviolet (UV) radiative energy absorbed by the atmospheric ozone layer. Space-based radiometers have measured directly changes in both the total (spectrally integrated) solar radiative output and in some important UV spectral regions,

NATO ASI Series, Vol. I 25
The Solar Engine and Its Influence
on Terrestrial Atmosphere and Climate
Edited by Elizabeth Nesme-Ribes

from the maximum to the minimum of an 11-year activity cycle (see review by Lean, 1991): When the level of solar activity is high (as indicated by sunspot numbers and cosmogenic isotopes) so too is solar radiative output (Fig. 1).

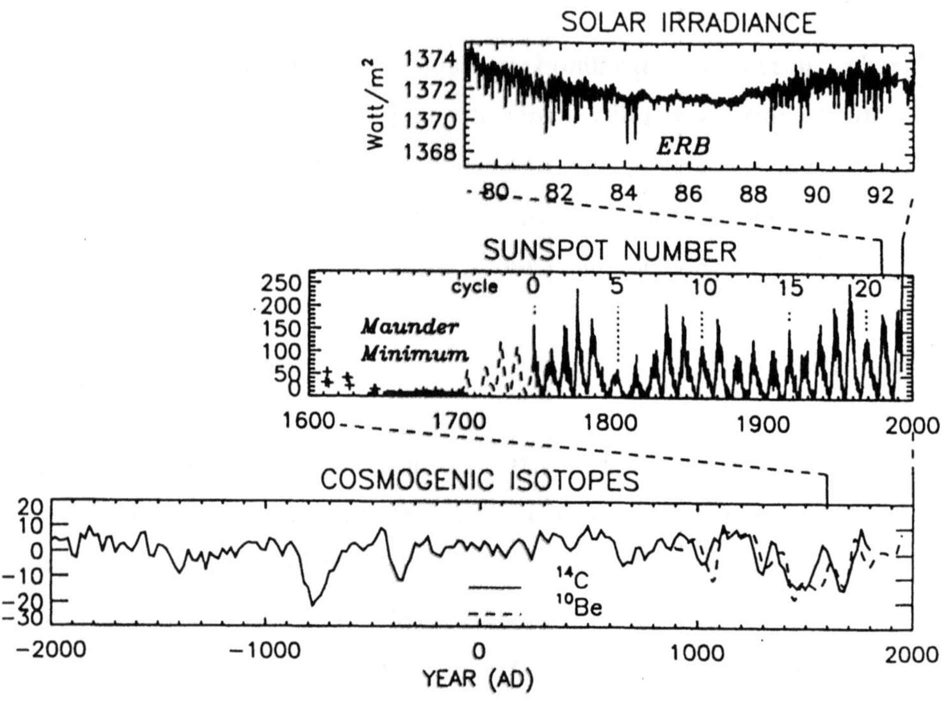

Fig. 1. Solar variability on time scales of years, decades and centuries, as evidenced in the total solar irradiance measured by the Earth Radiation Budget (ERB) cavity radiometer onboard the Nimbus-7 satellite (Hoyt et al., 1992), the sunspot number and the ^{10}Be and ^{14}C cosmogenic isotopes in, respectively, ice cores (McHargue and Damon, 1991) and tree rings (Stuiver and Reimer, 1993).

Semi-empirical solar variability models (Foukal and Lean, 1988, 1990) can account for a significant fraction of the 11-year cycle in total solar irradiance in terms of the competing effects of dark sunspots (which deplete the local emission) and bright faculae (where the local emission is enhanced, in the brightest active regions and the surrounding network). During high levels of solar activity the increased occurrence of sunspots and faculae, both manifestations of solar magnetic activity, modifies the emission from the "quiet" Sun (i.e., that seen during solar minimum). Models developed from the extant radiometry have been extended to climatological time scales, for example by postulating the effect on solar radiative output of reductions in both magnetic and non-magnetic solar emission, and in conjunction with observations of solar-like stars (Baliunas and Jastrow, 1990; Lean et al., 1992; Hoyt and Schatten, 1993; Nesme-Ribes et al., 1993). Scenarios for future solar forcing have been developed (Wigley and Kelly, 1990; Damon and Sonnet, 1991; Lean and Rind, 1994).

Progress has also been made in identifying changes in the global environment that may arise from solar forcing. For example, global surface temperature increases of ~0.5° C over the past century appear to correlate with decadal-scale increases in solar variability (Reid, 1991; Friis-Christensen and Lassen, 1991). Evidence for tropospheric temperatures warmer by ~1°C during solar maximum (Labitzke and van Loon, 1993) suggests larger effects may be occurring regionally in response to the 11-year solar cycle. Such correlations have prompted quantitative evaluation of the relative contributions of solar and anthropogenic forcings (greenhouse gases, aerosols, ozone depletion) to the climate record of the past 150 years (Kelly and Wigley, 1992; Schlesinger and Ramankutty, 1992; Hansen et al., 1993). General Circulation Models (GCMs) have calculated the response of the climate system to the newly determined solar forcing measurements and models (Rind and Overpeck, 1993; Nesme-Ribes et al., 1993). The observed Sun-climate correlations have, thus far, not been reproduced by climate simulations in which solar forcing is constrained to the changes measured by the space-based radiometers. Some significant uncertainties which remain in specifying the solar forcing and the climate and atmosphere responses are discussed below for both contemporary and paleo- scenarios.

Contemporary Solar Variability

Measurements

During 1980 and 1990, both years of activity maxima (in solar cycles 21 and 22, respectively), space-based radiometers observed total solar irradiance increases of about 0.1% relative to 1986, the intervening solar minimum. Superimposed on this solar cycle variation are larger changes (as much as 0.3%) related to the Sun's 27-day rotation. The solar cycle and rotation modulation are shown in Figure 2. Of the five radiometers that have contributed to this database only ACRIM/UARS remains operational. By virtue of their relatively high long term precision (0.005 to 0.01%), the radiometers have tracked the overall solar cycle variations in total radiative output (Willson and Hudson, 1991; Hoyt et al., 1992). However, differences in the absolute irradiance levels determined by the individual radiometers, traceable to calibration inaccuracies, can be as large as 6 W/m^2 (~ 0.4%). A critical aspect of the historical solar irradiance database is the redundancy provided by multiple observations, without which confidence in total solar irradiance variability would be considerably reduced. Not shown in Figure 2 are six months of data made by two separate instruments on Eureka (Romero et al., 1994).

The variations in total solar irradiance shown in Figure 2 arise from the integrated effect of solar spectral irradiance changes which, as indicated in Figure 3, are not a uniform function of wavelength (see Lean, 1991). Solar UV radiation varies considerably more than does the visible emission, and contributes disproportionately to changes in total irradiance (Lean, 1989). Reliable, continuous measurements of UV spectral irradiance variations are now being made by SOLSTICE on UARS (London et al., 1993), but spectrum changes at wavelengths longward of 400 nm, which may not necessarily be in phase with solar activity (e.g., Fig. 3, near 2000 nm), are not being monitored.

Proxy Variability Models

First attempts to explain the observed total solar irradiance variations focused on the coincidence between relatively large rotational-related depletions and the presence on the solar disk of large, dark sunspots. Thus sunspots were initially believed to be the dominant magnetic phenomenon responsible for changing total irradiance, and by including this effect (parameterized in terms of the fractional disk area of the sunspots, bolometric sunspot contrast and limb-darkening) the very first proxy models were able to reproduce much of the observed rotational modulation (Hudson et al., 1982; Sofia et al., 1982).

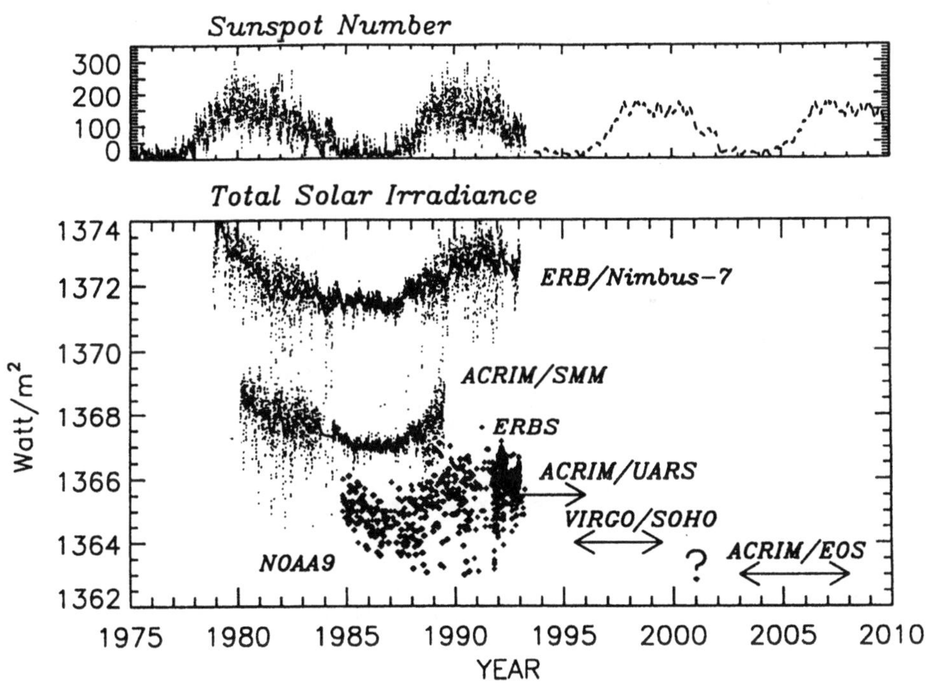

Fig. 2. Levels of solar activity according to the sunspot number and space-based radiometric monitoring of the total solar radiative output. Differences in irradiance values measured by the radiometers reflect inaccuracies in their absolute radiometric calibrations. A gap in solar monitoring is expected in the 1999-2003 time frame.

To account for the observed decrease in total irradiance from maximum to minimum of the 11-year solar activity cycle, it was subsequently shown that facular brightening dominates sunspot blocking over these longer time scales (sunspot blocking alone would cause increased irradiance at sunspot minimum) (Foukal and Lean, 1988). New proxy models were developed that included both sunspot blocking and facular brightening (assumed here to include all regions of enhanced emission on the solar disk, both the brightest active regions, and the more dispersed surrounding network); a critical aspect of their success in explaining both the short term rotational modulation and the longer term solar cycle variations was the use of certain global (full-disk) facular proxies such as the He 1083 nm equivalent width. In fact, it was shown that the emission from the brightest active regions alone (and proxies such as the Ca K plage index that

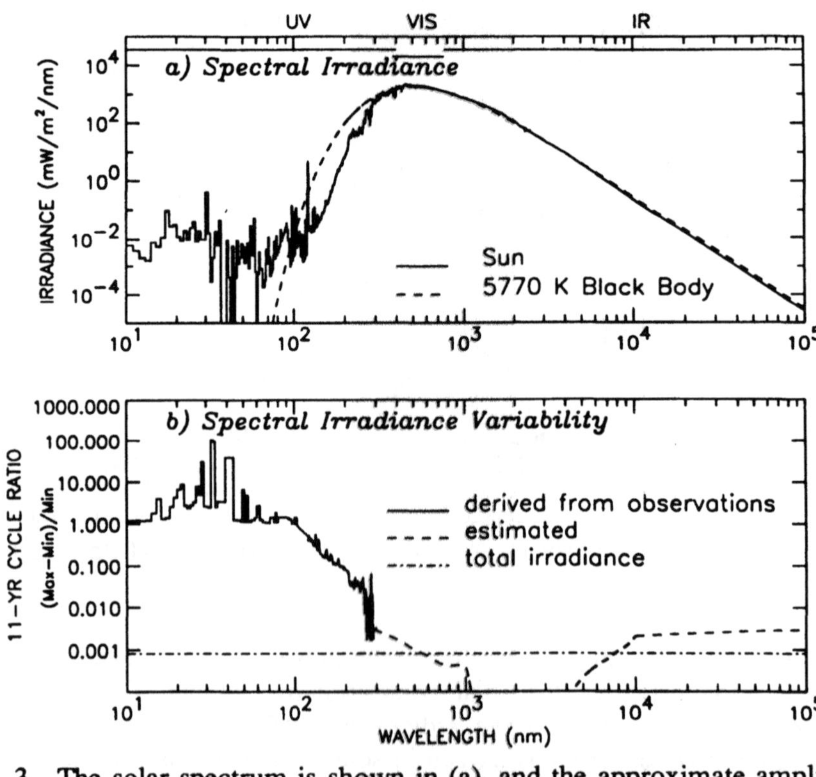

Fig. 3. The solar spectrum is shown in (a), and the approximate amplitude of the 11-year solar cycle variation in this spectrum is shown in (b) (Lean, 1991).

reflect this) underestimated by about a factor of two the actual variation of the total irradiance brightness source (Foukal and Lean, 1988; Lean, 1992).

Further modifications of proxy models are needed, and in progress. Improved sunspot blocking parameterizations that account for differences in sunspot darkness with size (Brandt et al., 1990) lead to reductions of 15% to 20% (on average) in the sunspot blocking effect, and hence in the compensating facular emission needed to account for the observed solar cycle irradiance variation. On the other hand, the inability of the current proxy models to match the high levels of solar irradiance measured by ERB in 1990 implies the need for a larger facular contribution (Fröhlich, 1993) or, instead, a completely different, non-magnetic mechanism (e.g., Kuhn and Libbrecht, 1991). Nor can instrumental effects be ignored. Consistent with a needed re-evaluation of the facular proxies are the differences between various proxies and the solar magnetic flux, shown in Figure 4. Solar magnetic flux generates the phenomena responsible for solar irradiance variability but differences between cycle 21 and cycle 22 maxima challenge the assumption of its linear relationships with the proxies.

Forcing of Global Change

The impact on the climate system of contemporary solar radiative output changes is uncertain. As indicated in Figure 5, the climate forcing arising from the 0.1% decrease in total solar radiative output from solar activity maximum in 1980 to the subsequent minimum in 1986 is approximately opposite and equal to the forcing by greenhouse gases over this same period (Hansen and Lacis, 1990). Solar forcing arising from a radiative output change of $\Delta S=0.1\%$ in $S=1370$ W/m^2 corresponds to an equilibrium global mean temperature change of $\Delta T=\kappa\times1370\times0.001\times0.7/4°C$ where κ is the temperature sensitivity and the factors 0.7 and 4 refer to the Earth's albedo and the geometrical conversion from spherical to disk area, respectively. For $\kappa=0.7°C/(W/m^2)$, $\Delta T=0.17°C$. This is

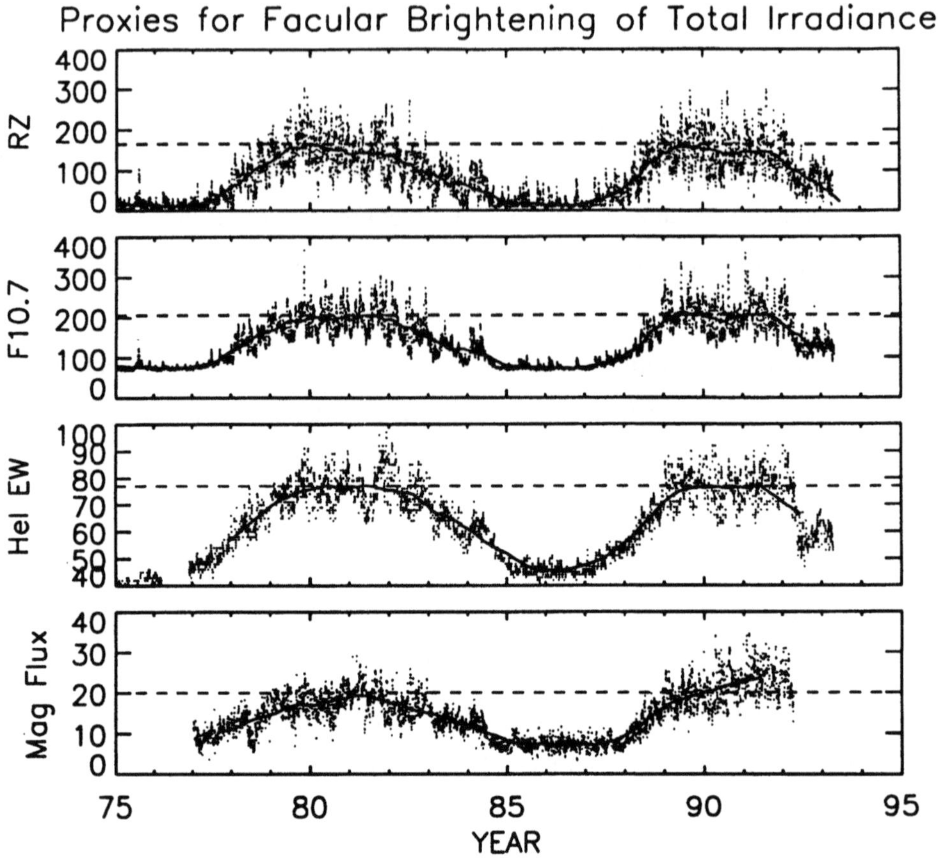

Fig. 4. Comparison of solar proxies used to parameterize the variations in the brightness source (assumed to be primarily active region and network faculae) of the total solar irradiance. Rz is the sunspot number, F10.7 is the 10.7 cm radio flux, Hel EW is the equivalent width of the He I infrared absorption feature at 1083 nm and Mag Flux is the disk-integrated absolute magnetic flux.

of the order of the ±0.2°C inherent natural variability of the climate system. Furthermore, due to the thermal inertia of the oceans the actual temperature change is thought to be some factor of two less than the equilibrium temperature change, which implies that the effect of direct solar radiative forcing over the 11-year cycle is unlikely to be distinguishable from other sources of variability in the climate record. This temperature change is also too small to account for observations of tropospheric temperatures warmer, on average, by 1°C at high levels of solar activity, relative to solar minimum conditions (Labitzke and van Loon, 1993). It is possible that these and other solar-terrestrial correlations arise from different, indirect physical mechanisms associated with, for example, the impact of changing solar UV radiation on ozone and the middle atmosphere, perhaps involving the ~26-month quasi-biennial oscillation (QBO) of the equatorial stratospheric winds.

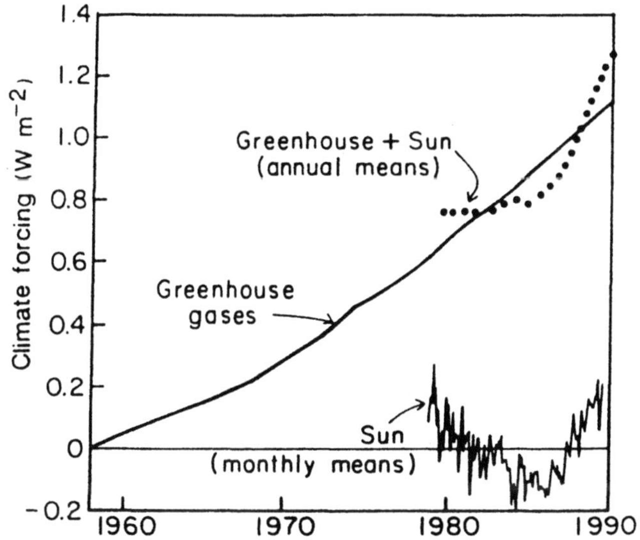

Fig. 5. Climate forcings arising from measured changes in greenhouse gases and solar irradiance. Zero point in solar forcing is the 1978-1989 mean (Hansen and Lacis, 1990).

The impact of solar UV irradiance variations on the ozone layer is relatively well established (Stolarski et al., 1991; Hood and McCormack, 1992), although some details remain uncertain. Global column ozone varies by 1.5% to 1.8% in response to the 11-year solar activity cycle. As illustrated in Figure 6, model studies indicate that larger variations occur in upper stratosphere ozone, although new observational evidence indicates larger than expected solar related ozone perturbations in the lower stratosphere (Hood et al., 1993). The implications of both solar and anthropogenic forcing of ozone for climate change are not yet understood, and depend critically upon the altitude of the perturbation (Fig. 6).

Solar Variability and Paleoclimate

Sun-Climate Correlations

The similarity over the past centuries between certain paleoclimate records and various indicators of historical solar activity, demonstrated by Eddy (1976) for ^{14}C and northern European temperatures, is also apparent during the most recent century. Figure 7 shows the apparent tracking of global sea surface temperature anomalies and the smoothed sunspot number, and of northern hemisphere temperature anomalies with another solar variability indicator, the length of the

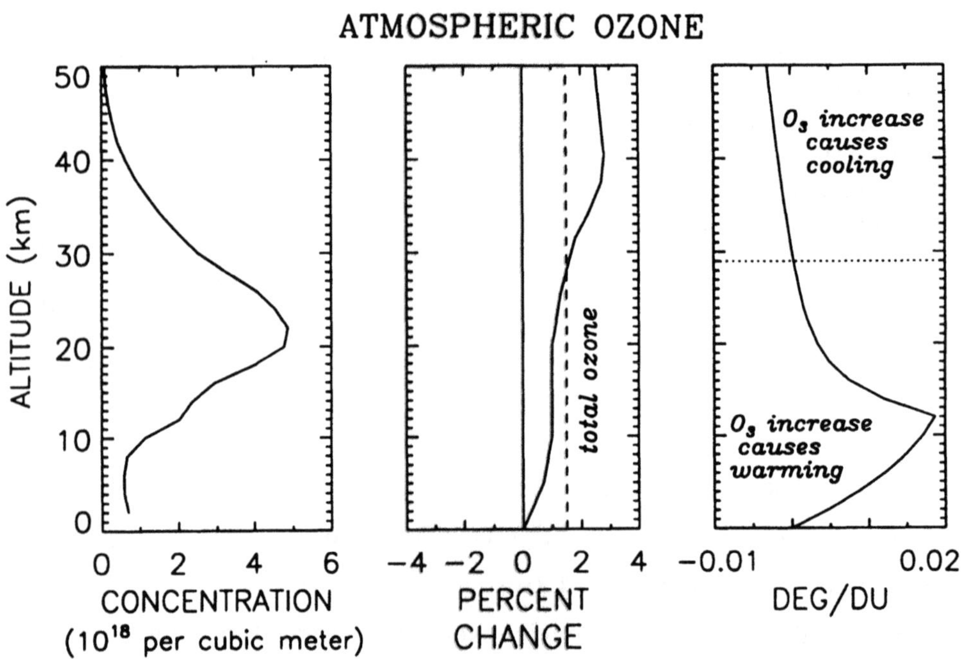

Fig. 6. The USSA76 atmospheric ozone profile is shown on the left and estimated solar cycle ozone changes are shown in the middle (Wuebbles et al., 1991). On the right is a one-dimensional radiative-convective equilibrium calculation of the vertical sensitivity of the radiative forcing of climate by changing ozone concentrations (in DEG/DU), indicating how the impact of altered ozone concentration (in Dobson Units) on surface temperature (in degrees C) depends on the altitude profile of the ozone changes (Lacis et al., 1990).

nominal 11-year cycle. In a simple ocean circulation model, to account for the sea surface temperature anomaly increase of ~0.5° C over the past 100 years by solar forcing requires a change in total solar irradiance of the order of ~0.5% (Reid, 1991), five times larger than the 0.1% variation that has been observed over the contemporary 11-year cycle. Larger solar irradiance variations are also inferred by climate simulations that better reproduce the observed surface temperature variations by including both solar and anthropogenic forcings (Kelly and Wigley, 1992; Schlesinger and Ramankutty, 1992).

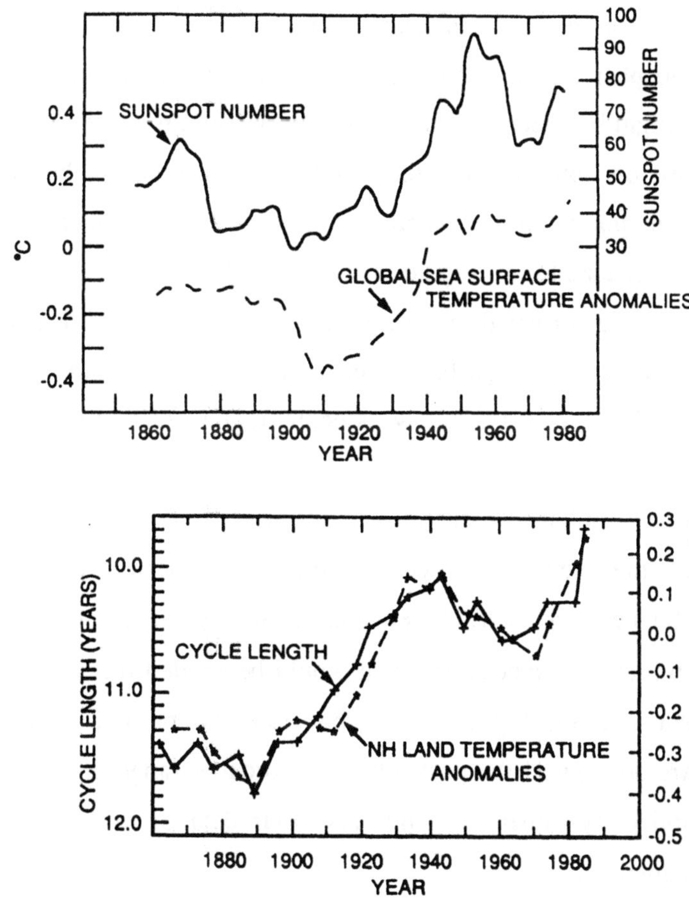

Fig. 7. Correlations between climate data and solar variability proxies. The upper figure is from Reid (1991) and the lower figure is from Friis-Christensen and Lassen (1991). Note the phase differences of as much as 20 years between solar cycle length and sunspot number, and between sea surface and NH land surface temperature anomalies.

Limits on Long-term Solar Variability Amplitudes

What is the likelihood of larger solar radiative output changes on longer time scales, unsampled by the limited, space-based, solar radiometric monitoring? Evaluating scenarios in which the facular emission (which controls the longer-term radiative output variations) is altered beyond its contemporary range can help provide limits on the amplitude of long term irradiance variations.

During contemporary solar minima a network of bright Ca K emission (a proxy for facular emission) remains over some 39% of the solar disk (Skumanich et al., 1975), associated presumably with residual magnetic flux in the solar atmosphere. Complete disappearance of this network of enhanced emission, such as might be speculated in a Maunder Minimum event, would deplete the facular emission and reduce the total solar irradiance below its level during contemporary solar minima (White et al., 1992; Lean et al., 1992). According to the correlation in Figure 8 between the Sun's Ca emission and the total irradiance corrected for sunspot blocking (i.e., the irradiance brightness source), removing all solar surface magnetic network reduces the total solar irradiance by ~1.5 W/m^2, or 0.1% below its contemporary solar minimum level.

Larger irradiance reductions might arise from depletion of the Sun's emission from the centers of the "cells" defined by the network, in addition to the removal of the network itself. Current solar theory cannot accommodate the (presumably) non-magnetic mechanism that would be needed for such a reduction in "basal" solar emission [Spruit, 1994], but observations of Ca emission from solar-like stars do suggest the possibility of a broader range of solar Ca emission than is implied by either contemporary solar data or even by removing all surface magnetism.

Shown in Figure 9 is the distribution of Ca emission from 13 solar-like stars.

Baliunas and Jastrow (1990) identify the upper two-thirds of this distribution (HK > 0.15) as stars having activity cycles analogous to the Sun's 11-year cycle whereas activity cycles have yet to be observed for stars in the lower ~⅓ of the distribution (HK < 0.15). The Sun's level of Ca emission places it amongst the most active of these solar-like stars. Even with zero surface network the Ca level of the Sun (HK=0.156, from Fig. 8) moves only to the low end of the distribution of the cycling stars, not into the lower part of the distribution comprised of the non-cycling stars. Matching the Ca emission of the solar-like

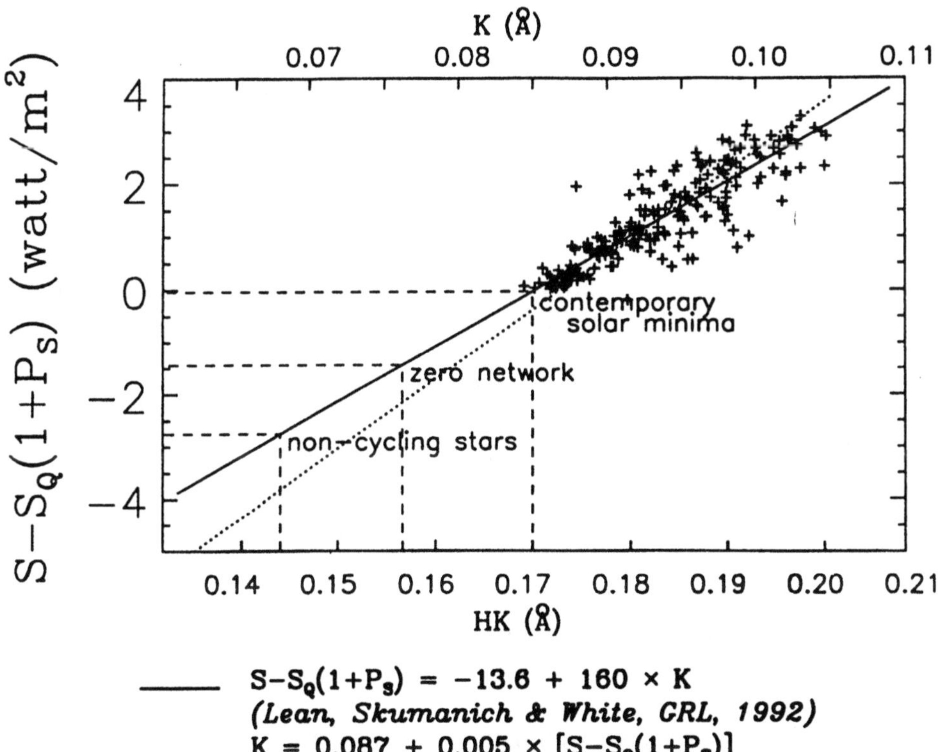

Fig. 8. Correlation between solar Ca K emission and total irradiance faculae (determined by correcting the measured total irradiance for sunspot blocking effects). A conversion between solar Ca K emission and stellar Ca H and K emission has been applied to provide equivalent stellar HK emission levels (White et al., 1992). Scenarios whereby solar Ca emission is reduced below its contemporary range are indicated (White et al., 1992; Lean et al., 1992). The dotted line provides an estimate of the uncertainty in this correlation; it was obtained by inverting the abscissa and ordinate in the regression.

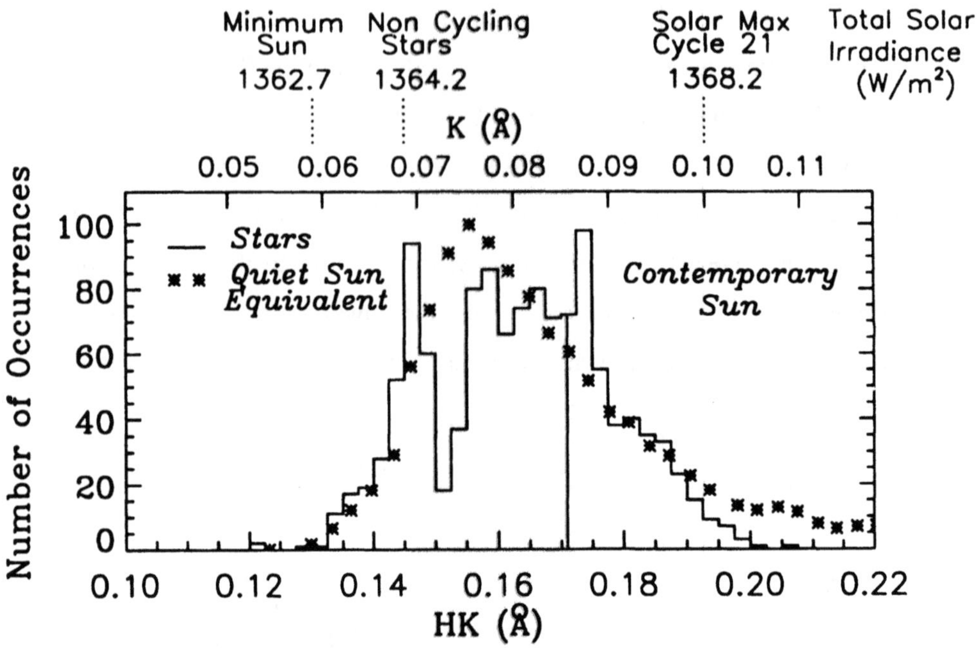

Fig. 9. Distribution of Ca emission from solar-like stars, compared with the distribution of Ca emission from the center-disk quiet Sun, and the contemporary range of solar Ca K (disk-integrated) flux. Levels of total solar irradiance are estimated for different levels of Ca emission (Lean et al., 1992).

non-cycling stars requires the solar Ca emission to be reduced to levels seen only in 15% of the quiet Sun area today (i.e., the darkest cells) (White et al., 1992). In this case, total solar irradiance is reduced to 0.2% below its value in the 1986 solar minima (0.24% below its mean value in solar cycle 21).

The stellar data are somewhat ambiguous. Not all non-cycling stars have Ca emission levels as low as the lower portion of the Figure 9 distribution. For example, solar-like, non-cycling HD10700 has Ca HK=0.17, and even though HD3651 has recently entered a non-cycling phase, its Ca emission is still HK=0.16 (i.e., in the portion of the distribution associated with cycling stars). Yet, the distribution of emission from the present day quiet Sun (asterisks in

Figure 9) provides limits of Ca emission independent of, but in agreement with, the stellar data. If the entire Sun were covered with emission levels seen only in the very darkest regions of the quiet Sun, the corresponding reduction in the disk-integrated Ca flux would place the Sun with the lowest levels of the Figure 9 distribution, consistent with the lowest Ca emission seen in solar-like stars. From the correlation in Figure 8, the total irradiance would be reduced to 0.35% below the contemporary solar minimum. This represents the minimum possible Sun. Note that this reduction is still less than the 0.4% reduction derived by Nesme-Ribes et al. (1993) for the Maunder Minimum.

Assuming that total solar radiative output variations in excess of the 0.1% 11-year cycle variation are possible, Figure 10 compares two historical reconstructions which each invoke a varying background irradiance component in addition to the solar cycle changes arising from active regions (sunspot and faculae) alone. Hoyt and Schatten's (1993) use of the number of umbral spots results in estimated background changes out of phase by as much as 20 years with a background component derived instead from the smoothed sunspot number, but that approximately match the solar cycle length data in Figure 7.

Climate Model Simulations

Motivated by the coincidence of the solar Maunder Minimum and the coldest temperatures of the Little Ice Age, and by the correlations between solar variability and temperature anomalies in the past century (Figure 7), recent studies have utilized climate models to investigate the likelihood that climate change arises from the response of known climate feedbacks to solar forcing.

Shown in Figure 11 is the geographical distribution of temperature change estimated by the Goddard Institute for Space Studies (GISS) GCM for a 0.25% reduction in total solar irradiance (Rind and Overpeck, 1993). The corresponding

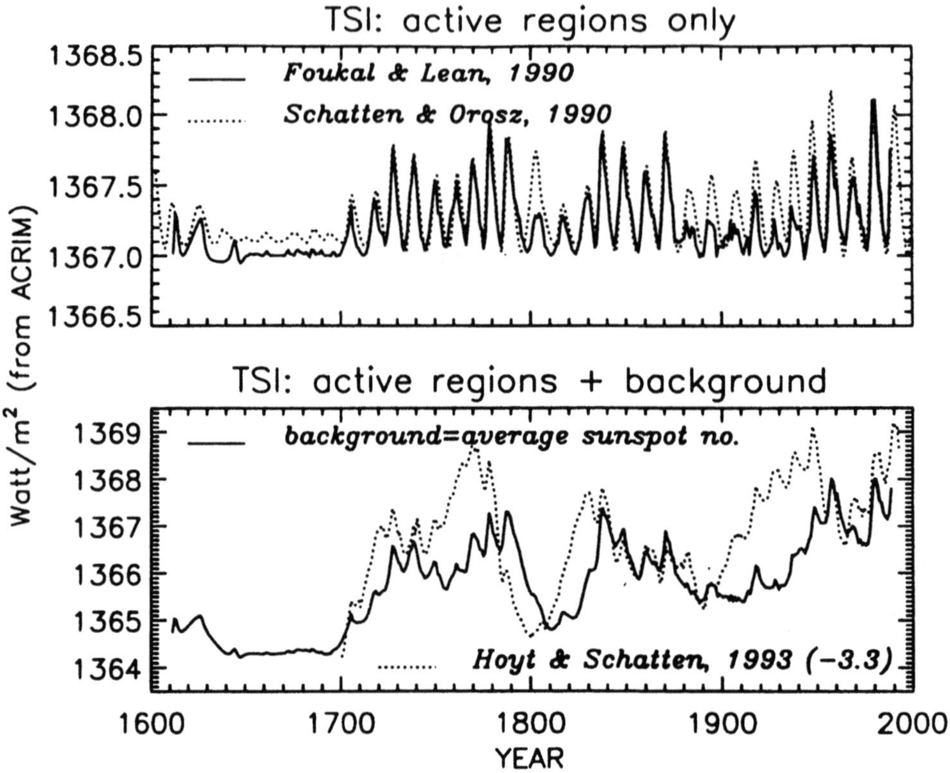

Fig. 10. Estimates of historical total solar irradiance (TSI) variations due to active regions alone (upper panel), and active regions plus an additional background component (lower panel) derived under two different scenarios.

global temperature reduction is 0.46°C, which is approximately half of the perhaps 1°C cooling that is thought to have occurred during the coldest period of the Little Ice Age, relative to modern times (although exact specification of Little Ice Age temperatures is difficult). Importantly, the GISS study indicates significant geographical inhomogeneities in response to this small reduction in solar forcing; some locations may actually warm in response to advective, rather than radiative effects. Figure 11 may thus provide clues for interpreting the regional paleoclimate data, only some of which indicate evidence for Little Ice Age cooling. Another GCM study predicts a more geographically-uniform cooling (Nesme-Ribes et al., 1993). However, neither of these GCM studies included spectral irradiance variations or the effects of ozone changes on the surface temperature.

Other climate simulations indicate that incorporating solar forcing (especially when parameterized by solar cycle length) in addition to greenhouse warming and aerosol cooling provides an improved reconstruction of surface temperature changes during the past 150 years (Kelly and Wigley, 1992; Schlesinger and Ramankutty, 1992; Hansen et al., 1993). In fact, based on simply fitting the climate data, solar forcing alone provides a very reasonable fit (e.g., Figure 7) but requires a climate sensitivity well in excess of current values. As well, attributing the entire warming of the past century to solar forcing would leave substantial greenhouse gas forcing unaccounted for in the temperature record.

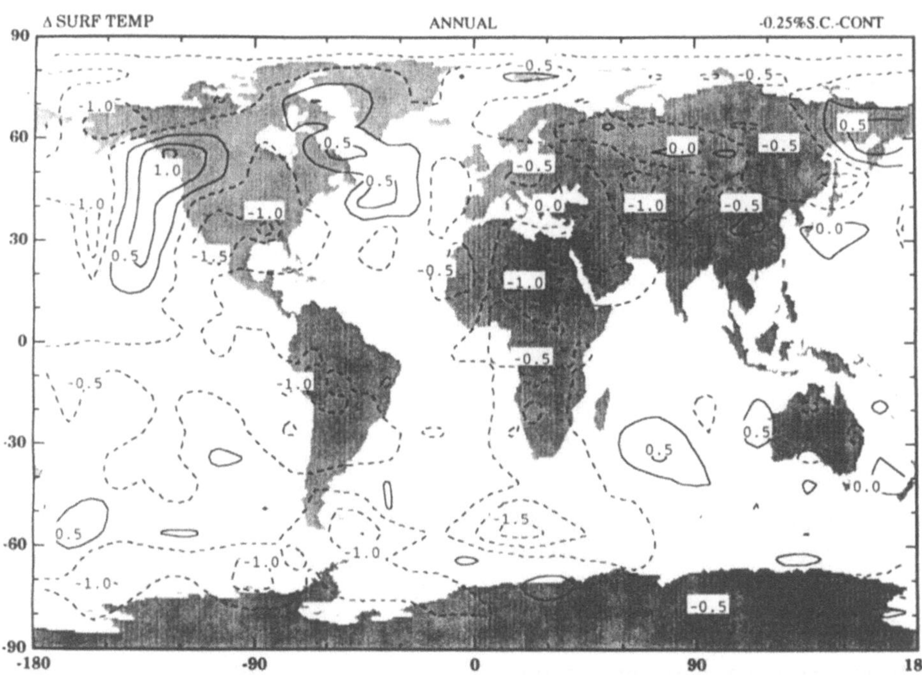

Fig. 11. Estimates by the GISS GCM of the temperature changes arising from a 0.25% spectrally-uniform, solar irradiance decrease (Rind and Overpeck, 1993).

Still, these climate simulations provide circumstantial evidence for non negligible solar forcing of climate change, as do the GCM results for Maunder Minimum forcing during the Little Ice Age.

Summary

Figure 12 compares the climate forcings thought to have been operable over the past 140 years. As demonstrated in Figure 13, current uncertainties in knowledge of solar forcing limit an assessment of both paleo- and future global change. Reducing these uncertainties will require a committed, co-ordinated, interdisciplinary effort of monitoring, assessment and interpretation of solar and climate data; without this, questions as to whether observed global changes are the result of natural or anthropogenic forcing will continue to be raised.

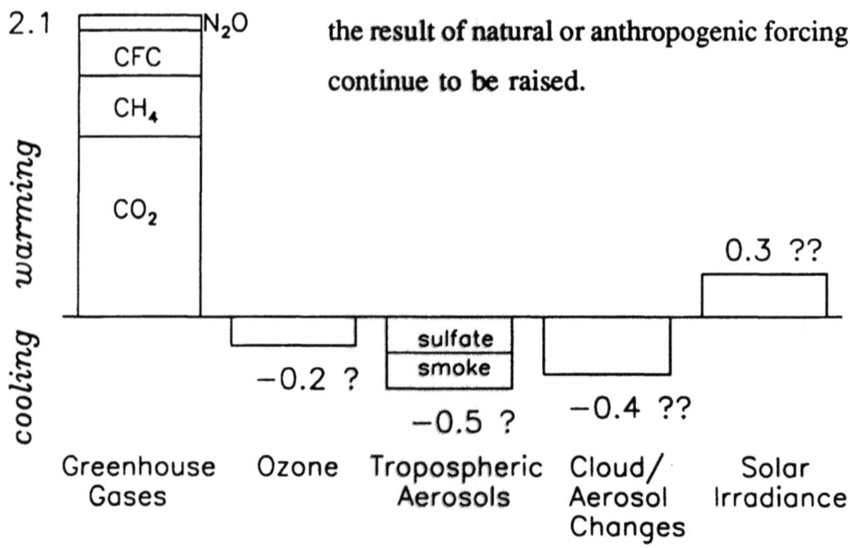

Fig. 12. Summary of climate forcings in the past 140 years (Hansen et al., 1993).

Current Uncertainties and Future Requirements

1) It is unknown whether, or with what amplitude, solar radiative output variations in addition to the measured 11-year cycle, are occurring, despite circumstantial evidence for larger changes on climatologically significant time

scales. Thus the likelihood of the Figure 13 scenario cannot, as yet, be estimated.

2) Radiometric inaccuracy and imprecision contribute uncertainties to long term solar forcing. For example, determining a solar irradiance change of 0.25% over 200 years (0.013% per solar cycle), as speculated in Figure 13, requires that the precision of the database be maintained at better than 0.005%, which may be difficult to achieve. Continuous, uninterrupted solar monitoring by redundant radiometers is critical. The gap projected in the 1999-2003 time frame between SOHO and EOS would prevent cross-calibration of past and future monitoring, thereby invalidating the decadal historical database for studies of solar forcing of global change.

3) Improvements are needed in proxy variability models of solar forcing. These models are largely empirical, exhibit some discrepancies with recent radiometric data and among themselves, and are not yet based on solar theory. Current solar theory precludes radiative output variations other than those associated with changes in magnetic flux; this would limit the long term variations to <0.2% which is less than various estimates of the Maunder Minimum reduction developed from empirical models and implied by observations of solar-like stars. Theoretically based models of both solar and stellar variability are needed.

4) The spectrum of solar radiative output variability should be properly defined at all wavelengths. This has yet to be achieved aside from preliminary estimates at the shortest UV wavelengths.

5) Impressively high correlations between solar variability and climate/atmosphere phenomena are not reproduced by climate simulations. However, there are sufficient uncertainties in the simulations and in the climate record that these correlations cannot yet be discounted entirely. Climate simulations of solar forcing should include the spectral dependence of the changing solar irradiance, feedbacks associated with solar-induced ozone changes and dynamical effects related to the QBO. Whether other physical feedback mechanisms are likewise absent must also be assessed. Reducing uncertainties

in the paleoclimate record may help provide an empirical determination of the limits of past, and perhaps future, solar forcing.

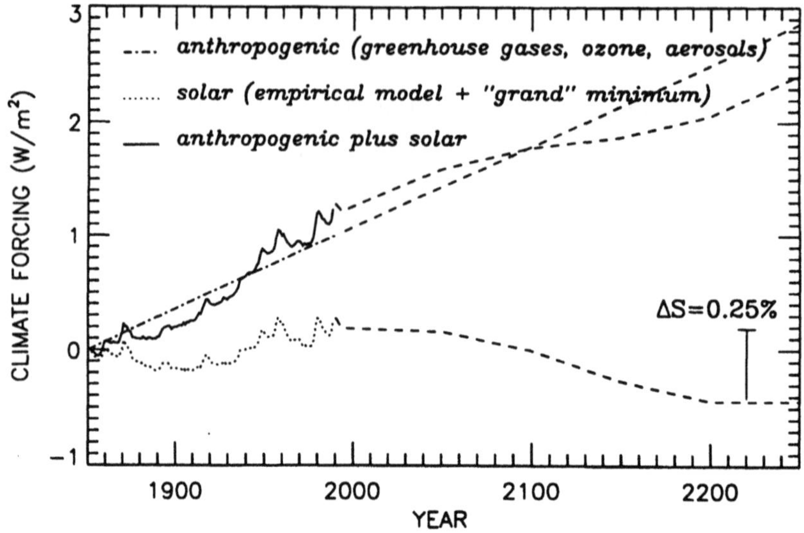

Fig. 13. A scenario for anthropogenic and solar forcing of future climate change; anthropogenic forcing continues to increase at a net rate of 1.1 W/m^2 per 140 years (Hansen et al., 1993) and the Sun enters a Maunder Minimum-type phase, commencing in the year 2000 and taking 200 years to reach a radiative output level 0.25% less than its current mean value (after Wigley and Kelly, 1990; Lean and Rind, 1994).

Acknowledgments. Data were kindly provided by R. Willson, J. Closs, M. Stuiver and H. Coffey. The He data are produced cooperatively by NSF/NOAO, NASA/GSFC and NOAA/SEL. Much appreciated are discussions with D. Rind. The SERDP provided partial support.

References

Baliunas, S., and Jastrow, R. (1990). Evidence for long-term brightness changes of solar-like stars, Nature, 348: 520-523

Beer, J., Raisbeck, G. M., and Yiou, F. (1991). Time variation of ^{10}Be and solar activity. In The Sun in Time, eds. C. P. Sonett, M. Giampapa and M. S. Matthews, The University of Arizona Press, Tucson: 343-359

Brandt, P. N., Schmidt, W., and Steinegger, M. (1990). On the umbra-penumbra ratio of sunspots, Solar Phys., 129: 191-194

Damon, P. E., and Sonett, C. P. (1991). Solar and terrestrial components of the ^{14}C variation spectrum. In The Sun in Time, eds. C. P. Sonett, M. Giampapa and M. S. Matthews, University of Arizona Press, Tucson: 360-388

Eddy, J. A. (1976). The Maunder minimum, Science, 192: 1189-1202
Foukal, P., and Lean, J. (1988). Magnetic modulation of solar luminosity by photospheric activity, Astrophys. J., 328: 347-357
Foukal, P., and Lean, J. (1990). An empirical model of total solar irradiance variation between 1874 and 1988, Science, 247: 505-604
Friis-Christensen, E., and Lassen, K. (1991). Length of the solar cycle: An indicator of solar activity closely associated with climate, Science, 254: 698-700
Fröhlich, C. (1993). Irradiance observations of the Sun, IAU Colloquium No. 143, June 20-25, Boulder, CO
Hansen, J. E., and Lacis, A. A. (1990). Sun and dust versus greenhouse gases: An assessment of their relative roles in global climate change, Nature, 346: 713-719
Hansen, J., Lacis, A., Ruedy, R., Sato, M., and Wilson, H. (1993). How sensitive is the world's climate?, National Geographic Research and Exploration 9(2): 142-158
Hood, L. L., and McCormack, J. P. (1992). Components of interannual ozone change based on Nimbus 7 TOMS data, Geophys. Res. Lett., 19: 2309-2312
Hood, L. L., Jirokowic, J. L., and McCormack, J. P. (1993). Quasi-decadal variability of the stratosphere: influence of solar-term solar ultraviolet variations, J. Atmos. Sci., 50: 3941-3958
Hoyt, D. V., and Schatten, K. H. (1993). A discussion of plausible solar irradiance variations, 1700-1992, J. Geophys. Res., 98: 18895-18906
Hoyt, D. V., Kyle, H. L., Hickey, J. R., and Maschhoff, R. M. (1992). The Nimbus 7 solar total irradiance: A new algorithm for its derivation, J. Geophys. Res., 97: 51-63
Hudson, H. S., Silva, S., Woodard. M., and Willson, R. C. (1982). The effects of sunspots on solar irradiance, Solar Phys., 76: 211-219
Kelly, P. M., and Wigley, T. M. L. (1992). Solar cycle length, greenhouse forcing and global climate, Nature, 360: 328-330
Kuhn, J. R., and Libbrecht, K. G. (1991). Nonfacular solar luminosity variations, Astrophys. J., 380: L35-L37
Labitzke, K., and van Loon, H. (1993). Some recent studies of probable connections between solar and atmospheric variability, Ann. Geophysicae, 11: 1084-1094
Lacis, A. A., Wuebbles, D. J., and Logan, J. A. (1990). Radiative forcing of climate by changes in the vertical distribution of ozone, J. Geophys. Res., 95: 9971-9981
Lean, J. (1989). Contribution of ultraviolet irradiance variations to changes in the Sun's total irradiance, Science, 244: 197-200
Lean, J. (1991). Variations in the Sun's Radiative Output, Rev. Geophys., 29: 505-535
Lean, J. (1992). The effect of surface inhomogeneities on total solar irradiance. In Surface Inhomogeneities on Late-Type Stars, eds. P. B. Bryne and D. J.

Mullan, Lecture Notes in Physics, 397, Springer-Verlag: 167-180
Lean, J., and Rind, D. (1994). Solar variability: Implications for global change, Eos, Transactions, American Geophysical Union, 75: 1
Lean, J., Skumanich, A., and White, O. (1992). Estimating the Sun's radiative output during the Maunder Minimum, Geophys. Res. Lett., 19: 1591-1594
London, J., Rottman, G. J., Woods, T. N., and Wu, F. (1993). Time variations of solar UV irradiance as measured by the SOLSTICE (UARS) instrument, Geophys. Res. Lett., 20: 1315-1318
McHargue, L. R., and Damon, P. E. (1991). The global Beryllium 10 cycle, Rev. Geophys., 29: 141-158
Nesme-Ribes, E., Ferreira, E. N., Sadourny, R., Le Treut, H., and Li, Z. X. (1993). Solar dynamics and its impact on solar irradiance and the terrestrial climate, J. Geophys. Res., 98: 18923-18935
Reid, G. (1991). Solar total irradiance variations and the global sea surface temperature record, J. Geophys. Res., 96: 2835-2844
Rind, D., and Overpeck, J. (1993). Hypothesized causes of decadal-to-century climate variability: climate model results, Quat. Sci. Rev., 12: 357-374
Romero, J., Wehrli, C., and Fröhlich, C. (1994). Solar total irradiance variability from SOVA 2 onboard EURECA, Solar Phys., 152: in press
Schlesinger, M. E., and Ramankutty, N. (1992). Implications for global warming of intercycle solar irradiance variations, Nature, 360: 330-333
Skumanich, A., Smythe, C., and Frazier, E. N. (1975). On the statistical description of inhomogeneities in the quiet solar atmosphere. I. Linear regression analysis and absolute calibration of multichannel observations of the Ca+ emission network, Astrophys. J., 200: 747-764
Sofia, S., Oster, L., and Schatten, K. (1982). Solar irradiance modulation by active regions during 1980, Solar Phys., 80: 87-98
Spruit, H. (1994). Theories of radius and luminosity variations. This volume
Stolarski, R. S., Bloomfield, P., McPeters, R. D., and Herman, J. R. (1991). Total ozone trends deduced from Nimbus-7 TOMS data, Geophys. Res. Lett., 18: 1015-1018
Stuiver, M., and Reimer, P. J. (1993). Extended ^{14}C database and revised CALIB radiocarbon calibration program, Radiocarbon, 35: 215-230
White, O. R., Skumanich, A., Lean, J., Livingston, W. C., and Keil, S. L. (1992). The Sun in a non-cycling state, Publ. Astronom. Soc. Pac., 104: 1139-1143
Wigley, T. M. L., and Kelly, P. M. (1990). Holocene climatic change, ^{14}C wiggles and variations in solar irradiance, in The Earth's Climate and Variability of the Sun over Recent Millennia: Geophysical, Astronomical and Archaeological Aspects, Phil. Trans. R. Soc. Lond. A, 330: 547-560
Willson, R. C., and Hudson, H. S. (1991). A solar cycle of measured and modeled total irradiance, Nature, 351: 42-44
Wuebbles, D. J., Kinnison, D. E., Grant, K. E., and Lean, J. (1991). The effect of solar flux variations and trace gas emissions on recent trends in stratospheric ozone and temperature, J. Geomag. Geoelectr., 43, Suppl: 709-718

THE OBSERVED SPATIAL NON-UNIFORMITY AND TEMPORAL VARIABILITY OF THE SOLAR LIMB-DARKENING FUNCTION

Ronald J. Kroll
Dept. of Physics, U. Arizona
Tucson, AZ 85721

Abstract

The solar radiation intensity was measured at selected heliographic latitudes by observing the continuum radiation at 550 nm over the outermost 32 arcseconds of the limb. At each observed limb position, solar edge locations are determined from the intensity profiles using the Finite Fourier Transform Definition with a series of scan amplitudes. The difference between edge locations for a fixed pair of scan amplitudes, the differential radius, provides information about the slope of the limb-darkening function. The results of the observations made from 1990 through 1992 indicate that a global change in the limb-darkening function occurred and that, for certain years, the limb-darkening function was not uniform over solar latitude. Since the energy flux through the photosphere determines the temperature gradient and hence the limb-darkening function, these observations suggest time variations in solar luminosity originating in the quiet photosphere. The latitudinal non-uniformity of the solar limb darkening is shown to be compatible with perturbations to the temperature gradient confined to a relatively thin sub-photospheric layer. Additionally, limb-darkening variations should bear directly on the interpretation of solar diameter observations.

Introduction

The variability of the total solar irradiance has been demonstrated by recent satellite observations, and considerable effort is underway to account for its origin. Models of total irradiance which incorporate effects of sunspots, active region and network faculae account for substantial variance, yet exhibit significant discrepancies with observations particularly during times of solar maxima (Pap et al., 1994). In spite of

NATO ASI Series, Vol. I 25
The Solar Engine and Its Influence
on Terrestrial Atmosphere and Climate
Edited by Elizabeth Nesme-Ribes

considerable efforts to refine these models, additional effects may be relevant, which some models have yet to incorporate. Among these are the possible luminosity changes originating in the quiet photosphere on a global scale, or changes in the distribution of radiation over the solar surface. One method of determining the reality of these effects is to monitor the solar limb-darkening function over time and heliographic latitude. Such ground-based observations can be a valuable compliment to both satellite observations and modeling efforts for understanding solar variability.

Limb-darkening observations made to detect variations over time have provided a variety of results (see, e.g., Petro et al., 1984; Brown, 1988; Beardsley et al., 1988; Kroll et al., 1990). Observations which employ the drift scan technique actually sample different solar latitudes through the year. If the limb-darkening function were different from one latitudinal region to another, then spatial variation may be interpreted as temporal variation. Therefore, it is important to determine the uniformity of the solar surface temperature and limb-darkening function. Kuhn et al. (1988) and Kroll (1994) demonstrate that there is spatial nonuniformity as well as temporal variability in these observables. Such results imply that the assumption of a uniform nonvarying background photosphere must be amended. With appropriate modifications to the existing models and additional data on photospheric variability, the problem of accounting for total irradiance variations may be more tractable.

The following sections briefly describe the method employed to determine the spatial uniformity of the solar limb-darkening function, the results, and a discussion of the implications of these results for total irradiance and solar diameter observations.

Observations and Data Analysis

The limb-darkening observations discussed herein were obtained with the SCLERA astrometric telescope (Oleson et al., 1974). The telescope is configured to project the solar image onto a horizontal focal plane while continuously tracking the sun. In the focal plane is a detector assembly consisting of six separate slits (Figure 1), each of which extend over 5.7° of the solar circumference and 1 arcsecond radially. The solar image is scanned sinusoidally across these slits so that about 32 arcseconds of the extreme solar limb is profiled in the radial direction. Throughout the day, the detector is rotated through a repeating sequence of angles resulting in the profiling of

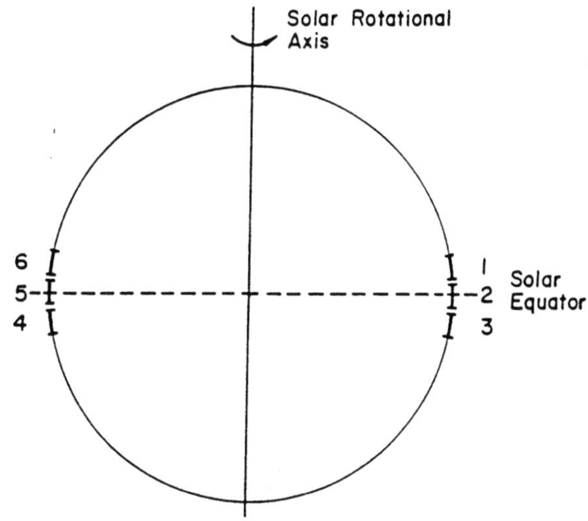

Figure 1. The configuration of detectors when the detector assembly is aligned with the solar equator. The detector assembly can rotate into alignment with any chosen diameter.

as many as 36 limb locations. The light is filtered to pass an 8 nm band of the continuum centered at 550 nm.

For each of the limb profiles obtained, it is possible to define the solar edge using the Finite Fourier Transform Definition (FFTD) (Hill et al., 1975). The edge location lies at a point along the intensity profile that depends on the amount of the limb profile incorporated into the FFTD integral. By limiting the limb profiles to what corresponds to 8- and 24-arcsecond scan amplitudes, the distance between the two edge locations is obtained and is a quantity which characterizes the slope of the solar limb-darkening function at the extreme limb. This quantity, termed the differential radius, is designated as $\Delta r(8,24)$ and is used to detect limb-darkening changes over time or differences from region to region.

The differential radius measurements to be discussed here were obtained during the spring observing seasons of 1990, 1991, and 1992. The 1990 observations contain data from only 24 limb locations, whereas the following two years include 36 limb locations. The data from any limb position when sunspot or faculae are present is excluded from the analysis, allowing a better determination of the state of the quiet photosphere. Average differential radius values at each limb position are determined for each of the three observing seasons.

Results and Discussion

The differential radius results for the seasons under discussion are shown in Figure 2. In 1990 the error bars are larger than in the following years due to fewer days of observations. Combined with fewer observed latitude positions, the results indicate a fairly uniform limb-darkening function around the limb. In 1991, however, the latitudes near 25°N and 30°S have limb-darkening functions different from those at nearby lower latitudes, which is statistically significant at the 99% level. It appears that the north and south polar limb-darkening functions are also different. Finally in 1992, the results are not so clear. There appears to be some significant variability at latitudes within ±15°, but there is no clear pattern. Further, a potentially important feature of these results over this period is that the average level of the differential radius is increasing with time. In addition to latitudinal variability, the average global limb-darkening function varied with time at least around the maximum of solar cycle 22.

The spatial non-uniformity of the limb-darkening function, as well as the observed distribution of differential radius values, suggests a connection to the results of Kuhn and Libbrecht (1991). They determined that the effective temperature of the quiet photosphere varied as a function of both latitude and time. The distribution of effective temperatures they observed during 1990, their last observing season, resembles the distribution of differential radius values present in 1991. A possible connection between these two results is investigated in Kroll (1994). Briefly, Kuhn (1991) suggests that magnetic fields at the base of the convection zone regulate the radiative flux into the convection zone. Modeling results indicate that variations in this flux can perturb the temperature gradient to a large degree in a shallow layer of the convection zone immediately below the photosphere, accounting for the observed variations in effective temperature. As modeled by Petro et al. (1985), the temperature perturbation in question will produce a different limb-darkening

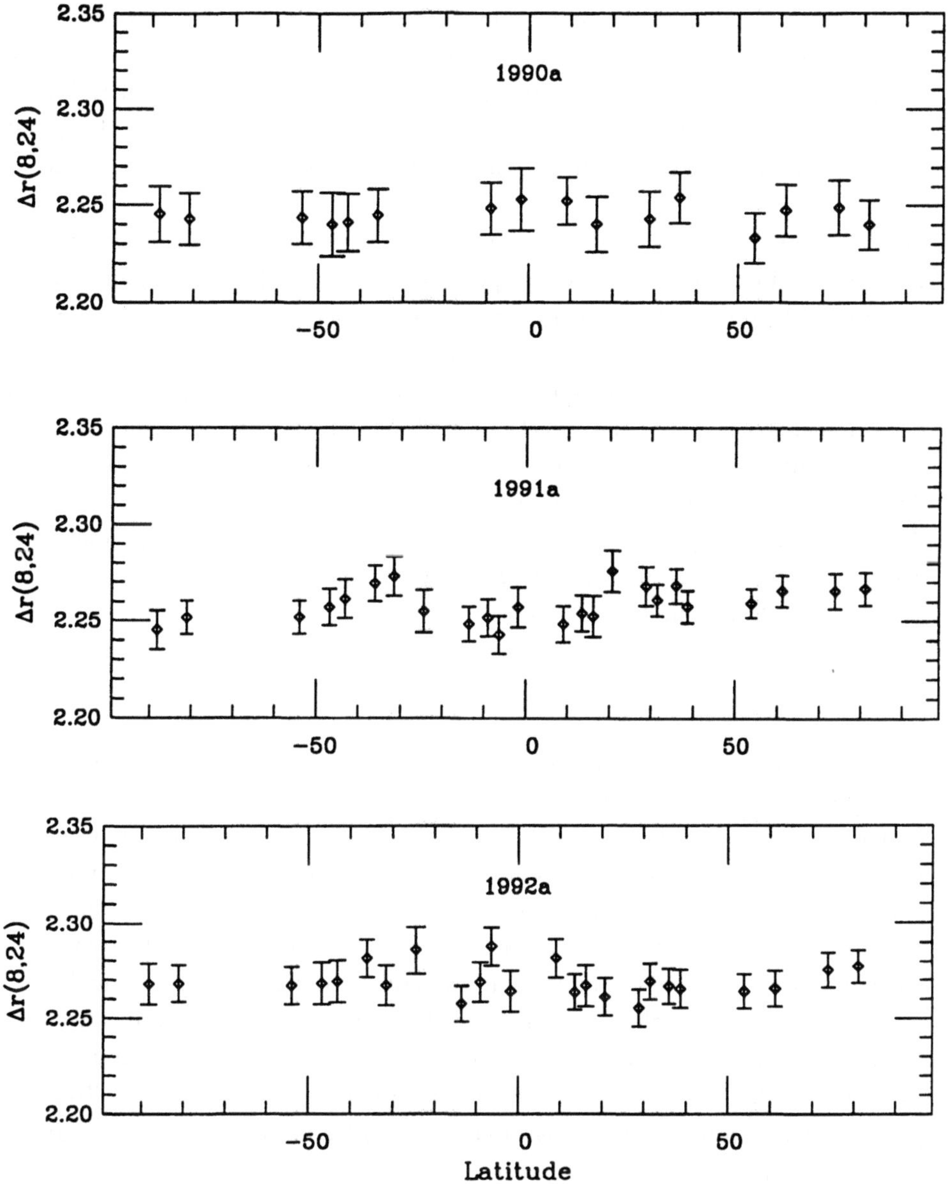

Figure 2. Differential radius $\Delta r(8, 24)$ measured at various solar latitudes for three observing seasons. Data have been corrected for offsets among the detectors, and data from slits 3 and 4 are not included.

function in a way that is consistent with the differential radius observations. These limb-darkening results, therefore, are consistent with the explanation Kuhn gives for his effective temperature results, and gives some additional weight to Kuhn and Libbrecht's (1991) total irradiance model, or alternative models, which incorporate variability of the quiet photosphere, in addition to effects of sunspots and faculae.

The observed change in the average level of the differential radius values in the 1990-1992 time period is relevant in two respects. First, this result signifies global change in the limb darkening of the quiet photosphere and perhaps an associated change in the radiative flux measured by satellite radiometers. Second, the location of the solar edge will depend on the shape of the limb-darkening function, so solar diameter measurements can show the effects of limb-darkening change. Solar diameter variability has been reported (Ribes et al. 1991), which is difficult to understand in light of our theortical understanding of physical radius changes (Spruit, 1991,1994). That limb-darkening variations could explain the changes in the apparent solar diameter without requiring a change in the actual solar radius becomes an attractive possibility (Delache and Kroll, 1994).

Conclusion

Measurements of the solar radiation intensity near the extreme solar limb reveal that from 1990 to 1992 the solar limb-darkening function was neither constant in time nor spatially uniform. In 1991 and 1992 the limb-darkening was not constant around the solar disk and the latitudinal pattern changed from one year to the next. Additionally, the average slope of the limb-darkening function changed during the 1991 to 1992 time period. These results suggest that the quiet photosphere cannot be taken to be either constant or uniform and should be considered as a potential source of total irradiance variability. The intensity profile at the solar limb affects the apparent location of the solar edge, and changes in this intensity distribution may contribute to the observed variability of the apparent solar diameter. Thus, the pursuit of such limb-darkening measurements may lead us to a better understanding of solar irradiance and diameter change.

References

Beardsley, B. J., Czarnowski, W. M., Kroll, R. K., Cornuelle, C. S., Oglesby, P. H., Yi, L., and Hill, H. A. (1988) Variations in solar luminosity since 1957 using differential radius observations. B.A.A.S. 20, No. 4: 1011.

Brown, T. M. (1988) Ground-based observations relating to the global state of the sun. 176-188 in *Solar Radiative Output Variation*, Foukal, P. (ed), Cambridge Research and Instrumentation, Inc. publication..

Delache, Ph., and Kroll, R. J. (1994) Panel discussion on solar diameter variations. This volume.

Hill, H. A., Stebbins, R. T., and Oleson, J. R. (1975) Finite Fourier Transform Definition of an edge on the solar disk. Ap. J. 200: 484-498.

Kroll, R. J., Hill, H. A., and Beardsley, B. J. (1990) A correlation between changes in solar luminosity and differential radius. 65-72 in *Climate Impact of Solar Variability*, Schatten, K. H., and Arking, A. (eds), NASA C.P. 3086,

Kroll, R. J. (1994) Latitudinal variation of the solar limb-darkening function. Solar Physics. In press.

Kuhn, J. R., Libbrecht, K. G., and Dicke, R. H. (1988) The surface temperature of the sun and changes in the solar constant. Science 242: 908-911.

Kuhn, J. R. (1991) Inferring solar structure variations from photometric and helioseismic observations. Adv. Space Res. 11 No. 4: 171-184.

Kuhn, J. R., and Libbrecht, K. G. (1991) Nonfacular solar luminosity variations. Ap. J. 381: L35-L37.

Oleson, J. R., Zanoni, C. A., Hill, H. A., Healy, A. W., Clayton, P. D., and Patz, D. L. (1974) SCLERA: an astrometric telescope for experimental relativity. Appl. Opt. 13: 206-211.

Pap, F., Willson, R. C., Fröhlich, C., Donnelly, R. F., and Puga, L. (1994) Long-term variations in total solar irradiance. Proceedings of *The Sun as a Variable Star*, June 20-25, 1993, Boulder, in press.

Petro, L. D., Foukal, P. V., Rosen, W. A., Kurucz, R. L., and Pierce, A. K. (1984) A study of solar photospheric limb-darkening variations. Ap. J. 283: 426-438.

Petro, L. D., Foukal, P. V., and Kurucz, R. L. (1985) Photospheric limb-darkening signatures of global structure variations. Solar Physics 98: 23-28.

Ribes, E., Beardsley, B., Brown, T. M., Delache, Ph., Laclare, F., Kuhn, J. R., Leister, N. V. (1991) The variability of the solar diameter. 59-97 in *The Sun in Time*, Sonnett, C. P. et al. (eds), Univ. Arizona Press, Tucson.

Spruit, H. C. (1991) Theory of luminosity and radius variations. 118-158 in *The Sun in Time*. Sonnett, C. P. et al. (eds), Univ. Arizona Press, Tucson.

Spruit, H. (1994) Theories of radius and luminosity variations. This volume.

Panel discussion on SOLAR DIAMETER VARIATIONS

Ph. Delache
Observatoire de la Côte d'Azur,
B.P. 229, F-06304, Nice Cedex 04, France
and
R. J. Kroll
University of Arizona, Dept. of Physics,
Tucson, AZ 85721, U.S.A.

1. Summary of results

1.1 Observations

The panel discussion had been to a large extent triggered by the CERGA astrolabe results. This is mostly due to the claim that variations have been detected with this instrument at a significant level, raising interest and stimulating discussions. A brief review of techniques was presented at the beginning of the session.

- The semi-diameter as a by-product of astrometric measurements.

The best data routinely obtained, now and in the past, come from measurement of transit times either through the meridian or through an elevation circle. In the first category, one finds the historical series from the Paris Observatory (La Hire 1683-1718) or from the automated meridian instrument which has been operated during several years in Boulder (Brown, 1988). The second type is the realm of astrolabes (Laclare, 1983) currently in operation at Observatoire de la Côte d'Azur, Malatya, Sao Paulo, Rio de Janeiro, San Fernando, Algiers.

NATO ASI Series, Vol. I 25
The Solar Engine and Its Influence
on Terrestrial Atmosphere and Climate
Edited by Elizabeth Nesme-Ribes

• The diameter as part of measurements of the solar limb figure.

One of the oldest interests lies in the search for the solar oblateness : for example, in classical newtonian dynamics, explaining Mercury's perihelion advance would require a substantial solar quadrupolar moment. So, as early as the XIXth century, the oblateness induced by solar rotation was an important parameter. One knows that it was already observed to be too small for explaining the Mercury phenomenon.

Later, the question of quadrupolar moment, and the search for gravity wave signatures on the solar limb have been continuously stimulating observations of the limb figure, e.g. Dicke *et al.*, (1985), Hill and Stebbins (1975), in the U.S., or more recently Rösch's heliometer project in France.

Finally, let us quote solar images already coming from outside the Earth atmosphere e.g. from the satellite Yokkoh. Sofia's solar sextant instrument (Sofia, 1993) has recently demonstrated its excellent capabilities in a balloon flight. In the near future, this instrument, or others like the solar imagers on board the SOHO mission will provide us with high precision permanent surveys of the limb figure, free from atmospheric blurring. However, the expected life times of such instruments are too short as compare to the time scales that we would like (at least one full solar cycle !). So ground based, high precision instruments will remain extremely valuable.

1.2. Variability of the radius

Before addressing the variability question, one must be sure that all the needed corrections have been made, such as the subtle effects of atmospheric refraction. For example Wittman has shown that a seasonal effect in meridional observations can be understood when differential refraction is considered in detail. See Wittman and Debarbat (1990).

Let us denote by R_{obs} the value of observed radius, calculated after having taken into account all relevant instrumental or atmospheric corrections. Still, we cannot be sure, at this step, that R_{obs} is a significant quantity. Indeed, actual observations have demonstrated the difficulties of disentangling latitudinal variations of the solar atmosphere structure and various solar blemishes from the effects of the terrestrial atmosphere. They have also occasionally shown that day to day dispersion on R_{obs} determinations can be of the order or even larger than the intrinsic instrumental uncertainty. Notice that this fact underlines the need for performing observations as continuously as possible. Unfortunately, a meridian type instrument can only observe at local noon. And once a day is probably not enough.

This is the case of the Boulder photoelectric instrument. It has shown no significant variations within its error bars. However, since astrolabe data sets, which have a better duty cycle, have shown convincing evidence of variations and correlations with other solar global parameters (Delache *et al.*, (1993a and b), it is necessary to investigate what is actually varying, which affects R_{obs}.

1.2.1. Visual observations

During the last decade, modern statistical techniques have been applied to historical data. Let us mention Gilliland, who confirms a fact which was already noticed in the XIXth century : the Sun seems smaller at the time of maximum activity. Secular trends have been discussed by Eddy and Boornazian (1979), and the unique Paris series at the time of the Maunder minimum has been studied by Ribes *et al.*, (1988).

The question of variability is best addressed when observations are homogeneous, which means that they should be made by a single observer and available over a long enough period of time (at least a solar cycle). Besides the La Hire data set, only the CERGA measurements obtained by Laclare have these important properties. A major interest of Laclare's data is due to the easy access to simultaneous determinations of other solar structural parameters such as irradiance, activity indices, p-modes frequencies, neutrino flux, etc..

Distinguishing signal from noise and assessing the significance of time series parameters such as periodicities and amplitudes has also been studied by Vigouroux and Delache (1993). They applied wavelet techniques to Laclare's time series and successfully determined radius variability.

1.2.2. Visual versus photoelectric methods

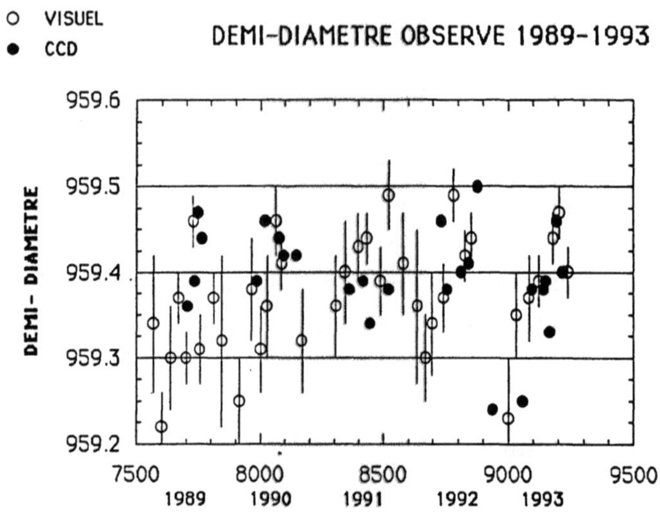

Figure **1** : Comparison of visual and CCD images determinations of the solar radius, 1989-1993. (courtesy F. Laclare)

Since 1988 Laclare has observed simultaneously with his eye and with a CCD camera. He has defined the photoelectric limb as the steepest part of the observed limb function along a portion of circle. The dispersion of individual measurements is, according to Laclare, slightly better with the photoelectric device. Results of both observations are in very good agreement, as shown in Fig. **1**. This provides a convincing argument in favor of the high quality of the past 16 years of visual data.

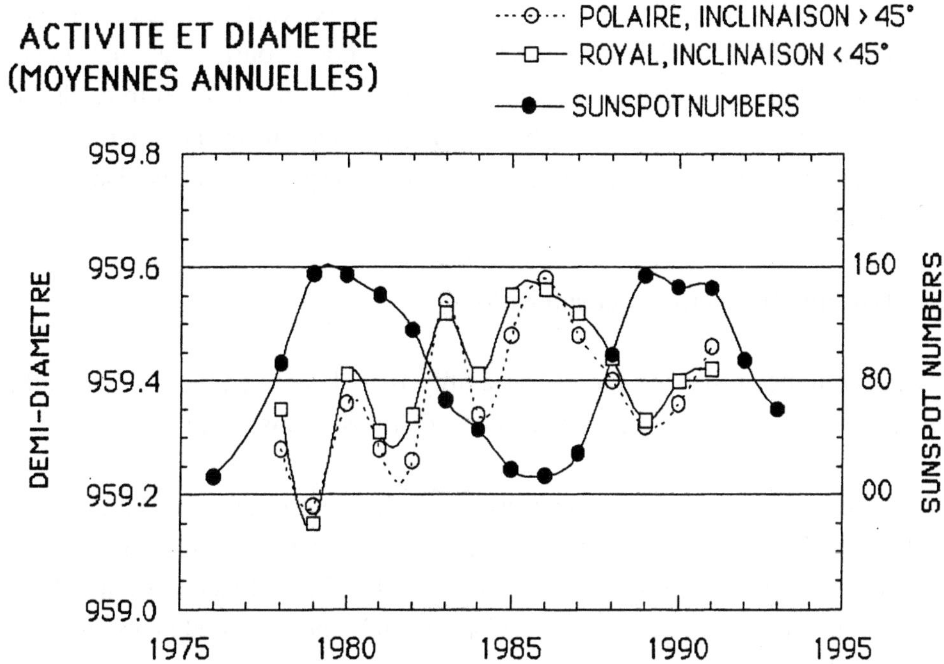

Figure **2** : Comparison of annual means of polar and radial determinations of the radius, 1975-1995. The annual sunspot number is shown for reference. (courtesy F. Laclare)

Another assessment of the reality of the observed variation has also been presented in the meeting by Laclare : Fig. **2** shows fourteen annual means of solar radius, separated in two independent series, one obtained towards the solar polar regions, the other obtained for inclination on the solar equator less than 45 degrees. For reference, the annual sunspot number is also shown in Fig. **2**.

2. Relevant theoretical considerations

Being convinced that the radius variability is real, how confident can we be that these variations reveal actual redistributions of mass in the Sun ?

Although not directly concerning people interested in Sun-climate relationship, it might be worthwhile to point out that a real measurement of radius change has profound theoretical implications. Indeed, as has been shown in Spruit's presentation (1994), significant radius changes are not expected with the standard ingredients of stellar structure : convection, magnetic fields and differential rotation. If theoretical predictions imply an upper limit of about 3 x 10^{-6} relative variation, and if the observed variations ($\delta R_{obs} \sim 10^{-5} \times R_{obs}$) correspond to real radius variations, would not we be faced with a problem potentially as serious as the neutrino discrepancy ?

The panel discussion has shown that it is indeed very difficult to escape from the schemes proposed by Spruit, even in going into magnetically inhomogeneous structure of the convection zone and upper photospheric layers. However, some recent papers, e.g. Kleeorin and Rogashevskii, (1994), present interesting contributions going in this direction. So, the rest of the discussion, quite naturally, had focused on the question of uncertainties and biases in the radius determination.

3. Origin of uncertainties

The rest of the panel session was devoted to examination of possible effects affecting the limb observational "definition" :

• From the earth atmosphere, e.g. stratospheric winds

The behaviour of the radius was suggesting a possible quasi-biennial variation, prior to 1986, that is no longer visible during recent years. It is thus likely that the correlation was of spurious origin. No recent attempt has been made to interpret the observed variability, or part of it, as induced by terrestrial atmospheric effects.

• From the solar active regions

The signature of active regions is sometimes visible as a local depression of the visual edge. The photoelectric images reveal it, and it is possible to avoid that part of the limb in the image processing. It is also detected by the observer's eye. In that case, Laclare does not make any measurement. It is thus unlikely that measurements would be seriously affected by isolated solar active regions. Furthermore, figure 2 shows quite similar variability for "royal" regions where most of the active regions lie and for polar regions where one observes very few of them. Although not completely excluded, it seems unlikely that active regions are responsible for the observed variations.

• From variations of the surface brightness or of the limb darkening function

Kuhn and Libbrecht (1985) have measured significant variability of the effective limb brightness from 1983 to 1990. The variation with solar latitude is strong, and highly dependent on the solar cycle phase. If there would be a direct effect on astrolabe determination, it should most likely show up as a significant difference between the two royal/polar yearly radius series, which is not observed.

Out of the discussion it appeared that the variation of the limb darkening function obtained in 1990, 1991 and 1992 by Kroll (1994) may provide a very interesting clue. Indeed, Kroll finds (see Figure 2 in his paper in this volume) a slight latitudinal variation on top of a significant change of the overall level of his differential radius Δr. Since this parameter is related to the steepness of the limb darkening function around the place where one locates the solar edge, it seems promising to investigate further along this direction.

- Variability of the solar granules horizontal sizes

This point was raised in the discussion and is worth mentioning. The fact that granules are significantly smaller at times of solar maximum activity is well documented but not adequately explained by standard theories of the convective zone. Should the vertical size be affected the same way as the horizontal size, while the granules were sitting on top of a constant radius surface, then the 11-year radius variability would be explained at the correct order of magnitude.

Conclusion

A brief review of planned missions, already quoted above, is made before the following tentative conclusions are proposed :

IF :
R_{obs} variability is entirely, or partially, of solar origin, as we think it is ,

AND IF :
solar surface structures, e.g. active regions, do not play any significant role,

AND IF :
limb darkening function and solar granulation would not suffice to explain the observations (which is not ruled out),

SINCE :
R_{obs} variability cannot be interpreted in the frame of classical solar modeling,

THEN :
correlations between diameter variations, p-mode frequency changes and neutrino production rates could suggest some direct connection between short term surface phenomena and the conditions in the core of the Sun affecting nuclear energy generation. But then a challenging question immediately arises : what mechanisms could adjust the core conditions on such short (11 year) timescales ?

References

Brown, T.M. (1988). Solar diameter measurements. Bull. Amer. Astron. Soc., 20: 376

Delache, Ph., Laclare, F., Gavryusev, V., Gavryuseva, E. (1993a). Signal and noise in long duration solar radius observations. Mem. S. A. It., 64: 237-243

Delache, Ph., Gavryusev, V., Gavryuseva, E., Laclare, F., Regulo, C., and Roca Cortes, T. (1993b). Time correlation between solar structural parameters : p-mode frequencies, radius, and neutrino flux. Astrophys. J., 407: 801-805

Dicke, R.H., Kuhn, J.R., and Libbrecht, K.G. (1985). Oblateness of the Sun in 1983 and relativity. Nature, 316: 687-690

Eddy, J.A., and Boornazian, A.A. (1979). Secular decrease in the solar diameter, 1863-1953. Bull. Amer. Astron. Soc., 11: 437 (abstract)

Hill, H.A., and Stebbins, R.T. (1975). The intrinsic visual oblateness of the Sun. Astrophys. J., 200: 471-483

Kleeorin, N., and Rogachevskii, I. (1994). A mechanism of 11-year oscillations of solar radius. Solar Phys., in press

Kroll, R.J. (1994). The observed spatial non-uniformity and temporal variability of the solar limb-darkening function. This volume

Kuhn, J.R., and Libbrecht, K.G. (1985). Observations of a solar latitude dependent limb brightness variation. Astrophys. J., 290: 758-764

Laclare, F. (1983). Mesures du diamètre solaire à l'astrolabe. Astron. Astrophys., 125: 200-203

La Hire, Ph., 1683-1718, Manuscripts D2.1-10, Archives Obs. Paris

Ribes, E., Ribes, J.C., and Barthalot, R. (1988). Size of the Sun in the seventeenth century. Nature, 332: 689-690

Sofia, S. (1993). Climate implications of the EUDOSSO experiment. Mem. S. A. It., 64: 207-220

Spruit, H. (1994). Theories of radius and luminosity variations. This volume

Vigouroux, A., and Delache, Ph. (1993). Fourier versus wavelet analysis of solar diameter variability. Astron. Astrophys., 278: 607-616

Wittman, A.D., and Debarbat, S. (1990). Der Sonnendurchmesser und Seine Variabilitat, Sterne und Weltraum, 7-8: 420-426

ATMOSPHERIC ^{14}C AS A PROXY OF SOLAR AND CLIMATIC CHANGE

Minze Stuiver
Quaternary Research Center and Department of Geology
University of Washington AK-60
Seattle WA 98195
USA

Summary

The complexity of the atmospheric $\Delta^{14}C$ record, as well as its relevance to the Sun's heliomagnetic modulation of the cosmic ray flux, is discussed. An overriding role of the Sun is unlikely as oceanic thermohaline circulation variability plays an important part as well. Further augmentation of our knowledge of century type solar variability (which is now limited to the past 11,400 yr due to sample availability) will have to come from ^{10}Be concentration measurements in ice cores.

Introduction

The atmospheric ^{14}C record contains important cosmic ray flux information. The record also is subject to climatic and geomagnetic influences. Although, as a first order, the century type variability of the record is usually considered to be all of cosmic (heliomagnetic) ray origin, climate and ocean variability may play a role as well. The evaluation of climatic and oceanic "noise" is an important part of this discussion because reliable heliomagnetic information depends on the removal of extraneous contributions.

Two primordial stable carbon isotopes, ^{12}C and ^{13}C, exist on earth. Primordial ^{14}C is not present anymore as it has a relatively short half life of 5730 years. Yet ^{14}C exists in nature because cosmic radiation from galactic sources produce neutrons in the atmosphere, which in turn interact with nitrogen to produce ^{14}C. Some ^{14}C is produced in trace amounts by the interaction of

NATO ASI Series, Vol. I 25
The Solar Engine and Its Influence
on Terrestrial Atmosphere and Climate
Edited by Elizabeth Nesme-Ribes

cosmic radiation with earth 'crustal' materials (e.g. basalt or polar ice) but these amounts are negligible compared to those produced in the atmosphere. Crustal ^{14}C also plays a minimal role in the geologically rapid carbon exchange between the atmosphere, oceans and biosphere (the carbon cycle) because, relative to its half life, ^{14}C is released very slowly, if at all, from crustal rock.

As the amount of ^{14}C produced by cosmic radiation is less than 100 kg per year, the $^{14}C/^{12}C$ ratio (~ 10^{-12}) of carbon participating in the carbon cycle is minute compared to the primordial $^{13}C/^{12}C$ ratio (about 1.12×10^{-2}). The carbon encountered in the atmosphere, oceans and biosphere also looses ^{14}C through radioactive decay at a rate of 1% every 83 years. Yet, although the $^{14}C/^{12}C$ ratio diminishes with time for a sample separated from newly produced ^{14}C (the basis for radiocarbon dating), the $^{14}C/^{12}C$ ratio in the earth system is determined (at least for steady state as a first order approximation) by an equilibrium between radioactive decay and cosmic ray ^{14}C production. Thus the total amount of cosmogenic ^{14}C on earth, as a first order approximation, equals 8300 years of production. The oceans are by far the largest reservoirs of exchangeable ^{14}C, with the atmosphere and the biosphere containing a little more than 1% each (about 100 years of ^{14}C production) of the total. Because the atmospheric ^{14}C reservoir is much smaller than the oceanic one a perturbation in the transfer of ^{14}C from the oceans to the atmosphere can have major consequences for atmospheric $^{14}C/^{12}C$ ratios, even when ^{14}C production stays constant.

The ^{14}C production in the atmosphere depends on cosmic ray intensity which, in turn, is influenced by the degree of change in solar wind magnetic properties and geomagnetic dipole intensity. The total amount of cosmogenic ^{14}C in the earth's exchangeable carbon reservoirs is defined by these parameters (external "forcing factors"). What is actually measured in samples, however, is not the globally integrated amount of ^{14}C but the $^{14}C/^{12}C$ ratio of carbon incorporated in specific reservoirs (e.g. the atmosphere or the ocean). Because changes in exchange rate and other physical properties (such as ocean circulation) can influence the ^{14}C distribution among the reservoirs, the atmospheric $^{14}C/^{12}C$ ratio depends not only on cosmic ray flux changes but also on mechanisms that cause ^{14}C re-distribution. Here climatic change (an internal "forcing factor") is the leading candidate for change. A discussion of the influence of climate change on carbon reservoir parameters requires a fair amount of simplification because, for instance, past physical changes in ocean carbon reservoir properties have rarely been measured directly. Changes that can be tied to climate are thermohaline circulation ("conveyer belt") change of the oceans and wind speed changes that influence the CO_2 exchange between the oceans and the atmosphere.

The $\Delta^{14}C$ record and geomagnetism

Past atmospheric (or oceanic) ^{14}C levels can be derived from materials that were initially in ^{14}C equilibrium with those reservoirs. Plants, through photosynthesis, are in isotopic equilibrium (usually with a fixed offset that can be corrected for by normalizing on a fixed stable carbon isotope ratio) with atmospheric (or oceanic) $^{14}CO_2$. The ^{14}C activity currently remaining in the sample can be measured in the laboratory. Paleoatmospheric ^{14}C activity can be calculated from such measurements by correcting for the radioactive decay (half life $T_{1/2}$ = 5730 yr) that took place since the death of the organism. This, of course, implies a precise age determination independent of the ^{14}C activity observation. The preferred methods for calibrated (cal) age determinations involve dendrochronology ("tree-ring counting") and ^{234}Uranium - ^{230}Thorium ratio estimations of corals. Here dendrochronology (with "zero" error claims) may yield a very precise time scale whereas Uranium/Thorium high precision cal ages (Edwards et al., 1993, Bard et al., 1993) have uncertainties approaching 1% around 10,000 cal yr BP (Before Present, 0 yr BP = AD 1950). The decay-corrected measured ^{14}C activity of the wood, after normalization on a fixed $^{13}C/^{12}C$ ratio to avoid biological fractionation, is expressed as a relative deviation $\Delta^{14}C$ (in percent or per mil) from the National Bureau of Standards (NBS) oxalic acid ^{14}C activity (Stuiver and Polach, 1977). The NBS standard approximates late 19 th century atmospheric $^{14}C/^{12}C$ ratios, and thus the $\Delta^{14}C$ term is based on a comparison of past atmospheric $^{14}C/^{12}C$ values to those near the end of the last century. For high-precision ^{14}C dating the calculated $\Delta^{14}C$ values will have an uncertainty (one standard deviation) of 2 - 4 per mil, with the larger uncertainty for sample ages near 10,000 cal year BP.

Wood $\Delta^{14}C$ values represent a well mixed global atmosphere whereas coral $\Delta^{14}C$ values usually yield a proxy record of $\Delta^{14}C$ change of the local (regional) mixed surface layer of the ocean. These records differ in detail because oceanic ^{14}C variations, when caused by production rate changes in the atmosphere, will be smoothed relative to those in the atmosphere. The ^{14}C content of the global surface ocean also is lower than the atmospheric one because upwelling deep ocean water that is deficient in ^{14}C (due to radioactive decay) lowers surface ^{14}C. Atmospheric $\Delta^{14}C$ values can be derived from coral records, but the correction for regional ^{14}C reservoir deficiency of the surface ocean complicates matters. The reservoir ^{14}C time dependency is usually unknown and approximate atmospheric data can only be obtained either by assuming a constant reservoir deficiency, or by calculating its time dependency.

The derivation of atmospheric and oceanic $\Delta^{14}C$ data sets has been, and is, a multi-laboratory (e.g. Belfast, Heidelberg, Hohenheim, Groningen, La Jolla, Seattle and Tucson) effort. Compilations of the wood ^{14}C results have yielded a bi-decadal $\Delta^{14}C$ data set spanning 11,390 cal yr, a decadal data set of 7950 cal yr, and a single year data set covering the AD 1510 - 1954 interval. The multi-laboratory effort and the computerized compilation of the bi-decadal and decadal atmospheric $\Delta^{14}C$ data sets (Stuiver and Reimer, 1993) are described in the Calibration 1993 issue of the journal Radiocarbon whereas the single year $\Delta^{14}C$ data has been reported elsewhere (Stuiver and Braziunas, 1993a). Beyond 11,390 cal yr BP atmospheric $\Delta^{14}C$ values are estimated by using coral ^{14}C data and deducting a fixed (400 ^{14}C yr) reservoir age (Bard et al, 1993) from the coral ^{14}C results. Because only a limited number of coral measurements is available the finer detail of the $\Delta^{14}C$ curve (one wood $\Delta^{14}C$ determination every twenty year for the 0 - 11390 cal yr interval) diminishes drastically beyond 11,390 cal yr BP.

A noticeable feature of the atmospheric record (Fig. 1) is the long-term $\Delta^{14}C$ decline over the past 25,000 cal yr. This decline is usually attributed to an increase of geomagnetic dipole intensity (and thus a decrease in atmospheric ^{14}C production) when advancing towards the present (Bard et al., 1990, Stuiver at al., 1991, Mazaud et al., 1991). Part of the long-term trend may also relate to changes in ocean circulation. Major postulated changes, such as a full turn off of deep water formation during the pre-Holocene portion, generate very large atmospheric $\Delta^{14}C$ changes in global carbon reservoir models (examples of some calculations are a 130 $^{o}/oo$ transient increase for a 1300 year cessation of deep water formation, or a 280 $^{o}/oo$ increase for an infinitely long steady sate switch off (Stuiver and Braziunas, 1993b). The same postulated changes, however, also strongly influence the ^{14}C age differences between the deep and surface ocean. These age differences can be derived from benthic and planktonic (b-p) foraminifera found in marine sediment (e.g. Andree et al., 1986, Shackleton et al., 1988).

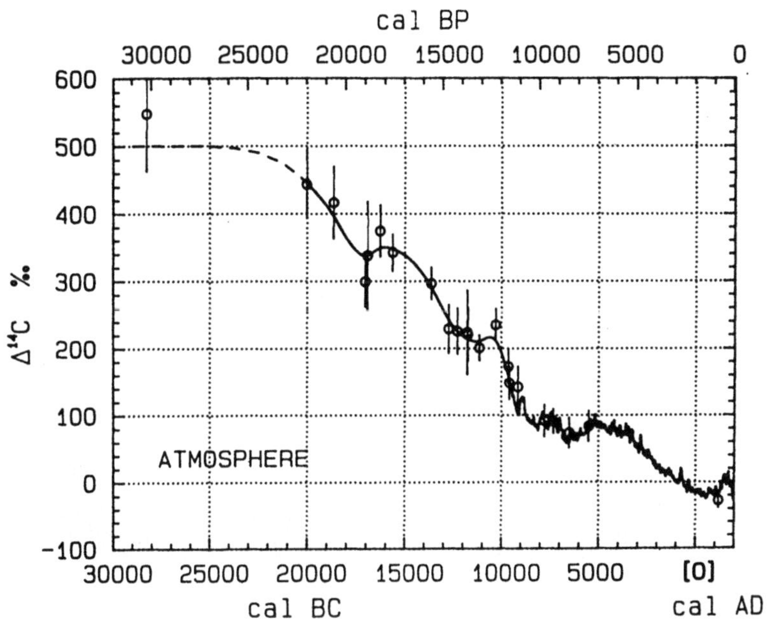

Fig. 1: The atmospheric $\Delta^{14}C$ record (Stuiver and Reimer, 1993, Bard et al., 1993) as derived from dendrochronologically calibrated bi-decadal tree-ring ^{14}C measurements (11,390 cal BP to 0 cal BP) and a cubic spline through U/Th calibrated coral ^{14}C averages corrected by 400 ^{14}C year. The coral measurements are shown as circles with 2σ error bars.

The b-p record, although variable, can not accommodate the large b-p changes required for a full turn off of deep water formation (calculated b-p age differences range from about 4000 yr for the purely "diffusive" ocean (eddy diffusivity = 0.2 cm^2/s) to a more acceptable 1000 yr for the same ocean with the addition of substantial rates (20 Sverdrups or 1.6 m/yr) of deep water formation). Therefore most of the long-term atmospheric $\Delta^{14}C$ trend has to be allocated to geomagnetic dipole change. As a first approximation the putative geomagnetic component of the $\Delta^{14}C$ record has been removed in the Fig. 2 residual record by deducting the Fig. 1 spline. The downward swing of $\Delta^{14}C$ prior to 11,200 cal yr BP (see Figure 2) may well relate to increased deep water formation (with its corollary of increased upwelling of ^{14}C deficient older waters) during the second half of the Younger Dryas interval (Stuiver and Braziunas, 1993a,b, 1994, Edwards et al., 1993).

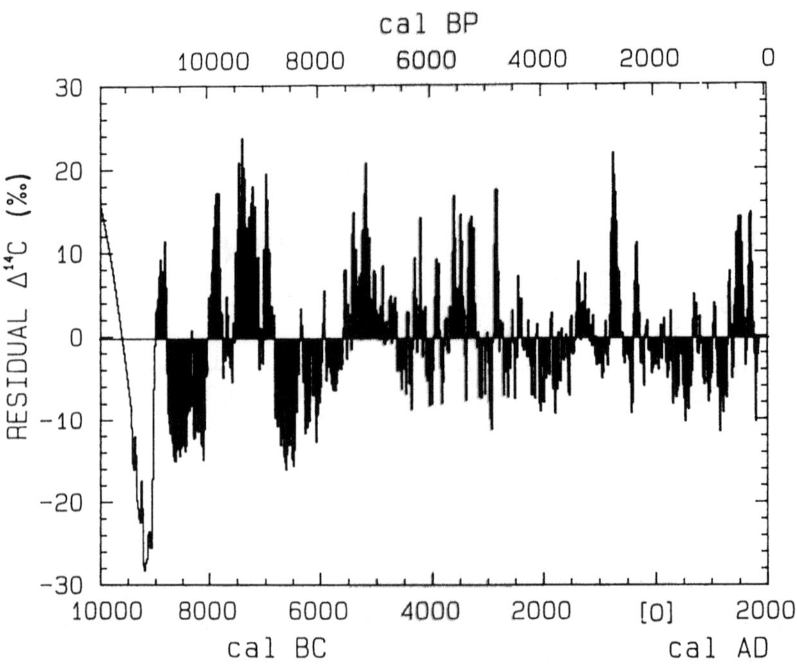

Fig. 2. Residual $\Delta^{14}C$ obtained by deducting the Figure 1 long-term trend. The oldest oscillation, which is partly (first 500 cal yr) derived from coral measurements has not been shaded. The remaining $\Delta^{14}C$ values were derived from tree-ring measurements of wood older than AD 1850. The early 18th - late 17th century $\Delta^{14}C$ maximum is associated with the Maunder sunspot minimum.

Solar and climate forcing of the bi-decadal $\Delta^{14}C$ record

The Fourier power spectrum of the Figure 2 residual $\Delta^{14}C$ record has periods exceeding the 2σ significance level at 512 yr, 206 yr, 148 yr, 87 yr and 46 - 49 yr (Stuiver and Braziunas, 1993a). A longer periodicity is tied to a 2200 - 2600 yr cycle discussed by Hood and Jirikowic (1990) for a 7200 yr long $\Delta^{14}C$ record. For the longer Figure 2 record this cycle is slightly below the 2σ significance level, and the period increases by a few hundred years.

The spectral power of the 512 yr cycle is concentrated in the earliest portion of the tree-ring record. This cycle can be tied to oceanic forcing (Stuiver and Braziunas, 1993a). The postulated mechanism for oceanic $\Delta^{14}C$ change is via the (unstable) salt oscillator in the North Atlantic where salinity lowering results in increased water stratification and reduced deep water formation. Reduced deep water formation in turn parallels reduced upwelling of ^{14}C deficient water elsewhere in the oceans thus increasing surface ocean and atmospheric $\Delta^{14}C$ levels.

Solar modulation of the cosmic ray flux must play a crucial role in the $\Delta^{14}C$ variance associated with the 208, 148 and 87 yr periodicities (e.g. Stuiver and Quay, 1980, Stuiver and Braziunas, 1989). The century type ^{10}Be oscillations in ice cores, which are mostly contemporary with the $\Delta^{14}C$ oscillations, also tend to favor this conclusion (Beer et al, 1988a, b, 1991). Yet, although it is unlikely that oceanic influences dominate this part of the spectrum, the possibility of an incremental oceanic contribution can not be dismissed out of hand. Increased oceanic circulation variance in the 320 yr range has been observed in the Hamburg Ocean General Circulation model when driven by an Atlantic white noise freshwater flux superimposed on climatological fluxes (Mikolajewicz and Maier-Reimer, 1990). Dr Martin Heimann, working with the same model but considering biospheric influences as well, noted at the 1993 Global Change Institute "The Carbon Cycle" (Snowmass, Colorado, 19-30 July, 1993) that the corresponding atmospheric $\Delta^{14}C$ changes of these century type model changes approach 10 ‰ (peak to trough). Thus, even though the peak to trough changes of the Figure 2 century type oscillations are more like 20 ‰, a portion of the $\Delta^{14}C$ variance may well relate to internal resonance of the ocean circulation system.

The dependence of cosmogenic ^{14}C production on solar wind properties (with increased ^{14}C production tied to lower sunspot numbers) is well established for the 11 yr sunspot cycle (Lingenfelter, 1963, O'Brien, 1979). A longer term production rate change evidently contributed to the late 17th century $\Delta^{14}C$ maximum of Figure 2. Apart for a small time lag, this Maunder $\Delta^{14}C$ maximum is contemporary with the AD 1650-1715 Maunder minimum (Eddy, 1978) in sunspot numbers. By using the 20th century observed relationship between the 11 yr cycles in sunspot numbers and atmospheric neutron production rate (which in turn is proportional to ^{14}C production rate) the changes in ^{14}C production rate during the Maunder Minimum can be calculated. This calculation leads to a Maunder $\Delta^{14}C$ increase of about half the observed one when assigning zero sunspot numbers to the entire 65 year interval (Stuiver and Quay, 1980). The Maunder $\Delta^{14}C$ amplitude discrepancy can be removed, however, when postulated long-term trends in solar modulation are taken into account (Stuiver and Quay, 1980). Solar modulation considerations of Lal (1992), however, would limit the maximum ^{14}C production rate change during the Maunder minimum to about 2/3 the increase needed to generate the observed $\Delta^{14}C$ augmentation in carbon reservoir models. Evidently solar modulation of the cosmic ray flux explains at least one half of the Maunder $\Delta^{14}C$ increase. The remaining half can either be attributed to additional cosmic ray flux increases when sunspots are absent (Stuiver and Quay, 1980), or to oceanic influences. For the latter scenario the solar signal would have been augmented by a synchronous oceanic response. Thus, if climate related ocean circulation change plays a role, it must have enhanced the Maunder $\Delta^{14}C$ signal. Here a postulated mechanism is reduced upwelling (and deep water formation) during episodes of climatic change induced by lower solar activity. This mechanism has been explored by Stuiver and Braziunas (1993a) as an alternative to the 100% solar origin for the Maunder $\Delta^{14}C$ maximum. This modification entails a hypothetical mechanism explaining postulated climate - ^{14}C relationships in which minor solar induced Maunder Minimum climate change is amplified by salinity effects on North Atlantic thermohaline circulation.

Single year $\Delta^{14}C$ variance

A nearly 450 year long $\Delta^{14}C$ record, derived from individual tree-rings, has been described by Stuiver and Braziunas (1993a). Figure 3 reproduces the results, with a detailed structure of the record given in the lower curve (3 yr moving average, with a 15 ‰ offset). The data series contain, in addition to random noise, cyclic components that can be spectrally analysed. Periodicities in a MEM (Maximum Entropy Method) power spectrum for the AD 1510 - 1954 interval are at 56, 26, 17, 13, 10.4, 6.4, 5.7, 3.5, and 2.7 - 2.1 yr (periodicities exceeding 80 yr have been omitted as they have been discussed in the previous section).

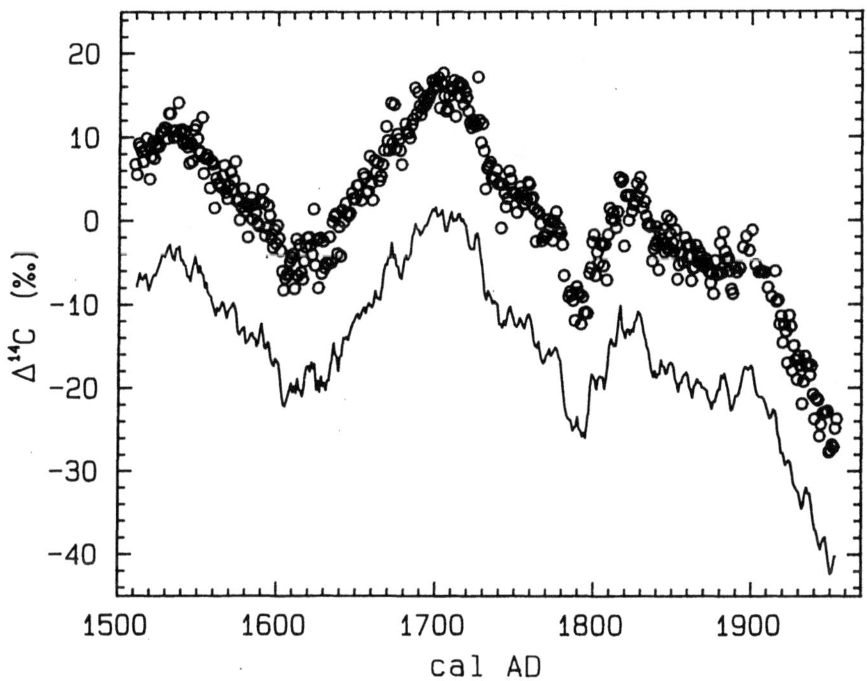

Fig. 3. Atmospheric $\Delta^{14}C$ in per mil (circles), as derived from single rings of trees grown in the Pacific Northwest. The fine structure of the data set is given in a 3 yr moving average (lower curve, offset by 15 ‰). The calibrated (cal) ages are based on ring counting. Approximate standard deviation in the measurements is 2 ‰.

Solar influence on ^{14}C production rate (11 yr sunspot cycle) relates to the 10.4 yr, and perhaps to some of the longer periodicities. Sunspot numbers are compared to a filtered (5 - 20 yr period spilt cosine bandpass) $\Delta^{14}C$ time series in Figure 4. In Figure 4 the $\Delta^{14}C$ cal ages have been reduced by 4 yr because the best match between inverted $\Delta^{14}C$ and sunspots is obtained for a 4 yr lag.

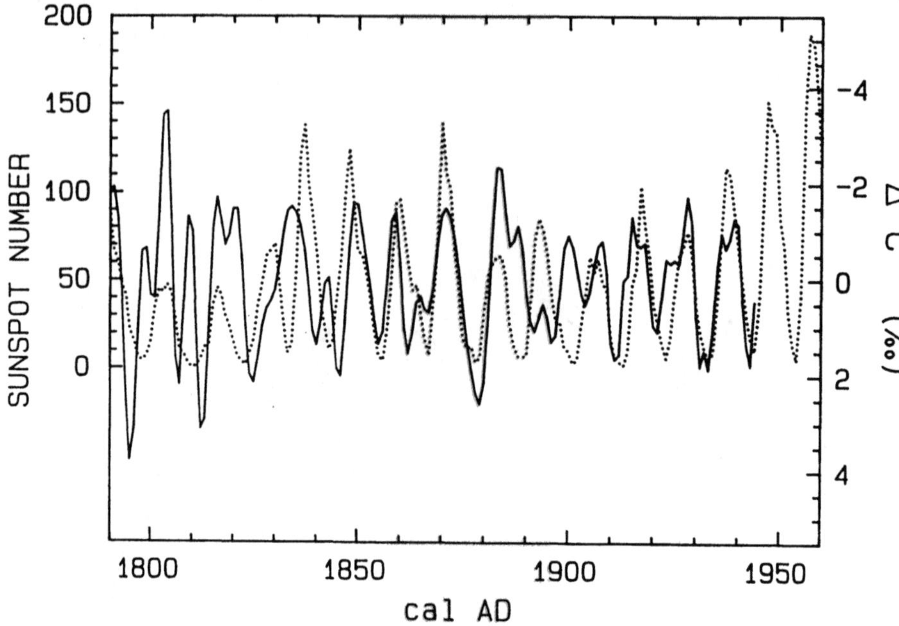

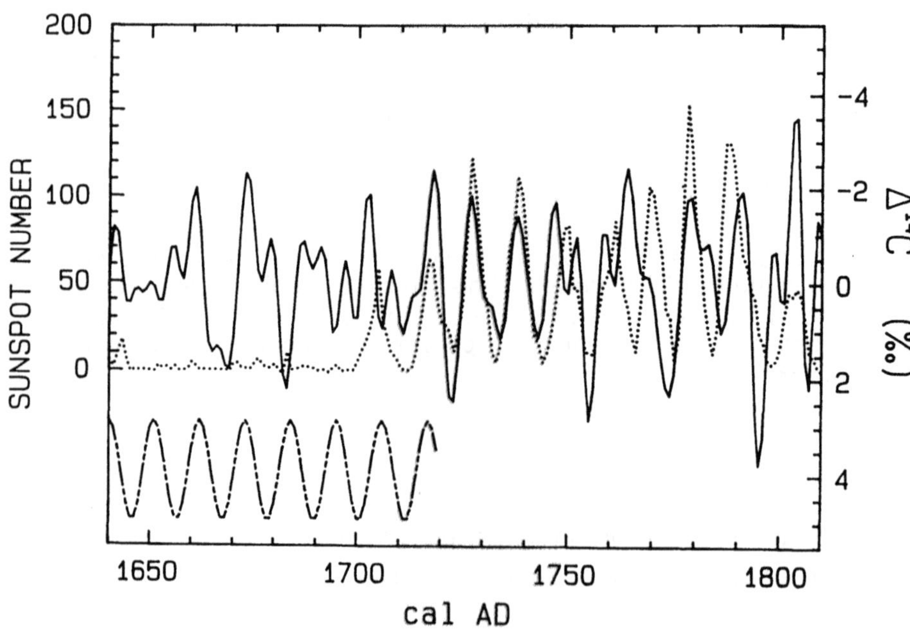

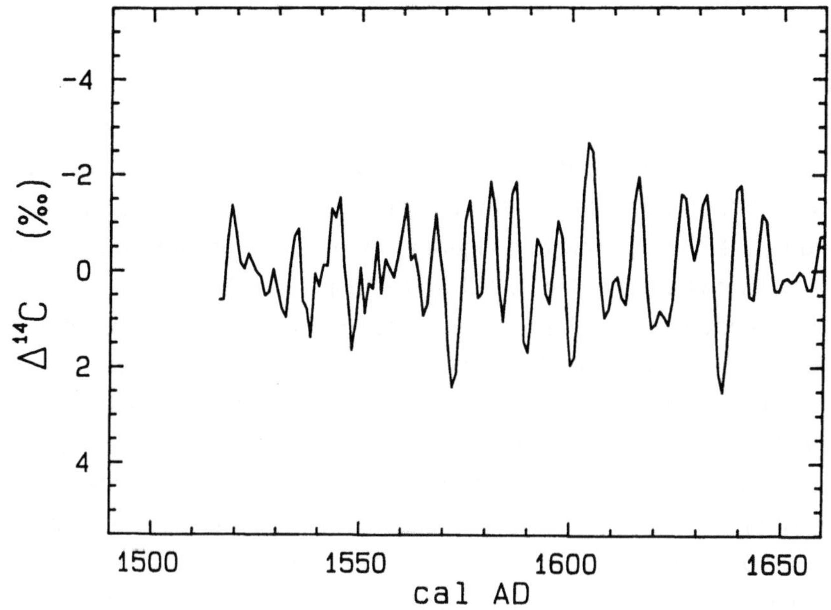

Figure 4 a, b and c. Filtered (5 - 20 yr bandpass) $\Delta^{14}C$ signal (solid line) derived from the Figure 3 data. The inverted $\Delta^{14}C$ signal is compared to unfiltered sunspot numbers (dotted line). For comparison purposes the $\Delta^{14}C$ cal ages were reduced by 4 yr (see text). The vertical scales were arranged such that the post- AD 1715 means (and variances) of the respective time series plot equivalently. An artificial 11 yr cycle (lower dashed curve) was generated for the Maunder Minimum (AD 1650 - 1715). The 4 yr cal age shift was not applied in Figure 4c above.

Linear cross correlation between AD 1715 - 1984 sunspot numbers and the Figure 4 filtered $\Delta^{14}C$ signal (with cal ages reduced by 4 yr) yields a correlation coefficient r = -0.39. This r value corresponds to a 0.5% probability of the parameters not being correlated when autocorrelation is taken into account. A 3 yr lag leads to an r value of - 0.32 with a 2.3% probability of not being correlated. Such a lag is close to the 2.5 yr lag calculated for 11 yr production rate changes in carbon reservoir models (Houtermans et al, 1973, Braziunas, 1990).

A perfect agreement between $\Delta^{14}C$ and sunspot numbers can not be expected because the precision of the individual $\Delta^{14}C$ determinations approximates the amplitude of the investigated signal (average amplitude of the 10.4 yr signal is 1.4 ‰). Furthermore, there is the potential generation of atmospheric ^{14}C (creating inverse $\Delta^{14}C$ minima in Figure 3) by extremely energetic solar flares. Agreement between inverted $\Delta^{14}C$ and sunspot numbers can be seen for several individual cycles, and, even though the low sunspot numbers associated with the Maunder Minimum suggest minimal 11 yr modulation, the magnitude of the filtered $\Delta^{14}C$ oscillations during this interval does not differ appreciably from those observed for the preceding and following years. Although this (and the comparison with the artificial 11 yr cycle) is very suggestive of continuing 11 yr cycle modulation during the Maunder minimum, absolute proof is elusive as only a limited portion of the $\Delta^{14}C$ variance (16%(r=-0.39) for the AD 1715 - 1948 interval) relates directly to sunspot numbers.

Cosmogenic isotope production rate change can also be monitored by ^{10}Be concentrations in ice. The 11 yr production rate variations for both ^{14}C and ^{10}Be are relatively large (in the range of 25%, peak to trough), but the ^{14}C signal recorded in tree-rings is strongly attenuated by the large atmospheric reservoir containing about 100 yr of ^{14}C production. In our global carbon reservoir model (Braziunas, 1990, Stuiver and Braziunas, 1988) the peak to trough change in atmospheric $\Delta^{14}C$ is 2.0 ‰ for such 11 yr changes which compares favorably with the 2.8 ‰ variations found in the filtered $\Delta^{14}C$ signal (Figure 4). Production rate changes for ^{10}Be, with an atmospheric residence time of 1 to 2 yr, are much less attenuated and the ^{10}Be concentration of polar ice (Dye 3, Greenland) has 11 yr concentration changes in the 15 to 25% range (Beer et al., 1990). These production rate changes are compared in Figure 5 to the ^{14}C production rate changes (Q), calculated from the Figure 3 $\Delta^{14}C$ record through carbon reservoir modeling. Variations are found to be approximately equal for measured ^{10}Be and calculated ^{14}C production Q, as would be expected for a common (solar modulation of the cosmic ray flux) origin. However, the Q - ^{10}Be correlation coefficient does not exceed 0.42, a value reached when ^{10}Be lags Q by 1.5 yr. The lag is fully compatible with the 1 to 2 yr atmospheric residence time of ^{10}Be. Nevertheless, an r value of 0.42, although confirming a common denominator in both records, allows for only

18% of the variance in the annual records to be tied to cosmogenic production rate change. Other variables, such as climate and continental dust, evidently play a major role in either the ^{10}Be or ^{14}C variance, or both.

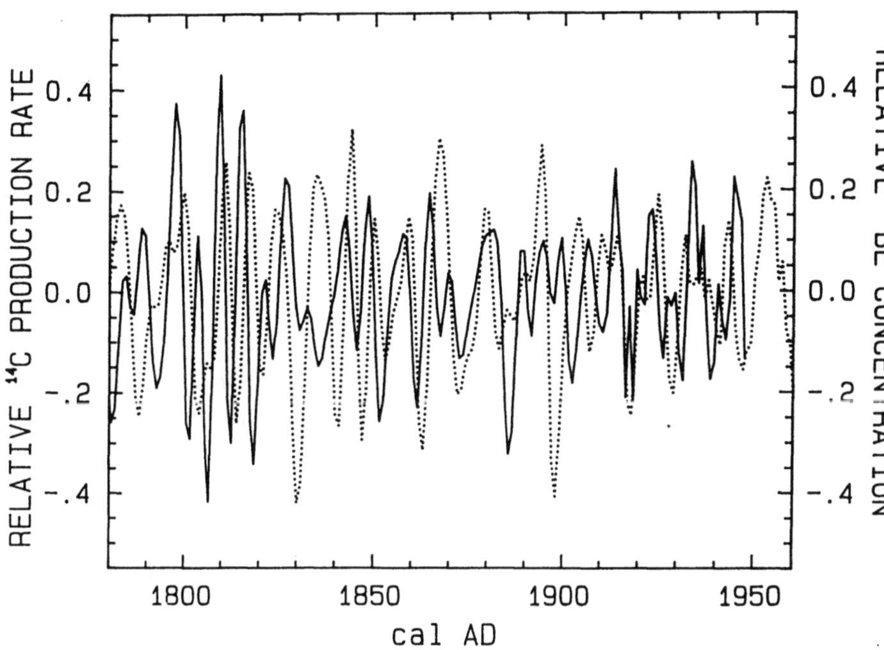

Fig. 5. Relative ^{14}C production rate Q (solid line), calculated from the Figure 3 $\Delta^{14}C$ record through carbon reservoir modeling, compared to observed (no lag was applied) ^{10}Be concentration in polar ice (Beer et al, 1990). Each time series is plotted relative to its mean value.

^{10}Be concentrations of the aerosols found in atmospheric precipitation show a distinct latitudinal gradient. Thus the climatic influences on the ^{10}Be concentration record in ice will be related to air mass transport mixing changes (and continental dust transport). Climatic influences on the $\Delta^{14}C$ record include variable rates of upwelling associated with EL Nino - Southern Oscillation (ENSO) events as well as North Atlantic thermohaline circulation change (Stuiver and Braziunas, 1993a). The 2 to 6.4 yr periodicities in the $\Delta^{14}C$ record correspond in duration to the irregular cycle of ENSO events (Quinn and Neal,

1992) whereas irregular thermohaline oscillations of approximately 40 - 50 yr were noted by Delworth et al. (1993) in a coupled ocean atmosphere model. And an 18.6 year cycle of air pressure, tied to lunar-solar tides (O'Brien and Currie, 1993), is sufficiently close to the ~17 yr periodicity in the $\Delta^{14}C$ record to suggest a possible causal relationship.

Causal diversity

Solar modulation of cosmic rays and climate induced (or related) oceanic changes are major players in generating atmospheric $\Delta^{14}C$ variance. The oceans play a role in generating a damped 512 yr oscillation in the bi-decadal $\Delta^{14}C$ record during the early Holocene, and also may generate annual $\Delta^{14}C$ variance on ENSO time scales. Internal oceanic oscillations and associated thermohaline circulation changes even may be responsible for some of the century type $\Delta^{14}C$ variability. Yet, despite this oceanic imprint, evidence for a strong solar component in the $\Delta^{14}C$ record is formidable. Century type $\Delta^{14}C$ and ^{10}Be variability are highly correlated, pointing to a common cosmic origin. The Maunder minimum $\Delta^{14}C$ increase coincides closely with the Maunder Minimum in sunspot numbers, and heliomagnetic modulation on an 11 yr time scale is part of the $\Delta^{14}C$ record. The existence of many harmonics in the $\Delta^{14}C$ record (Stuiver and Braziunas, 1989, Damon and Sonett, 1992) also suggests major contributions of a single body, most likely the Sun, with substantial internal oscillations.

A complicating factor are the suggestions (e.g. Sonett and Suess, 1984, Wigley and Kelly, 1990, Anderson, 1992, Taylor et al., 1992, Magny, 1993, Cook et al., 1992, 1993 and Scuderi, 1993) of other broader relationships between climate and $\Delta^{14}C$ such as a joint solar - climate effect on $\Delta^{14}C$. If such close relationships exist they influence conceptual ideas underlying climate and $\Delta^{14}C$ forcing. Amplification of a weak solar radiation signal is needed to explain such correlations, perhaps involving UV radiation and North Atlantic thermohaline circulation change (Stuiver and Braziunas, 1993a). Such a hypothetical mechanism, for example, could lead to an oceanic contribution towards a solar Maunder minimum type atmospheric $\Delta^{14}C$ change of up to one third to one half of the ultimate $\Delta^{14}C$ signal.

Acknowledgements

The ^{14}C research of the Quaternary Isotope Laboratory is supported by National Science Foundation grant ATM-9310121.

References:

Anderson RY (1992) Possible connection between surface winds, solar activity and the Earth's magnetic field. Nature 358: 51-53.

Andrée M, Beer J, Loetscher HP, Moor E, Oeschger H, Bonani G, Hofmann HJ, Suter M, Woelfli W, and Peng TH (1986) Limits on the ventilation rate for the deep ocean over the last 12000 years. Climate Dynamics 1: 53-62.

Bard E, Arnold M, Fairbanks RG, and Hamelin B (1993) ^{230}Th-^{234}U and ^{14}C Ages Obtained by Mass Spectrometry on Corals. Radiocarbon 35: 191-199.

Bard E, Hamelin B, Fairbanks RG, and Zindler A (1990) Calibration of the ^{14}C timescale over the past 30,000 years using mass spectrometric U-Th ages from Barbados corals. Nature 345: 405-410.

Beer J, Blinov A, Bonani G, Finkel RC, Hofmann HJ, Lehmann B, Oeschger H, Sigg A, Schwander J, Staffelbach T, Stauffer B, Suter M and Wölfli W (1990) Use of ^{10}Be in polar ice to trace the 11-year cycle of solar activity. Nature 347: 164-166.

Beer J, Raisbeck GM and Yiou F (1991) The variations of ^{10}Be and solar activity. In: The Sun in Time, eds CP Sonett, MS Giampapa and MS Matthews, The University of Arizona Press, Tucson: 343-359.

Beer J, Siegenthaler U, Bonani G, Finkel RC, Oeschger H, Suter M and Wölfli W (1988a) Information on past solar activity and geomagnetism from ^{10}Be in the Camp Century ice core. Nature 331: 675-679.

Beer J, Siegenthaler U and Blinov A (I988b) Temporal ^{10}Be variations in ice: Information on solar activity and geomagnetic field intensity. In Secular Solar and Geomagnetic Variations in the last 10,000 years, Stephenson FR & Wolfendale AW (edts) Kluwer Dordrecht : 297-314.

Braziunas TF (1990) Nature and origin of variations in late-glacial and Holocene atmospheric ^{14}C as revealed by global carbon cycle modeling. Ph.D. dissertation, Univ. of Washington.

Cook ER, Bird T, Peterson M, Barbetti M, Buckley B, D'Arrigo R and Francey R (1992) Climatic change over the last millennium in Tasmania reconstructed from tree-rings. The Holocene 2: 205-217.

Cook ER, Buckley BM and D'Arrigo RD (1993) Decadal-scale oscillatory modes in a millennia-long temperature reconstruction from Tasmania: Proceedings of the National Academy of Sciences Workshop on "The natural variability of the climate system on 10-100 year time-scales", Irvine, Cal. Sept. 21-24, 1992, in press.

Damon PE and Sonett CP (1992) Solar and terrestrial components of the atmospheric ^{14}C variation spectrum. In: The Sun in Time (CP Sonett, MS Giampapa and MS Matthews edts), University of Arizona, Tucson: 361-387.

Delworth T, Manabe S and Stouffer RJ (1992) Interdecadal Variations of the Thermohaline Circulation in a Coupled Ocean-Atmosphere Model. Journal of Climate, in press.

Eddy JA (1976) The Maunder Minimum. Science 192:1189-1202.

Edwards RL, Beck JW, Burr GS, Donahue DJ, Chappell JMA, Bloom AL, Druffel ERM and Taylor FW (1993) A large drop in atmospheric $^{14}C/^{12}C$ and reduced melting in the Younger Dryas, documented with ^{230}Th ages of corals. Science 260: 962-968.

Hood LL and Jirikowic JL (1990) Recurring variations of probable solar origin in the atmospheric $\Delta^{14}C$ time record. Geophys. Res. Lett. 17: 85-88.

Houtermans JC, Suess HE and Oeschger H (1973) Reservoir models and production rate variations of natural radiocarbon. J. Geophys. Res. 78: 1897-1907.

Lal D (1992) Expected secular variations in the global terrestrial production of radiocarbon. In: NATO ASI Series, vol 12. The Last Deglaciation: Absolute and Radiocarbon Chronologies. Edts E Bard and WS Broecker. Springer Verlag Heidelberg: 113-126.

Lingenfelter RE (1963) Production of carbon 14 by cosmic-ray neutrons. Rev. Geophys. 1: 35-55.

Magny M (1993) Solar and Geomagnetic Influences on Holocene Climatic Changes. Quat. Res. 40: 1-9.

Mazaud A, Laj C, Bard E, Arnold M and Tric E (1991) Geomagnetic field control of ^{14}C production over the last 80 ky: implications for the radiocarbon time-scale. Geophys. Res. Lett. 18: 1885-1888.

Mikolajewicz U and Maier-Reimer E (1990) Internal secular variability in an ocean general circulation model. Climate Dynamics 4:145-156.

O'Brien DP and Currie RG (1993) Observations of the 18.6-year cycle of air pressure and a theoretical model to explain certain aspects of this signal. Climate Dynamics 8: 287-298.

O'Brien K (1979) Secular variations in the production of cosmogenic isotopes. J. Geophys. Res. 84: 423-431.

Quinn WH and Neal VT (1992) The historical record of El Nino events. In Climate since AD 1500, R. S. Bradley and P. D. Jones, edts, Routlidge, Chapmann and Hall: 623-648.

Scuderi LA (1993) A 2000-Year Tree Ring Record of Annual Temperatures in the Sierra Nevada Mountains. Science 259: 1433-1436.

Shackleton NJ, Duplessy J-C, Arnold M, Maurice P, Hall MA and Cartlidge J (1988) Radiocarbon age of last glacial Pacific deep water. Nature 335: 708-711.

Sonett CP and Suess HE (1984) Correlation of bristlecone pine ring widths with atmospheric ^{14}C variations: a climate-Sun relation. Nature 307: 141-143.

Stuiver M and Braziunas TF (1988) The solar component of the atmospheric ^{14}C record. In Secular Solar and Geomagnetic Variations in the last 10,000 years, Stephenson FR & Wolfendale AW edts, Dordrecht Kluwer: 245-266.

Stuiver M and Braziunas TF (1989) Atmospheric ^{14}C and century-scale solar oscillations. Nature 338: 405-408.

Stuiver M and Braziunas TF (1993a) Sun, Ocean, Climate and Atmospheric $^{14}CO_2$ An evaluation of causal and spectral relationships. The Holocene 3: 289-305.

Stuiver M and Braziunas TF (1993b) Modeling Atmospheric ^{14}C Influences and Radiocarbon Ages of Marine Samples to 10,000 BC. Radiocarbon 35: 137-189.

Stuiver M and Braziunas TF (1994) Isotope and Carbon Cycle Inferences. OIES Global Change Institute, the Carbon Cycle July 19-30, Snowmass, CO, in press.

Stuiver M, Braziunas TF, Becker B and Kromer B (1991) Climatic, solar, oceanic, and geomagnetic influences on Late-glacial and Holocene atmospheric $^{14}C/^{12}C$ change. Quat. Res. 35: 1-24.

Stuiver M and Polach HA (1977) Discussion: Reporting of ^{14}C data. Radiocarbon 19: 355-363.

Stuiver M and Quay PD (1980) Changes in atmospheric carbon-l4 attributed to a variable sun. Science 207: 11-19.

Stuiver M and Reimer PJ (1993) Extended ^{14}C Data Base and Revised CALIB 3.0 ^{14}C Age Calibration Program. Radiocarbon 35: 215-230.

Taylor K, Rose M and Lamorey G (1992) Relationship of solar activity and climatic oscillations on the Colorado plateau. Journ. of Geophys. Res. 97: 15803-15811.

Wigley TML and Kelly PM (1990) Holocene climatic change, ^{14}C wiggles and variations in solar irradiance. Phil. Trans. R. Soc. Lond. A 330: 547-560.

^{10}Be as an indicator of solar variability and climate

J. Beer[1], F. Joos[3], Ch. Lukasczyk[1], W. Mende[2], J. Rodriguez[3], U. Siegenthaler[3] and R. Stellmacher[2]

1) Swiss Federal Institute for Environmental Science
and Technology (EAWAG),
CH-8600 Dübendorf,
Switzerland

Abstract

The atmospheric production rate of the cosmogenic isotope ^{10}Be is modulated by the solar wind. Spectral analysis of a 560 year long ^{10}Be concentration record from a Greenland ice core shows that this modulation process was active at all times even during quiet sun periods (Maunder minimum). To separate production effects from the effects of transport and deposition processes, three ^{10}Be records of different length and time resolution were used to calculate the expected ^{14}C variations. Comparisons with measured ^{14}C variations show that during the Holocene, for time scales larger than 3 years, the mixing and transport processes played only a minor role. Measurements of these isotopes in terrestrial archives therefore provide information about past solar variability.
Satellite measurements reveal a positive relationship between solar activity and solar irradiance. Comparisons between ^{10}Be in polar ice samples and ^{14}C in tree rings with temperature records show common features which confirm this relationship over the last few centuries.

---

2) Institute of Meteorology, Freie Universität Berlin, Germany
3) Physics institute, University of Bern, CH-3012 Bern

NATO ASI Series, Vol. I 25
The Solar Engine and Its Influence
on Terrestrial Atmosphere and Climate
Edited by Elizabeth Nesme-Ribes

Introduction

The sun is the engine which drives the climate system of the earth. We can distinguish between three main causes of changes in the external forcing. The largest of these is the increase in solar luminosity during the evolution of the sun over the last 4.5 billion years. Although this increase is very large (about 30%), it is so slow that it can be neglected when considering time scales on the order of millennia. The second cause is associated with changes in the orbital parameters of the earth as described by the theory of Milankovich. This leads to small variations in the absolute irradiance and to changes in the distribution of the solar irradiation on the earth's surface with time scales of 20, 40 and 100 ky. The third cause is related to turbulent and magneto-hydrodynamic processes in the convective zone of the sun. Measurements from satellites during solar cycle 21 (1978-1989) revealed a change in solar irradiance of about 1‰, which, though small, cannot be neglected when considering its long-term effects on climate. This change is the result of fluctuations in the energy transport in the convective zone, and not of fluctuations in energy production in the core. If we compare the energy content of the convective zone (10^{45} erg) with today's luminosity, we obtain a time of 10^4 years. In principle it would therefore be possible for convective effects to produce changes in solar irradiance with time scales up to the thermodynamic time scale of about 10'000 years.
The fundamental question of course is whether the sun ever made or will make use of this theoretical potential of variability of the irradiance. Based on observations of sun-like stars, it seems that the solar variability on a decadal time scale is at present rather low compared to other sun-like stars.
Beside the external forcing factors the climate is strongly determined by the atmosphere, biosphere, hydrosphere and the couplings between them. Of special interest is the expected global warming caused by anthropogenic greenhouse gases. During the last decades, a tremendous amount of information on the solar terrestrial relationships has been continuously gathered. This very short period resembles a snapshot. It does not provide any information about long-term solar dynamics. Such information can only be gained by carefully collecting all the available data from direct observations and by investigating natural archives such as ice, sediment cores and tree rings. The main basic problems are that the signal-to-noise ratio in these systems is rather low and that the signals are influenced by different local and global processes. In the following we discuss the potential and the limitations of cosmogenic isotopes for the reconstruction of solar variability and its possible relationship to changes in the earth's climate.

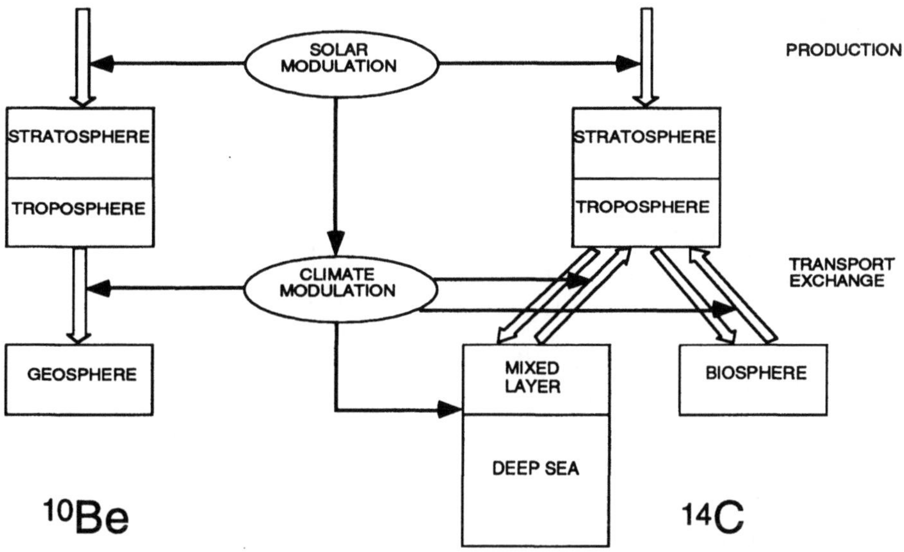

Fig. 1: Schematic diagram of the production and distribution of the cosmogenic isotopes ^{10}Be and ^{14}C.

Cosmogenic Isotopes

Cosmogenic isotopes like ^{10}Be and ^{14}C are produced continuously by the interaction of cosmic ray particles with the atmosphere. Although the nuclear processes involved in the production of ^{10}Be and ^{14}C are quite different, a change in the cosmic ray flux leads to very similar changes in the production rates of both isotopes. However, their geochemical behaviour is completely different. After production, ^{10}Be becomes attached to aerosols, whilst ^{14}C is oxidised to $^{14}CO_2$. ^{10}Be is removed from the atmosphere after a mean residence time of 1 to 2 years and is stored in the geosphere (Fig. 1). $^{14}CO_2$ on the other hand is invoked in exchange processes between atmosphere, biosphere and the ocean. The two main differences between the ^{10}Be system and the ^{14}C system are: 1. The short atmospheric residence time of ^{10}Be results in a very direct recording of production rate changes. But it also produces a local component in the signal because tropospheric mixing can be slow compared to the tropospheric residence time. 2. ^{14}C is globally well mixed in the atmosphere but does not reflect short-term production variations well (e.g., the 11 year Schwabe cycle) because of the large sizes of the reservoirs, which lead to strong attenuations of short-term variability (a factor of 100 for the 11 year cycle) and rather large phase lags (e.g. 10 yr for a 100 yr cycle).

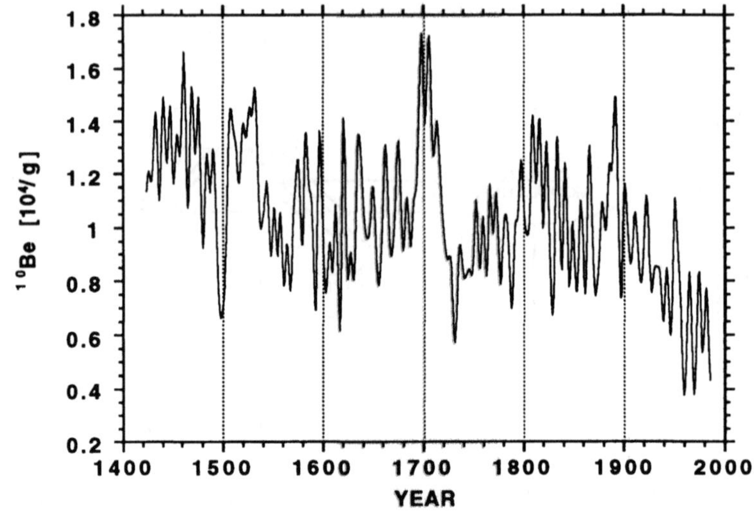

Fig. 2: ^{10}Be concentrations measured on annual ice samples from Dye 3, Greenland (Beer et al. 1994), after applying a low-pass filter (6 y).

In a very simple model we can therefore assume that measurements of ^{10}Be in ice and ^{14}C in tree rings tell us something about the production and the transport and exchange processes between the different reservoirs. From direct observations (e.g. neutron monitors, satellites) and models it is known that the production rates of cosmogenic isotopes are modulated by the magnetic properties of the solar wind, which itself is related to the complex processes taking place in the convective zone of the sun (Fig. 1). The transport and exchange processes, on the other hand, are influenced by the climate. The situation is even more complicated by the fact that the climate itself is also influenced by solar variability, and that many of the processes involved, and their interactions, are not yet well understood.

Based on the very simple model of Figure 1 and the above discussion, we expect changes in the production rate to be reflected almost equally in the ^{10}Be and ^{14}C data. However, ^{14}C production variations are much more dampened. On the other hand climatic changes are not expected to affect both systems the same way. Therefore, by comparing ^{10}Be and ^{14}C records it should be possible to distinguish between production and climate effects.

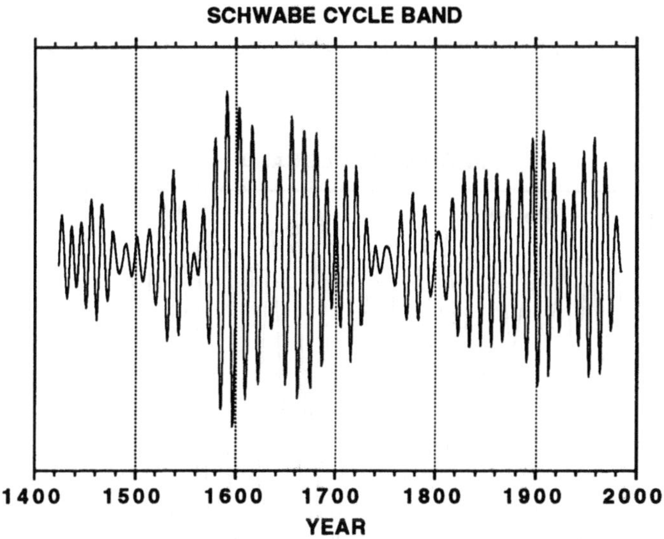

Fig. 3: ^{10}Be data of Fig. 2 after applying a spectral bandpass filter (9-13.7 y).

^{10}Be as a tool to reconstruct solar variability

The application of ^{10}Be as a tool to reconstruct solar variability has been discussed several times (Raisbeck and Yiou 1980), (Beer et al. 1990), (Beer et al. 1991), (Beer et al. 1994). Therefore in the following we concentrate on a comparison of the ^{10}Be with the sunspot record in the Schwabe and the Gleissberg cycle band. In Fig. 2 the low-pass (6 year cut off) filtered ^{10}Be data from a 300 m long ice core (Dye 3, Greenland) are shown. They cover the period 1423-1985. During the Spoerer minimum (1420-1580) and the Maunder minimum (1645-1715) the ^{10}Be concentrations are relatively high; during the 20th century a decrease in the values indicates increasing solar activity. In Fig. 3 the filter output of the band filtered (spectral band filter with a window of 9 - 13.7 yrs) ^{10}Be data are shown. The filter output around the Schwabe cycle frequency is large for the periods 1580 - 1720 and 1820 to 1960. Obviously the reflection of the Schwabe cycle is weak during that part of the Spoerer minimum which is covered by the data (1423 - 1580) and strong throughout the Maunder minimum (1645-1715), when the sunspot cycle seemed to have been interrupted. This indicates that the solar modulation of cosmic rays and the sunspot numbers are two different aspects of solar variability. A comparison between the band-filtered ^{10}Be and sunspot data (Fig. 4) shows good agreement with regard to phases and amplitudes except around 1900. The main discrepancy arises between 1750 and 1780, when ^{10}Be shows three cycles and the sunspots four.

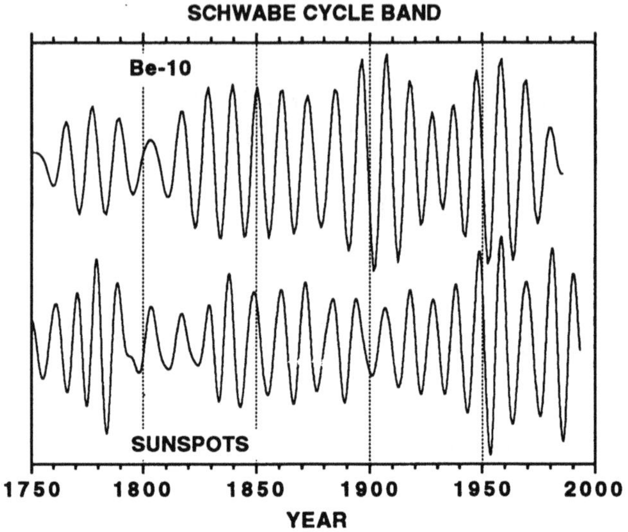

Fig. 4: Comparison of part of the ^{10}Be data from Fig. 3 with the corresponding bandpassed record of sunspots.

In a next step we want to show that the observed ^{10}Be cycle around the Schwabe cycle band is indeed mainly the result of production modulation and not of climate effects (Fig. 1). For this purpose we compare three ^{10}Be records covering different time periods with the corresponding $\Delta^{14}C$ records.

Comparison of ^{10}Be and ^{14}C

For this comparison we use a procedure which we have already applied earlier (Beer et al. 1988b), (Siegenthaler and Beer 1988). The ^{10}Be concentration record from Dye 3 is used as input function to a carbon cycle model, assuming that ^{10}Be directly describes the production rate. This assumption is a strong simplification because it neglects all climatic effects on the ^{10}Be record. The carbon cycle model used is an improved version of the one shown in Fig. 1 consisting of two atmospheric plus four biospheric reservoirs and a box-diffusion ocean (Siegenthaler and Oeschger 1987). The result of the model calculation for the record of Fig. 2 is displayed in Fig. 5. The general agreement between the measured (Stuiver and Braziunas 1993) and the calculated curve is good except for two periods. 1. During the Maunder minimum (1645-1715) the calculated peak is, for unknown reasons, too narrow. 2. The decrease in the measured $\Delta^{14}C$ (the relative deviation in ^{14}C activity from the standard after correction for decay and isotope fractionation) during the 20th century is steeper than that calculated from ^{10}Be. This discrepancy is to be expected. The combustion of fossil fuel leads to a dilution of the atmospheric carbon reservoir by dead carbon ("Suess effect"). However, Fig. 5 shows that only part of the observed ^{14}C decrease is due to the Suess effect.

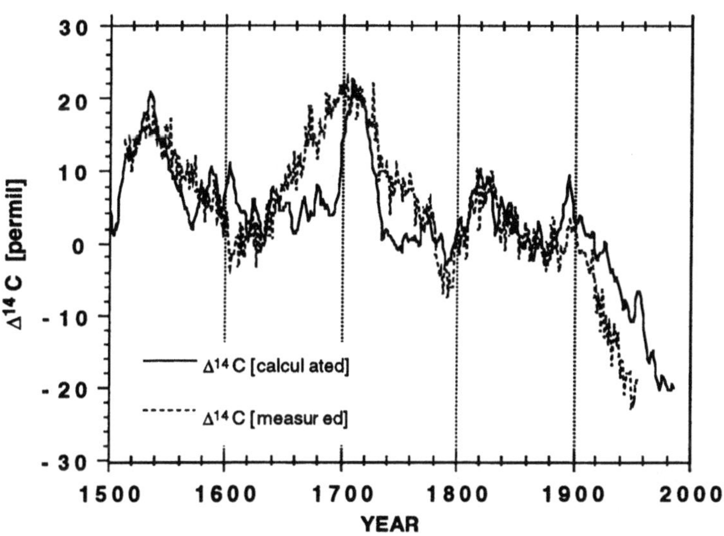

Fig. 5: Comparison of measured $\Delta^{14}C$ (the relative deviation in ^{14}C activity from the standard after correction for decay and isotope fractionation) data (Stuiver and Braziunas 1993) with calculated tropospheric $\Delta^{14}C$ using the ^{10}Be data from Fig. 2 as input function to a carbon cycle model (Siegenthaler and Oeschger 1987).

One way to get rid of the local ^{10}Be component is to combine ^{10}Be records from different sites. In Fig. 6 the model calculations are repeated for a ^{10}Be data set which was obtained by combining the Milcent record from Greenland with the South pole record from Antarctica (Beer et al. 1991) analysed by Raisbeck and Yiou. The combined data set covers the period 1200-1800 AD. The agreement of the measured $\Delta^{14}C$ with the calculated $\Delta^{14}C$ based on the combined ^{10}Be record is better than with the individual records not shown here. This indicates that combining records removes local disturbances.

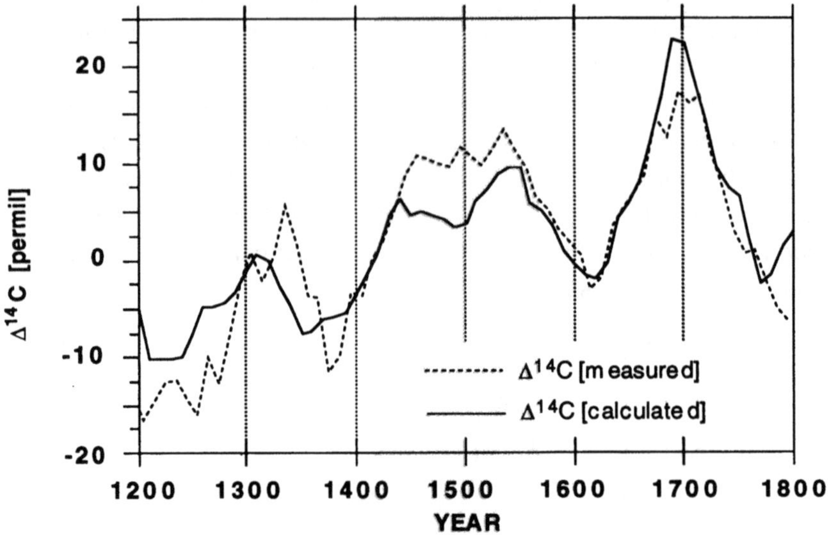

Fig. 6: Comparison of measured $\Delta^{14}C$ data (Stuiver and Quay 1980) with calculated $\Delta^{14}C$ using the combined ^{10}Be records of Milcent and South pole (Beer et al. 1991) as input function of a carbon cycle model (Siegenthaler and Oeschger 1987).

Finally we used a combined ^{10}Be data set from Camp Century, Milcent, South pole and Dye 3 to calculate the $\Delta^{14}C$ record over the last 4000 years (Fig. 7). Again we find a good agreement with regard to the long-term trend and the main short-term features. Based on this good agreement with respect to trends and shapes of the main peaks, we conclude that a large part of the observed signals are common to both isotopes and therefore most probably caused by changes in the production rate. The differences may be due to local ^{10}Be components and to climate-induced system effects. The prominent peaks at 300 and 700 BC are therefore probably also of solar origin (like the Maunder and Spoerer minima) (Stuiver and Braziunas 1989).

^{10}Be and climate

The ^{10}Be concentration measured in an archive such as ice depends not only on the production rate, but also on several atmospheric processes such as latitudinal

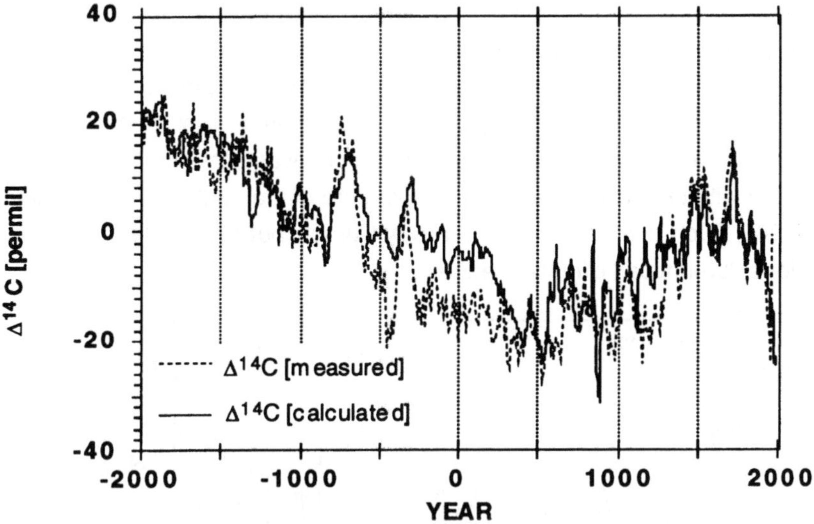

Fig. 7: Comparison of measured $\Delta^{14}C$ data (Stuiver et al. 1986) with calculated $\Delta^{14}C$ using a combination of the ^{10}Be records from Camp Century (Beer et al. 1984), Milcent-South pole and Dye 3 as input function of a carbon cycle model (Siegenthaler and Oeschger 1987).

mixing, exchange between stratosphere and troposphere, and wet and dry removal from the troposphere. All these processes are connected to the climate. A clear relationship between ^{10}Be concentration and $\delta^{18}O$, reflecting temperature, has been found during glacial times in polar ice cores (Raisbeck et al. 1981). Changes in the ^{10}Be concentration by up to a factor of 2 during glacial time could be explained by corresponding changes in the accumulation rate. During the Holocene, however, climate variations were comparatively small and there is no clear relationship between ^{10}Be and accumulation rate (Beer et al. 1988a). This does not mean that ^{10}Be does not reflect climate changes. However, the generally good agreement with the $\Delta^{14}C$ data suggests that production effects during the last few millennia are the main cause for variations on time scales of more than 2 - 3 years. On shorter time scales, climate effects seem to become dominant. This important result still has to be confirmed by further high-resolution records of different cosmogenic isotopes from other sites.

Recent direct observations of solar irradiance from satellites have revealed a clear positive correlation between solar activity expressed in terms of sunspot numbers and solar irradiance (Willson and Hudson 1988). If this relationship holds true for the past, we should find an anticorrelation between irradiance and ^{10}Be. In Fig. 8 the surface air temperature in central England (Manley 1974) is plotted together with the inverse ^{10}Be concentration. The agreement is good, in view of all the uncertainties and simplifications involved. A similarly good result was obtained by comparing the ^{10}Be record with the mean surface air temperature of the northern hemisphere (Beer et al. 1994). Such correlations alone are not proof, but they justify continuing the search for solar-induced climate changes.

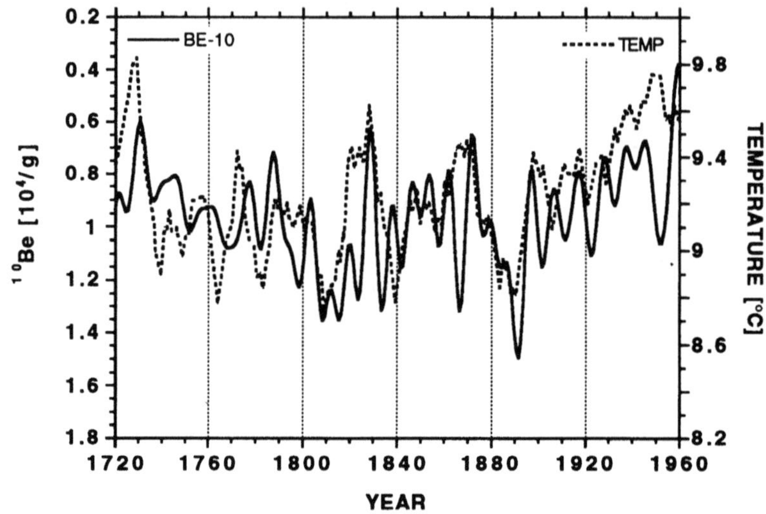

Fig. 8: Comparison of the low-pass filtered ^{10}Be data from Dye 3 (Fig. 2) with the temperature record from central England (Manley 1974).

Recently several authors (Friis-Christensen and Lassen 1991), (Hoyt 1993) working with sunspot numbers as indicator of solar luminosity argued that it is more appropriate to use the cycle length instead of the amplitude. This seems to be true for sunspots, but not necessarily for ^{10}Be. Further evidence for a connection between ^{10}Be, ^{14}C and temperature records can be gained by investigating the Gleissberg cycle around the 90 year period. In Fig. 9 three data sets have been analysed using a spectral band filter (bandwidth: 30 years): 1. the ^{10}Be record of Fig. 2., 2. the ^{14}C record from tree rings (Stuiver and Quay 1980) and 3. a combined temperature record (Groveman and Landsberg 1979),(Jones et al. 1986). The phase lag of 10 years induced by the carbon system has been taken into account by shifting the ^{14}C results correspondingly. The good agreement of the spectral peak frequencies in these bands (89.1 yrs for ^{10}Be, 88.5 yrs for ^{14}C and 90.7 yrs for temperature) indicates a common source of the frequency components in the Gleissberg band. The phase coincidence near 1780 is not the result of any kind of normalisation procedure; it is a true result of the filter output. The amplitudes are also not normalised. The good phase coherence between the three curves is again in favour of a common source, most likely the solar activity and the solar irradiance. All arguments taken together provide strong evidence for a causal relationship between these three records and emphasize the climatic relevance of the Gleissberg cycle (Mende et al. 1993).

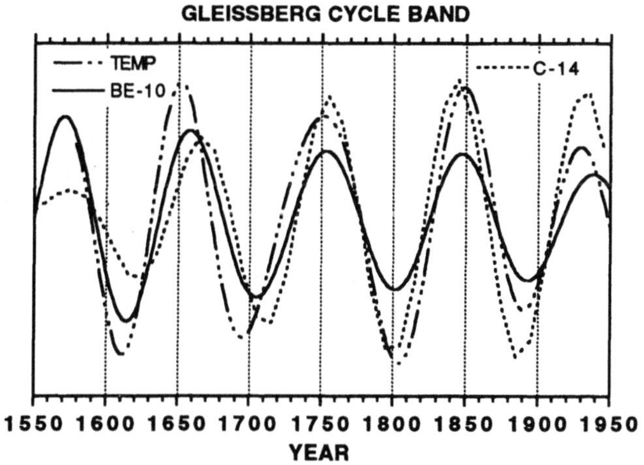

Fig. 9: Comparison of the records of ^{10}Be (Dye 3), ^{14}C (Stuiver et al. 1986) and temperature (Groveman and Landsberg 1979), (Jones et al. 1986). The records are band-filtered around the Gleissberg periodicity of 90 yrs. The $\Delta^{14}C$ curve is shifted by 10 years (carbon system lag).

Conclusion

The comparison of records of ^{10}Be, ^{14}C, sunspots and temperature indicates that the main cause of the observed isotope variations occurring during the last few centuries are changes in the production rate induced by solar wind modulation of the cosmic ray flux. The effects caused by various mixing and exchange processes within the ^{10}Be and the carbon systems seem to play only a minor role for time scales larger than a few years. This justifies low-pass filtering of the ^{10}Be data. The common features existing between the temperature and the isotope records provide unique information not only about the history of turbulent and magneto-hydrodynamic processes in the convective zone of the sun, but also about terrestrial climate changes induced by changes in the solar irradiance. Careful calibrations over the last 2-3 centuries should allow quantification of the effect of solar forcing on time scales of decades and longer. This is also of great importance within the framework of the discussion about global change because it allows us to distinguish between anthropogenic effects caused by greenhouse gases and natural variability due to changes in the solar engine.

References

Beer, J., Andrée, M., Oeschger, H., Siegenthaler, U., Bonani, G., Hofmann, H., Morenzoni, E., Nessi, M., Suter, M., Wölfli, W., Finkel, R. and Langway Jr., C. (1984). The Camp Century ^{10}Be Record: Implications for Long-Term Variations of the Geomagnetic Dipole Moment. Nucl. Instrum. Meth. B5: 380-384.

Beer, J., Baumgartner, S., Dittrich-Hannen, B., Hauenstein, J., Kubik, P., Lukasczyk, C., Mende, W., Stellmacher, R. and Suter, M. (1994). Solar Variability Traced by Cosmogenic Isotopes in The Sun as a Variable Star: Solar and Stellar Irradiance Variations , Cambridge University Press: 291-300.

Beer, J., Blinov, A., Bonani, G., Finkel, R. C., Hofmann, H. J., Lehmann, B., Oeschger, H., Sigg, A., Schwander, J., Staffelbach, T., Stauffer, B., Suter, M. and Wölfli, W. (1990). Use of ^{10}Be in polar ice to trace the 11-year cycle of solar activity. Nature 347: 164-166.

Beer, J., Raisbeck, G. M. and Yiou, F. (1991). Time Variations of ^{10}Be and Solar Activity. The Sun in Time. Tucson, Univ. of Arizona press. 343-359.

Beer, J., Siegenthaler, U. and Blinov, A. (1988). Temporal ^{10}Be Variations in ice: information on solar activity and geomagnetic field intensity. Secular Solar and Geomagnetic Variations in the Last 10,000 Years. Kluwer Academic Publishers. 297-313.

Beer, J., Siegenthaler, U., Bonani, G., Finkel, R. C., Oeschger, H., Suter, M. and Wölfli, W. (1988). Information on Past Solar Activity and Geomagnetism from ^{10}Be in the Camp Century Ice Core. Nature 331: 675-679.

Friis-Christensen, E. and Lassen, K. (1991). Length of the Solar Cycle: An Indicator of Solar Activity Closely Associated with Climate. Science 254: 698-700.

Groveman, B. S. and Landsberg, H. E. (1979). Reconstruction of Northern hemisphere temperature. 1579 - 1880. University of Maryland, College Park, MD, : Publication 79-181.

Hoyt, D. V. (1993). A Discussion of Plausible Solar Irradiance Variations, 1700-1992. Journal of Geophysical Research 98, A11: 18895-18906.

Jones, P. D., Raper, S. C. B., Bradley, R. S., Diaz, H. F., Kelly, P. M. and Wigley, T. M. L. (1986). Northern Hemisphere surface air temperature variations, 1851-1984. J. Clim. Appl. Met. 25: 161-179.

Manley, G. Q. (1974). Central England Temperatures: Monthly Means 1659 - 1973. Quart. J. R. Met. Soc. 100: 389-405.

Mende, W., Stellmacher, R. and Olberg, M. (1993). Relationships between relative sunspot numbers and air temperature. Meteoro. Zeitschrift N.F. 2: 121-126.

Raisbeck, G. M. and Yiou, F. (1980). ^{10}Be in polar ice cores as a record of solar activity. Proc. Conf. on Ancient Sun. Pergamon Press, New York. 185-190.

Raisbeck, G. M., Yiou, F., Fruneau, M., Loiseaux, J. M., Lieuvin, M., Ravel, J. C. and Lorius, C. (1981). Cosmogenic ^{10}Be concentrations in Antarctic ice during the past 30,000 years. Nature 292: 825-826.

Siegenthaler, U. and Beer, J. (1988). Model Comparison of ^{14}C and ^{10}Be Isotope Records, Proc. NATO-Workshop. Secular Solar and Geomagnetic Variations in the Last 10,000 Years. Durham 1987, Kluwer Academic Publishers. 315-328.

Siegenthaler, U. and Oeschger, H. (1987). Biospheric CO_2 emissions during the past 200 years reconstructed by deconvolution of ice cores. Tellus 39B: 140-154.

Stuiver, M. (1980). Solar Variability and Climatic Change During the Current Millennium. 286: 868-871.

Stuiver, M. and Braziunas, T. F. (1989). Atmospheric ^{14}C and Century-Scale Solar Oscillations. 338: 405-408.

Stuiver, M. and Braziunas, T. F. (1993). Sun, ocean, climate and atmospheric $^{14}CO_2$, an evaluation of causal and spectral relationships. The Holocene. 3: 289-305.

Stuiver, M., Kromer, B., Becker, B. and Ferguson, C. W. (1986). Radiocarbon age calibration back to 13,000 years B.P. and the ^{14}C age matching of the German oak and US Bristlecone pine chronologies. Radiocarbon 28: 969-979.

Stuiver, M. and Quay, P. D. (1980). Changes in Atmospheric Carbon-14 Attributed to a Variable Sun. Science 207: 11-19.

Willson, R. C. and Hudson, H. S. (1988). Solar luminosity variations in solar cycle 21. Nature 332: 810-812.

Panel Discussions on Total Solar Irradiance Variations and the Maunder Minimum

J. M. Pap[1], and O.R. White[2]

I. Measurements and models of total solar irradiance

I.1. Summary of the results on irradiance variations: J.M. Pap and R.C. Willson, JPL

For more than a decade, total solar irradiance has been monitored from several satellites, namely the Nimbus-7, Solar Maximum Mission (SMM), the NASA ERBS, NOAA9 and NOAA10, EURECA, and the Upper Atmospheric Research Satellite (UARS) (e.g. Willson and Hudson, 1991; Hoyt et al., 1992; Mecherikunnel et al., 1988; Romero et al., 1994). These observations have revealed variations in total irradiance ranging from minutes to the 11-year solar cycle (Figure 1, from Fröhlich 1994). The very small, rapid irradiance fluctuations are due to solar oscillations (Woodard and Hudson, 1983; Fröhlich, 1992). The short-term variations (from days to months) are directly related to the evolution of active regions via the combined effect of dark sunspots and bright faculae (Chapman, 1987). The most important discovery of irradiance observations is the 0.1% peak-to-peak variation in total solar irradiance over the solar cycle (Willson and Hudson, 1991). This solar-cycle-related variation of total irradiance is attributed to the changing emission of bright magnetic elements, including faculae and the magnetic network (Foukal and Lean 1988). This solar cycle variability may also be related to changes in the photospheric temperature; however it is not clear as yet whether this change can be linked to the bright network component (Kuhn et al., 1988).

Although considerable information exists about the variations in total solar irradiance, the underlying physical mechanisms, especially that of its long-term variations, are not well understood as yet. Furthermore, several theories for explaining the reradiation mechanism of the missing energy in the sunspot-related irradiance dips have been put forward. One particular part of the theories suggests that the reduced radiation results in storage of the blocked heat flux as thermal and potential energy of the convection zone around a spot, and this energy is re-radiated over the long radiative relaxation time of the layers effected by the spot (Foukal, 1981; Spruit, 1982a, b; Foukal et al., 1983; Spruit, 1994). Recent numerical convection models of energy and material

[1] Jet Propulsion Laboratory, California Institute of Technology, 4800 Oak Grove Drive, Pasadena, CA 91109

[2] High Altitude Observatory, National Center for Atmospheric Research, P. O. Box 3000, Boulder, CO 80307

NATO ASI Series, Vol. I 25
The Solar Engine and Its Influence
on Terrestrial Atmosphere and Climate
Edited by Elizabeth Nesme-Ribes

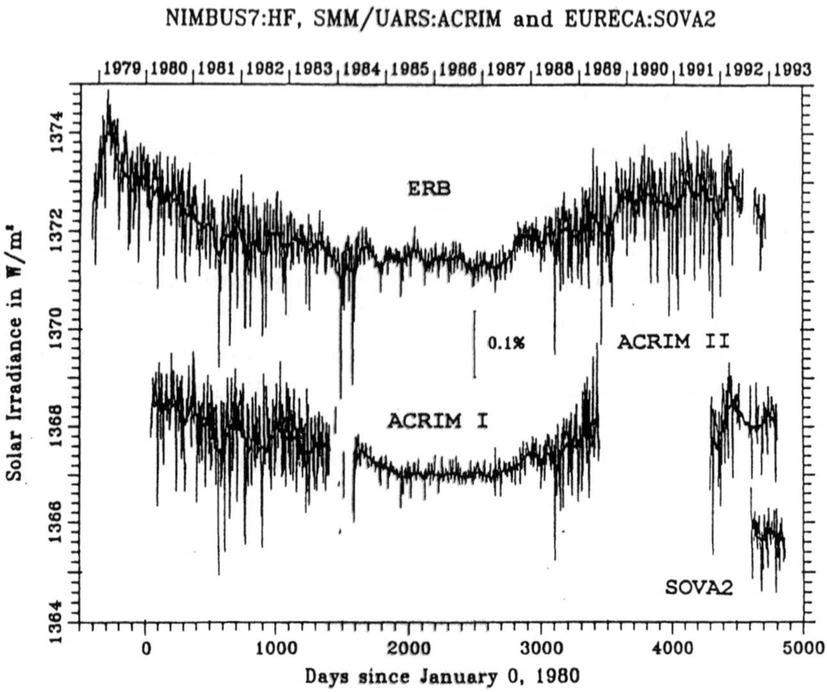

Fig. 1. Observations of total solar irradiance by the Nimbus-7/ERB, SMM/ACRIM I, UARS/ACRIM II, and EURECA/SOVA2 radiometers are plotted.

flowing around a sunspot do not indicate long-term storage of the heat flux, but show that most of the heat flux blocked by sunspot-like objects (magnetic perturbations) reappears at the surface, although some fraction of the blocked flux is carried horizontally far from the object (Fox et al., 1992; Fox and Sofia, 1994). Another explanation suggests that the energy blocked by sunspots is stored in the magnetic fields of the spots, in which case sunspots should cause dips in total irradiance during the early stage of their development (Wilson, 1981), as it is shown by some of the statistical analyses (Pap, 1985; Fröhlich and Pap, 1989). A slow down flow is also supposed around the spots during their growing phase that is capable of transporting the blocked energy to surrounding regions where it emerges as facular enhancement (Schatten et al., 1986). Some calculation indicates that the facular excess flux can compensate the spot deficit on active regions time scale (Chapman, 1987).

I.2. History of irradiance observations: R.C. Willson and J. M. Pap, JPL

Since the radiative output of the Sun is the main driver of the physical processes within the Earth's atmosphere, the precise and continuous determination of the value of total solar irradiance and its variation is extremely important. There are indications that changes in the solar energy output influence the Earth's climate on time scales ranging from the Gleissberg cycle (Reid, 1987; Friis-Christensen and Lassen, 1991) up to the Maunder type of climate anomalies (Lean et al., 1992). Furthermore, it is shown that persistent variations in total solar irradiance (as small as 0.5%) over a century could have explained a wide range of climatic changes that occurred in the past (Eddy, 1977). Therefore, one of the main questions to be addressed is:

What do we need to achieve long-term high precision irradiance database for climatic studies?

The first attempts to detect variations in total solar irradiance started as early as the beginning of this century. The first continuous observational program of total solar irradiance was performed at the Smithsonian Institution over the first half of the 20th century (Abbot, 1925; Hoyt, 1979). During the following 30 years, experiments with the same objective were conducted using aircraft, high altitude balloons, sounding rockets, and space flight platforms (Fröhlich, 1977). All these experiments were unable to detect variations in total irradiance that were unambiguously solar in origin. On the one hand, the intra- atmospheric experiments were highly effected by the selective absorption of the terrestrial atmosphere that masked solar induced irradiance variations from even high flying aircraft and balloons. On the other hand, failure of extra-atmospheric experiments on rockets and early satellites was due to the lack of sufficient radiometric accuracy.

Significant development in radiometry in the late sixties led to the development of the electrically self-calibrating active cavity irradiance detectors at different institutions: at the Jet Propulsion Laboratory (Willson, 1984), Eppley Laboratory (Hickey and Karoli, 1974), World Radiation Center at Davos, Switzerland (Brusa and Fröhlich, 1975), and at the Royal Meteorological Institute of Belgium (Crommelynck, 1973). These new radiometers are capable of measuring total solar irradiance with an absolute accuracy of +/-0.2%. The development of these new type of radiometers and long-term flight opportunities from space led to the discovery of changes in total irradiance related to solar activity.

The first long-term solar irradiance monitoring experiment was the Earth Radiation Budget (ERB) on the Nimbus-7 satellite that provided irradiance database from late 1978 to early 1993 (Kyle et al., 1993). The Nimbus-7/ERB experiment was followed by the ACRIM I and ACRIM II on aboard the Solar Maximum Mission (SMM) and Upper Atmospheric Research Satellite (UARS), respectively (Willson, 1994); and by the SOVA1 and SOVA2 experiments on EURECA (Crommelynck et al., 1994, Romero et al., 1994). The demonstration that these experiments (ERB, ACRIM I and II, SOVA1 and SOVA2) showed the large excursions due to passage of large sunspots across the center of the Sun's disk convinced the skeptics that the darkening effects of sunspots on the total irradiance had indeed been detected. Furthermore, combination of ERB, ACRIM I, and ACRIM II give the solar cycle amplitude of the irradiance change from 1978 to 1993, covering two solar maxima and one solar minimum. The different observations of total solar irradiance are summarized in Figure 2 (from Willson, 1994).

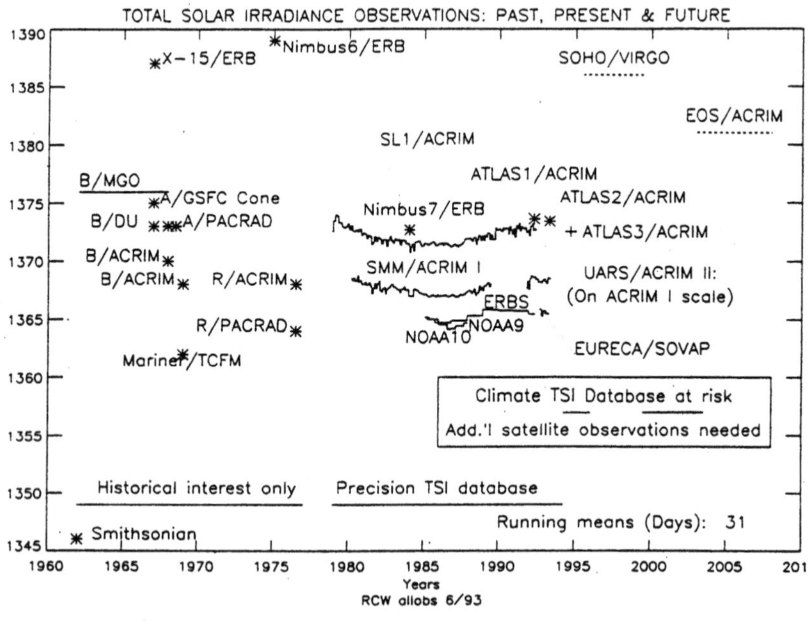

Fig. 2. Summary of the past and current irradiance observations is presented. Forthcoming irradiance observations of NASA and ESA are also indicated.

However, the current absolute accuracy of the operating active cavity radiometers (+/-0.2% SI) is still not adequate to provide long-term irradiance database for climatic studies. Therefore, the main goal is to build high-precision irradiance database that covers several solar cycles. For this purpose continuous observations of total solar irradiance with the same type of instrument with an overlapping strategy is needed to provide the maximum precision due to the smaller number of possible degrading variables. This approach can provide a long-term irradiance database with precision better than 50 ppm over decades that is sufficient to reveal variation as small as 0.5% over a century.

Since the overlap requirement cannot always be achieved (e.g. failure of satellites and/or lack of flight opportunities), data sets gathered by different instruments should be used as a backup. For this purpose, extensive comparison of different flight instruments during the overlapping portion of their missions is required. In the case of the gap between SMM/ACRIM I and UARS/ACRIM II data, their adjustment is achieved through their numerical intercomparisons with the Nimbus-7/ERB data. Systematic variations arising due to differing inherent responses and flight environments must be removed to make maximum quality comparisons. With these adjustments, the relationship between ACRIM I and ACRIM II should be established to better than 50 ppm (Willson, 1994).

A new type of radiometers: the cryogenic radiometers can achieve a higher accuracy than the operating active cavity radiometers (Martin and Fox, 1994). However, cryogenic radiometers operating near the temperature of liquid He ($\leq$ 20 K) can provide an order of magnitude smaller SI uncertainty in the laboratory environment than ambient temperature devices. Their SI uncertainty (probably not less than +/- 200 ppm SI in flight experiments) is still above that required by the long-term climate data base. Further, the small apertures required, due to the need for power by the Stirling cycle He coolers, makes them much more vulnerable to contamination than larger ambient temperature radiometers. They may not provide the consistent measurement ability of ambient temperature instrumentation that facilitates the precision overlap capability.

I.3. Overview of the available surrogates for irradiance models: O.R. White, HAO

The detection of total irradiance variations by satellite-based experiments during the last 15 years stimulated modeling efforts to help identify their causes and to provide estimates of irradiance data for time intervals when no satellite observations exist. Since there is no adequate quantitative physical model for the variations in total solar irradiance, one has to rely on empirical irradiance models based on 'proxy' indicators of solar activity.

There are two basic sources of our measures of solar activity as far as solar irradiance is concerned: (1) Full disk measurements of solar output, i.e., the Sun as a star (direct irradiance measurements), and (2) Images of the Sun showing the presence of dark sunspots, bright faculae, and bright network. Historically, images were used in Wolf's and Schwabe's work first to define the solar cycle on the basis of sunspot counts.

With the discovery of solar radio emission and development of methods for accurately measuring radiative energy from the Sun at radio wavelengths, we see Covington's use of the 10.7 cm radio flux in 1947 as a measure of activity on the Sun as a star. With the development of precision spectral photometers and solar magnetometers, several long-term experiments began in about 1975 to measure solar magnetic fields in the photosphere and chromospheric activity as shown by HeI, MgII, and CaII lines. The four principal full-disk radiometric indices are:

Type est. precision	Wavelength/line	Time Period	Measurement
Radio flux	10.7 cm	1947-1993	absolute flux 1%
Spectral line	CaII K	1974-1993	relative flux 1%
Spectral line	HeI 10830	1975-1993	relative flux 2%(?)
Spectral line	MgII h&k	1978-1993	relative flux 2%(?)

I stress that the line indices are chosen because these emissions show the bright structures associated with sunspots on the solar disk with much higher contrast than in the photospheric spectrum. This is precisely why they are chosen as indicators of solar activity.

The principal indices derived from solar images are the Zürich sunspot number, plage index, magnetic field index, and the photometric sunspot index (PSI). Each index has its own virtue of length of measurement, completeness, or precision. The magnetic and photometric sunspot indices are the only ones based on quantitative measurements of the solar magnetic field or sunspot size and brightness. The time series are summarized as follows:

Type est. precision	Wavelength/line	Time Period	Measurement
sunspot no.	white light	1610-1993	sunspot counts ?
plage index	CaII K	1958-1987	area, intensity ?
magnetism	FeI lines	1975-1993	zeeman effect 15%
PSI	white light	1980-1993	sunspot area, contrast ?

With the advent of high quality electronic cameras, a new generation of solar imaging experiments produce new indices for the facular excess - the bright features - and the sunspot deficit - the dark features - on the solar disk. These measures are photometric quantities usable directly in analysis of the sources of variability in total irradiance measurements, but they do not explicitly distinguish between individual sunspots and plages on the visible solar disk. The analysis is statistical and requires very stable image recording through the full optical train.

The source of these new types of measures is as follows:

Observatory	Wavelength/line period	Observer time
San Fernando	CaII K & 6723A	Chapman 1990-1993
NSO (U.S.)	CaII K	Harvey 1992-1993
NAO(Japan)	5450,7770A	Nishikawa 1988-1989, 1992

The time base is still quite short because the technique is new, but these experiments are the basis for development of precision photometric telescopes in the U.S. SunRise program. Chapman (1994) shows that relative variations in the total irradiance can be recovered well for large sunspots that are easily identified. Early use of the statistical image analysis technique by both Chapman (1994) and Nishikawa (1994) also gives reasonable estimates of the relative variation of the total irradiance.

In addition to contemporary measurements, we have historical sets of solar spectroheliograms from Mt. Wilson Observatory, Meudon Observatory, and Kodiakanal Observatory extending back to about 1910. Both the Mt. Wilson and Meudon images are being digitized and will be analyzed to give new sets of CaII K line data defining variability of plages over the last 80 years. The Meudon program is of particular interest since three wavelengths sampling three levels from the photosphere to the upper chromosphere, i.e., core and wing of the CaII K line and the core of the HI alpha line.

I emphasize that all of these ground-based measurements are narrow band or 'point' measurements in the solar spectrum from which we are trying to infer the area under the solar energy distribution curve as well as its shape. Furthermore, most of our available 'points' in the spectrum are spectral lines (CaII K, MgII k, HeI 108.3) or radio emission (10.7 cm) more sensitive to solar activity than the lower levels where the total solar irradiance originates.

The accuracy of our estimates of total solar irradiance is limited by the inaccuracy of the 'mapping spectral irradiances from these wavelengths to the total irradiance.

I.4. Summary of current irradiance models: J.M. Pap, JPL

Considerable effort has been made to combine the 15-year record of total solar irradiance from space with a variety of solar measurements more sensitive to solar activity in an effort to extend the absolute output record backward in time as far as is possible. However, the fundamental question is:

Whether the empirical models of total solar irradiance based on the described surrogates can reasonably predict solar irradiance changes with the long-term precision required by climatic studies?

On the one hand, the current empirical models of total irradiance, developed from various solar activity indices, such as the He-line equivalent width (Foukal and Lean, 1988), several Fraunhofer-lines (Livingston et al., 1988), the 10.7 cm radio flux (Brandt et al., 1994; Fröhlich, 1994), and the MgII h & k core-to-wing ratio (Pap et al., 1994), disagree significantly with the irradiance observations at the maximum of solar cycles 21 and 22 (Figure 3). Results of multivariate cross-spectral analysis show that considerable variation remains unexplained in total irradiance after removing the effect of sunspots and bright magnetic elements, and the residual variability changes with the phase of solar cycle (Pap and Fröhlich, 1992; Figure 4).

The lack of agreement between the measured total irradiance and its model estimates from other full-disk solar indices is a current outstanding problem in the study of solar-cycle-related irradiance variability. One of the largest uncertainties in the irradiance models comes from the lack of knowledge of the effect of faculae, principally because of the lack of accurate long-term synoptic data. The full disk observations of faculae started only in 1988 by Nishikawa and recently by Chapman in 1990; however, these results are not yet incorporated into the irradiance models. On the other hand, previous studies have shown that besides the faculae, the magnetic network contributes to the changes in total irradiance, especially on long time scales (Foukal and Lean, 1988; Willson and Hudson, 1988). It should be mentioned, however, that these new photometric observations do not include the bright magnetic network because of the difficulty of the precise measurement of its area.

Therefore, most of the irradiance variations related to the bright magnetic features (faculae and the magnetic network) are modeled by chromospheric surrogates such as the CaII K index, the full disk equivalent width of the He-line at 1083 nm, MgII h & k core-to-wing ratio, and the 10.7 cm radio flux, although more than 90% of the total radiative flux of the Sun is emitted from the photosphere. However, the conversion factor between the area and intensity of photospheric and chromospheric plages is not known, the center-to-limb behavior of the contrast of white light faculae and that of the Ca K plages is quite different (Chapman et al., 1992; Schatten and Mayr, 1992). The

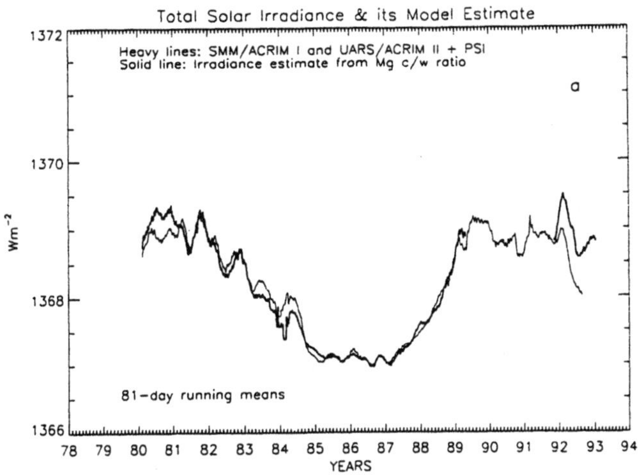

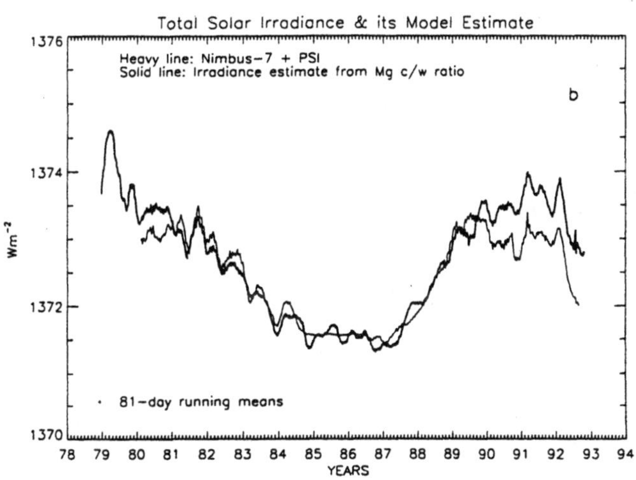

Fig. 3. The heavy lines show the 81-day running means of the Nimbus-7/ERB (a) as well as the SMM/ACRIM I and UARS/ACRIM II (b) total solar irradiance corrected for sunspot darkening. The solid lines represent the 81-day running means of the empirical model of total solar irradiance corrected for sunspot darkening.

current spatially resolved plage data set is combined from three different and incoherent data sets (obtained at the McMath Observatory from December 1970 until September,

1979, Mt. Wilson Observatory from October 1979 to August 1982, and Big Bear Solar Observatory from September 1982 to November 1987); however, most modelers are not aware of this fact. Since the Big Bear data reduction ended in November of 1987 because of budget cuts, no intensity and area data of CaII K plages are measured and published on a routine basis (Marquette, 1992). Furthermore, it should also be mentioned that these plage data exclude the remnants of plages, which still cause a significant variation in both total and UV irradiances (Pap et al., 1991).

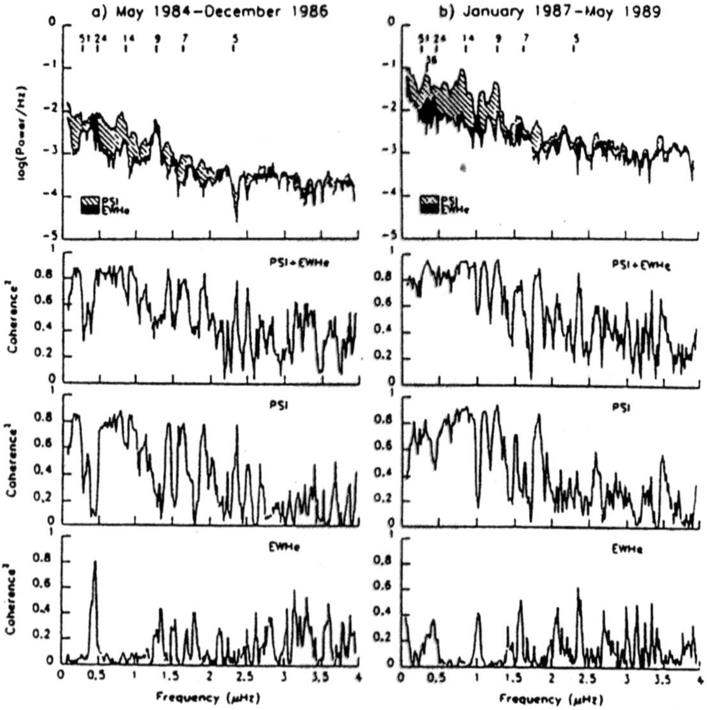

Fig. 4. The upper panels show the power spectra of the SMM/ACRIM I total solar irradiance for the time intervals of 1984-86 and 1987-89, respectively. The shaded areas give the portion of the ACRIM variance explained by the PSI and the He-line equivalent width at 1083 nm, respectively. The lower panels show the total and partial coherences squared.

Variability of the HeI equivalent width index and the 10.7 cm radio flux is rather complex: variation of both indices has a coronal contribution that is not present in the lower solar atmosphere where the total irradiance originates (Harvey, 1984; Tapping, 1987). While most of the variation in the HeI line equivalent width is related to plages, about 15% of its variability is related to dark filaments that do not effect total irradiance (Harvey and Livingston, 1994). This effect of filaments is also present in the indices derived from optically thick lines such as CaII K , MgII k, and H-alpha. Furthermore, it is not clear whether the irradiance effect of the quiet chromospheric network contributes the same as the plage-related variation of the HeI line (Harvey and Livingston, 1994). The 10.7 cm radio flux has contributions from both free-free (bremsstrahlung) emissions (related to the weaker fields of plages and the network) and gyro resonance emission (related to the strong magnetic fields of sunspots). A more recent analysis suggests that the free-free contribution from the low chromosphere dominates its variability, except at the time of very high solar activity (Tapping and Harvey, 1994). It should also be mentioned that the lack of good synoptic data sets for sunspots, especially during the last 15 years when high precision irradiance observations exist, introduce additional uncertainties into the irradiance models. The publication of high precision area and position of sunspots in the Greenwich Catalogue ended in 1976. Since then the sunspot observations are reported in the NOAA World Data Center Solar Geophysical Data catalogue with much less precision. The current precision of the sunspot area measurements is 20-25% for large spots, but it can be as high as 50% for small spots (Sofia et al., 1982). Fröhlich et al. (1994) have recently demonstrated that the irregularities in the sunspot area cause the largest uncertainty in the PSI. The results of sunspot photometry show that the contrast of sunspots depends on the area of sunspots (Steinegger et al., 1990; Chapman et al., 1992) and the current PSI model overestimates the sunspot effect on total irradiance with about 40% (Brandt et al., 1990). Moreover, the change of the contrast of sunspots over the solar cycle (Maltby et al., 1986) is not taken into account in the irradiance models.

It has been shown that the evolution of active regions plays an important role in the changes of total irradiance (Willson, 1982; Pap, 1985; Fröhlich and Pap, 1989). However, neither the evolution of sunspots or plages is taken into account in the current irradiance models; only mean values of contrast for these structures go into the regression calculations. No broad statistical study of the change of radiation from the sunspots and plages due to their evolution has been done to give us the needed data to improve the empirical transformations. Full disk measures of solar activity, such as the He-line equivalent width, Mg core-to-wing ratio, and the 10.7 cm radio flux contain the evolutionary changes but in an undifferentiated form where sunspot, plage, and network contributions cannot be separated.

As has been described above, the principal sources of the total solar irradiance variability are known to be sunspots and bright faculae and plage regions. However, we still need to separate a quiet-Sun component from a chromospheric network component to account for the total amount of radiation from layers above the photosphere. Thus, there is the possibility of slow secular changes both in the network and globally in the Sun that remain unmeasured at this time. In addition to the unmeasured long-term changes, we also see the underestimate of total irradiance by surrogates of chromospheric origin at the times of solar maximum in the last two solar cycles.

We cannot assume that the difficulty lies in an instrumental problem: we must also consider that the empirical models fail at solar maximum because of a change in how the solar atmosphere responds to the presence of strong magnetic fields sustained over the 2-3 years of solar maximum. The empirical regression models derived from the full disk surrogates are controlled by the ascending and descending phases of the solar cycle and not by the mutual behavior of the various measurements during solar maximum. Thus, it may be that we simply do not have the same empirical transformations between total irradiance and surrogates such as CaII K, 10.7 cm, HeI 1083, and MgII h & k at solar maximum as during the periods when the emerging magnetic flux is both low and changing steadily at the beginning and end of a solar cycle.

Since the physics of the formation of the total irradiance and our common solar activity indices is not identical, our empirical models give only estimates with a precision of a few percent for the variability of the total solar irradiance due to both sunspots and faculae (plages). This inherent limitation of the current simple empirical models of total irradiance must be taken into account when irradiance models are used in climatic studies. Considering the significance of irradiance variations as a potential natural source of climatic changes, continuous observations of total irradiance from space are required to maintain a long-term, high precision irradiance database for climatic studies. In parallel with the direct irradiance observations, advanced theoretical and statistical studies are necessary to understand why, how and on what time scale the total radiative output of the Sun changes and thus, to reconstruct and predict the solar induced climatic changes.

I.5. Discussions led by J.M. Pap

Schlesinger: This is all very depressing. Do we really have to wait 50 years for the data to accumulate before we can answer this question of solar influence? Are there any simplifying insights or pivotal experiments that can be used to speed up our understanding of solar change and its effect in the atmosphere?

Parker: Given the complexity of both the solar atmosphere and the Earth's climate system, it will probably take 50 years to really understand how and why they vary as they do.

Schlesinger: Cannot we clarify the existence of long-term irradiance changes from observations prior to the space observations?

Pap: The current absolute accuracy of radiometers is low (+/- 0.2%), therefore individual measurements from the ground, balloons, aircraft, rockets, and from the space shuttle cannot reveal the small amplitude changes in total irradiance. In case of the measurements from the ground, balloons, and aircraft, the atmospheric effects cannot be neglected, the accuracy of these measurements is not better than 1%. Therefore, long-term, high precision irradiance observations from space with overlapping strategy are needed to reveal the climatically significant irradiance changes.

Lean: It is not yet determined whether the discrepancy between irradiance observations and models is solar, rather than instrumental, effect, especially in the late seventies and in 1980 considering the degradation of the instruments.

Pap: The degradation of the SMM/ACRIM I instrument was 600 ppm over its 9.5 years of operational interval. The degradation of the radiometers are caused by the high energy solar fluxes, such as the UV and particle fluxes. Because of this, the ACRIM instruments contain three radiometers: one of them measures the solar energy flux continuously, the second one once a month, and the third one is exposed to solar radiation only in every second or third month. With this operational mode, the change of total irradiance due to the instrument degradation was corrected to a 50 ppm level. The difference between ACRIM I total irradiance and its model estimate in 1980 was 300 ppm and it is well above the degradation level. Furthermore, the Nimbus-7/ERB instrument which started to operate in late November of 1978 showed the same trend as the ACRIM I instrument in 1980.

Fröhlich: Furthermore, the degradation of the SOVA instrument on EURECA during the first month of its operation was the same as that of the ACRIM I.

Hoyt: There is a discrepancy between the solar irradiance models and the Nimbus-7 measurements for 1979-1980 and 1987-1988. The measured irradiance is higher than the model values and modelers often ask what is wrong with the measurements. In order to force the measurements to match the model, the algorithm to reduce the counts to irradiances could be changed. Changes in the pointing, temperature sensitivity, electronic gain, or contamination could be made, but there is no experimental evidence to make the algorithm changes. Even if the changes could somehow be made, the results after 1980 would probably be altered such that a new model/measurement disagreement would occur. For 1987 and later, the Nimbus-7 radiometer was operating under increasingly favorable conditions as other platform instruments were turned off. Any differences then are even harder to explain by measurements problems.

Pap: It should be added that the UARS/ACRIMII total solar irradiance is also well above the irradiance estimates computed from our standard surrogates at the maximum of solar cycle 22. This difference strongly suggests that irradiance values cannot be estimated accurately during solar maximum. Research at the U.S. National Solar Observatory and at the Jet Propulsion Laboratory will determine if the magnetic activity during the maxima of both cycles 21 and 22 provides a mechanism to explain the difference. We should keep in mind that our common surrogates are radiations formed in atmospheric layers higher than the photosphere where more than 90% of total irradiance is emitted. Therefore, these indices represent different physical conditions than in the photosphere. Because of the lack of a physical theory to describe the radiation emitted by different solar structures, we are unable to map radiative losses from higher layers to those lower down accurately and, thus, explain the difference between observation and our current empirical models.

Livingston: The observations at the He I line at 1083 nm were essentially performed to study coronal holes. It has been a surprise that the correlation between the variations in the He-line equivalent width and total as well as UV irradiances is so good.

White: To improve the surrogates used in the irradiance models, digitization of different features (sunspots, faculae, network) is needed from a new set of CaII K images and magnetograms. As a further step, computation of the solar spectral energy distribution is necessary, taking into account the new photometric results from Chapman and Harvey using image analysis.

II. Solar Output in The Maunder Minimum: O.R. White, HAO

a. How well do we know the Sun's output in the Maunder Minimum?

We have three methods for estimating the Sun's total irradiance in the Maunder Minimum: 1) decrease in solar energy required for the amount cooling in the terrestrial atmosphere in the 17th century, 2) extrapolation of empirical models back in time, and 3) estimates from stars in a 'Maunder Minimum' state. The estimate of the decrease of irradiance from contemporary values ranges from 0.2% to 0.7% with a consensus value at about 0.3%.

Discussion led by O.R. White

Schlesinger: What is the range of variation in the solar input do we have to consider realistically?

White: Zero to 1 percent in the radiative output. This estimate allows for factor of 2X on both the low and high side of current estimates from extrapolation of the regression formula back 300 years. The statistical realities in this extrapolation are horrible because we go well outside the fitted range. I also included the low values obtained from the analysis of solar-type stars (Zhang, et al., 1993) in this guess as well as the bias to the high side from Damon's early climate model analysis (Jirikowic and Damon, 1994). Other sources for estimates of the solar output in the Maunder Minimum are Lean, et al. (1992), Nesme-Ribes and Mangeney (1992), Nesme-Ribes, et al. (1993), and Hoyt and Schatten (1993).

Concerning the plasma output, we have information on the variability of the plasma output of the Sun; but have little notion of how it may couple to the lower atmosphere and cause sensible changes. Geomagnetic activity shows a steady increase over the last fifty years, and this reflects both the level of activity in the solar wind and variability of Earth's magnetosphere. Radionucleide data clearly show variation in cosmic ray flux caused by changes in the output of solar plasma over the last seven thousand years. These data show characteristic periodicities of 208, 88, and 11 years. Conversion of these radioisotope variabilities to sources and properties of the solar wind and transient plasma ejections has not been done; therefore, the detailed connection to activity as measured by sunspot occurrence remains in a qualitative state at this time.

Fröhlich: It should also be kept in mind that the correlation between total irradiance and its surrogates depends on the phase of the solar cycle. Models give different results for different phases of solar cycle since the value of the slope and intercept calculated from linear regression changes with the time during the solar cycle, even on yearly time scale. How do we know what is the right value of the intercept and slope when we extend the models back to the 17th century?

Lean: Current irradiance models extended back to the time of the Maunder Minimum use the a, b linear regression coefficients determined from the decreasing phase of solar cycle 21.

White: One basic limitation in the current extrapolations is the lack of data covering many solar cycles: the data used in these estimates covers only two solar maxima and one solar minimum, at best. As such, the regression formulae describes only the 'strength' of two recent solar activity cycles and not any long-term variation in its zero point, i.e., variability of the quiet stellar atmosphere undisturbed by emergence of strong magnetic fields at the visible surface.

b. Did the Maunder Minimum really occur?

This point is now moot because of the careful work of Dr. Ribes and her colleagues in the study of the Paris Observatory archive for the period from 1640 to 1720. This study of the daily observations by Picard, La Hire, Cassini, and their successors clearly establishes the basic character of the solar cycle in the period from AD1645 to AD1715, within which no sunspots were recorded during extended periods despite regular daily observation. Ribes, et al. (1987), Ribes, et al. (1989) and Ribes and Nesme-Ribes (1993) give details of this analysis of the Paris Observatory archive. In particular, between AD1690 and AD1702 only one spot was observed while an average of about 15 daily observations per month were recorded. This work gives the solar theorist a quantitative description of the rate and location of magnetic flux emergence and the change in solar rotation rate during this unique episode of solar history. Continued examination of the historical record only strengthens the original arguments made by Eddy (1976) for the existence and importance of the Maunder Minimum episode. Its identification in the 10Be and 14C paleoclimate records points to similar occurrences in the past and suggests its reality as a naturally occurring irregularity in the cyclic variation of stars of the same age and mass as the sun.

Discussion:

White: Hoyt's current re-examination of the quality of sunspot observations from 1610 to the present shows that the Maunder Minimum period was well covered by other observers throughout western Europe. A lack of solar observations did occur in the mid 18th century, well after the resumption of the solar cycle in AD1710 (see Hoyt, Schatten, and Nesme-Ribes, this volume).

c. What are the implications for the solar dynamo?

Clearly the lack of sunspots for several decades and their occurrence in a narrow latitude band near the solar equator (about +/-18 deg latitude) is an extremum that has to be allowed by a credible dynamo model. More important may be the observation that between 1660 and 1710 almost all of the sunspots emerged in the southern solar hemisphere. The apparent increase in the degree of differential rotation on the Sun also gives another observational constraint to the theory.

Discussion:

White : Throughout this workshop, several colleagues (Nesme-Ribes, Krause, Sokoloff, Kremliovsky) mentioned the potential importance of 'mixed parity' dynamos in understanding the occurrence of sunspots in only one solar hemisphere. See Sokoloff and Nesme-Ribes (1994) and papers given at this workshop by these colleagues. Periods of large N-S asymmetry correspond to times of minimum total magnetic energy produced by the alpha-omega dynamo described by Brandenburg, Krause, and Tuominen (1989, Turbulence and Nonlinear Dynamics, (Meneguzzi, et al. editors)).

White: We have to ask whether the toroidal solar dynamo operating at low latitudes actually ceased generating magnetic field during this time, or were conditions in the torus and the convection zone such that flux tubes did not rise to the visible surface of the Sun?

d. Did the Maunder Minimum really contribute to the Little Ice Age?

Since the Maunder Minimum lasts only about 70 years, the question arises about its causal influence in the 300 year Little Ice Age, which lasted from AD1550 to AD1850. In his discussion, Bradley cautions not to oversimplify the Little Ice Age as a global phenomena involving only the mean temperature or glacial advance. Our view is biased by the picture in Europe where the most complete records exist. These indicate that this local climate was much harsher than today, a condition which did not apply worldwide. He also points out the need to consider the change in volcanic activity and aerosol production throughout this period and into the 20th century as another major factor in climate modulation.

Discussion:

Ribes: The deepest part of the Maunder Minimum corresponded to periods of extreme low temperature in Western Europe.

Parker: The Maunder Minimum was not the only solar excursion in this period. Both the Dalton and Spörer Minima occurred in this time interval, i.e., at AD1800 and AD1500, respectively.

e. What about solar influence in the Medieval Warm Epoch (AD1000-AD1250)?

Discussion:

White: Is there any question about the occurrence of the Medieval Warm Epoch?

Bradley: In the report of the workshop devoted to this period, Malcolm Hughes concludes that the moderated climate observed in western Europe and eastern North America may not have occurred on a global scale. Thus, it is not possible to judge that the mean global temperature increased significantly.

White: In the 14C and 10Be time series shown at this meeting, I notice that none of them yet extend earlier than about AD1500. Is there any doubt about the so-called 'Grand Maximum" of solar activity in the period from AD1100 to AD 1300 in the radionucleide record?

Raisbeck: The earlier data are available: we did not show them because the analysis is not complete. Yes, the Grand Maximum is present in our new time series.

White: This period is the example of a 'positive' extremum in solar output as opposed to the 'negative' extrema shown by Maunder Minima episodes of very low solar activity. Although the Maunder Minima episodes appear to be easier to identify and, therefore, study as a isolated phenomenon; the positive extrema should not be ignored. The 10Be level today is approaching the high levels seen in the 12th century as pointed out by Jirikowic and Damon (1994); therefore, characteristics of the AD1100-AD1300 period may indicate the direction of climate forcing by the sun when the level of solar activity remains high for many decades. However, Bradley's caution at this meeting about the Little Ice Age also applies here, i.e., we may not be dealing with a global changes described simply by, say, the mean global temperature or a single forcing function strong enough to drive the mean climate state above the natural variability of the Earth's atmosphere.

Acknowledgements: This paper was completed at the Jet Propulsion Laboratory, California Institute of Technology under a contract with the National Aeronautics and Space Administration (J.M.P.) and at the High Altitude Observatory, National Center for Atmospheric Research, with partial support under Contract NASW 4721 to the Solar Physics Research Corporation (O.R.W.). The National Center for Atmospheric Research is sponsored by the U.S. National Science Foundation. The authors thank D. Hoyt and Drs. R.C. Willson and C. Fröhlich for their insight and helpful comments.

References

Abbot CG (1925) Transactions of the International Astronomical Union 2:32.

Brandenburg A, Krause F, Tuominen I, (1989) Parity Selection in Non-linear Dynamos, Turbulence and Nonlinear Dynamic Effects in MHD Flows edited by Meneguzzi, Pouquet, and Sulem, Elsevier.

Brandt PN, Schmidt V, Steinegger M (1990) On the Solar Physics Area Ratio of Sunspots, Solar Physics 129:191-194.

Brandt PN, Stix M, Weinhardt W (1994) Modelling Solar Irradiance Variations with Area Dependent Photometric Sunspot Index, Solar Physics, in press.

Brusa RW, Fröhlich C (1975) Scientific Discussions IPC IV, World Radiation Center, Davos, Switzerland.

Chapman G (1987) Variations of Solar Irradiance due to Magnetic Activity, Annual Review of Astronomy and Astrophysics 25:633-667.

Chapman G (1994) Photometric Observations of the Sun, in Proceedings of IAU Colloquium 143 "The Sun as a Variable Star: Solar and Stellar Irradiance Variations", edited by J.M. Pap, C. Fröhlich, H.S. Hudson and S. Solanki, Cambridge Univ. Press, in press.

Chapman G, Lawrence JK, Hudson HS (1992) Photometric Measurements of Sunspot Deficits and Facular Excess, in Proceedings of Solar Electromagnetic Radiation Study for Solar Cycle 22 edited by R.F. Donnelly, NOAA/ERL/SEL, Boulder, CO., p. 135.

Crommelynck D (1973) Instrumentation Theory of Absolute Radiometry, Pub. A. 81, Royal Meteorological Institute of Belgium, Brussels, Belgium.

Crommelynck D, Domingo V, Fichot A, Lee R (1994) Irradiance Observations from the EURECA and ATLAS Experiments, in Proceedings of IAU Colloquium 143 The Sun as a Variable Star: Solar and Stellar Irradiance Variations, edited by J.M. Pap, C. Fröhlich, H.S. Hudson and S. Solanki, Cambridge Univ. Press, in press.

Eddy JA (1976) The Maunder Minimum, Science 192:1189-1202.

Eddy JA (1977) Climate and the Changing Sun, Climatic Change 1:173-190.

Foukal P (1981) Sunspots and Changes in the global output of the Sun, in Physics of Sunspots, edited by L. Cram and J. Thomas, SPO Publication p. 391-423.

Foukal P, Lean J (1988) Magnetic Modulation of Solar Luminosity by Photospheric Activity, Astrophysical Journal 328:347-357.

Foukal, P., Fowler, LA, Livshits, M (1983) A Thermal Model of Sunspot Influence on Solar Luminosity, Astrophysical Journal 267:863-871.

Fox P, Sofia S (1994) Convection and Irradiance Variations, in Proceedings of IAU Colloquium 143 The Sun as a Variable Star: Solar and Stellar Irradiance Variations, edited by J.M. Pap, C. Fröhlich, H.S. Hudson and S. Solanki, Cambridge Univ. Press, in press.

Fox P, Sofia S, Chan K (1992) Numerical Simulations of the Surface Energy Balance of Sunspot-Like Objects, in Proceedings of Solar Electromagnetic Radiation Study for Solar Cycle 22, edited by R.F. Donnelly, NOAA/ERL/SEL, Boulder, CO., p. 136-141.

Friis-Christensen E, Lassen K (1991) Length of the Solar Cycle: An Indicator of Solar Activity Closely Associated with Climate, Science 254:698-700.

Fröhlich C (1977) Contemporary Measurements of the Solar Constant, in The Solar Output and its Variations edited by O.R. White, Colorado Assoc. Univ. Press, p. 93-110.

Fröhlich C (1992) Solar Irradiance Variability, in Proceedings of Solar Electromagnetic Radiation Study for Solar Cycle 22, edited by R.F. Donnelly, NOAA/ERL/SEL, Boulder, CO., p. 1-10.

Fröhlich C (1994) Irradiance Variations of the Sun, in Proceedings of IAU Colloquium 143 The Sun as a Variable Star: Solar and Stellar Irradiance Variations, edited by J.M. Pap, C. Fröhlich, H.S. Hudson and S. Solanki, Cambridge Univ. Press, in press.

Fröhlich C, Pap J (1989) Multi-Spectral Analysis of Total Solar Irradiance Variations, Astronomy and Astrophysics 220:272-280.

Fröhlich C, Pap J, Hudson HS (1994) Improvement of the Photometric Sunspot Index and Changes of the Sunspot Contrast within the Solar Cycle, Solar Physics, in press.

Harvey J (1984) He 10830A Irradiance: 1975-1983, in Proceedings of Solar Irradiance Variations on Active Regions Time Scales, edited by B.J. LaBonte, G.A. Chapman, H.S. Hudson, and R.C. Willson, NASA CP-2310, 197-211.

Harvey JW, Livingston WC (1994) Variability of the Solar He I 10830A Triplet, in Proceedings of IAU Symposium 154: Solar Infrared Astronomy, edited by D. Rabin, J. Jeffreys, and C. Linsey, Kluwer Academic Publishers, p. 59-64.

Hickey JR, Karoli AR (1974) Radiometric Calibrations for the Earth Radiation Budget Experiment, Applied Optics 13:523-533.

Hoyt DV (1979) The Smithsonian Astrophysical Observatory Solar Constant Program, Reviews of Geophysics and Space Physics 17:427-458.

Hoyt DV, Kyle HL, Hickey JR, Maschhoff RH (1992) Nimbus-7: Over Twelve Years of Solar Total Irradiance Measurements, in Proceedings of Solar Electromagnetic Radiation Study for Solar Cycle 22, edited by R.F. Donnelly, NOAA/ERL/SEL, Boulder, CO., p. 43-48.

Hoyt DV, Schatten KH (1993) Solar Irradiance Variations, Journal of Geophysical Research 98:18,895-18,906.

Hoyt DV, Schatten KH, Nesme-Ribes E (1994) New Reconstruction of Solar Activity: 1610-1993. This volume.

Jirikowic JL, Damon PE (1994) The Medieval Solar Activity Maximum, Climatic Change, in press.

Kuhn J, Libbrecht KG, Dicke R (1988) The Surface Temperature of the Sun and Changes in the Solar Constant, Science 242:908-911.

Kyle HL, Hoyt DV, Hickey JR, Maschhoff RH, Vallette BJ (1993) Nimbus-7 Earth Radiation Budget Calibration History - Part I: The Solar Channels, NASA Reference Publ. 1316, GSFC, Greenbelt.

Lean J, Skumanich A, White OR (1992) Estimating the Sun's Radiative Output during the Maunder Minimum, Geophysical Research Letters 19:1591-1594.

Livingston W, Wallace L, White OR (1988) Spectrum Line Intensity as a Surrogate for Solar Irradiance Variations, Science 240:1765-1767.

Maltby P, Avrett EH, Carlsson M, Kjedseth-Moe, O, Kurucz RL, Loeser R (1986) A New Sunspot Umbral Model and its Variation with the Solar Cycle, Astrophysical Journal 306:284-303.

Marquette WH (1992) Calcium Plage Observations at Big Bear Solar Observatory, in Proceedings of Solar Electromagnetic Radiation Study for Solar Cycle 22, edited by R.F. Donnelly, NOAA/ERL/SEL, Boulder, CO., 154-159.

Martin JE, Fox NP (1994) Cryogenic Absolute Radiometer - CSAR, Solar Physics, in press.

Mecherikunnel AT, Lee RB, Kyle HL, Major ER (1988) Intercomparison of Solar Total Irradiance Data from Recent Spacecraft Measurements, Journal of Geophysical Research 93:9503-9509.

Nesme-Ribes E, Ferriera EN, Sadourny R, Le Treut H, Li L (1993) Solar Dynamics and its Impact on Solar Irradiance and the Terrestrial Climate, Journal Geophysical Research 98:18,923-18,935.

Nesme-Ribes E, Mangeney A (1992) On a Plausible Mechanism Connecting the Maunder Minimum to the Little Ice Age, Radiocarbon 34:263-273.

Nishikawa J (1994) On the Cause of Total Irradiance Variations by the CCD Solar Surface Photometer, Solar Physics, in press.

Pap J (1985) Activity of Sunspots and Solar Constant Variations during 1980, Solar Physics 97:21-33.

Pap J, Fröhlich C (1992) Multi-Variate Spectral Analysis of Short-Term Irradiance Variations, in Proceedings of Solar Electromagnetic Radiation Study for Solar Cycle 22, edited by R.F. Donnelly, NOAA/ERL/SEL, Boulder, CO., p. 62-77.

Pap J, Marquette WH, Donnelly RF (1991) Modelling Solar Irradiances Using Ground-Based Measurements, Advances in Space Research 11:271-274.

Pap J, Willson RC, Fröhlich C, Donnelly RF, Puga L (1994) Long-Term Variations in Total Solar Irradiance, Solar Physics, in press.

Reid GC (1987) Influence of Solar Variability on Global Sea Surface Temperature, Nature 329:142-143.

Ribes E, Ribes J-C, Barthalot R (1987) Evidences for a large Sun and a Slower Rotation during the Seventeenth Century, Nature 326:52-54.

Ribes E, Ribes J-C, Merlin P, Barthalot R, (1989) Absolute Periodicities in the Solar Diameter Derived from Historical and Modern Data, Annales Geophysicae 7:321-329.

Ribes J-C, Nesme-Ribes E (1993) The Solar Sunspot Cycle in the Maunder Minimum AD1645 to AD1715, Astronomy and Astrophysics 276:549-563.

Romero J, Wehrli C, Fröhlich C (1994) Solar Total Irradiance Variability from SOVA 2 on Board EURECA, Solar Physics, in press.

Schatten K, Mayr H (1992) On the Structure and Contrast of Sunspots and Faculae, in Proceedings of Solar Electromagnetic Radiation Study for Solar Cycle 22, edited by R.F. Donnelly, NOAA/ERL/SEL, Boulder, CO., p. 142-153.

Schatten K, Mayr H, Omidvar K, Maier M (1986) The Hillock and Cloud Model for Faculae, Astrophysical Journal 311:460-473.

Sofia S, Oster L, Schatten K (1982) Solar Irradiance Modulation by Active Regions during 1980, Solar Physics 80:87-98.

Sokoloff D, Nesme-Ribes E (1994) The Maunder Minimum, a Mixed-Parity Dynamo Mode?, Astronomy and Astrophysics, in press.

Spruit H (1982a) Effect of Spots on a Star's Radius and Luminosity, Astronomy and Astrophysics 108:348-356.

Spruit H (1982b) The Flow of Heat Near a Spot, Astronomy and Astrophysics 108:356-360.

Spruit H (1994) Theory of Irradiance Variations, in Proceedings of IAU Colloquium 143, The Sun as a Variable Star: Solar and Stellar Irradiance Variations, edited by J.M. Pap, C. Fröhlich, H.S. Hudson and S. Solanki, Cambridge Univ. Press, in press.

Steinegger M, Brandt P, Pap J, Schmidt, W (1990) Sunspot Photometry and the Total Solar Irradiance Deficit Measured in 1980 by ACRIM, Astrophysics and Space Science 170:127-133.

Tapping KF (1987) Recent Solar Radio Astronomy at Centimeter Wavelengths, the Temporal Variability of the 10.7 cm Flux, Journal of Geophysical Research 92:829-838.

Tapping KF, Harvey, K (1994) Coronal Microwave Emissions, in Proceedings of IAU Colloquium 143, The Sun as a Variable Star: Solar and Stellar Irradiance Variations, edited by J.M. Pap, C. Fröhlich, H.S. Hudson and S. Solanki, Cambridge Univ. Press, in press.

Willson RC (1982) Solar Irradiance Variations and Solar Activity, Journal of Geophysical Research 86:4319-4326.

Willson RC (1984) Measurements of Solar Total Irradiance and its Variability, Space Science Reviews 38:203-242.

Willson RC (1994) Irradiance Observations from the SMM, UARS, and ATLAS Experiments, in Proceedings of IAU Colloquium 143, The Sun as a Variable Star: Solar and Stellar Irradiance Variations, edited by J.M. Pap, C. Fröhlich, H.S. Hudson and S. Solanki, Cambridge Univ. Press, in press.

Willson RC, Hudson HS (1988) Solar Luminosity Variations in Solar Cycle 21, Nature 332:810-812.

Willson RC, Hudson HS (1991) The Sun's Luminosity over a Complete Solar Cycle, Nature 351:42-44.

Wilson P (1981) Theories of Sunspot Structure and Evolution, in Physics of Sunspots, edited by L. Cram and J. Thomas, SPO Publication p. 83-97.

Woodard M, Hudson HS (1983) Frequencies, Amplitudes and Linewidths of Solar Oscillations from Solar Total Irradiance Observations, Nature 305:589-593.

Zhang Q, Soon WH, Baliunas SL, Lockwood GW, Skiff BA, Radick RR (1993) The Brightness Variation of the Sun in Past Centuries, Nature submitted.

THE QUASI-BIENNIAL OSCILLATION IN THE EARTH'S ATMOSPHERE AND ITS LINKS TO LONGER PERIOD VARIABILITY

James R. Holton
Atmospheric Sciences Department
University of Washington
Seattle WA 98195
U. S. A.

Summary

The dynamical theory of the quasi-biennial oscillation (QBO) of the zonal winds in the equatorial stratosphere is reviewed, and the role of the QBO in producing interannual variability in the extratropical stratosphere is discussed. It is argued that the extratropical winter stratosphere has two basic climate regimes; one in which the eddy motions are strong and polar temperatures are warm, and a second in which the eddy motions are weak and polar temperatures are cold. The equatorial QBO influences the extratropics through the mechanism of changing the probability that the winter stratosphere lies in either the strong eddy or weak eddy regime. This influence appears to undergo a quasi-decadal variation (QDO), which may or may not be related to the solar cycle. It is speculated that if the QDO is related to solar forcing, the mechanism is one in which the net effect of weak solar cycle forcing alters the probability density function of the climate regimes in such a way that the strong eddy regime becomes more probable in the westerly QBO phase during sunspot maximum, than during sunspot minimum.

Introduction

In the extratropical stratosphere variations of the zonally averaged wind and temperature fields are dominated by the annual cycle; in winter the flow is westerly (from the west) and temperatures decrease from equator to pole, in summer the flow is easterly (from the east) and temperatures increase from equator to pole. This annual cycle in the extratropics can be understood in general terms as a linear response to the annual cycle in solar heating, which in the stratosphere has a maximum at the summer pole and a minimum at the winter pole. In the equatorial stratosphere, however, the annual cycle is quite weak. The major oscillation in wind and temperature in that region is the quasi-biennial oscillation (QBO). This oscillation features zonally symmetric easterly and westerly wind regimes that alternate regularly with period varying from about 24 to 30 months. Due to the variation of the QBO period, and the absence of any significant external forcing corresponding to this variable period, the source of the QBO must be sought in the internal dynamics of the atmospheric system. The QBO is

NATO ASI Series, Vol. I 25
The Solar Engine and Its Influence
on Terrestrial Atmosphere and Climate
Edited by Elizabeth Nesme-Ribes

a notable example of a low frequency oscillation generated by the nonlinear dynamics of the atmosphere.

Although the QBO is most prominent in the tropics, there is evidence for significant variability in the temperature, wind and ozone distributions in the extratropics that is correlated with the equatorial wind QBO (Holton and Tan, 1982)). The Northern Hemisphere winter polar stratosphere tends to have colder temperatures and stronger zonal winds during years when the equatorial QBO winds in the lower stratosphere are westerly than in years when the equatorial QBO winds are easterly. Labitzke and van Loon (see paper by Labitzke in this volume for details and references) found that this association between cold polar stratospheric temperatures and the westerly phase of the equatorial stratospheric QBO is modulated on a quasi-decadal time scale. Their work suggests that there is a significant correlation between polar stratospheric temperatures and the 11-year solar cycle if the data is stratified according to phase of the equatorial QBO. Some workers have argued that this correlation is not statistically significant [Baldwin and Dunkerton (1989), Hamilton (1990), and Salby and Shea (1991)]. But, whatever the merits of the various statistical studies of this phenomenon, dynamical theory must still explain how the equatorial QBO can influence winds and temperatures in the extratropical stratosphere, and why this influence undergoes a quasi-decadal oscillation (QDO).

Many of the past attempts to link solar variations with atmospheric responses have involved calculating the *linear* response of the atmosphere to altered solar forcing. However, with the exception of the diurnal and annual cycles and their harmonics, most atmospheric variability can only be understood through *nonlinear* dynamical thinking. Due to the small perturbations in forcing associated with the solar cycle, any model that attempts to linearly link variations in solar forcing to atmospheric dynamical responses is almost certain to fail. An understanding of the quasi-decadal variations in the atmosphere discovered by Labitzke and van Loon will surely require a nonlinear dynamical model. This will be true whether or not the QDO ultimately turns out to be related to solar forcing or not.

In this paper we briefly review the dynamical theory of the equatorial QBO and our current understanding of the mechanisms by which it may influence the extratropical circulation. We will then be in a position to speculate on the nature of the QDO and how it might be related to the solar cycle.

Overview of the Dynamics of zonal mean flows

Before discussing the dynamical theory of stratospheric motions it is useful to review the basic meteorological balances involved in large-scale atmospheric motions. We are here primarily concerned with the longitudinally averaged dynamical equations. Since our purpose is to illustrate the qualitative aspects of wave driving of the zonal mean flow as simply as possible, we make a number of simplifications in the dynamical equations (see, e. g., Andrews, et al., 1987). We first make the hydrostatic approximation, which allows us to eliminate explicit reference to the variable density of the atmosphere by replacing geometric altitude by the "log pressure" coordinate z defined as

$$z = -H \ln(p / p_s)$$

where H is a reference density scale height, p is pressure, and p_S is a standard reference pressure, usually taken to be 1000 hPa. Surfaces of constant z are thus constant pressure surfaces, and geopotential replaces pressure as a dependent variable. For simplicity we replace the spherical coordinates by Cartesian coordinates (x, y) directed eastward and northward, respectively. The zonal and meridional components of the momentum equation can then be written as

$$Du/Dt - fv + \Phi_x = X \tag{1a}$$

$$Dv/Dt + fu + \Phi_y = Y \tag{1b}$$

Here, (u, v) are the (x, y) components of velocity, f is the Coriolis parameter, Φ is the deviation of geopotential from a standard profile $\Phi_0(z)$, subscripts x and y denote partial derivatives on a constant pressure surface, (X, Y) are components of unspecified mechanical forcing (e. g., friction), and

$$\frac{D}{Dt} \equiv \frac{\partial}{\partial t} + u\frac{\partial}{\partial x} + v\frac{\partial}{\partial y} + w\frac{\partial}{\partial z}$$

is the rate of change following the motion.

The geopotential and temperature fields are related by the hydrostatic equation, which can be expressed as

$$\Phi_z = H^{-1}RT \tag{1c}$$

where T designates the deviation of temperature from a standard reference profile, $T_0(z)$, and R is the gas constant for dry air. [Note that density does not appear in (1c).] The thermodynamic energy equation states simply that temperature changes following the motion are caused only by adiabatic expansion (compression) and diabatic heating (cooling):

$$DT/Dt + N^2HR^{-1}w = J / c_p \tag{1d}$$

where N is the buoyancy frequency (a measure of static stability), c_p is the specific heat at constant pressure, and J is the diabatic heating per unit mass. Finally, the mass continuity equation may be expressed as

$$u_x + v_y + \rho_0^{-1}(\rho_0 w)_z = 0 \tag{1e}$$

where $\rho_0(z) = \rho_{00}\exp(-z/H)$ is a standard atmosphere density profile, and ρ_{00} is the standard density at $z = 0$.

Low frequency variations, such as the QBO, involve the interaction of longitudinally varying disturbances (referred to as eddies, and denoted by primed variables) with the longitudinally averaged flow (referred to as the mean-flow, and denoted by overbars). This sort of average is an *Eulerian mean* since it is evaluated at fixed latitude, pressure, and time. In the extratropical stratosphere eddy motions are dominated by planetary-scale waves for which the velocities are in approximate geostrophic balance with the pressure field (i. e., the Coriolis force due to the rotation of the earth and the horizontal pressure gradient force are in approximate balance). For such eddies the horizontal eddy fluxes of heat and momentum are much larger than the vertical eddy fluxes. The Eulerian mean zonal momentum, thermodynamic energy, and continuity equations, obtained by taking zonal averages of (1a, 1d, and 1e) can then be written in the following approximate form:

$$\bar{u}_t - f_0\bar{v} - \bar{X} = -\left(\overline{u'v'}\right)_y \tag{2a}$$

$$\bar{T}_t + \bar{w}N^2HR^{-1} = \bar{J}/c_p - \left(\overline{v'T'}\right)_y \tag{2b}$$

$$\bar{v}_y + \rho_0^{-1}(\rho_0\bar{w})_z = 0 \tag{2c}$$

where we have neglected advection by the mean meridional circulation $(\bar{w},\bar{v})$ in (2a,b) consistent with the fact that $|\bar{u}| \gg |\bar{v}|$. The quadratic terms involving primed variables on the right hand side in (2a) and (2b) represent the meridional fluxes of zonal momentum and heat, respectively.

The zonal mean meridional momentum equation is approximated by geostrophic balance

$$f\bar{u} \approx -\bar{\Phi}_y$$

which can be combined with the hydrostatic relationship to give the very important *thermal wind* relation

$$f_0\bar{u}_z + RH^{-1}\bar{T}_y = 0 \tag{3}$$

This relationship between the zonal mean wind and temperature distributions imposes a strong constraint on the mean meridional circulation $(\bar{v},\bar{w})$. Any process that changes the zonal mean temperature distribution must also change the vertical wind shear. That is, the wind and temperature are uniquely coupled through the pressure distribution. In the absence of a mean meridional circulation the eddy momentum flux divergence in (2a) and eddy heat flux divergence in (2b) would tend separately to change the mean zonal wind and temperature fields, and hence would destroy thermal wind balance. The pressure gradient force that results from any small departure of the mean zonal wind from geostrophic balance will, however, drive a mean meridional circulation, which adjusts the mean zonal wind and temperature fields so that (3) remains valid. In many situations this compensation allows the mean zonal wind to remain unchanged even in the presence of large eddy heat and momentum fluxes. Thus, it is rather inefficient to attempt to diagnose the net eddy forcing of the mean flow from the conventional Eulerian mean equations.

An alternative approach that provides clearer diagnosis of eddy forcing is the transformed Eulerian mean (TEM) formulation introduced by Andrews and McIntyre (1976). This transformation takes account of the fact that in (2b) there tends to be a strong cancellation between the eddy heat flux divergence or convergence and the adiabatic heating or cooling due to vertical motion, so that the net adiabatic heating (cooling) balancing diabatic cooling (heating) is a small residual.

The TEM equations can be obtained from (2) by defining a residual circulation by

$$\bar{v}^* = \bar{v} - \rho_0^{-1} R H^{-1}\left(\rho_0 \overline{v'T'}/N^2\right)_z, \quad \bar{w}^* = \bar{w} + R H^{-1}\left(\overline{v'T'}/N^2\right)_y \tag{4}$$

Substituting into (2) then yields

$$\bar{u}_t - f\bar{v}^* = \bar{X} + \rho_0^{-1}\nabla \bullet \mathbf{F} = \bar{G} \tag{5a}$$

$$\bar{T}_t + N^2 H R^{-1} \bar{w}^* = \bar{J}/c_p \tag{5b}$$

$$\bar{v}^*{}_y + \rho_0^{-1}\left(\rho_0 \bar{w}^*\right)_z = 0 \tag{5c}$$

where f is the Coriolis parameter. Here $\mathbf{F}$, the Eliassen-Palm flux (EP flux), is a vector in the meridional (y, z) plane, which for quasi-geostrophic eddies has the form

$$\mathbf{F} \equiv \left[0,\ -\rho_0 \overline{u'v'},\ \rho_0 f R \overline{v'T'}/\left(N^2 H\right)\right] \tag{6}$$

and $\bar{G}$ designates the sum of the zonal momentum forcing by the EP flux divergence and by unresolved mechanical forcing.

Eqs. (5) show that the eddy heat and momentum fluxes do not act separately on the zonal mean circulation, but in the combination given by the divergence of the EP flux. According to the TEM view the basic role of the eddy fluxes is to provide a zonal force, and thus to tend to accelerate or decelerate the mean zonal flow.

We now consider the response of the stratosphere to a small change in solar heating in the absence of eddy forcing ($\bar{G}$=0). In this case it turns out that the actual temperature is quite close to radiative equilibrium so that we can parameterize the diabatic heating as proportional to the departure of the zonal mean temperature $\bar{T}(y,z,t)$ from its radiative equilibrium value $T_r(y,z,t)$:

$$\bar{J} / c_p = -\alpha_r[\bar{T} - T_r(y,z,t)] \tag{7}$$

where α_r is the Newtonian cooling rate. Substituting from (7) into (5) and combining to eliminate $(\bar{u}, \bar{v}^*, \bar{w}^*)$ we find that the temperature tendency $(\partial T/\partial t)$ satisfies the equation

$$\left[\frac{\partial^2}{\partial y^2} + \frac{\partial}{\partial z}\left(\rho_0^{-1}\frac{\partial}{\partial z}\rho_0\varepsilon\right)\right]\bar{T}_t + \frac{\partial}{\partial z}\left\{\rho_0^{-1}\frac{\partial}{\partial z}\left[\alpha_r\rho_0\varepsilon(\bar{T} - T_r)\right]\right\} = 0 \tag{8}$$

where $\varepsilon(z) \equiv f^2 / N^2$. Here the forcing of the zonal mean temperature tendency is approximately proportional to the departure of temperature from its radiative equilibrium value. The temperature response will, however, lag the forcing (due to the finite relaxation time) and the response will be smoothed in space due to the elliptic operator on the left hand side in (8).

A simple scale analysis of (8) can be used to illustrate the difficulty of driving a significant atmospheric oscillation by thermal forcing in the tropics. For motions with vertical scale $D \sim 10$ km and horizontal scale $L \sim 1000$ km an order of magnitude estimate of the terms in (7) shows that

$$\left(1 + f^2L^2/N^2D^2\right)\frac{\partial T}{\partial t} \sim \alpha_r(T - T_r)\left(f^2L^2/N^2D^2\right)$$

in the extratropics where $f^2L^2/N^2D^2 \sim 1$ the temperature tendency will be of the same order of magnitude as the radiative heating perturbation. But in the tropics, where $f^2L^2/N^2D^2 \sim 10^{-2}$, the temperature tendency is much smaller than the radiative heating perturbation (by a factor of at least 100). Thus, it is unlikely that the QBO could be driven directly by thermal forcing. Rather, mechanical eddy forcing due to the EP flux divergence term as shown in (5a) is far more effective as a direct drive for mean flow changes such as occur in the QBO. The inefficiency of thermal forcing in the tropics can be understood by noting that the thermal wind relation (3) is valid in the zonal mean even quite close to the equator. Thus, any change in the mean temperature distribution must be accompanied by a change in the mean zonal wind distribution. But in the absence of mechanical eddy forcing (5a) shows that the mean zonal wind can only be changed by the Coriolis force associated with the residual meridional circulation. Since the Coriolis force goes to zero at the equator, this term can provide only a very weak zonal flow acceleration. Thus, it must be concluded that eddy forcing is essential for driving mean wind oscillations in the equatorial stratosphere.

Eddy forcing is also required to explain the large departures of the zonal-mean temperature distribution from radiative equilibrium in the winter stratosphere. Although, as stated earlier, temperatures in the winter stratosphere decrease from equator to pole, the observed polar temperatures remain much above radiative equilibrium. This may be understood by considering quasi-steady conditions in which we can neglect the time derivatives in (5a, b). In that case the residual circulation is simply related to the eddy forcing and diabatic heating rate as:

$$-f_0\bar{v}^* = \overline{G}, \qquad N^2HR^{-1}\bar{w}^* = \bar{J}/c_p \tag{9a,b}$$

These relations state that the Coriolis force due to the residual meridional drift is balanced by the zonal force due to large-scale and small-scale eddies; and that the adiabatic heating (cooling) caused by the residual vertical motion is balanced by radiative cooling (heating). The relationships in (9) are not independent, but are connected by the continuity equation (5c), which implies that

$$-\overline{G}_y + f_0\rho_0^{-1}\left(\rho_0\bar{J}\kappa/N^2H\right)_z = 0 \tag{10}$$

where $\kappa \equiv R/c_p$. Eq. (10) shows how the diabatic heating rate must be related to the eddy forcing in this simple steady state model. If the eddy forcing vanishes ($\overline{G} = 0$) the residual meridional drift must vanish. The continuity equation then shows that if $\bar{w}^*$ is zero at the lower boundary it will be zero everywhere and so the diabatic heating must also be zero everywhere and the zonal mean temperature is in a steady state radiative balance.

We conclude that it is necessary to include the effects of eddy forcing to model the observed departure of the mean temperature from its radiatively determined state. For variations on the seasonal time scale the steady balance given in (10) is reasonably accurate. Thus, substituting from (7) into (10) we find that the departure from radiative equilibrium $\delta\overline{T} \equiv (\overline{T} - T_r)$ is given by

$$\left(\frac{\rho_0 R\alpha_r}{N^2H}\delta\overline{T}\right)_z = \frac{\rho_0}{f_0}\overline{G}_y \tag{11}$$

which implies that

$$\delta\overline{T} \sim \tau_r\left(\frac{N^2H^2}{f_0R}\right)\int_z^\infty \overline{G}_y dz$$

where τ_r designates the inverse of the radiative relaxation rate α_r. Eq. (11) shows that radiation behaves like a spring that relaxes temperature toward an equilibrium state (i. e., $\delta\overline{T} \to 0$ as $\tau_r \to 0$), while the zonal force $\overline{G}$ caused by the eddies drives temperature away from equilibrium.

There is much observational evidence to indicate that the large-scale circulation of the stratosphere is driven primarily by the zonal forces exerted by eddies that are generated in the troposphere and propagate into the stratosphere. The eddy forcing $\overline{G}$ cannot, however, be represented simply as an external forcing of the stratosphere, because wave propagation from the troposphere is very sensitive to the zonal wind distribution in the stratosphere, and hence there is a complex nonlinear interaction between mean zonal flow and the eddies. This nonlinear interaction can generate variability in the stratosphere even in the absence of any variations of the eddies in the troposphere.

Dynamics of the equatorial QBO

An elegant example of the generation of stratospheric oscillations by nonlinear forcing due to waves propagating from the troposphere is given by the equatorial wind QBO. As earlier mentioned this oscillation is characterized by zonally symmetric easterly and westerly wind regimes that alternate regularly with periods varying from about 24 to 30 months. Such regimes first appear above 30 km, but propagate downward at a rate of order 1 km month^{-1}. The downward propagation occurs without loss of amplitude between 30 and 23 km, but there is rapid attenuation below 23 km. The oscillation is symmetric about the equator with a maximum amplitude of about 20 m s^{-1}, and an approximately Gaussian distribution in latitude with a half-width of about 12°.

This oscillation is best depicted by means of a time-height section of the zonal wind speed at the equator as shown in Fig. 1. It is apparent from the figure that the vertical shear of the wind is quite strong at the level where the westerly regime is replacing the easterly regime, or vice versa. The observed magnitude of vertical shear of the mean zonal wind at the equator is ~ 5 m s^{-1} km^{-1}. The main factors that a theoretical model of the QBO must explain are the approximate biennial periodicity, the downward propagation without loss of amplitude, and the occurrence of zonally symmetric westerlies at the equator. Because a zonal ring of air in westerly motion at the equator has an angular momentum per unit mass greater than that of the surface of the earth at the equator, no known zonally symmetric advection process could account for the westerly phase of the oscillation. Therefore, there must be a ransfer of momentum by eddies (transferring momentum from the troposphere to the stratosphere, or from higher latitudes to the equator) to produce the westerly accelerations in the downward-propagating shear zone of the QBO.

Observational and theoretical studies have confirmed that vertically propagating equatorial waves provide a substantial portion of the zonal momentum sources necessary to drive the QBO. Equatorial stratospheric waves are subject to thermal damping by infrared radiation,

---

Figure 1: Time-height section of departure of monthly mean zonal winds (m s^{-1}) from the long-term average for each month at the equator. Westerly winds are shaded. (Updated from Naujokat (1986), courtesy of B. Naujokat.)

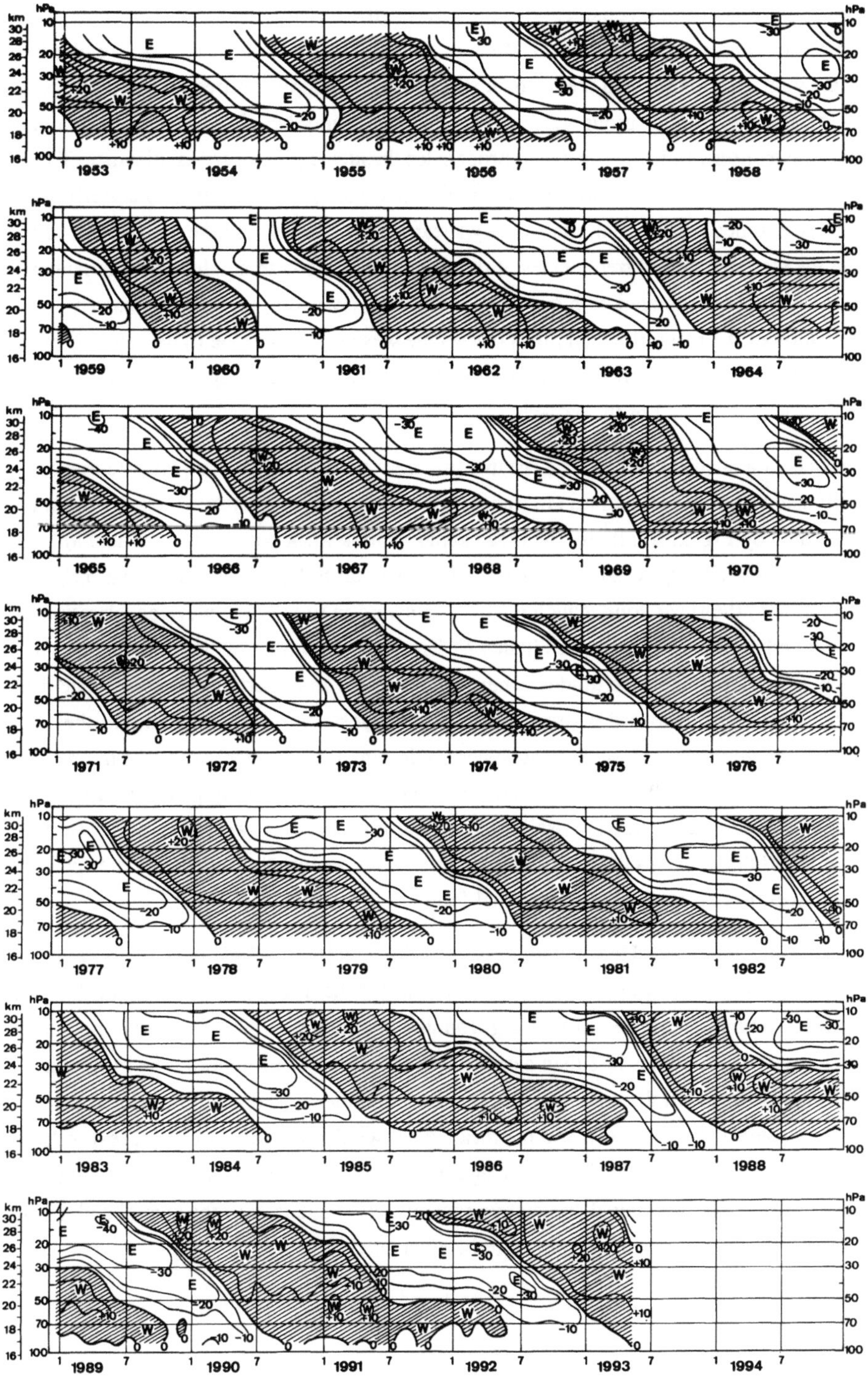
km
hPa
1953
1954
1955
1956
1957
1958
1959
1960
1961
1962
1963
1964
1965
1966
1967
1968
1969
1970
1971
1972
1973
1974
1975
1976
1977
1978
1979
1980
1981
1982
1983
1984
1985
1986
1987
1988
1989
1990
1991
1992
1993
1994

and to both thermal and mechanical damping by small scale turbulent motions. Such damping is strongly dependent on the Doppler shifted frequency of the waves (i. e. the frequency relative to the local mean zonal wind). As the Doppler shifted frequency decreases, the vertical component of group velocity also decreases, and there is a longer time available for the wave energy to be damped as it propagates through a given vertical distance. Thus, westerly waves tend to be damped preferentially in westerly shear zones, and easterly waves tend to be damped preferentially in easterly shear zones. In each case the damping occurs where the Doppler shifted frequencies of the waves decrease with height. The momentum flux convergence associated with damping of the westerly waves provides a westerly acceleration of the mean flow, and thus causes the westerly shear zone to descend. Similarly, the easterly waves are damped in easterly shear zones, thereby causing an easterly acceleration and descent of the easterly shear zone.

How such a mechanism can cause a mean zonal flow oscillation when equal amounts of easterly and westerly momentum are transferred upward across the tropopause by the waves is illustrated by Fig. 2. If initially the mean zonal wind is weak and westerly (panel a), westerly waves will be preferentially damped at lower altitude, and produce a westerly acceleration in the shear zone, which moves downward in time as the wave acceleration is concentrated at ever lower altitudes (panel b). The easterly waves, on the other hand, will initially penetrate to higher altitudes, where they produce an easterly shear zone (panel b), which also moves downward in time. Eventually the westerlies are damped out as they approach the tropopause, and the stratosphere is dominated by easterlies so that the westerly waves can again penetrate to high altitude and produce a new westerly phase. In this manner the mean zonal wind is forced to oscillate back and forth between westerlies and easterlies with a period that depends primarily on the vertical momentum transport and other properties of the waves, not on an oscillating external forcing.

Numerical models (e. g., Holton and Lindzen, 1972, and Plumb, 1977) and a laboratory analog (Plumb and McEwan, 1978) have demonstrated that this sort of oscillation should be regarded as a nonlinear instability arising from the interaction of the waves with the mean flow. For wave amplitudes less than a critical value there is no oscillation, the mean wind varies only slowly with the annual cycle. But, for amplitudes exceeding the critical value the stationary solution breaks down and the mean zonal wind undergoes continual reversals. In the earth's atmosphere the vertically propagating waves in the tropics are thought to be generated primarily by tropical convective systems, but the details of the spatial and temporal distribution of the wave sources remain to be elucidated.

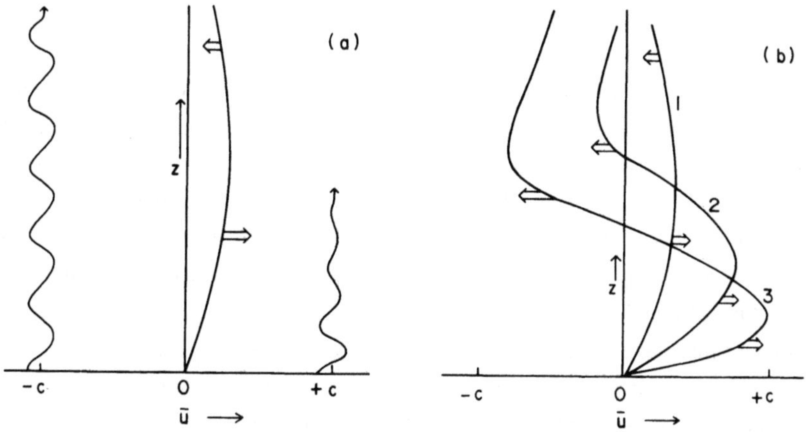

Figure 2: Schematic representation of the wave driven accelerations that lead to the zonal wind QBO. Equatorial waves of zonal phase speeds $+c$ and $-c$, respectively, propagate upward and are dissipated at rates dependent on the Doppler shifted frequency. (a) Initial weak westerly current selectively damps the westerly wave and leads to westerly acceleration at lower levels and easterly acceleration at higher levels. (b) Descending westerly shear zones block penetration of westerly waves while easterly waves produce descending easterlies aloft. Broad arrows show locations and direction of maxima in mean wind acceleration. Wavy lines indicate relative penetration of waves.(From, Plumb and Bell, 1982.)

Relation of equatorial QBO to extratropics

Holton and Tan (1982) composited a 16 year set of monthly mean geopotential height and temperature data for several pressure surfaces in the Northern Hemisphere stratosphere according to the phase of the equatorial QBO. They found that in years corresponding to the westerly phase of the equatorial QBO at the 50-mb level , the geopotential heights and temperatures during Northern Hemisphere winter are lower in the polar region and higher in midlatitudes than for years during the easterly phase. This result partly reflects the fact, as shown by Labitzke (1987), that sudden polar stratospheric warmings, which are associated with enhanced propagation of large-scale wave activity into the polar stratosphere, tend to occur more frequently during the easterly phase of the equatorial QBO than during the westerly phase.

These results may be partly understood by recognizing that the refractive index for propagation of wave activity in latitude and height in the stratosphere is strongly dependent on the distribution of the zonal mean wind. In fact stationary planetary waves can propagate into the stratosphere only when the mean winds are westerly, but less than some critical speed (see, e. g., Andrews, et al., 1987). This fact explains why the winter stratosphere (where winds are predominately westerly) has much stronger wave activity than the summer stratosphere (where winds are predominately easterly). It is through this sensitivity to the mean wind distribution that the equatorial QBO is apparently able to remotely affect the meteorological conditions of the winter polar region. When the equatorial QBO is in its westerly phase wave activity can propagate deep into the tropics; when the equatorial QBO is in its easterly phase wave activity is ducted toward higher latitudes, and has a greater impact on the polar circulation. Thus, the QBO provides a sort of resistive valve for wave propagation in the lower stratosphere. Numerical simulations by Holton and Austin (1991) and O'Sullivan and Salby (1990) suggest that when the wave activity is very weak the QBO valve effect is not significant since the polar night jet becomes very strong and limits the propagation of planetary waves. For medium levels of wave activity the resistance of the valve has an important influence on wave propagation, and the sensitivity of the model results to the phase of the QBO is very large. For large values of wave activity the stratospheric waves become saturated, and resistance of the valve is easily overcome, leading to reduced sensitivity to the phase of the QBO.

Stratospheric climate regimes

A number of studies in addition to those cited in the previous section have confirmed the strong interannual variability in the Northern Hemisphere winter stratosphere (e. g., Labitzke, 1982, Geller, et al., 1984, Boville and Randel, 1986) This variability can be regarded as an example of *climate regime* behavior. Apparently the Northern winter stratosphere has two preferred modes, or regimes. In the first mode the polar temperatures are very cold, the polar night jet is strong, and the eddy motions are relatively weak. In the second mode the polar temperatures are relatively warm, the polar night jet is weak, and the eddy motions are strong. This association of strong eddies and warm temperatures is, of course, consistent with (11), which indicates that the departure of temperature from radiative equilibrium is proportional to the zonal forcing by the eddies. We will refer to the weak eddy forcing regime as the *undisturbed regime* and the strong eddy forcing regime as the *disturbed regime*.

The existence of regime behavior is one of the fundamental aspects of nonlinear dynamical systems. The behavior of climate regimes, from a nonlinear perspective, is treated elegantly in a recent essay by Palmer (1993a), [see also Palmer (1993b)]. Palmer suggests that shifts in climate may occur primarily not through changes in the character of the individual regimes, but rather through changes in the relative probability that the climate system lies in a given regime. He illustrates this hypothesis with the famous Lorenz "chaos" model, for which it can be shown that "climate" changes are entirely due to changes in the probability density functions for the two regimes, not to changes in the characteristics of either regime. In the atmosphere regime shifts are likely controlled by very subtle and even rather indirect interactions involving both external forcing and internal dynamics. But they are likely to be most pronounced in those local regions of the climate phase space where the system is most unstable.

In the stratosphere the undisturbed and disturbed Northern Hemisphere winter circulations may be regarded as two natural modes of the circulation, i. e. as climate regimes. Even without the influence of the equatorial QBO there would likely be some shifts between these two regimes both on a sub seasonal and an interannual basis. The equatorial QBO, through its influence on the planetary wave propagation, acts to drive the system preferentially into the disturbed or undisturbed regime depending on whether the QBO is easterly or westerly in phase. The role of the QBO here is not to change the nature of the regimes --but only to change the probability that the system resides primarily in either the disturbed or undisturbed regime in a particular year.

Solar cycle influence?

Of course the major concern of the current volume is the connection between changes in the solar input and changes in the atmosphere. Simple scaling arguments show that it is very unlikely that changes over the course of solar cycle would alter the climate regimes of the stratosphere directly through radiative forcing because the solar cycle related radiative heating changes are simply too small. However, significant changes in the ozone distribution in the upper stratosphere are known to occur as a result of variations in ozone photochemistry caused by variations in ultraviolet radiation associated with the solar cycle (e. g., Angel, 1989). These ozone variations in turn have a minor impact on radiative equilibrium temperature as well as on the radiative relaxation rate, and hence will affect the zonal mean temperature distribution in the upper stratosphere. Through the thermal wind balances changes in the meridional temperature gradient will alter the vertical shear of the zonal mean wind.

Thus, solar cycle related changes in radiative heating may influence the zonal mean wind in the upper stratosphere, which in turn may influence the amount and distribution of upward propagating wave activity. As we have earlier seen, it is the distribution of vertically propagating wave activity that determines whether the winter stratosphere is in a disturbed or undisturbed state. Thus, any solar cycle related changes in the stratospheric wave guide should, like QBO related changes, have some influence on the climate regime behavior of the winter stratosphere.

That this sort of behavior may occur has been demonstrated in the general circulation model (GCM) experiment by Kodera, et al. (1991). In this experiment the perturbation to the radiative forcing was specified to be at least an order of magnitude greater than that associated with the actual solar cycle in order that a clear signal could be obtained from limited duration model runs. According to linear thinking, if the forcing in the Kodera et al. (1991) experiment were of realistic amplitude, the response would be reduced by a factor of 10, and would not account for the observed quasi-decadal oscillation. However, viewed from a nonlinear system perspective, it is possible that the net effect of weak solar cycle forcing may be to alter the probability density function of the climate regimes in such a way that the disturbed regime becomes more probable in the westerly QBO phase during sunspot maximum, than during sunspot minimum. Of course other processes may also influence the regime transition frequency. Any process that alters the mean wind and/or radiative spring of the stratosphere, and any process that alters the amount and distribution of upward propagating wave activity could play a role.

In conclusion, viewed as a chaotic system the atmosphere may have oscillations of a wide variety of time scales without the necessity of having a linear relationship with forcing of any particular period; a QDO does not require existence of an 11 year forcing cycle. Nevertheless, it is possible that forcing --even very weak forcing--may be important in climate variability not through altering the nature of climate regimes directly, but through altering the time spent in one or another regime. If the system is near a transition only a small forcing is required to push it into an alternative regime.

Although such ideas must be regarded as speculative (they are confirmed directly only for the highly simplified Lorenz model, which is more of a parable than paradigm), they do offer a possible path through which small solar cycle changes can have a significant impact, and thus may provide some guidance for research strategies in the search for physical links between the well documented changes of solar output, and the apparently statistically robust, but difficult to explain, QDO in the earth's atmosphere.

References

Andrews, D. G., J. R. Holton and C. B. Leovy, 1987: Middle Atmospheric Dynamics. Academic Press, Orlando, 489 pp.

Andrews, D. G. and M. E. Mc Intyre, 1976: Planetary waves in horizontal and vertical shear: the generalized Eliassen-Palm relation and the mean zonal acceleration. J. Atmos. Sci., **33**, 2031-2048.

Angel, J. K., 1989: On the relation between atmospheric ozone and sunspot number. J. Climate, **2**, 1404-1416.

Baldwin, M. P. and T. J. Dunkerton, 1989: Observations and statistical simulations of a proposed solar cycle/QBO/weather relationship. Geophys. Res. Lett., **16**, 863-866.

Boville, B. A. and W. J. Randel, 1986: Observations and simulation of the variability of the stratosphere and troposphere in January. J. Atmos. Sci., **43**, 3015-3034.

Geller, M. A., M. Wu and M. E. Gelman, 1984: Troposphere-stratosphere (surface-55 km) monthly winter general circulation statistics for the Northern Hemisphere -interannual variations. J. Atmos. Sci., **41**, 1726-1744.

Hamilton, K., 1990: A look at the recently proposed solar-qbo relationship. J. Clim., **3**, 497-503.

Holton, J. R. and J. Austin, 1991: The influence of the equatorial QBO on sudden styratospheric warmings. J. Atmos. Sci., **48**, 607-618.

Holton, J. R. and R. S. Lindzen, 1972: An updated theory for the quasi-biennial cycle of the tropical stratosphere. J. Atmos. Sci., **29**, 1076-1080.

Holton, J. R. and H. Tan, 1982: The quasi-biennial oscillation in the Northern Hemisphere lower stratosphere. J. Meteorol. Soc. Jpn., **60**, 140-148.

Kodera, K., M. Chiba and K. Shibata, 1991: A general circulation model study of the solar and QBO modulation of the stratospheric circulation during the Northern Hemisphere winter. Geophy. Res. Lett., **18**, 1209-1212.

Labitzke, K., 1982: On the interannual variability of the middle stratosphere during the nortyhern winters. J. Meteorol. Soc. Jpn., **60**, 124-139.

Labitzke, K. B., 1987: Sunspots, the QBO, and the stratospheric temperature in the north polar region. Geophys. Res. Letts., **14**, 535-537.

Naujokat, B., 1986: An update of the observed quasi-biennial oscillation of the stratospheric winds over the tropics. J. Atmos. Sci., **43**, 1873–1877.

O'Sullivan, D. and M. L. Salby, 1990: Coupling of the quasi-biennial oscillation and the extratropical circulation in the stratosphere through planetary wave transport. J. Atmos. Sci., **47**, 650-673.

Palmer, T. N., 1993a: A nonlinear dynamical perspective on climate change. Weather, **48**, 314-326.

Palmer, T. N., 1993b: Extended-range atmospheric prediction and the Lorenz model. Bull. Amer. Meteorol. Soc., **74**, 49-65.

Plumb, R. A., 1977: The interaction of two internal waves with the mean flow: implications for the theory of the quasi-biennial oscillation. J. Atmos. Sci., **34**, 1847-1858.

Plumb, R. A. and R. C. Bell, 1982: A model of the quasi-biennial oscillation on an equatorial beta-plane. Q. J. R. Meteorol. Soc., **108**, 335-352.

Plumb, R. A. and A. D. McEwan, 1978: The instability of a forced standing wave in a viscous stratified fluid: a laboratory analogue of the quasi-biennial oscillation. J. Atmos. Sci., **35**, 1827-1839.

Salby, M. and D. Shea, 1991: Correlations between solar activity and the atmosphere: An unphysical explanation. J. Geophys. Res., **96**, 22,579-22,595.

THE QUASI-BIENNIAL OSCILLATION OBSERVED IN THE SEMIDIURNAL GROUND PRESSURE DATA AND IN THE IONOSPHERIC F LAYER

H. Teitelbaum
LMD/CNRS, Ecole Polytechnique
91128 Palaiseau Cedex, France

Abstract

Hamilton (1983) and Hamilton and Garcia (1984) pointed out that the semidiurnal tidal pressure variations at ground level (S2(p)) exhibit amplitude modulations which appear to be related to the stratospheric Quasi-Biennial Oscillation (QBO). Ground pressure observations made at Macao (22° N, 113° E) from 1953 to 1989 have been analyzed and compared with the stratospheric QBO data obtained during the same time interval. The periods of the two phenomena and their time evolutions are found to be close to each other. Furthermore, time series of the stratospheric winds (QBO) and the S2(p) QBO signature are highly correlated, thus confirming the earlier analysis. On this basis, pressure measurements obtained at Batavia (now Djakarta: 6° S, 107° E) from 1870 to 1944 have been used in order to trace back the QBO phenomena before the advent of systematic stratospheric balloon measurements. The inferred period which varies between 25 and 32 months, suggest that the QBO is present in the atmosphere at least since 1870. Furthermore we shall show that the QBO signature exists also in the stratospheric F region and we shall explain its presence in this region. Finally, we shall analyze some results concerning the correlation of temperature, ozone and atmospheric electricity with the solar cycle when the data are stratified according to the phase of the QBO.

NATO ASI Series, Vol. I 25
The Solar Engine and Its Influence
on Terrestrial Atmosphere and Climate
Edited by Elizabeth Nesme-Ribes

1. Introduction

The Quasi-Biennial Oscillation (QBO) is a very strong oscillation of the mean zonal wind and temperature of the tropical stratosphere, characterized by a somewhat irregular period averaging, for the last 30 years, at about 27 months.

The equatorial QBO was discovered in 1960 by Reed (Reed *et al.*, 1961) and independently by Veryard and Ebdon (1961), from systematic wind and temperature observations in the stratosphere performed since the beginning of the 1950's. The problem of extracting the QBO signature prior to this date was taken up by Hamilton (1983) and Hamilton and Garcia (1984) through the use of hourly surface pressure data obtained at Batavia (now Djakarta : 6° S, 107° E). The ground pressure in the tropical regions exhibits a semidiurnal variation, which is, in fact, the sum of two components : the first is induced by the direct solar heating of the stratosphere, the second by a tropospheric semidiurnal variation in latent heat release (Lindzen, 1978 ; Hamilton, 1981). The first component peaks at 0900 and 2100 LST and the second at 0000 and 1200 LST. Hamilton and Garcia (1984) showed that the 0900 LST component is modulated over a 27 month period in the time interval 1905-1944, but such a modulation is absent from the 1200 LST data. Such a behavior is consistent with the fact that the semidiurnal tide induced by the solar heating is generated in the ozone layer at about 30 to 40 km and propagates through the QBO region before reaching the ground. The propagation through this region has the effect of modulating its amplitude (Hamilton 1983, 1984 ; Hagan *et al.*, 1992). It is interesting to note that Hamilton and Garcia do not find the 27 month period before 1905. They conclude that the Batavia data do not allow one to determine whether the QBO is absent or exists with a different period. We will show that it is possible to determine that the QBO was present prior to 1905 with a different period averaging at about 30 months.

In this study we use ground pressure data obtained at Macao (22° N, 113° E) from 1953 to 1989 together with stratospheric winds obtained in the

equatorial region (Naujokat, 1986) at 30 Hpa level during the same period. To show that the QBO was present in the eighteenth century we use the same data as Hamilton (1983) and Hamilton and Garcia (1984).
In section 2, following some considerations concerning the method used in the data processing, we analyse the ground pressure data obtained at Macao together with the stratospheric wind data at 30 Hpa. We compare the period variation existing in both data and calculate the correlation between the time series.

In the light of these results, in section 3, the analysis of the ground pressure data obtained at Batavia is extended in time back to 1870.

In section 4 we shall show the result concerning the ionospheric F layer and in section 5 we shall analyse the effect of the stratification of the data on the correlation between some atmospheric parameters and the solar cycle.

2. Comparison between ground pressure and stratospheric data

Due to the "quasi periodic" nature of the QBO, caution must be exerted in the processing of the data and, in particular, Fourier transform analysis should be used with care. A method which is particularly efficient for processing signals which amplitude and period vary, is the wavelet transform technique (Grossman *et al.*, 1989). An application of wavelets is described in the work of Andrew *et al.*, (1993). The wavelet analysis is related to the windowed Fourier transform but with an adjustable window. The window contracts at small periods and dilates at large ones. In particular, we use the Morlet's wavelet which roughly consists in a sinusoid which amplitude is modulated by a Gaussian curve. A sample of the signal is integrated, each point multiplied by the wavelet which is centered at the middle of the sample. After trying different periods for the sinusoid, we take as the local period corresponding to the center of the sample, the one which gives the maximum value for the integral. The length of the sample depends on the period, increasing the number of periods averages the result. In general two or three periods are enough. To avoid the loss of points at the ends of the time series, we use a part of the wavelet with a different normalization in approaching the first and the last point. This last part of the technique allows to use all the data, but unfortunately deteriorates the results near the limits.

In this section we analyse the ground pressure measured at Macao (22° N, 113° E). The data consist in the monthly mean of each hour L. S. T. (local solar time) from 1953 to 1989. The stratospheric data are the zonal wind at 30 Hp as in Naujokat (1986) but up to 1989. It consists of a wind velocity value for each month. The first step is to perform the Fourier transform of the ground pressure data for each month, to obtain monthly values of the amplitude and phase of S2(p) (semidiurnal ground pressure). From this result we take the S2(p) at 0090 LST, according to the fact that at this time the maximum component corresponds to the semidiurnal tide, generated in the stratosphere. Before the use of the wavelet technique both the ground pressure and the stratospheric data are filtered with a band pass filter which spans the range between 20 to 40 months. On this step we used a fourth order Butherworth filter. The variation of the period of the zonal stratospheric wind is compared with that of the amplitude of S2(p) after the use of the wavelet technique as shown in Figure 1. A minimum of the period

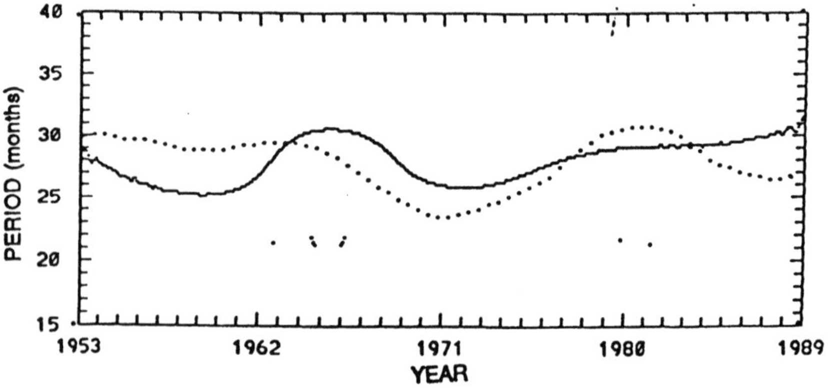

Figure 1 : Period variations of the QBO deduced from S2(p) at 0090 LST from data obtained at Macao (22° N, 113° E) (. . . . .), compared with the period variation deduced from stratospheric wind data at 30 Hpa (_______).

is clearly seen around the year 1971 in the S2(p) amplitude as well as in the stratospheric zonal wind. Then, both periods increase down to 1963 and up to 1980. Near both extremities of the time series the agreement deteriorates according to what has been stated above. Wavelet transforms give not only the period but also the amplitude and phase.

Figure 2 displays both time series after use of the wavelet technique. We can remark that the amplitude of the stratospheric zonal wind remains almost whereas the amplitude of the S2(p) variations changes. Nevertheless for the whole period the correlation is very good with a coefficient as high as 0.75 for stratospheric wind lagging S2(p) amplitude by two months.

To confirm the results obtained with the wavelet, we used another method which cannot give the period variations but permits to compare the stratospheric time series with the S2(p) one. It consists in performing the inverse Fourier transform. Figure 3 shows both time series after the application of this method. The Fourier components are taken with periods between 20 and 40 months. The result is very similar to that obtained with the wavelet technique : the correlation coefficient is of 0.73 for the same lag.

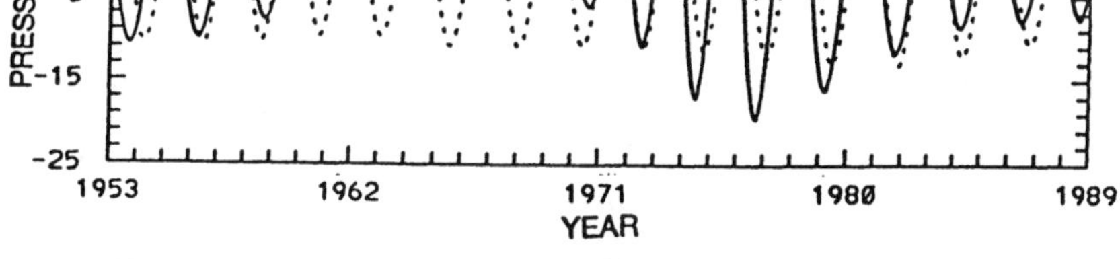

Figure **2** : Time series of the S2(p) at Macao (______) and stratospheric wind data (. . . . .) after processing with the Wavelet method. The correlation of the series coefficient is 0.75 for stratospheric wind lagging S2(p) amplitude by two months. The stratospheric wind has been normalized to be compared with pressure data in μbar.

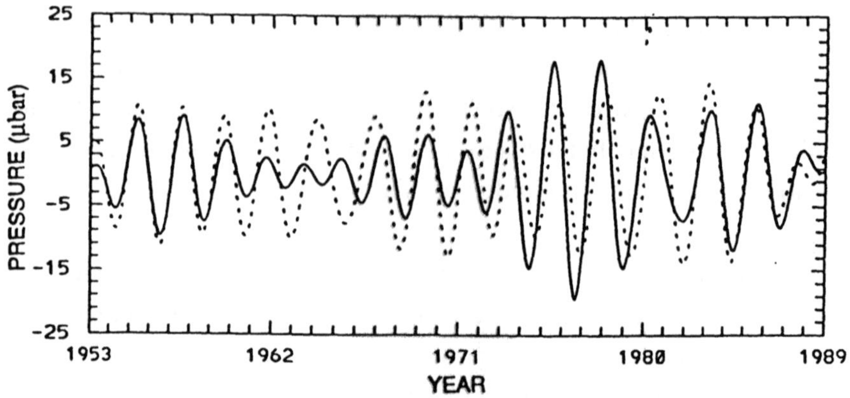

Figure 3 : The same as in Figure 2 but after processing the data wih the inverse Fourier transform. The correlation coefficient is of 0.73 with the same lag.
(______) S2(p) data ; (. . . . .) stratospheric data. The stratospheric wind has been normalized as in Figure 2.

The above results confirm the fact that the 0090 LST component of S2(p) carries out the signature of the stratospheric QBO as it has been already stated by Hamilton (1983) and Hamilton and Garcia (1984). Furthermore, they provide the basis for applying the wavelet method to ground pressure data at early times, even when stratospheric data were not available.

3. Analysis of the Batavia ground pressure data

Ground pressure data have been obtained continuously at Batavia from 1866 until 1944. Nevertheless, due to the irregular character of the first four years we start the analysis from 1870. Following the previous processing of Macao data, the semidiurnal amplitude at 0900 LST has been extracted. As before, a band pass filter which spans the range between 20 and 40 months period was performed. Figures 4a and 4b show the period and the amplitude variations obtained using the wavelet technique. The QBO seems to be present all the time but with an important change of the period around 1905, which is the earliest year at which Hamilton and Garcia (1984) found the

QBO. In Figure 4b we note that it is also around 1905 that the amplitude is the smallest, suggesting that some changes in the driving forces could have existed in this period. The amplitude that we find is roughly the same obtained by Hagan *et al.*, (1992) with a numerical model, specially after 1910. Their model provided an amplitude of the order of 15μbar in the equatorial region. We notice some gaps in the result between 1910 and 1915. These gaps correspond to large variations in the amplitude of the 0090 LST S2(p). The amplitude variation has been found by Hamilton and Garcia (1984) which suggested some explanations. We do not enter in this subject here. Figure 5 displays the time series deduced from the wavelet analysis. There are same gaps as in the time variation of the period. Furthermore, it is possible to see that the mean period is shorter after 1905 than before this year.

This proves that the wavelet technique is able to describe the period variation, whereas Fourier transform is not. Furthermore, before 1905 as shown in Figure 4a the period varies roughly between 30 and 32 months which is a less important variation that after this year. This encouraged us to try another type of spectral analysis, say the multitaper method of Thompson (1982). This method uses an optional set of tapers instead of a single one. The tapers are calculated so as to maximize the spectral power inside a given frequency bandwidth and to minimize it outside. This leads to a high resolution and provides a statistical F-test on the location of each spectral peak. In addition the statistical significance of a given peak is practically independent of its amplitude. The method is used after the application of a low pass filter to the data, to eliminate all periods shorter than 15 months. This step is necessary due to the presence of the annual and semi-annual periods which are very important. Figure 6a and 6b show results of the multitaper method together with the confidence level obtained by the Fisher's test.

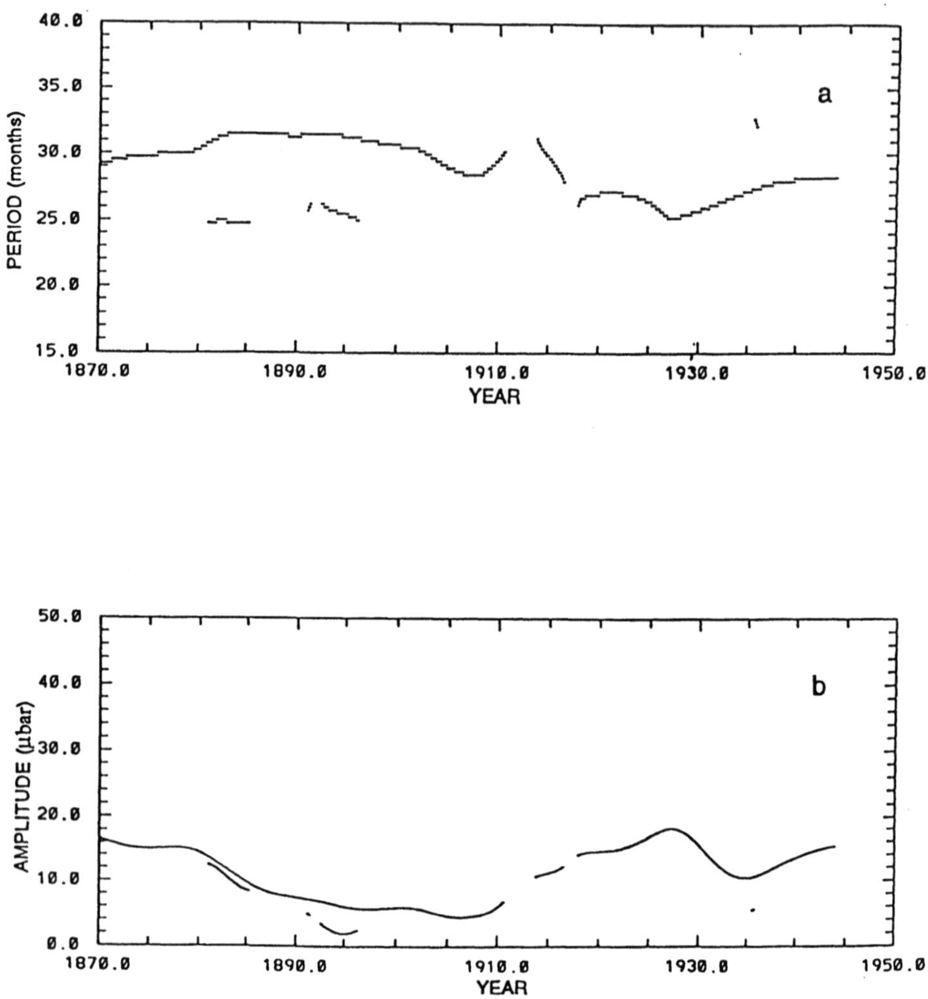

Figure **4** : Results of the wavelet analysis of S2(p) at 0090 LST from data obtained at Batavia (6í S, 107í E). a) QBO period variation ; b) amplitude variation.

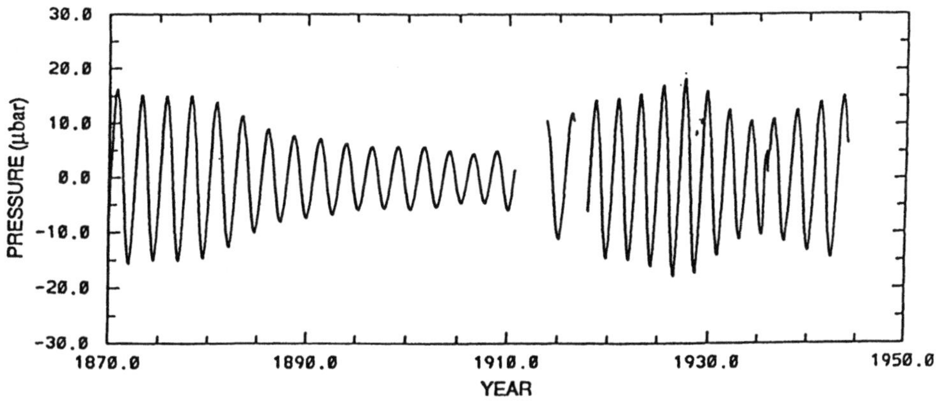

Figure **5** : Time series of the S2(p) Batavia data after using the wavelet method.

Figure 6a corresponds to the data taken between 1870 and 1905. The 31. 3 months period appears with a confidence level of 0. 995, whereas the 27 months period which appears between 1905 and 1944 (Figure 6b) has a confidence level of 0. 96. This difference in the confidence level is due to the fact that after 1905 the QBO period is much more irregular than before this year as seen on Figure 4a.

4. The QBO in the ionospheric F layer

After Forbes *et al.* (1982) and Fesen *et al* (1986)., semidiurnal tide in the F region has un important ingredient which comes from below 100 km. We can ask on the possibility that the semidiurnal tide carry out the signature of the QBO from the stratosphere up to the F region.

In this study we used ionosonde data obtained at Dourbes, Belgium (50° lat., 5° long.). Thermospheric meridional winds were derived from ionosonde data by the method of Miller *et al.* (1986), that is to say using the height of the F2-layer electron density peak.

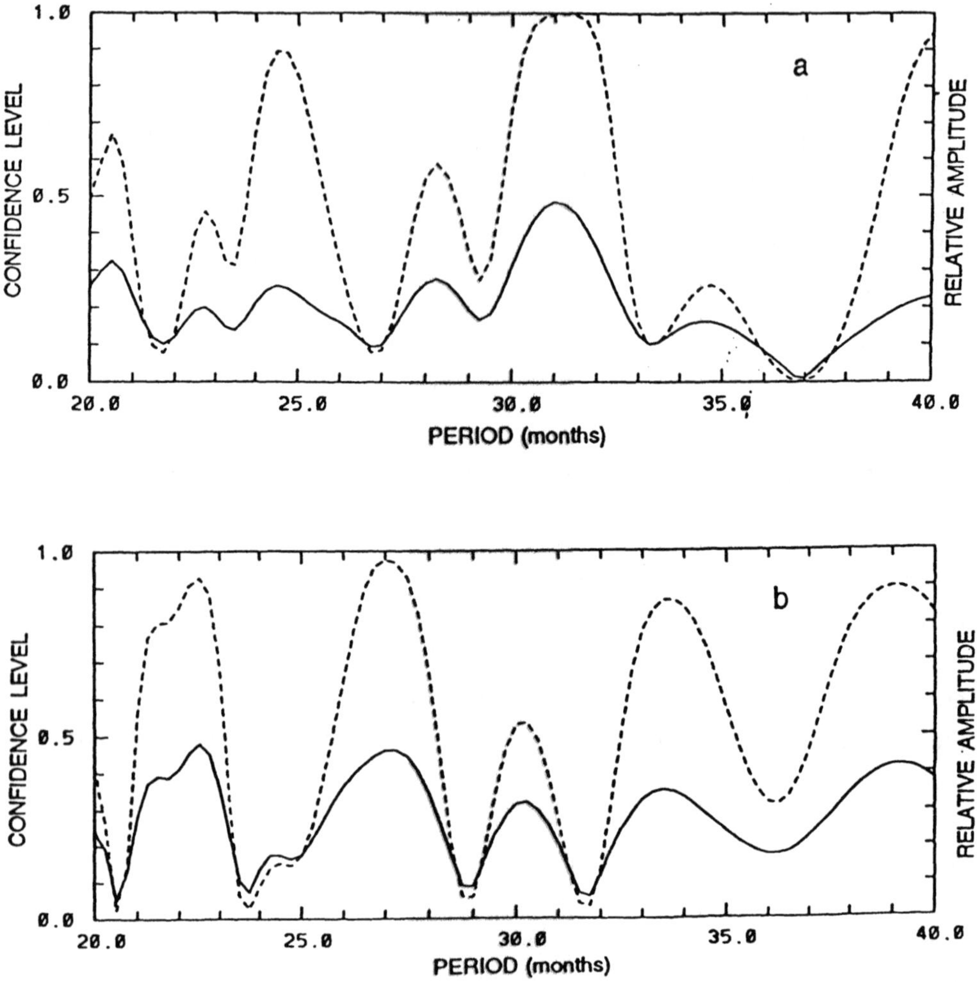

Figure **6** : Multitaper spectral analysis of the S2(p) Batavia data. (______) relative amplitude ; (. . . . .) confidence level. a) Between 1870 and 1905 ; b) Between 1905 and 1945.

The first step in the data processing is to calculate the monthly mean of daily variation. Then a Fourier transform is performed to calculate the diurnal and semidiurnal tides. Figure 7 displays the amplitude variation of tides. We can see that the eleven year solar cycle is present in both tides, but the most striking aspect is the difference concerning the annual and semi-annual oscillations. It is clear that beside the eleven year solar cycle, the most important period apparent in the variation of the diurnal tide amplitude is the annual oscillation, whereas for the semidiurnal tide the semi-annual oscillation is much more important. The phenomenon is more clearly seen in the spectrum (Figure 8). This is due to the origin of the F2 layer semidiurnal tide which comes partly from lower levels. We can try to determine the time at which the most important part of the tide comes from the stratosphere with the help of the ratio of the semi-annual oscillation and the eleven year solar cycle. Figure 9 shows that the time of the maximum is 3 hours local time. Taking the amplitude of the semidiurnal tide at 3 hours LT and after the use of the wavelet technique we can compare the period variation of the semidiurnal F2 amplitude with that of the stratospheric wind. Figure 10 displays the results showing that the temporal variation of both periods are very symilar.

5. Apparent correlation between the eleven year solar cycle and some atmospheric parameters.

Several authors come to the conclusion that there exists a correlation between the eleven year solar cycle and some atmospheric parameters like stratospheric temperature (Labitzke and Van Loon, 1988), ozone (Marcz, 1990) and even atmospheric electricity near the ground (Varotsos 1989). In all of these results the correlation is obtained whenever the data for January and February are sorted out according to the phase of the QBO zonal wind at 50 mb. We tried to find something similar in the ground pressure data obtained at Batavia from 1870 to 1944. For that we used the amplitude variation of the semidiurnal tide which as we know, carried out the signature of the QBO.

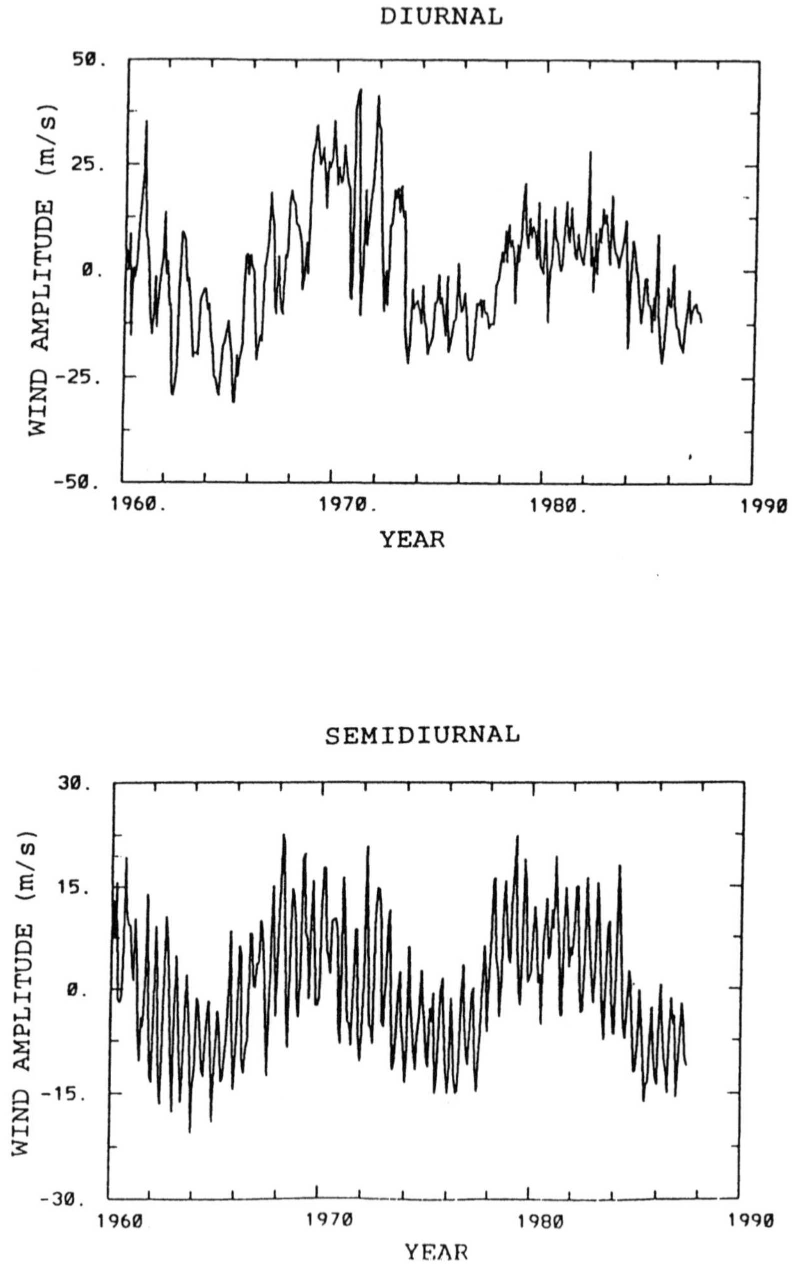

Figure **7** : Amplitude variation of the tides as a function of time, after elimination of the mean value.

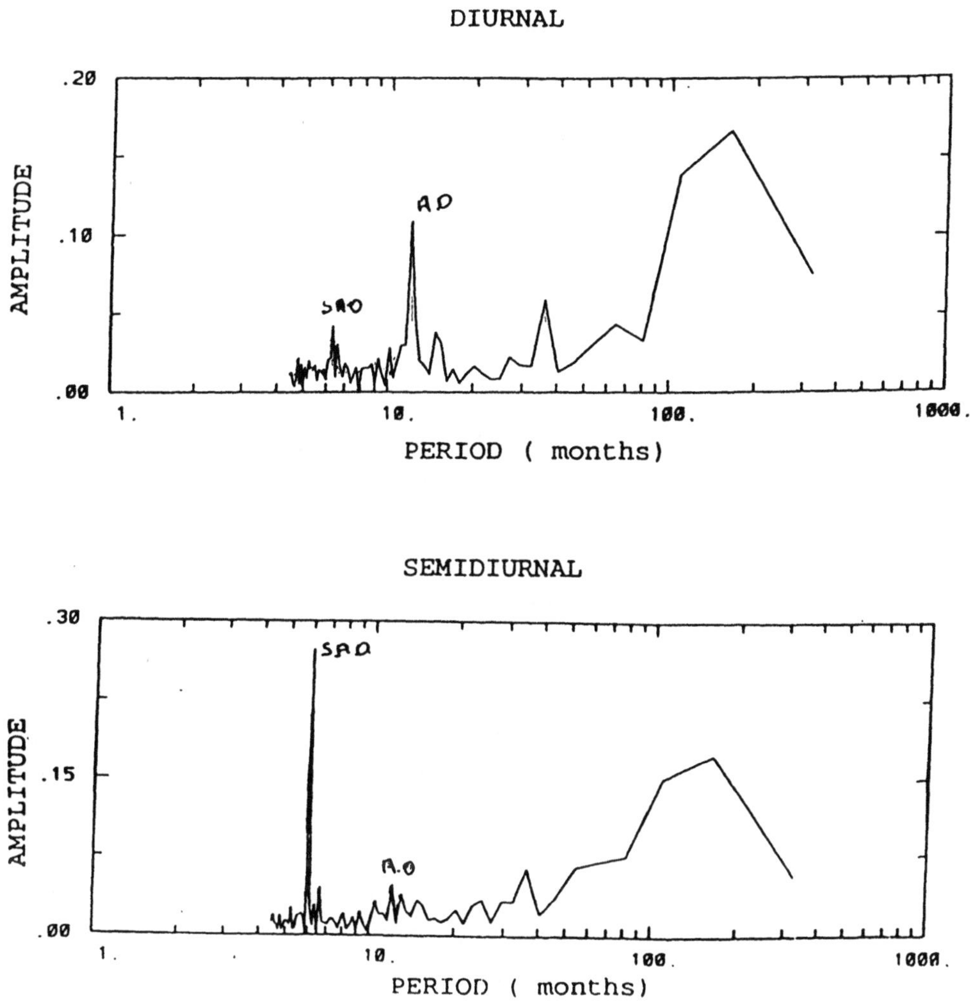

Figure **8** : Spectral analysis of the amplitude variation of tides as a function of the period.

The Fourier transform of the result shows a decadal variation as seen in figure 11. For the whole data the decadal variation disappears as seen in the same figure. We conclude that in this case the decadal and other periods which appear in the stratified data are a consequence of aliasing. We think that the other correlations with temperature, ozone and atmospheric electricity are produced or at least enhanced by this kind of data processing as shown by Teitelbaum and Bauer (1990)

6. CONCLUSION

We have analysed tropical ground pressure semidiurnal amplitude oscillation and the equatorial zonal wind at the 30 Hpa level from 1953 to 1989. The aim was to search the QBO signature in the semidiurnal ground pressure oscillation. In the data processing we used the wavelet method, capable of finding local periods and amplitudes.Our results show that the period variations in the S2(p) and stratospheric zonal wind are similar. Furthermore, from these results and by using the inverse Fourier transform we compare both time-series. The two methods give comparable results: both time-series are highly correlated with the same lag. This first part of our study confirms earlier analysis and suggest the use of the wavelet method for data obtained in the last century. In this second part of our study we found that the QBO exits at least since 1870 but with a mean period of 31 months up to 1905 which is different from the 27 months mean period found after. At the present time, we cannot give an explanation of this difference and only suggest that perhaps the driving force of the QBO was different.

The QBO signature has been found in the F2 ionospheric layer. This period is carried out by the stratospheric ingredient of the semidiurnal tide.

Some correlations found between the eleven year solar cycle and atmospheric temperature, ozone and electricity are probably due to or higly enhanced by a special data processing.

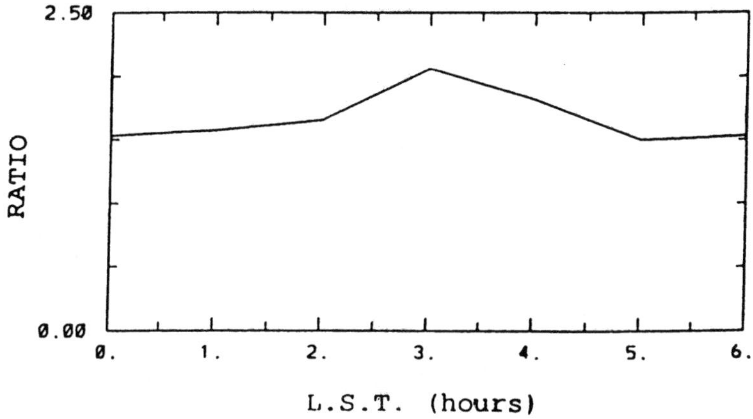

Figure 9 : Variation of the ratio between the semi-annual and the eleven year solar cycle as a function of local solar time.

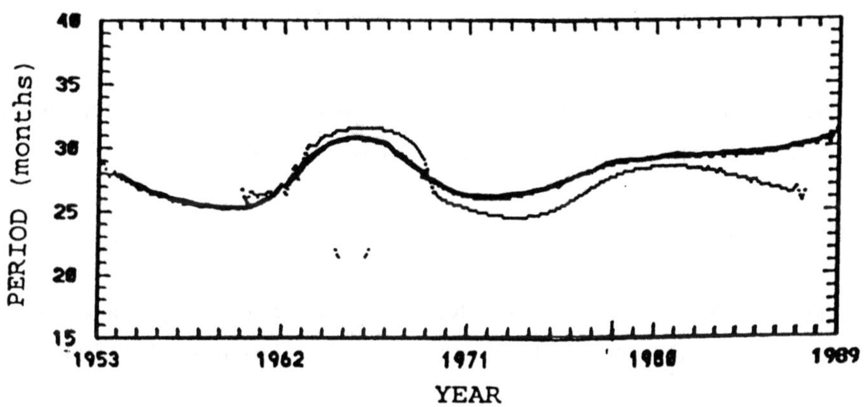

Figure 10 : Period variation of the QBO at Dourbes (50° lat., 5° long.)deduced from the meridional S2 wind(-----), compared with the period variation of the zonal wind at 30 Hpa level.

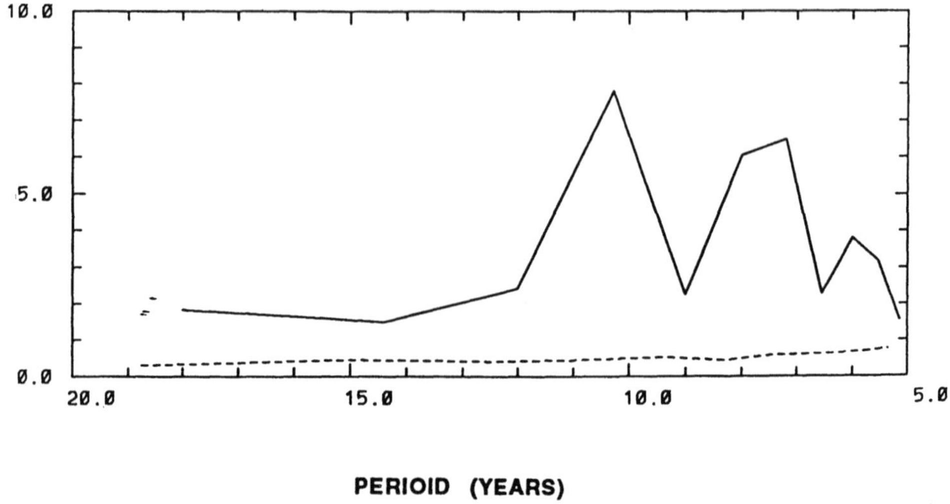

Figure **11** : Power spectra of the Batavia data (see the text) for (_____) stratified and (. . . . .) unstratified data.

Acknowledgements

We wish to warmly thank R. Garcia for providing us the Batavia hourly data, and S. Pawson for the stratospheric data.

References

Andrew, R., Ochadlik, Jr., Kritikos, N., and Giegengack, R. (1993). Variations in the period of the sunspot cycle. Geophys. Res. Let., 20: 1471-1474

Fesen, C.G., Dickinson, R.E., and Roble, R.G. (1986). Simulation of thermospheric tide at equinox with the NCART GCM. J. Geophys. Res., 91: 4471-4489

Forbes, J.M., and Gillette, D.F. (1982). A compendium of theoretical atmospheric tidal structure. Environment Research Papers N° 780. AFGL-TR

Grossman, A., Kronland-Martinet, R., and Morlet, J. (1989). Reading and understanding continuous wavelet transforms, in Wavelet, Springer-Verlag

Hagan, M.E., Vial, F., and Forbes, J.M. (1992). Variability in the upward propagating semidiurnal tide due to effects of QBO in the lower atmosphere. J. Atmos. Terr. Phys., in press

Hamilton, K. (1981). Latent heat release as a possible forcing mechanism for atmospheric tides. Mon. Weather Rev., 109: 3-17

Hamilton, K. (1983). Quasi-Biennial and other long-period variations in the solar semidiurnal barometric oscillation : Observations, theory and possible application to the problem of monitoring changes in global ozone. J. Atmos. Sci., 40: 2432-2443

Hamilton, K. (1984). Calculation of the effect of stratospheric mean wind variations on the solar semidiurnal barometric oscillation. Atmosphere-Ocean, 22: 48-66

Hamilton, K., and Garcia, R.R. (1984). Long-period variations in the solar semidiurnal atmospheric tide. J. Geophys. Res., 89: 11705-11710

Labitzke, K., and Van Loon, H. (1988). Association between the 11-year solar cycle, the QBO and the atmosphere ; Part 1: the troposphere and the stratosphere in the northern hemisphere in winter. J. Atmos. Terr. Phys., 50: 197-206

Lindzen, R.S. (1978). Effects of daily variations of cumulonimbus activity on the atmospheric semidiurnal tide. Mon. Weather rev., 106: 526-533

Märcz, F. (1990). Atmospheric electricity and the 11-year solar cycle associated with QBO. Ann. Geophys., 8: 525-530

Miller, K.L., Torrand, D.G., Richards, P.G. (1986). Meridional winds in the thermosphere derived from measurements F2 layer height. J. Geophys. Res., 91: 4531-4535

Naujokat, B. (1986). An update of the observed quasi-biennial oscillation of the stratospheric winds over the tropics. J. Atmos. Sci., 43: 1873-1877

Reed, R.J., Campbell, W.J., Rasmussen, L.A., and Rogers, D.G. (1961). Evidence of downward-propagating annual wind reversal in the equatorial stratosphere. J. Geophys. Res., 66: 813-818

Teitelbaum, H., and Bauer, P. (1990). Stratospheric temperature eleven years variations : solar cycle influence or stroboscopic effect ?. Ann. Geophys., 8: 239-242

Thomson, D.J. (1982). Spectrum estimation and harmonic analysis. Proc. IEEE, 70: 1055-1196

Varotsos, C. (1989). Comment on connections between the 11-year cycle, the QBO and total ozone. J. Atmos. Terr. Phys., 51: 367-370

Veryard, R.G., and Ebdon, R. A. (1961). Fluctuations in tropical stratospheric winds. Meteorol. Mag., 90: 125-143

THE QUASI-BIENNIAL OSCILLATION (QBO): SOME POINTS ABOUT THE TERRESTRIAL QBO AND THE POSSIBILITY OF RELATED PHENOMENA IN THE SOLAR INTERIOR.

Michael E. McIntyre
Centre for Atmospheric Science,
Dept of Applied Mathematics and Theoretical Physics,
University of Cambridge, Silver St, CB3 9EW, U.K.
[M.E.McIntyre@amtp.cam.ac.uk]

Introduction

As Professor Holton has reminded us (*q.v.*, this Proceedings), the terrestrial quasi-biennial oscillation or QBO is a remarkably persistent phenomenon. It has been clearly seen in tropical meteorological radiosonde data since the early 1950s, and there is indirect but persuasive evidence that it has been present in varying degrees since at least the 1880s (*e.g.* Hamilton 1983; Hamilton and Garcia 1984,6; Hagan *et al.* 1993; Teitelbaum *et al.* 1994). I suggest we postpone the interesting question of the possible modulation of solar-terrestrial effects by the terrestrial QBO until after the presentation by Professor Labitzke. I thought, however, that it might be interesting to consider one or two very fundamental points about the fluid dynamics of the terrestrial QBO and its robustnesses and sensitivities, and then to consider briefly the possibility that some aspects of solar variability — some of the many aspects that we have been hearing about — might be related to partially similar mechanisms in the Sun's interior.

There is a wider motivation as well. It seems overwhelmingly likely that certain facts now established about the fluid dynamics of the Earth's stratosphere will, in one way or another, turn out to have direct relevance to understanding the fluid dynamics of the Sun's interior, regardless of whether QBO-like phenomena exist there. The Earth's stratosphere and the Sun's interior are dynamically similar systems to the extent that both are rapidly rotating, with strong Coriolis effects, and both are stably stratified with large values of the fundamental dynamical aspect ratio, Prandtl's ratio $N/2\Omega$. Here Ω is a typical angular velocity, and N a typical buoyancy frequency of the stable stratification, or Schwarzschild–Milch–Hesselberg–Brunt–Väisälä frequency. We have

$$N/2\Omega \sim 10^2 \ , \tag{1}$$

NATO ASI Series, Vol. I 25
The Solar Engine and Its Influence
on Terrestrial Atmosphere and Climate
Edited by Elizabeth Nesme-Ribes

numerically about 1.5×10^2 in the Earth's stratosphere and about twice that value in the outer parts of the present-day Sun's radiative interior, the buoyancy periods $2\pi/N$ being about five minutes and one hour respectively.

There is promise here of new keys to understanding primeval spindown and lithium depletion in the Sun and other stars. For instance, observation and numerical modelling of the dynamics and chemistry of the Earth's extratropical stratosphere have confirmed the likely effects of so-called 'geostrophic' or layerwise-two-dimensional turbulence. In particular, such turbulence has no tendency to produce shellular solid rotation. I shall explain why, then argue that if the outer parts of the Sun's radiative interior away from the poles are actually nearer solid rotation than is the overlying convection zone — a state of things indicated by current helioseismic results (*e.g.* Goode *et al.* 1991) — then there has to be a sufficiently strong poloidal magnetic field $\mathbf{B}_\mathrm{p}$ in the interior. This is consistent with the observed north–south asymmetry at the Maunder Minimum (*e.g.* Nesme–Ribes *et al.* 1994). The suggestion of such an interior field is hardly new in itself; and reasonable estimates of magnetic torsional periods range all the way down to tens of years or less (*e.g.* Gough 1990 & refs.). Layerwise-two-dimensional turbulence could still be present, affecting the transport of atomic species including lithium (Spiegel and Zahn 1992, Zahn 1992, & refs.) and possibly having weak dynamo action as well.

Again, observation and numerical modelling of the Earth's stratosphere have repeatedly confirmed the importance of Rossby waves. In dynamical systems of the kind we are dealing with, Rossby waves tend to occur whenever layerwise-two-dimensional turbulence occurs; and Rossby waves can be powerful transporters of angular momentum, having a ratchet-like, robustly one-signed effect. This has long been recognised as important in the Earth's extratropical winter stratosphere, and there has been a suspicion that Rossby waves of extratropical origin also play a part in the QBO. That suspicion is now being confirmed by, among other things, the recent recognition of a new form of Rossby-wave breaking that is always important in some latitude band surrounding the equator. The same phenomenon can occur in the Sun also, and could have formed an internal equatorial channel for the vertical leakage of lithium and other atomic species. Professor Gough and I are looking into this. It is worth noting that Rossby waves originating near the polar tachocline, from nonlinear baroclinic or other local shear instabilities (*e.g.* Thorncroft *et al.* 1993 & refs.), would have a robust tendency to propagate not only downward but also equatorward for the same reasons that they propagate upward and equatorward in the Earth's winter stratosphere. The one-signedness is in the right sense to help explain, also, the puzzling helioseismic result about the sign of the vertical differential rotation below the equatorial convection zone.

An important modelling failure

What is our most secure piece of knowledge about the mechanism of the terrestrial QBO? There is a clear answer: the terrestrial QBO is a wave-driven mean flow. The alternating deficits and excesses of angular momentum observed in the tropical stratosphere, reversing sign at roughly 14-month intervals — the full cycle time being variable between about 24 and 36 months — owe their existence and timing to the irreversible transport of angular momentum associated with the generation, propagation and dissipation of internal waves, of one sort or another, in the Earth's atmosphere.

There is little room for doubt about this. It is not only that many suitable types of internal waves are available: the atmosphere is observed to be full of internal gravity waves, inertia-gravity waves, Rossby waves and various other hybrid, buoyancy–Coriolis types including equatorially-trapped Rossby–gravity and Kelvin waves (*e.g.* Gill 1982). It is not only that transport of momentum and angular momentum is a robust and almost inevitable accompaniment to wave propagation in fluid media, as is well known and as was illustrated by the simple experimental demonstration at the Workshop (Fig. 1)

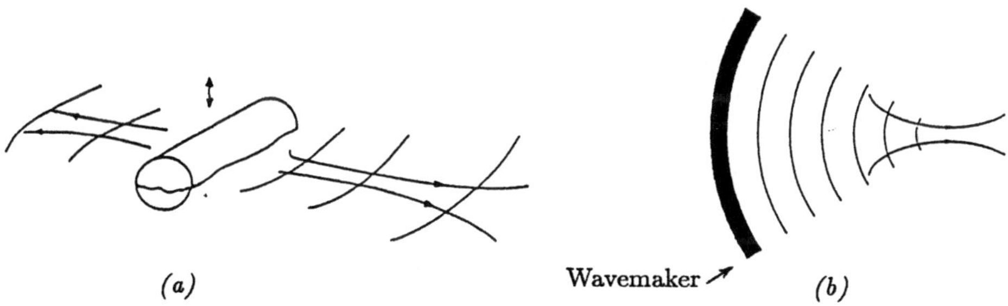

Fig. 1. Simple experiments with 5 Hz water waves, illustrating the irreversible wave-induced momentum transport that accompanies the generation of waves in one place and their dissipation in another. The strong mean flows thereby generated can be made conspicuous by sprinkling a little powder such as chalk dust on the surface of the water. Configuration (*a*) works well with a cylindrical wavemaker whose diameter is of the order of 2 to 4 cm, and can be demonstrated in a transparent dish on an overhead projector. Configuration (*b*) works well with a larger wavemaker whose radius of curvature is of the order of half a metre. From McIntyre and Norton (1990) and McIntyre (1992,3); *q.v.* for further discussion.

— noting also that, as hinted above and discussed in two recent reviews (McIntyre 1992,3), wave-induced angular momentum transport is now recognised to be of primary importance for understanding almost all global-scale aspects of the Earth's atmospheric circulation including those aspects that depend on what used to be called 'negative eddy viscosity'. It is not only that the internal waves usually thought to be responsible for the QBO, which are hybrid upward-propagating equatorially-trapped waves, have been observed to exist in the real atmosphere and in some cases to have amplitudes of the right order of magnitude. This is often true, at least, of the equatorial Kelvin waves (e.g. Wallace and Kousky 1968, Kousky and Wallace 1971, Ziemke and Stanford 1994 & refs.). It is also, above all, the clear failure of serious, highly skilled and intelligent attempts to model the QBO *without* invoking wave-induced angular momentum transport that compels us to deduce that the QBO is a wave-driven mean flow.

This modelling failure is an example of an important negative result in science. I am thinking especially of the work of Wallace and Holton (1968), which Professor Holton, with typical modesty, has not emphasised. There, an attempt to model the QBO in a dynamically self-consistent manner — consistent also with the known orders of magnitude of relevant and well-understood physical processes such as radiative heat transfer, and consistent with available observational estimates of eddy fluxes of angular momentum — showed that the observed angular momentum changes could not be explained without assuming the existence of a mysterious additional eddy flux of angular momentum. This had to be quite unlike the fluxes suggested by the observational estimates. It also had to be quite unlike any flux by quasi-diffusive processes such as the eddy viscosity often assumed, rightly or wrongly, to be associated with unresolved fluctuating motions.

The convergence of this additional eddy flux of angular momentum had to correspond to a time-dependent, axisymmetric mean torque, acting over a certain fairly narrow range of altitudes, a few kilometres, that systematically descended as time progressed. The sign of the torque had to be such that it systematically tended to reinforce, not reduce, the local angular momentum anomaly. If one tried to describe it in terms of eddy viscosity one would be compelled to admit nonsensical negative and infinite values. The descent had to begin at altitudes around 30km and end around 15 km, the same pattern being repeated with opposite sign and some temporal overlap and the whole cycle taking, of course, about 24 to 36 months. The time-altitude contour plot of the torque had to look a bit like half a sunspot butterfly pattern, but with thinner 'wings'.

Despite being unable to say what physical process or processes might produce such a time-dependent mean torque, Wallace and Holton reported their negative result, concluding that the torque must exist despite its strange character and despite the lack of any direct observational confirmation — but also bearing in mind that the available observational estimates were based on sparse radiosonde data and therefore tentative. It was only subsequently recognised (Lindzen and Holton 1968, Holton and Lindzen 1972; see also *e.g.* Plumb 1977, Saravanan 1990) that exactly this qualitative kind of mean torque, with exactly this qualitative kind of time-dependence, can be expected, on theoretical grounds, to result from wave-induced angular momentum transport. This is because of the refractive effect of the differential rotation associated with the QBO upon internal wave propagation, giving rise to a crucial feedback.

The differential rotation selectively refracts and Doppler-shifts the internal waves in much the same way as an ordinary bathplug vortex selectively refracts and Doppler-shifts randomly incident surface waves (tending to deceive the eye as to the direction of rotation). In the tropical stratosphere, under plausible assumptions about the types of internal waves and how they might be dissipated, the net effect turns out to be a time-dependent mean torque whose spacetime pattern is qualitatively the same as the required thin-half-butterfly pattern, with the correct phase and sign.

Mean upwelling in the Earth's tropical stratosphere

With hindsight, one can add that if Wallace and Holton had known in 1968 what we know today about upward mean motion in the tropical stratosphere, then their conclusion would have been further strengthened. For reasons connected with the stratosphere's response to *extratropical* wave-induced angular momentum transport, which latter is dominated by Rossby waves and related eddy motions, there is a persistent global-scale upwelling or upward mean motion in the tropical stratosphere. It is of the order of a fraction of a millimetre per second, or 10 km per year, in the altitude range of interest. This can be taken as a fact well established both directly from many observations — including global-scale, space-based spectroscopic observations of long-lived chemical tracers — and indirectly from strong lines of theoretical argument and numerical modelling as discussed, for instance, in the two recent reviews I have mentioned and in the many publications cited therein. See especially the remarks about the consequences of the Earth's rotation and the one-signedness of Rossby-wave dynamics. The extratropical stratosphere acts as a gentle but persistent one-way, global-scale pump, extracting air from the tropical stratosphere and feeding it down into the extratropical troposphere. This pump acts persistently all year round, despite some seasonal variability (Yulaeva *et al.* 1994). To satisfy mass continuity, the circulation has to be completed by mean upwelling in the tropical stratosphere. The resulting adiabatic temperature tendencies are balanced by infrared radiative heat transport at low to moderate opacity. The net effect is qualitatively like Newtonian cooling to space; in this respect the stratosphere is more like the Sun's photosphere than the Sun's interior.

One consequence of the mean upwelling in the Earth's tropical stratosphere is that the torque required to account for the QBO must still conform to the thin-half-butterfly spacetime pattern but, in the presence of the upwelling, must have a greater magnitude than would otherwise be required. The torque now has to overcome the upward advection of angular momentum, represented by the $\bar{w}\,\partial/\partial r$ term in the Eulerian angular momentum equation where $\bar{w}$ is the mean upwelling velocity. The numerical magnitudes imply a torque greater by nearly a factor 2 than would be required in the absence of upwelling. From a Lagrangian viewpoint, the persistent upward motion of the air through the (descending) range of altitudes where the torque is significant shortens the time available to accelerate a given ring of air. It shortens the time by nearly a factor 2 because, by chance, the upwelling velocity has nearly the same magnitude as the typical descent velocity of the thin-half-butterfly pattern.

Model studies of the effects of mean upwelling on the QBO can be found for instance in the papers by Saravanan (1990) and Dunkerton (1991). Saravanan points out that the much stronger mean upwelling in the tropical troposphere can explain why the QBO is observed to be confined to the stratosphere.

Wave dissipation, Doppler shifting and the 'critical-layer effect'

It seems likely that the internal waves that drive the QBO in the tropical stratosphere dissipate mainly by infrared radiative damping, or wave breaking, or both. The notion of 'wave breaking' needs to be understood in a suitably generalised way, relevant to its part in making the wave-induced momentum transport irreversible (McIntyre and Palmer 1984,5). The waves of interest could include almost any non-acoustic internal waves. Many such waves propagate slowly enough to be Doppler shifted significantly by the differential rotation, whose characteristic relative velocities are of the order of tens of metres per second. Such waves become increasingly vulnerable to breaking and other forms of dissipation whenever the differential rotation Doppler-shifts them toward small intrinsic frequencies and group velocities. This is the relevant aspect of what is sometimes, for historical reasons, called the 'critical-layer effect', but which could more plainly be called wave dissipation enhanced by Doppler shifting, more specifically Doppler downshifting. Call it what we may, it is an important part of the feedback that gives rise to the QBO.

(There is a critical-layer *myth*, still occasionally found in the literature, saying or tacitly assuming that there is a separate critical-layer effect independently producing convergence of the wave-induced angular momentum flux, additional to the convergence produced by the various forms of wave dissipation. However, this comes from too literal an interpretation of linearised steady-state wave theory. For real, finite-amplitude waves the real 'critical-layer effect', wave dissipation enhanced by Doppler shifting, represents merely one subset — albeit an important subset — of the circumstances in which the waves must dissipate.)

If the Doppler shift goes far enough toward small group velocities while dissipation remains negligible, then the waves will inevitably break, however small their initial amplitude may be. Linearised, nondissipative wave theory predicts its own failure, by predicting a pileup of wave activity. In this respect, a real critical layer behaves like the surf zone of an ocean beach. The edge of the beach corresponds in the simplest examples to the place where the group velocity of the internal waves is Doppler shifted to zero. In the absence of other dissipation mechanisms, waves of non-infinitesimal amplitude must break *before* reaching this notional edge. How long before reaching it depends on the amplitude. These points were well explained in a useful review by Fritts (1984, Fig. 9), for the case of ordinary internal gravity waves in the usual, large-Richardson-number parameter régime, to which these ideas apply in their simplest form. For Rossby waves, the surf-zone analogy still applies, with some qualifications. There can be more backscattering — Rossby-wave surf zones tend to be more reflective than ocean beaches — and there is no simple characterisation of edge position, which is more sensitive to wave amplitude and to details of the differential rotation than to group-velocity values.

Uncertainties about the terrestrial QBO

Even though we can be practically certain that the QBO is wave-driven, there is great uncertainty about exactly which types of waves are most important. It was suggested by Dickinson (1968), and strongly argued by Andrews and McIntyre (1976) on the basis of a radiosonde data study by Lindzen and Tsay (1975) that, in addition to the upward-propagating equatorial Kelvin and Rossby–gravity waves traditionally considered, there could also be a significant contribution from planetary-scale Rossby waves, coming in obliquely or nearly horizontally from the extratropical winter hemisphere.

The possibility of driving a realistic-looking QBO entirely by such Rossby waves together with an equatorial Kelvin wave, and with no other waves at all, was demonstrated in an idealised QBO model simulation by Dunkerton (1983). The Rossby waves were assumed to provide a retrograde torque near the equator, and the Kelvin wave a prograde torque. This also proved to be one way of producing space-time patterns of torque and differential rotation more realistic than traditional simulations using Kelvin and Rossby–gravity waves alone, albeit not the only way (*e.g.* Plumb and Bell 1982, Saravanan 1990). However, while the Kelvin wave is confined near the equator and has long been known to be able to dissipate there in an appropriate way, both by infrared radiative damping and by wave breaking (Kousky and Wallace 1971), it was less clear how the Rossby waves might dissipate in suitable locations near the equator.

It therefore seems significant that recent work on 'asymmetric inertial instabilities' (Ciesielski *et al.* 1989, O'Sullivan and Hitchman 1992, Dunkerton 1993, Clark and Haynes 1994) has produced theoretical and observational evidence for the existence of a previously unrecognised form of Rossby-wave breaking that depends on smallness of the Coriolis parameter and is confined to some neighbourhood of the equator. The latitudinal width of that neighbourhood is of the same order as the latitudinal particle displacement amplitude of the Rossby waves. Although quantitative estimates have yet to be made, a significant contribution to the retrograde torque is indicated by order-of-magnitude considerations, applied to Rossby-wave theory together with wave–mean interaction theory expressed in terms of the Rossby–Ertel potential vorticity (see eq. (8)ff. below, also McIntyre and Norton 1990 & refs., McIntyre 1993 & refs., Bühler *et al.* 1994). A good qualitative understanding is available from these theories, and it is clear in particular that there is a robust tendency for real, finite-amplitude Rossby waves to produce a retrograde torque when they dissipate — an aspect of the 'one-signedness' already alluded to.

In any case, the work of Lindzen and Tsay (*op. cit.*), and a large range of observational and modelling studies since then (*e.g.* Takahashi and Boville 1992 & refs.), have been suggesting more and more clearly that some such extra contribution is needed to account for the required torque during the phases of the QBO when the torque has to be retrograde. The observed equatorially-trapped waves to which the retrograde torque is usually attributed, the Rossby–gravity waves, typically seem to be too weak and too sporadic to provide all the torque, even before allowance is made for the effect of mean upwelling. Besides this, the newly-recognised inertial-instability-related form of wave breaking near the equator is practically certain to affect Rossby–gravity waves, tending to limit their amplitudes regardless of source strength. This is because, like Rossby waves, Rossby–gravity waves involve latitudinal particle displacements. Unlike the breaking of Kelvin waves, which resembles the breaking of ordinary internal gravity waves and has a fairly definite finite-amplitude threshold — at least for idealised theoretical models of Kelvin waves with zero latitudinal particle displacements — the inertial-instability-related breaking of Rossby and Rossby–gravity waves can be expected to take place at all wave amplitudes and at all altitudes. It should not switch off below a threshold amplitude, but should merely be confined nearer the equator. This may tend to prevent large-amplitude Rossby–gravity waves, and indeed other equatorially trapped waves apart from the Kelvin wave, from propagating very far in the vertical, while not, of course, preventing Rossby waves of extratropical origin from reaching the neighbourhood of the equator by propagating obliquely or horizontally.

There is also great uncertainty about the contribution from vertically-propagating high-frequency gravity waves, although vertical soundings by radars and lidars in the tropics might, in the coming few years, tell us more about this contribution, by observing the waves directly, and similarly for the equatorially-trapped waves. For all these wave types there is extreme uncertainty, in addition, about wave generation mechanisms and wave source strengths. The sources all involve nonlinear fluid dynamics, among other things the dauntingly complex, multi-scale phenomena associated with large ensembles of tropospheric cumulonimbus convection cells. To quote briefly from my 1993 review — and this applies to all the relevant wave types — "we have even less knowledge, either observational or theoretical, of possible wave generation mechanisms. The real QBO seems to involve highly complicated, chaotic, nonlinear wave generation processes (mainly in the troposphere, both tropical and extratropical) to which there is a robustly non-chaotic response in the tropical stratosphere."

The latitudinal scale of the terrestrial QBO

The angular momentum changes associated with the terrestrial QBO are observed to take place mainly within a band extending to about 10 to 15 degrees of latitude to either side of the equator. Let us say that the width of this band is something like π times the latitudinal length *scale*, so that the scale itself is of the order of 10^3km. This localisation of the angular momentum changes, together with our ignorance, until recently, of the abovementioned inertial-instability-related form of wave breaking, may have been part of the reason why many researchers interested in the QBO seem to have paid little attention to wave types other than the equatorially-trapped Kelvin and Rossby–gravity waves, despite the increasingly conspicuous problems with Rossby–gravity wave amplitudes.

The equatorially-trapped waves have latitudinal scales that depend on the Doppler shift but, fairly typically, are such that the waveguide structure fits within the same 10-to-15-degree latitudinal band. When one looks more carefully at the effects of Doppler shifting toward small intrinsic frequencies, and considers what is now known about wave breaking, one gets somewhat narrower latitudinal scales on both counts; but then one can reasonably call on processes like sideways or layerwise-two-dimensional shear instability to broaden the scale, giving something like that observed. So it may still be tempting to assume that dominance by equatorially-trapped waves is part of the reason why the angular momentum changes occur mainly in the 10-to-15-degree latitude band around the equator, and not in a band of broader latitudinal scale.

However, in an important recent investigation Haynes (1994) has thrown new light on the problem by demonstrating that the terrestrial QBO could not, in any case, occur in a band of much broader latitudinal scale, no matter how broad the latitudinal distribution of wave-induced torque. The localisation of the angular momentum changes to a 10-to-15-degree latitude band around the equator can be accounted for purely in terms of the dynamical and thermal nature of the axisymmetric response to an axisymmetric wave-induced mean torque, in a spherical stratified rotating system having large Prandtl's ratio (1) and large Richardson number. This removes any possible basis for confining attention to equatorially-trapped waves alone. Such waves may still of course contribute; and the importance of equatorially-trapped Kelvin waves, in particular, is not in question.

Haynes' results show in detail how the different characters of the tropical and extra-tropical responses to wave-induced angular momentum transport fit together. Consider the axisymmetric response to a given time-dependent torque, of shallow vertical scale D, say, and broad latitudinal scale comparable to the radial distance R to the centre of the Earth or the Sun. Thus the torque might be proportional, for example, to something like the cosine squared of the latitude. Its time variation is assumed to be characterised by the timescale τ_{torque}, let us say quasi-periodic with period of order $2\pi\tau_{\mathrm{torque}}$, so that for the terrestrial QBO τ_{torque} is several months. The temperature and buoyancy perturbations associated with the evolving differential rotation, taken to be in hydrostatic and cyclostrophic balance, are subject to a thermal relaxation process characterised by timescale τ_{thermal}. In the relevant altitude range in the Earth's tropical stratosphere, τ_{thermal} can be taken to be of the order of weeks. Because of the stratosphere's low infrared opacity, already mentioned, τ_{thermal} is less strongly dependent on vertical scales of temperature variation than the diffusive thermal relaxation found in the Sun's interior, going more like $D^{1/2}$ than D^2 for vertical scales of order D (Fels 1984, Haynes and Ward 1993). For the terrestrial QBO, then,

$$\tau_{\text{thermal}}/\tau_{\text{torque}} \sim 10^{-1} \ . \qquad (2)$$

Define

$$\widehat{D}^2 \ = \ \min\left(D^2, \ HD\right) \ , \qquad (3)$$

where H is the e-folding pressure scale height, $\sim$ 7km for the Earth's stratosphere and $\sim$ 10^5km for the outer parts of the Sun's interior. Then the unsteady changes in differential rotation are found to be localised to a band around the equator with latitudinal scale

$$L_{\text{lat}} \ \sim \ \left(\frac{N\widehat{D}R}{2\Omega}\right)^{1/2} \left(\frac{\tau_{\text{thermal}}}{\tau_{\text{torque}}}\right)^{1/4} \ . \qquad (4)$$

For the terrestrial QBO, the second factor in this expression is about 0.6, from (2). The quarter power makes it insensitive to the exact timescales assumed. From (1)ff. and with $\widehat{D} \sim D \sim$ 4km and $R \sim$ 6000km we have, consistent with what is observed, $L_{\text{lat}} \sim$ 1000km.

Further poleward, the response to the torque consists mainly of the extratropical 'pumping' action already mentioned. There the essential feature, which depends on the Earth's rapid rotation, is a poleward or equatorward flow whose Coriolis force comes into balance with the applied torque on a timescale less than τ_{torque}. For a retrograde torque, such as that due to dissipating Rossby waves, the flow is poleward; this is what occurs in the real extratropical stratosphere, as already mentioned, although there we should take $2\pi\tau_{\text{torque}} = 1$ year. The two regimes merge smoothly into one another near latitudes of order some number of order unity times $\pm L_{\text{lat}}/R$ radian. All these results can be made plausible via scaling considerations related to those given in Professor Holton's paper in this Proceedings. Haynes' very thorough investigation includes, also, precise quantitative modelling using both analytical and numerical techniques.

Could there be a QBO-like phenomenon in the Sun's interior?

Since, as (1) reminds us, the Sun's radiative interior is fluid-dynamically very like the Earth's stratosphere, the question that inevitably arises is whether any of the observed unsteadinesses in solar properties are connected, directly or remotely, with QBO-like phenomena in the Sun's interior. To be sure, dynamos can exhibit unsteadinesses by themselves, but is that the whole story?

In the relatively opaque solar interior, τ_{thermal} becomes simply a thermal diffusion time,

$$\tau_{\text{thermal}} = D^2/\kappa \ , \tag{5}$$

where a typical thermal diffusivity would be $\kappa \sim 10^7 \text{cm}^2\text{s}^{-1}$. Thus, with $\widehat{D} \sim D \lesssim H$, (4) is replaced by

$$\frac{L_{\text{lat}}}{D} \sim \left(\frac{NR}{2\Omega}\right)^{1/2} (\kappa\tau_{\text{torque}})^{-1/4} \ . \tag{6}$$

With $R \sim 5 \times 10^5$km we have $NR/2\Omega \sim 1.5 \times 10^8$km. If we take $2\pi\tau_{\text{torque}} = 22$ years, then $\tau_{\text{torque}} \sim 10^8$s, so that $(\kappa\tau_{\text{torque}})^{1/2} \sim (10^{15}\text{cm}^2)^{1/2} \sim 3 \times 10^7\text{cm} \sim 3 \times 10^2\text{km}$, whence $L_{\text{lat}}/D \sim 7 \times 10^2$. If we take $2\pi\tau_{\text{torque}} \sim 2000$ years, then L_{lat}/D is only a factor 3 smaller, that is, $L_{\text{lat}}/D \sim 2 \times 10^2$.

If we take this at face value, there are two possibilities. The first is that the phenomenon is very shallow, of order 10^3km or less, in which case, if it is also very close to the Sun's centre or to the base of the convection zone, then N might be an order of magnitude smaller hence L_{lat}/D several times smaller, which, however, does not change the picture very much. The second possibility is that the phenomenon is significantly deeper, in which case the whole of the Sun's interior is, for present purposes, dynamically like the Earth's tropics. QBO-like phenomena would be possible on a broad latitudinal scale if suitable waves were present. More precisely, if the vertical scale D is significantly greater than the scale implied by substituting R for L_{lat} in (6), then there is little high-latitude pumping action like that operating in the Earth's extratropical stratosphere. Such pumping is then largely choked off by the stable stratification in combination with the far higher opacity to heat transport implied by typical values of κ. In other words thermal diffusion, though possibly crucial to wave dissipation (Gough 1977, Press 1981) can be altogether neglected in the dynamics of the differential-rotation changes. The latter can then be modelled simply as a frictionless, time-varying adiabatic response to the time-varying torque.

The next question is whether suitable waves are excited at sufficient amplitude. In fact there are two questions, first what amplitudes might be needed to drive some kind of QBO-like phenomenon, and second whether sufficiently strong wave sources might actually exist. I now argue that the answers to both questions turn critically upon the interpretation of recent helioseismic results, which give a new twist to the story.

Some implications of recent helioseismic results

As hinted earlier, the helioseismic evidence (*e.g.* Goode *et al.* 1991, Dziembowski and Goode 1993) suggests that the actual differential rotation of the present-day Sun is remarkably different from what might have been expected from purely fluid-dynamical considerations. Specifically, the evidence suggests that

(a) the angular velocity Ω decreases with decreasing R below the base of the Sun's equatorial convection zone, and

(b) the outer parts of the Sun's radiative interior away from the poles are nearer shellular solid rotation than the convection zone; and the entire convection zone has approximately the same latitude-dependent Ω as the photosphere.

By shellular solid rotation I mean constancy of Ω on the nearly-spherical stratification surfaces $\mathcal{S}$, say, of the radiative interior. If the helioseismic inversions are correct to the extent that statements (a) and (b) are true, then there are far-reaching implications for the problem under consideration. I shall provisionally assume that (a) and (b) are both true and consider the implications.

Statements (a) and (b) are both, at first sight, very surprising indeed. They contradict expectations from any fluid-dynamically-credible stratified, rotating primeval spindown model. In particular, statement (b) cannot be explained as an effect of layerwise-two-dimensional turbulence, for the very good reason that layerwise-two-dimensional turbulence is potential-vorticity-transporting.

For present purposes, this puts it at an opposite extreme from the momentum-transporting turbulent viscosities usually postulated in models of the Sun's differential rotation (*e.g.* Mestel *et al.* 1988, Zahn 1992, & many refs.). The potential-vorticity-transporting character of real layerwise-two-dimensional turbulence is clear from many observational, theoretical and numerical-modelling studies of the Earth's stratosphere and similar fluid-dynamical systems. It represents another of our most secure pieces of knowledge about such systems (see for instance §§8–9 & pp.362–5 of my 1992 review);

and on top of this the whole picture is now being strongly confirmed by a wealth of new observational data, obtained as a result of concerns about stratospheric ozone-layer chemistry (*e.g.* Lahoz *et al.* 1993, Waugh *et al.* 1994, Manney *et al.* 1994, & many refs.).

The potential-vorticity-transporting character of layerwise-two-dimensional turbulence means that, if such turbulence has a significant effect at all, then the effect is to reduce large-scale gradients not of momentum, but of the scalar quantity Q known as the potential vorticity — more precisely the Rossby–Ertel potential vorticity (Rossby 1936, 1940; Ertel 1942) — on each stratification surface $\mathcal{S}$. Q values tend to be carried advectively along each surface $\mathcal{S}$ and hence to undergo two-dimensional chaotic advection and mixing. This is just what we routinely see happening in the Earth's stratosphere and in similar systems. Q may be defined by

$$Q = b^{-1}\boldsymbol{\omega}\cdot\mathbf{n} \ , \tag{7}$$

where $\boldsymbol{\omega}$ is the absolute vorticity vector, or curl of the fluid-dynamical velocity field in an inertial reference frame, and $\mathbf{n} = \nabla s/|\nabla s|$, the local unit normal to the surface $\mathcal{S}$, where s is any thermodynamic variable, such as specific entropy, that labels the stratification surfaces $\mathcal{S}$. These surfaces, by definition, are material surfaces when the motion is adiabatic. The scalar field b^{-1} is defined as $|\nabla s|$ divided by the mass density. It is a strictly positive quantity for a stably stratified system, as is its reciprocal b, which latter can be regarded as a sort of stratification-modified mass density. More precisely, $b\,ds$ is the mass per unit area between neighbouring stratification surfaces, so that if dA is the surface area element then $b\,dA\,ds$ is the mass element.

The notation has been chosen to make conspicuous the fact that the surface area integral of bQ over each closed, quasi-spherical surface $\mathcal{S}$ must vanish, by Stokes' theorem:

$$\iint_{\mathcal{S}} bQ\,dA \; = 0\ . \tag{8}$$

The implication is that there is only one possible effect of strong layerwise-two-dimensional turbulence, if by strong we mean effective in eliminating the isentropic gradient of Q, *i.e.* the gradient of Q on the surface $\mathcal{S}$. The only possible such effect allowed by (8) is to make Q *zero* everywhere on $\mathcal{S}$, since b is strictly positive. The resulting state might be called 'shellular stagnation': it is a state having zero absolute angular velocity Ω everywhere on $\mathcal{S}$, apart from any small oscillations due to acoustic and other wave motions. It is the only state of shellular solid rotation that can be produced by layerwise-two-dimensional turbulence. It is a fantastically improbable state, for the real Sun's interior and the real Earth's stratosphere, if only because its realization would require implausibly large, or persistent, retrograde torques to have been applied to $\mathcal{S}$ by whatever was exciting the turbulence.

It can therefore be safely assumed that any layerwise-two-dimensional turbulence actually present cannot be strong in anything like the foregoing sense. It cannot greatly reduce the overall, pole-to-pole contrast in Q values. There are then only two likely situations. The first is that the turbulence is so weak or sporadic as to have a negligible effect on large-scale isentropic gradients of Q, hence a negligible effect on the distribution of Ω. The second is that the turbulence is stronger but inhomogeneous, effective in reducing isentropic gradients of Q but over limited regions only. The Earth's stratosphere, and numerical models of it, routinely and repeatedly provide examples of both situations. At solstice, between altitudes $\sim$ 30–40km, for instance, we see the first situation in the summer hemisphere, to good approximation, and we see the second situation in the winter hemisphere. The typical character of the fluid motion in the second situation is well illustrated in a recent numerical model study by Norton (1994).

It might well be asked where the implied torques come from, in this second situation. Though far smaller than the fantastically large torques that, as just shown, are part of the necessary conditions for exciting strong layerwise-two-dimensional turbulence, nevertheless the torques cannot generally be zero when layerwise-two-dimensional turbulence is effective. The answer has already been hinted at: the torques come from other altitudes or latitudes — which could be far away or nearby — via Rossby-wave-induced angular momentum transport or, if you prefer, Rossby-wave radiation stress, associated with propagating or diffracting Rossby waves. In particular, the surfaces $\mathcal{S}$ undulate gently but significantly, allowing pressure fluctuations to exert torques from other altitudes. Some of the consequences have already been mentioned in the case of the Earth, where for instance most of the retrograde torque on the stratosphere is exerted from the troposphere below in just this way.

The implied relation between layerwise-two-dimensional turbulence and Rossby-wave radiation stress is a fundamental reason why, in fluid-dynamical systems of the type we are dealing with, layerwise-two-dimensional turbulence tends to occur together with Rossby waves of one sort or another, as was suggested in the introduction. Indeed, it often makes sense to regard the layerwise-two-dimensional turbulence as having been excited by the breaking of Rossby waves, provided that 'breaking' is understood in a relevant and suitably generalised way, as already mentioned. The reader interested in the mathematical underpinning may consult my reviews and references therein to a large literature on 'vortex forces', 'potential-vorticity inversion', 'wave–mean interaction' and Rossby-wave critical layers, also Bühler *et al.* (1994). The resulting body of theory shows how the ratchet-like, robustly one-signed character of Rossby-wave-induced angular momentum transport and all its consequences are related to the likelihood that dissipative processes, including layerwise-two-dimensional turbulence, will locally reduce rather than increase the large-scale isentropic gradients of Q. This produces retrograde rather than prograde torques on the regions where the Rossby waves are being dissipated.

The expectation that any layerwise-two-dimensional turbulence that is effective will also be inhomogeneous, mixing Q on some regions of $\mathcal{S}$ but not on others, is reinforced by considering the nature of 'Rossby-wave elasticity'. This is the peculiar fluid-dynamical quasi-elasticity to which Rossby-wave propagation owes its existence, and against which work must be done if we want to excite Rossby waves or to initiate layerwise-two-dimensional turbulence. This quasi-elasticity depends on the magnitude of the isentropic gradient of Q, to which Rossby-wave intrinsic frequencies are proportional. If isentropic gradients of Q are reduced over some region of a stratification surface $\mathcal{S}$, then the gradients will tend to be increased in adjacent regions of the same $\mathcal{S}$. This is a positive feedback since the reduced gradients have weaker Rossby elasticity and facilitate further layerwise-two-dimensional mixing, while the increased gradients have stronger Rossby elasticity and inhibit such mixing. The Rossby elasticity is only part of the picture because it turns out that horizontal shear also has an important role; but the end result is exactly what the argument about positive feedback suggests, namely an extreme inhomogeneity with some regions dominated by undular, quasi-elastic behaviour adjacent to other regions dominated by layerwise-two-dimensional turbulence. The effects of this highly inhomogeneous 'wave-turbulence jigsaw puzzle' are seen, for example, in the various phenomena associated with the Antarctic ozone hole. There, the chemistry depends on the inhibition of layerwise-two-dimensional mixing by an undular, Rossby-elastic 'potential-vorticity barrier' marking the edge of the winter stratospheric polar vortex.

An immediate corollary is that homogeneous or weakly-inhomogeneous turbulence theory is inapplicable, indeed profoundly misleading for present purposes. The same conclusion can be reached by a careful consideration of the way in which homogeneous turbulence theory predicts *its* own breakdown, *e.g.* via Fjørtoft's theorem on 'upscale energy cascading', equivalently the well known tendency of nearby vortices to merge inelastically, seen in a whole range of theoretical models (*e.g.* Legras and Dritschel 1993, Dritschel 1993 & many refs.). The tendency toward inhomogeneity is very powerful, invalidating among other things the scale-separation, or weak-inhomogeneneity, assumption on which the usual notion of 'turbulent viscosity' is based.

What then is the main point? It is that layerwise-two-dimensional turbulence drives the system away from, not toward, shellular solid rotation. How then are we to make sense of statement (*b*) above? All laminar spindown models give latitude-dependent Ω in the Sun's interior, contradicting statement (*b*); and layerwise-two-dimensional turbulence will, if anything, make Ω still more strongly latitude-dependent, as in fact it does in the winter stratosphere. The clear implication is that some further physical effect must be pulling the outer parts of the Sun's radiative interior toward shellular solid rotation, except perhaps near the poles where helioseismology gives no information. There is one and only one obvious, and well known, candidate: a sufficiently strong poloidal magnetic field $\mathbf{B}_\mathrm{p}$.

The question of the origin of interior magnetic fields is outside the scope of this paper (see, *e.g.*, Mestel and Weiss 1987, & refs.). Suffice it to mention first that the usual argument against an interior $\mathbf{B}_\mathrm{p}$ from north-south symmetry of the sunspot butterfly pattern has been vitiated by the north-south *asymmetry* found at Maunder Minimum (*e.g.* Nesme–Ribes *et al.* 1994, this Proceedings), second that magnetic diffusivities are small enough to allow persistence of a primeval $\mathbf{B}_\mathrm{p}$ in a non-turbulent interior, as is well known, and third that if, on the other hand, layerwise-two-dimensional turbulence is significant in the interior, then helicity and dynamo action are possible through the enhanced magnetic buoyancy of field lines locally wrapped up by the turbulent eddies (E. A. Spiegel, personal communication). What can, with practical certainty, be said is that if statement (*b*) stands scrutiny, then the fluid-dynamical facts will compel us to accept the existence of the poloidal field $\mathbf{B}_\mathrm{p}$, whatever its origin.

The implications, which evidently include major implications for helioseismic inversion strategies as well as for our understanding of primeval solar spindown, will be discussed more adequately elsewhere (paper with Goode and Gough in preparation); but let me close with a very brief sketch. It will include the beginnings of an answer, of an unexpected kind, to the original question about the possibility of QBO-like phenomena.

$\mathbf{B}_p$ lines as 'reinforcing rods', and helioseismic inversions

The picture we have arrived at has the $\mathbf{B}_p$ lines of a well-contained interior field acting as if they were rigid 'reinforcing rods', to good approximation frozen into the fluid and spanning a substantial range of latitudes in the outer parts of the interior, albeit not necessarily as far as the poles. The 'reinforcing rods' may be twisted toward the toroidal, in a quasi-static way, by torques due to waves and turbulence. They are nonetheless effectively rigid, to the extent that they stop differential rotation, if we assume that magnetic torsional oscillations have been damped out. This last assumption has a strong theoretical justification, to be mentioned below; and the picture is fully consistent with statement (*b*). It tells us, furthermore, that the differential rotation throughout most of the interior must conform not to simple shellular solid rotation but to what might be called toroidally-shellular or doughnut-shellular solid rotation, otherwise known as Ferraro's law of isorotation: Ω must be constant not on the surfaces $\mathcal{S}$, but on the toroidal shells of $\mathbf{B}_p$. At present this cannot be distinguished helioseismically from shellular solid rotation, since present helioseismic results give us no more than a low-resolution view of the outer parts of the radiative interior away from the poles.

It therefore appears that there is an exceedingly strong case for assuming toroidally-shellular solid rotation, including the implied jump in Ω at the tachocline, when doing helioseismic inversions — unless and until further observational information forces us to do otherwise by, for instance, overturning statement (*b*). Occam's razor suggests that first attempts should use simple $\mathbf{B}_p$ configurations somewhat like the gravest modes obtained from classic calculations of primeval magnetic diffusion, with a single neutral ring or limiting torus on the equatorial plane. The boundary conditions on $\mathbf{B}_p$ at the tachocline will need reconsideration, if only because of the requirement for a well-contained field to give a reinforcing-rod effect consistent with statement (*b*). Any such procedure will usefully constrain the Ω structure not only near the tachocline but also in the deep interior, where the scope of helioseismic information is most severely limited.

The same picture also promises a natural and convincing explanation for statement (*a*), consistent with the idea that the convection zone is relatively vigorous dynamically and has a latitude-dependent differential rotation $\Omega = \Omega_{CZ}$ for dynamical reasons of its own, and that the tachocline is the necessary transition layer between Ω_{CZ} and the interior, $\mathbf{B}_p$-constrained Ω field. This predicts first of all that the tachocline must be as thin as nonlinear fluid dynamics or magnetohydrodynamics permit it to be. Specifically, we can expect the tachocline to be the seat of three-dimensional, shear-related instabilities one of whose effects, to a first approximation, is likely to be a local, quasi-frictional torque across the tachocline far larger than the net torque associated with magnetic solar-wind braking (Gough 1985, Brown *et al.* 1989, Dziembowski and Goode 1993).

An inevitable by-product of such local shear instabilities will be radiation, into the interior, of various kinds of waves — some of them with far larger amplitudes, and with radial wavelengths far less subject to drastic thermal dissipation, than the weak internal gravity waves generated directly by convection zone eddies (*e.g.* Gough 1977, Press 1981, García López and Spruit 1991, Schatzman 1993 & refs.). The timescales probably favour Rossby waves; and Rossby waves from the tachocline should be emitted mostly into the high-latitude interior, because it is at high latitudes that the interior has prograde Ω relative to the tachocline. The one-signed Rossby-wave dynamics and dispersion properties then allow long-range propagation over substantial depths and latitudes. Rossby-wave propagation will hardly feel $\mathbf{B}_\mathrm{p}$ and should be far less drastically limited by thermal dissipation than gravity-wave propagation (Gough 1977, Press 1981); and Rossby-wave dispersion properties favour not only downward, but also equatorward propagation, of waves originating in high latitudes as can easily be shown. It seems possible that the Rossby waves in question could affect all the toroidal shells of $\mathbf{B}_\mathrm{p}$, especially if at some stage of spindown the shells were all rotating at the same Ω as, or faster than, the outermost shells relevant to statement (*b*).

This puts the high-latitude tachocline into close analogy with the Earth's winter troposphere, which emits Rossby waves into the relatively prograde winter stratosphere. In just the same way, the relatively prograde motion, Ω larger, in the winter stratosphere favours long-range Rossby propagation; and, again, the Rossby-wave dynamics and dispersion properties favour not only upward but also equatorward propagation, so that some Rossby-wave activity affects the tropical as well as the extratropical winter stratosphere. As argued earlier, it probably contributes thereby to the terrestrial QBO.

In the case of the Sun's Rossby waves, any wave dissipation, whether by thermal damping or by various modes of wave breaking, will produce a retrograde torque on the toroidal shells of $\mathbf{B}_\mathrm{p}$. This will slow them down until their Ω values are nearly outside the range of Rossby-wave phase speeds being produced by the tachocline. This too is consistent with statement (*a*), and it suggests moreover that future helioseismic deductions of Ω values on the shells of $\mathbf{B}_\mathrm{p}$ will contain key information about Rossby-wave phase speeds. The whole picture makes sense provided that the timescale for the retrograde torque to act is shorter than primeval spindown timescales. In all likelihood it is very much shorter; I shall refer to this as the 'strong Rossby-wave hypothesis'.

If the Rossby waves reach the Sun's equatorial plane — and in some circumstances they might even do so more than once, after propagating to polar regions and back hence reaching greater depths — then there are two further consequences of interest. The first is the inertial-instability-related wave breaking that could provide the lithium channel near the equator. This, too, merits further investigation. The second is the possibility of a QBO-like phenomenon near the equator. Why near the equator, in view of what was said above about the scale L_{lat} being likely to be broad, a conclusion that might seem at first sight unlikely to be altered by the presence of a broad-scale $\mathbf{B}_{\mathrm{p}}$ field? To understand this, we must first return to the question of the damping of magnetic torsional oscillations.

The peculiarly high damping of magnetic torsional oscillations

The hypothesis of toroidally-shellular solid rotation requires that magnetic torsional oscillations be sufficiently damped. Now it appears that typical magnetic torsional oscillations are much more highly damped, even without help from shear or magnetohydrodynamic instabilities, than one might guess from the magnetic diffusion time τ_{I} for the whole interior ($\tau_{\mathrm{I}} \lesssim 10^{10}$ years) and from the corresponding viscous diffusion time, which is several orders of magnitude greater still. Typical magnetic torsional damping times τ_{mag} turn out to be considerably shorter even than the Kelvin–Helmholtz thermal time $\sim 3 \times 10^7$ years.

Recent unpublished results independently obtained by Professor Gough and myself show that τ_{mag} depends on the magnitude $|\nabla \ln \tau_{\mathrm{A}}|$ of the logarithmic gradient, across the toroidal shells of $\mathbf{B}_{\mathrm{p}}$, of the Alfvén time τ_{A} around those shells in a meridional plane:

$$\tau_{\mathrm{mag}} \sim {\tau_{\mathrm{A}}}^{2/3} \{\eta(|\nabla \ln \tau_{\mathrm{A}}|)^2\}^{-1/3} , \tag{9}$$

where η ($\sim 10^5 \mathrm{cm}^2\mathrm{s}^{-1}$) is the magnetic diffusivity. The gradient causes an inexorable scale-shrinkage as t^{-1}, giving enormously enhanced dissipation rates. If the cross-shellular variation of the reciprocal Alfvén time ${\tau_{\mathrm{A}}}^{-1}$ is of the same order as ${\tau_{\mathrm{A}}}^{-1}$ itself, then very roughly $\tau_{\mathrm{mag}} \sim {\tau_{\mathrm{A}}}^{2/3} {\tau_{\mathrm{I}}}^{1/3}$. If for instance $\tau_{\mathrm{A}} \lesssim 10^4$ years, then in this simplest case $\tau_{\mathrm{mag}} \lesssim 10^6$ years. Magnetic-buoyancy or other instabilities could, at finite amplitude, shorten this still further (Gough, personal communication).

The theory is a nontrivial generalization of the theory of Kelvin sheared disturbances (Thompson 1887), with τ_{A} taking the place of the single-wavelength advection time but with the unequal magnetic and viscous diffusivities complicating the analytical and behavioural details. When the diffusivities are equal, the theory becomes mathematically the same as Kelvin's. When they are unequal, as in the real Sun, it is the larger, magnetic, diffusivity that governs the damping.

The possibility of a 'watch-spring QBO' near the neutral ring

Should we expect there to be, or have been, a solar QBO-like phenomenon, specifically, a QBO-like fluctuation in the interior differential rotation field Ω? If statement (*b*) is true and $\mathbf{B}_\mathrm{p}$ exists, and if furthermore statement (*a*) and the strong Rossby-wave hypothesis are true, then at least some of the ingredients are present. Indeed the analogy between the tachocline and the Earth's troposphere, already discussed, and the indication from helioseismic data that typical tachocline shears are far stronger than required by primeval spindown dynamics alone, point toward Rossby-wave amplitudes that could easily be far above the minimum necessary. Some order-of-magnitude estimates are being attempted, to gain an idea of the retrograde torques that might result.

On the other hand, it is still a moot point whether any wave types capable of exerting prograde torques are available, with sufficient amplitudes. To the extent that the partial analogy between the Sun's tachocline and the Earth's troposphere is valid, the tachocline must be a far more efficient generator of Rossby waves than of gravity waves and equatorial hybrid types. This seems unlikely to be altered by Lorentz forces of plausible magnitude and timescale in the tachocline. The direct wave source from the convection zone seems weak, also, for this purpose, as already mentioned, with current estimates pointing to rates of change of Ω not much faster than primeval spindown (*e.g.* Schatzman 1993 & refs.).

So it is worth asking first whether we could get a QBO-like phenomenon with only one type of wave. The answer, albeit with certain caveats, is "perhaps yes" — in interesting contrast to the terrestrial QBO. Moreover, the requirement on wave amplitude is far less severe. The reason is the Alfvénic quasi-elasticity, or watch-spring mechanism, provided by $\mathbf{B}_\mathrm{p}$. The waves need only do work on the torsional oscillations once per cycle; and the work per cycle need only be a small fraction of the energy of the oscillation. The wave-induced angular momentum transport is required to be significant not on the timescale τ_A of a single oscillation, but only on the damping timescale τ_mag, which even though far shorter than τ_I may well be far longer than τ_A.

In order for the wave-induced angular momentum transport or radiation stress to do positive work on an incipient torsional oscillation, the waves need only satisfy two of the conditions already familiar from studies of the terrestrial QBO. First, the waves must have nonzero angular phase speeds relative to the mean Ω values of the toroidal magnetic shells in question. Second, their dissipation must be enhanced by Doppler downshifting. Then it is not too hard to see that the resulting torque, whether prograde or retrograde, from any waves that propagate past the relevant places, will always do positive work on the incipient torsional oscillation because such waves dissipate most strongly, exerting the greatest torques, when and where the fluctuation in Ω is in the same sense as, albeit of smaller magnitude than, the waves' intrinsic phase speeds — just as in the terrestrial QBO. So whether the magnetic torsional oscillations are excited or not is partly a question of the competition between their own damping rate τ_{mag}^{-1} and the strength of the excitation by the waves, in turn dependent on radiation stress, approximately proportional to wave amplitude squared, and on the sensitivity of wave dissipation to Doppler downshifting.

The main caveat is that such an oscillation could not persist by itself, because if nothing else were happening there would inevitably be a time-mean retrograde torque that would gradually take Ω out of the effective range of available Rossby-wave phase speeds. This is the same 'strong Rossby wave' effect already invoked to account for statement (a). It would be necessary for some other agency, perhaps a weaker prograde wave, to stop Ω drifting completely out of range. An intriguing possibility, in no way contradicting statements (a) and (b), is that the field $\mathbf{B}_{\mathrm{p}}$ might have been expelled from a rapidly (and progradely) rotating core, or axial vortex, in which case an inner 'spindown tachocline' would exist and might provide enough prograde torque either in a local quasi-frictional sense or, for instance, through short-range gravity-wave propagation, on parts of the toroidal shells of $\mathbf{B}_{\mathrm{p}}$, keeping their Ω values in range.

Where should we first look for a solar QBO-like phenomenon? One reasonable answer is to look first where magnetic torsional oscillations are least damped. The estimate (9) tells us clearly where that is, even though the estimate itself obviously becomes invalid there: the place to look is near the neutral ring, or limiting torus, of $\mathbf{B}_{\mathrm{p}}$, where τ_{A} is stationary and $|\nabla \ln \tau_{\mathrm{A}}|$ vanishes. Here we are again invoking the Occam's-razor assumption that $\mathbf{B}_{\mathrm{p}}$ is simple, with just one neutral ring on the equatorial plane.

Quantitative estimates are beyond the scope of this discussion, and indeed will demand much detailed work. But it might not be accidental that helioseismic results, in a recent interpretation reported by Goode and Dziembowski (1991), give a hint of possible time-dependence of Ω in a deep region below the solar equator, which could indeed be part of the neighbourhood of the neutral ring of a plausible $\mathbf{B}_p$. If this putative time-dependence is real, then the timescale is comparable to that of the solar cycle and, as the authors point out, could only be a magnetic torsional oscillation on a $\mathbf{B}_p$ of kilogauss strength.

Acknowledgements — It is a special pleasure to thank Professors D. O. Gough and E. A. Spiegel for sharing their many interesting insights and for many stimulating discussions, over the past two decades, including discussions of the possible rôles of wave-induced angular momentum transport and layerwise-two-dimensional turbulence in stellar interiors. Many of the foregoing ideas arose from those discussions, and some of them were originally recorded in an essay that shared the 1981 Adams Prize in the University of Cambridge. Dr P. H. Haynes has shared many insights about the terrestrial QBO and kindly allowed me to communicate his important new results on the latitudinal scales of QBO-like phenomena, in advance of publication. I thank also Dr Nesme-Ribes and all the Workshop organisers and participants for stimulating discussions during the Workshop that have further helped to shape my thinking, and Dr Nesme-Ribes especially warmly for the original invitation to the Workshop and for encouraging me to write this paper. My own research has been generously supported by grants from the UK Natural Environment Research Council through the UK Universities' Global Atmospheric Modelling Programme and the British Antarctic Survey, from the European Community through contract CHRX-CT92-0001, from the Innovative Science and Technology Program through grants N00014-92-J-2009 and N00014-93-1-G029 administered by the US Naval Research Laboratory, and from the UK Science and Engineering Research Council in the form of a Senior Research Fellowship.

References

Andrews, D. G., McIntyre, M. E. (1976) Planetary waves in horizontal and vertical shear: the generalized Eliassen–Palm relation and the mean zonal acceleration. J. Atmos. Sci., 33: 2031–2048. [See pp.2043–5.]

Brown, T. M., Christensen-Dalsgaard, J., Dziembowski, W. A., Goode, P. R., Gough, D. O., Morrow, C. A. (1989) Inferring the Sun's internal angular velocity from observed *p*-mode frequency splittings. Ap. J., 343: 526–546

Bühler, O., McIntyre, M.E., Mo, R., Warner, C. D. (1994) On potential-vorticity evolution and the inefficiency of momentum transport by low-frequency inertia-gravity waves. In preparation.

Ciesielski, P. E., Stevens, D. E., Johnson, R. H., Dean, K. R. (1989) Observational evidence for asymmetric inertial instability. J. Atmos. Sci., 46: 817–831

Clark, P. D., Haynes, P. H. (1994) Inertial instability of an asymmetric low-latitude flow. Q. J. Roy. Meteorol. Soc., submitted.

Dickinson, R. E. (1968) Planetary Rossby waves propagating vertically through weak westerly wind waveguides. J. Atmos. Sci., 25: 984–1002

Dritschel, D. G. (1993) Vortex properties of two-dimensional turbulence. Phys. Fluids, A 5: 984–997

Dunkerton, T. J. (1983) Laterally-propagating Rossby waves in the easterly acceleration phase of the quasi-biennial oscillation. Atmos.-Ocean, 21: 55–68

Dunkerton, T. J. (1991) Nonlinear propagation of zonal winds in an atmosphere with Newtonian cooling and equatorial wavedriving. J. Atmos. Sci., 48: 236–263

Dunkerton, T. J. (1993) Inertial instability of nonparallel flow on an equatorial β-plane. J. Atmos. Sci., 50: 2744–2758

Dziembowski, W. A., Goode, P. R. (1993) The Sun's internal angular momentum from seismology. In: GONG 1992: Seismic investigation of the Sun and Stars. Ed. T. M. Brown ASP Conference Series vol 42, San Francisco, Astron. Soc. Pacific. 225–227.

Ertel, H. (1942) Ein Neuer hydrodynamischer Wirbelsatz. Meteorol. Zeitschr., 59: 271–281

Fels, S. B. (1984) The radiative damping of short vertical scale waves in the mesosphere. J. Atmos. Sci., 41: 1755–1764

Fritts, D. C. (1984) Gravity wave saturation in the middle atmosphere: a review of theory and observations. Rev. Geophys. Space Phys., 22: 275–308. [Note that the formula for the effective momentum flux due to inertia-gravity waves needs correction when Coriolis forces are important; see *e.g.* Bühler *et al.* & refs.]

García López, R. J., Spruit, H. C. (1991) Li depletion in F stars by internal gravity waves. Ap. J., 377: 268–277

Gill, A. E. (1982) Atmosphere-Ocean Dynamics. Academic Press, 662pp

Goode, P. R., Dziembowski, W. A., Korzennik, S. G., Rhodes, E. J. (1991) What we know about the Sun's internal rotation from solar oscillations. Ap. J., 367: 649–657

Goode, P. R., Dziembowski, W. A. (1991) Solar-cycle dependence of the Sun's deep internal rotation shown by helioseismology. Nature, 349: 223–225

Gough, D. O. (1977) Random remarks on solar hydrodynamics. In: The Energy Balance and Hydrodynamics of the Solar Chromosphere and Corona (Proc. IAU Colloq. 36), ed. R. M. Bonnet, P. Delache; Clermont-Ferrand, G. de Bussac, 3–36

Gough, D.O. (1985) Future missions in solar, heliospheric and space plasma physics. Ed. E. J. Rolfe and B. Battrick, Proc. European Space Agency ESA SP-235 Nordwijk 183–197.

Gough, D. O. (1990) On possible origins of relatively short-term variations in the solar structure. Phil. Trans. Roy. Soc. London, 330: 627–640

Hagan, M. E., Vial, F., Forbes, J. M. (1993) Variability in the upward propagating semidiurnal tide due to effects of QBO in the lower atmosphere. J. Atmos. Terrest. Phys., 54: 1465–1474

Hamilton, K. (1983) Quasi-biennial and other long period variations in the solar semidiurnal barometric oscillation: observations, theory, and possible application to the problem of monitoring changes in global ozone. J. Atmos. Sci., 40: 2432–2443

Hamilton, K., Garcia, R. R. (1984) Long-period variations in the solar semidiurnal atmospheric tide. J. Geophys. Res., 89: 11705–11710

Hamilton, K., Garcia, R. R. (1986) Theory and observations of the short-period normal mode oscillations of the atmosphere. J. Geophys. Res., 91: 11867–11875

Haynes, P. H. (1994) The latitudinal structure of the quasi-biennial oscillation. In preparation.

Haynes, P. H., and McIntyre, M. E. (1987) On the evolution of vorticity and potential vorticity in the presence of diabatic heating and frictional or other forces. J. Atmos. Sci., 44: 828-841

Haynes, P. H., McIntyre, M. E. (1990) On the conservation and impermeability theorems for potential vorticity. J. Atmos. Sci., 47: 2021–2031

Haynes, P. H., Ward, W. E. (1993) The effect of realistic radiative transfer on potential vorticity structures, including the influence of background shear and strain. J. Atmos. Sci., 50: 3431–3453

Holton, J. R. (1994) The quasi-biennial oscillation in the Earth's atmosphere and its links to longer period variability. (This Proceedings.)

Holton, J. R. and Lindzen, R. S. (1972) An updated theory for the quasi-biennial cycle of the tropical stratosphere. J. Atmos. Sci., 29: 1076–1080

Kousky, V. E., Wallace, J. M. (1971) On the interaction between Kelvin waves and the mean zonal flow. J. Atmos. Sci., 28: 162–169

Lahoz, W. A., Carr, E. S., Froidevaux, L., Harwood, R. S., Kumer, J. B., Mergenthaler, J. L., Peckham, G. E., Read, W. G., Ricaud, P. D., Roche, A. E., Waters, J. W. (1993) Northern hemisphere mid-stratospheric vortex processes diagnosed from H_2O, N_2O and potential vorticity. Geophys. Res. Lett., 20: 2671–2674

Legras, B., Dritschel, D. (1993) Vortex stripping and the generation of high vorticity gradients in two-dimensional flows. In: Advances in Turbulence IV, ed. F. T. M. Nieuwstadt (Applied Scientific Research Vol 51); Dordrecht, Kluwer, 445–455

Lindzen, R. S., Holton, J. R. (1968) A theory of the quasi-biennial oscillation. J. Atmos. Sci., 25: 1095–1107

Lindzen, R. S., Tsay, C. Y. (1975) Wave structure of the tropical stratosphere over the Marshall Islands area during 1 April – 1 July 1958. J. Atmos. Sci., 32: 2008–2021

Manney, G. L., Zurek, R. W., O'Neill, A., Swinbank, R. (1994) On the motion of air through the stratospheric polar vortex. J. Atmos. Sci., 51: in press. [Upper Atmosphere Research Satellite Special Issue.]

McIntyre, M. E. (1992) Atmospheric dynamics: some fundamentals, with observational implications. In: Proc. Internat. School Phys. "Enrico Fermi", CXV Course, ed. J. C. Gille, G. Visconti (ISBN 0-444-89896-4); Amsterdam, Oxford, New York, Toronto, North-Holland, 313–386. [A list of updates and minor corrigenda is available from the author, M.E.McIntyre@amtp.cam.ac.uk or mem2@phx.cam.ac.uk.] See also Haynes and McIntyre (1987, 1990).

McIntyre, M. E. (1993) On the rôle of wave propagation and wave breaking in atmosphere ocean dynamics. Sectional Lecture from "Theoretical and Applied Mechanics 1992" (Proc. XVIII Int. Congr. Theor. Appl. Mech., Haifa), ed. S. Bodner, J. Singer, A. Solan, and Z. Hashin. Amsterdam, New York; Elsevier, 281–304

McIntyre, M. E., Norton, W. A. (1990) Dissipative wave-mean interactions and the transport of vorticity or potential vorticity. J. Fluid Mech., 212: 403–435; Corrigendum 220: 693

McIntyre, M. E., Palmer, T. N. (1984) The "surf zone" in the stratosphere. J. Atm. Terr. Phys., 46: 825–849

McIntyre, M. E., Palmer, T. N. (1985) A note on the general concept of wave breaking for Rossby and gravity waves. Pure Appl. Geophys., 123: 964–975

Mestel, L., Weiss, N. O. (1987) Magnetic fields and non-uniform rotation in stellar radiative zones. Mon. Not. Roy. Astron. Soc., 226: 123–135

Mestel, L., Moss, D. L., Tayler, R. J. (1988) The mutual interaction of magnetism, rotation and meridian circulation in stellar radiative zones. Mon. Not. Roy. Astron. Soc., 231: 873–885

Nesme-Ribes, E., Sokoloff, D., Ribes, J.-C., Kremliovski, M. (1994) The Maunder Minimum and the solar dynamo. This Proceedings.

Norton, W. A. (1994) Breaking Rossby waves in a model stratosphere diagnosed by a vortex-following coordinate system and a technique for advecting material contours. J. Atmos. Sci., 51: 654–673 [Video available from the author]

O'Sullivan, D. J., Hitchman, M. H. (1992) Inertial instability and Rossby wave breaking in a numerical model. J. Atmos. Sci., 49: 991–1002

Plumb, R. A. (1977) The interaction of two waves with the mean flow: implications for the theory of the quasi-biennial oscillation. J. Atmos. Sci., 34, 1847–1858

Press, W. H. (1981) Radiative and other effects from internal waves in solar and stellar interiors. Ap. J., 245: 286–303

Plumb, R. A., Bell, R. C. (1982) A model of the quasi-biennial oscillation on an equatorial beta-plane. Q. J. Roy. Meteorol. Soc., 108: 335–352

Rossby, C. G. (1936) Dynamics of steady ocean currents in the light of experimental fluid mechanics. Mass. Inst. of Technology and Woods Hole Oc. Inst. Papers in Physical Oceanography and Meteorology, 5(1): 1–43 [See Eq. (75)]

Rossby, C. G. (1940) Planetary flow patterns in the atmosphere. Q. J. Roy. Meteorol. Soc., 66(Suppl.): 68–97

Saravanan, R. (1990) A multiwave model of the quasi-biennial oscillation. J. Atmos. Sci., 47: 2465–2474

Schatzman, E. (1993) Transport of angular momentum and diffusion by the action of internal waves. Astron. Astrophys., 279: 431–446

Spiegel, E. A., Zahn, J.-P. (1992) The solar tachocline. Astron. Astrophys. 265: 106–114

Takahashi, M., Boville, B. A. (1992) A three-dimensional simulation of the equatorial quasi-biennial oscillation. J. Atmos. Sci., 49: 1020–1035

Teitelbaum, H., Vial, F., Bauer, P. (1994) The stratospheric quasi-biennal oscillation observed in the semidiurnal ground pressure data. J. Climate, submitted. [Preprints available from Dr François Vial, Lab. Météorol. Dynamique, École Normale Supérieure, Paris]

Thompson, W. (Lord Kelvin) (1887) Stability of fluid motion — rectilineal motion of viscous fluid between two parallel planes. Phil. Mag., 24: 188–196

Thorncroft, C. D., Hoskins, B. J., McIntyre, M. E. (1993) Two paradigms of baroclinic-wave life-cycle behaviour. Q. J. Roy. Meteorol. Soc., 119: 17–55

Wallace, J. M., Holton, J. R. (1968) A diagnostic numerical model of the quasi-biennial oscillation. J. Atmos. Sci., 25: 280–292

Wallace, J. M., Kousky, V. E. (1968) Observational evidence of Kelvin waves in the tropical stratosphere. J. Atmos. Sci., 25: 900–907

Waugh, D. W. (1993) Subtropical stratospheric mixing linked to disturbances in the polar vortices. Nature, 365: 535–537

Waugh, D. W., Plumb, R. A., Atkinson, R. J., Schoeberl, M. R., Lait, L. R., Newman, P. A., Loewenstein, M., Toohey, D. W., Avallone, L. M., Webster, C. R., May, R. D. (1994) Transport of material out of the stratospheric Arctic vortex by Rossby wave breaking. J. Geophys. Res., 99: 1071–1088

Yulaeva, E., Holton, J. R., Wallace, J. M. (1994) On the cause of the annual cycle in tropical lower stratospheric temperatures. J. Atmos. Sci., 51: 169–174

Zahn, J.-P. (1992) Circulation and turbulence in rotating stars. Astron. Astrophys., 265: 115–132

Ziemke, J. R., Stanford, J. L. (1994) Kelvin waves in total column ozone. Geophys. Res. Lett., 21: 105–108

SENSITIVITY OF CLIMATE TO VARIATIONS OF THE SOLAR OUTPUT: RADIATIVE PROCESSES AND THEIR PARAMETRIZATION IN NUMERICAL MODELS

Jean-Jacques Morcrette
European Centre for Medium range Weather Forecasts
Shinfield Park
Reading Berkshire RG2 9AX
United Kingdom

Summary

We first stress the need for an accurate representation of the radiative processes in numerical models of the atmosphere used for climate studies and medium-range weather prediction. We then briefly illustrate some of the approximations and simplifications required for maintaining a good trade-off between computational efficiency and accuracy of the results. We then discuss uncertainties in some aspects of the radiative parametrizations likely to be important for modelling the sensitivity of the climate to variations of the solar output. Finally, we review the recent effort at studying the sensitivity of the climate to variations of the solar output from the viewpoint of the radiative transfer and of its parametrization in the general circulation models used for these studies. Depending whether the forcing is of the top-down or bottom-up kind, we argue that the recently made simulations of the Maunder minimum may not be adequate.

1. Introductory remarks

1.1 Radiation balance

Radiation transfer is the only physical process by which energy can be exchanged between the Earth's atmosphere-surface system and the outside space. A zeroth order model of the Earth's climate follows the global radiation balance equation

NATO ASI Series, Vol. I 25
The Solar Engine and Its Influence
on Terrestrial Atmosphere and Climate
Edited by Elizabeth Nesme-Ribes

$$\frac{S_0}{4} (1 - \alpha_p) = \sigma T_E^4$$

The previous equation simply states the balance between the solar radiation received by the Earth at shorter wavelengths and the radiation emitted at longer wavelengths (following the Stefan-Boltzmann law). S_0 is the so-called "solar constant", the factor 4 appears since the area of the Earth's surface that emits is 4 times the area of the Earth's disk which intercepts sunlight, α_p is the planetary albedo, which defines the fraction of solar radiation reflected back to space, σ is Stefan's constant (5.67 x 10^8 W m^{-2} K^{-4}) and T_E is the Earth's effective blackbody radiative temperature (in K, as seen from the top of the atmosphere). With S_0 = 1365 W m^{-2} , α_p = 0.3 , T_E is approximately 255 K.

1.2 Need for an accurate radiation transfer parametrization

It is well known that the meridional gradient in the deposition of the net radiative energy drives the atmospheric and oceanic circulations. Seen from the top of the atmosphere, the solar (shortwave SW) radiation absorbed at low latitudes exceeds the longwave (LW) radiation lost to space, where the contrary applies at high latitudes, so that some transport mechanism must be active to transfer this excess of energy from the low to the high latitudes (see Stephens et al., 1981; their Fig. 3). However, the amount of energy actually available for driving the general atmospheric and oceanic circulation is the small residual of a delicate balance between the absorbed SW and emitted LW radiation that affects differently the surface and the atmosphere (Ramanathan, 1987). In the tropics, the maximum net annual heating of 50 to 60 W m^{-2} is only roughly 20 % of the total absorbed SW radiation, and 80 % of this radiation contributes to the warming of the tropical surface and atmosphere, which in turn will radiate the energy back to space in terms of LW radiation. Therefore, any systematic error of the determination of one of the SW or LW components of model column net heating in the tropics can potentially induce an error 5 times larger in the required poleward transport of heat. Similarly, for globally averaged conditions, 70 % of the absorbed SW radiation is deposited at the surface, whereas a net radiative cooling prevails within the atmosphere as the atmospheric cooling by emission of LW radiation is roughly 2.5 larger than the atmospheric heating by absorption of SW radiation (see Ramanathan, 1987; his Fig.2). The surface is subject to a net radiative heating of about 100 W m^{-2} while the troposphere is subject to a net radiative cooling of the same magnitude, a pattern responsible for the destabilization of the troposphere and the development of convective processes. Here again, a systematic error in the model representation of the partitioning of the deposited radiation between the surface and the troposphere will have a direct impact on the model simulated atmospheric structure.

The actual accuracy of the model radiation fields will strongly depend on the quality of the representation of the cloud fractional cover and cloud water loading. However, a number of issues also directly relate to the representation of the radiative processes in clear-sky conditions or to the details of the radiative transfer within the clouds. Some of them are discussed in section 3.

1.3 Some examples of the importance of a proper representation of the cloud-radiation interactions

Recently, the cloud-radiation interactions have been proposed to be central in explaining a number of interactions with the hydrological cycle and atmospheric dynamics.

Slingo and Slingo (1988) and Randall et al. (1989) described the positive feedback loop between the radiative warming/cooling gradients associated with the high clouds produced by deep convection, and the large-scale rising motion associated with the convection. However, on a global scale, a negative feedback loop is also active, in which a stronger convection leads to more high-level cloud, which reduces the atmospheric net radiative cooling (obtained by combining the net radiative energy flux at the top of the atmosphere with the net radiative energy at the Earth's surface), which reduces the precipitation rate and the level of convective activity. On a global and annual mean scale, the net atmospheric cooling is balanced by the turbulent heat fluxes. To a first approximation, this latter example links the globally averaged precipitation to the radiative processes.

On a smaller than global scale, there are numerous mechanisms that link the radiative processes to the other physical processes, hydrological in particular, and the atmospheric dynamics. Betts and Ridgway (1988) proposed an idealized energy balance model of the steady state tropical circulation over the uniform ocean surface that links the tropospheric net radiative cooling, the surface fluxes, and the mean subsidence in the descending branch of the Hadley circulation (the regions of deep precipitating convection occupy only a small fraction of the tropics). In this simplified description, which neglects horizontal advection and the atmospheric export of heat to higher latitudes, the surface evaporation is coupled to the ascending mass flux in the ITCZ and the compensating subsidence in the descending branch where the surface fluxes compensate the integrated radiative cooling. This simplified framework has been very operational at explaining the changes in the tropical circulation undergone by the ECMWF forecast model when different parts of the package of physical parameterizations have been successively changed (Morcrette, 1990, 1991).

On an even smaller scale, the anvils and cirrus clouds associated with deep convection in the tropics have a net radiative effects near zero when seen from the top of the atmosphere (Ramanathan et al., 1989) as the cooling by increased reflection of solar radiation and the warming by increased greenhouse effect under these high-level clouds nearly cancel. However, the SW cooling is felt mainly at the surface and the LW heating with in the atmosphere. Ramanathan and Collins (1991) have proposed that these bright upper clouds act as a "thermostat" which mitigates the warming of the tropical oceans and help to moderate the El Niño events.

From these few examples, the actual role of the cloud and radiative processes in their interactions with the other physical processes (hydrological cycle, in particular) and with the dynamics is far from completely understood specially because it encompasses very different scales (global, regional, local, to limit oneself to the scales explicitly accounted for in the present GCMs) and because the feedback loops might appear to change sign with the spatial scale considered.

2. A GCM-type parametrization: A compromise between accuracy and efficiency

Different steps are necessary in evaluating the radiative transfer in the atmosphere:

* Formal solution of the radiative transfer equation
* Integration over the vertical coordinate taking into account the inhomogeneities in T, and absorber amounts
* Integration over wavenumber to get total radiances
* Integration over zenith angle to get total fluxes
* Differentiation of fluxes with respect to vertical coordinate to get heating rate

Then for each of these steps, there is a need for simplifications and approximations in order to allow the radiation transfer parametrization to remain computationally efficient. Below, we illustrate, on the derivation of the downward longwave radiation, some of the approximations, currently made, which allow to maintain a good trade-off between computational efficiency and accuracy of the resulting radiative fluxes and cooling rates.

The most explicit form for the downward longwave radiation is

$$F^-(z) = \int_0^\infty d\nu \int_0^1 \mu \, d\mu \int_z^\infty B(\nu, T(z')) \frac{d}{dz'} t(\nu, z', z'') \, dz'$$

with ν the wavenumber, μ the cosine of the zenith angle, B the Planck function at temperature T(z'), and t the transmission function

$$t(\nu, z', z'') = \exp\left\{ -\frac{1}{\mu} \int_{z'}^\infty \sum_i S_i(z'') \, f_i(\nu, z'') \, \rho(z'') \, dz'' \right\}$$

with S_i the line intensity, f_i the line shape factor, ρ the density of absorber

A simplication generally made when integrating over the optical path is to replace the integration over z" by one over a discrete number of homogeneous layers, using either a one-parameter (or scaling) approximation

$$\delta_\nu = k_\nu(p_0, T_0) \int \frac{\phi(p, T)}{\phi(p_0, T_0)} \, dm$$

$$\delta_\nu = k_\nu(p_0, T_0) \, m_e$$

$$m_e = \int \left(\frac{p}{p_0}\right)^n \left(\frac{T_0}{T}\right)^m dm$$

or a two parameter approximation

$$\overline{pm} = \int p \, dm$$

$$\overline{m} = \int dm$$

usually applied in a band model of the transmission function (e.g., Goody band model)

$$T_{Goody}(m) = \exp\left\{ -\frac{\overline{S}}{\delta}\overline{m} \left(1 + \frac{\overline{Sm}}{\pi\, \alpha_{LO}\frac{\overline{pm}}{\overline{m}}}\right)^{-\frac{1}{2}} \right\}$$

Then the integration over the wavenumber is usually performed using a band model of the transmission function

$$F^{-}(z) = \sum_j \Delta\nu_j \int_0^1 \mu \, d\mu \int_z^{\infty} B(\nu_j, T(z')) \frac{d}{dz'} T(\Delta\nu_j, \frac{m_e}{\mu}, \overline{p}) \, dz'$$

Finally the diffusivity approximation is often used to replace the integration over the zenith angle:

$$F^{-}(z) = \sum_j \Delta\nu_j \int_z^{\infty} \pi \, B(\Delta\nu_j, T(z')) \, \frac{d}{dz'} T(\Delta\nu_j, r\overline{m}, \overline{p}) \, dz'$$

3 Outstanding issues in radiation transfer modelling of relevance for climate modelling

The errors in the model atmospheric radiative divergence are often assigned to a deficient distribution of the cloud parameters. Some of these parameters are clearly the outputs of the model cloud scheme, as the cloud fraction, cloud height, cloud water loading. They will not be discussed here. A sizeable amount of the errors is also directly linked to the representation of the clear-sky radiative processes or to the cloud overlap assumption. Both topics are addressed below.

3.1 Accuracy of clear-sky radiative transfer parametrizations: Water vapour continuum absorption

As illustrated in section 2, the development of a GCM-type radiation scheme requires a number of approximations and simplifications to make the problem tractable and to solve it in a reasonably computer-efficient way. The InterComparison of Radiation Codes in Climate Models (ICRCCM; Ellingson and Fouquart, 1991; Ellingson et al., 1991; Fouquart et al., 1991) was a concerted effort on the part of the atmospheric radiation community to intercompare the various computational techniques used for representing the radiation transfer in numerical models. The principal objectives of ICRCCM were to develop an understanding of the various RT algorithms, to investigate the causes and nature of the differences in the results among the various codes (including their accuracy with respect to benchmark computations), and to ascertain the sensitivity of the algorithms to a number of factors (including spectral parameters, constitutents amounts, and meteorological conditions).
One of the main results of the ICRCCM was to pinpoint the large contribution of the water vapour continuum in explaining differences between the longwave radiation schemes (Ellingson et al., 1991). The agreement of the more parametrized schemes with the line-by-line models was generally reasonable. For a set of more than 30 radiation schemes applied to 5 standard clear-sky atmospheric profiles (tropical, mid-latitude summer and winter, sub-arctic summer and winter), the percentage error computed as the standard deviation over the average was better than 3 % for the downward flux at the surface, the net flux at the tropopause, and the outgoing longwave flux at the top of the atmosphere, better than 7 % for the tropospheric divergence of longwave radiation and around 20 % for the stratospheric divergence of longwave radiation, when absorption by the main atmospheric absorbers (H_2O lines and continuum, CO_2 and O_3) was accounted for. But much poorer agreements were found when absorption by one constituent only was considered, or when considering the sensitivity to a ± 25 % change in water vapour, or to a doubling of the CO_2 concentration. Similarly, the ICRCCM exercises shed light on how the parametrization of the water vapour continuum absorption was compensating for deficiencies in the H_2O line absorption parametrization.
In the shortwave part of the spectrum, even for the simplest case of pure water vapour absorption (without any scattering), standard deviation ranging from 1 to 3 % were found on the downward radiation at the surface, and from 6 to 11 % for the total atmospheric

absorption (Fouquart et al., 1991). In the longwave part of the spectrum, line-by-line model simulations of the radiation transfer have been available for about 10 years over large part of the longwave spectrum because of the need for reference calculations when developing radiative transfer algorithms used for operationally inverting satellite-measured channel radiances into temperature profiles (e.g., McMillin et al., 1976, 1979; Scott and Chédin, 1981). There was no such operational pressure for the shortwave part of the spectrum, and the first shortwave line-by-line models accounted for absorption only (Chou, 1982). Results from line-by-line models taking into account the full interaction between the absorption and scattering processes are now becoming available (Ramaswamy and Freidenreich, 1992), which should allow a better understanding of the requirements for a proper parametrization of these interactions in GCM-type shortwave radiative schemes (Freidenreich and Ramaswamy, 1993). In the shortwave part of the spectrum, the presence and amplitude of a water vapour continuum absorption is still uncertain and debated (Tsay and Stephens, 1990; Nakajima et al., 1991; Taylor, 1992). Figure 1 compares the absorption coefficent for water vapour derived from the HITRAN'92 compilation of line parameters. In panel 1a, the absorption coefficient is derived as is customary when using statistical models of the transmission function (Rodgers and Walshaw, 1966) by summing up the intensities of all the lines whose centre falls within the given 10 cm^{-1} wide spectral interval and assuming that the lines are only foreign broadened (i.e., through collisions with N_2 molecules). In panel 1b, the absorption coefficients are computed accounting only for the fraction of the line intensity actually corresponding to the spectral interval containing the line centre. Results are shown assuming that the lines can be either self- or foreign broadened (collisions H_2O - H_2O, or H_2O - N_2). The panel 1c corresponds to the contribution of the line intensity outside the spectral interval containing the line centre, again assuming that the line can be self- or foreign-broadened. This approach introduces a continuum-type absorption (seen in those parts of the spectrum that would be otherwise transparent in either Fig. 1a or 1b). Furthermore, the contribution from the self-broadened lines is always slightly larger than the one from the foreign-broadened lines. In contrast with the usual way of parametrising the absorption, the continuum absorption (presented here as the contribution of self- and foreign-broadened line wings) does not depend on an independently specified continuum absorption, and is therefore fully consistent with the rest of the line spectroscopic characteristics. The impact on the water vapour absorption over the whole shortwave spectrum for different atmospheric amounts of precipitable water is presented in Table 1. The main results are that i/ the new approach leads to systematically higher water vapour absorption by 1 or 2 percent when averaged over the whole solar spectrum, and ii/ that the actual shape assumed for the line (Lorentz or sub-Lorentz in the computations presented in Table 1) is of secondary importance.

3.2 Cloud overlap assumptions : One-dimensional radiative computation

The most common cloud overlap assumptions (COAs) in the radiation schemes presently used in GCMs are the random, the maximum-random and the maximum overlap assumptions, neither of which is completely justifiable on a global scale from observations of the actual atmosphere. Tian and Curry (1989) have tested the validity of the various cloud overlap assumptions by applying them to the U.S. AirForce 3DNEPH in January 1979 over the north Atlantic Ocean. This data set is believed to be the best available information on the vertical distribution of clouds as it integrates both satellite and conventional observations over a relatively data rich area. They compared the observed total cloud amount to the computed total cloudiness obtained with the different overlap assumptions. Their conclusion is somewhat in favour of the maximumrandom overlap assumption first proposed by Geleyn and Hollingsworth (1979): Adjacent levels containing cloud may be combined by using maximum overlap to form a discrete cloud layer. Random overlap of discrete cloud layers, separated by clearsky, performed the best, although this assumption results in a systematic underestimation of the total cloud cover for the large horizontal resolutions currently used in climate oriented GCMs.

In a model simulation of the cloud and radiative fields, we are interested not only in the total cloud cover and radiation at the top of the atmosphere, but also to the vertical profile of these quantities. Therefore the definition of the different COAs is given from the point of view of the cloudiness CC_{IJ} encountered between any two levels I and J in the atmosphere.

Let C_K be the cloud fraction of layer K located between levels K and K+1; the maximum overlap assumption gives

$$CC_{I\,J} = \max\ (C_I\ ,\ C_{I+1}\ ,\ \ldots\ ,\ C_{J-1}\ )$$

the random overlap assumption gives

$$CC_{I\,J} = 1 - \prod_{K=I}^{J-1} (1-C_K)$$

and the maximum-random overlap assumption gives

$$CC_{I\,J} = 1 - \prod_{K=I}^{J-1} \{[(1 - \max(C_K, C_{K-1})] / (1-C_{K-1})\}$$

Results of one-dimensional radiative computations are presented first to confirm the validity of the modifications made to the radiation scheme to allow the handling of the different cloud overlap assumptions, and second to illustrate the direct impact of a change in COA on the radiative profiles. The radiative fluxes and heating/cooling rate profiles are computed for a standard midlatitude summer atmosphere (McClatchey et al., 1971), with a surface SW albedo of 0.175 and a solar zenith angle of 23.5 ° corresponding to noon insolation at 45 °N on a 15 July and chosen to maximise the differences in SW heating. Provision for the LW and SW optical properties of water, ice and mixedphase clouds is made in the radiation scheme. Ice cloud optical properties are derived from Ebert and Curry (1992). Water cloud optical properties follow Smith and Shi (1992) in the longwave and Fouquart (1987) in the shortwave.

Calculations are performed for different cloud configurations, often encountered in 3-dimensional simulations. The first one correspond to a typical convective system and includes a convective tower of fractional cover 0.15 from 820 to 290 hPa topped by stratiform anvil of cover 0.4 and 70 hPa thick. The second case includes three layers of stratiform cloud of cover 0.3 between 890 and 820 hPa (low level), 640 and 545 hPa (middle level), and 290 and 220 hPa (high level), respectively. The profiles of the cloud longwave, shortwave and net radiative forcing for these two cases are presented in Figs. 2 and 3, respectively. The cloud forcing term is simply the difference between the heating obtained for the cloudy atmosphere minus the clear-sky heating.

The different COAs are treated exactly in the LW part and only approximately in the SW part of the radiation scheme. Therefore, the LW MAXRAN and MAXIMUM results are exactly the same for case 1, and the same holds for the LW MAXRAN and RANDOM results for case 2. In the SW, the agreement on the SW forcing profiles is very good but some small discrepancies can be found in the SW fluxes at the boundaries, with the SW MAXRAN results lying very close to the SW MAXIMUM results for case 1, and between MAXIMUM and RANDOM SW results for case 2.

In the first case, the anvil produces a strong LW cooling whereas a relative heating is found in the rest of the convective tower below. Compared to MAXRAN/MAXIMUM, the LW RANDOM shows a smaller cooling peak in the anvil, smaller heating below within the tower, and larger warming in the clear PBL. All these features are consistent with the larger effective cloudiness that the RANDOM assumption produces between two cloudy layers. This larger effective cloudiness prevents the radiation from the lower atmospheric layers from escaping to space, thus the heating in the PBL, decreases the possible radiative exchanges within the tower thus the smaller radiative heating, and decreases the upward LW flux at the base of the anvil, thus the smaller LW divergence in the anvil. In the SW, the strong heating is found lower down within the upper part of the convective tower. A relative cooling is found below the cloud.

These one-dimensional calculations show that, although the cloud overlap assumption is usually considered an intrinsic part of a GCM-type parametrisation of the cloud-radiation interactions, it is likely to have a large impact on the radiative impact of the clouds both within the atmosphere and at the surface. Furthermore, any attempt at validating the cloudiness produced by a model against cloud observations should consider the uncertainties brought by the differences between the actual overlapping of clouds and the model assumption. Both questions have up to now received little attention.

3.3 Insufficient observational data for validating the cloud-radiative parametrizations

It is still difficult to know how much SW and LW radiation is actually available within the atmosphere. Whereas the Earth Radiation Budget Experiment (ERBE) has provided good quality measurements of the SW and LW radiative fluxes at the atmosphere, for both cloudy and clear-sky conditions (at least on a monthly mean basis), we currently have very few data on the climatology of surface radiation over the oceans (or even over land). Efforts to produce such a climatology from satellite measurements are now underway, particularly for the SW (Darnell et al., 1992; Laszlo and Pinker, 1993; Li and Leighton, 1993). However, the preliminary results do not allow yet an accurate estimation of the extra SW absorption introduced by the presence of clouds within the atmosphere. The observed global annual-mean solar energy disposition is 98.1 W m^{-2} (estimated from ground measurements), 83 W m^{-2} (from ERBE measurements; Li and Leighton, 1993), 76.3 W m^{-2} (from ISCCP data; Darnell et al., 1992), 68.9 W m^{-2} (from ISCCP data; Laszlo and Pinker, 1993), 67.1 W m^{-2} (from ISCCP data; Rossow, 1993) whereas the SW radiation absorbed at the surface varies from 142 to 171.2 W m^{-2}. This large range of uncertainties explains why the first two estimates of the global annual-mean effect of clouds on the SW atmospheric absorption does not even agree on sign: +1 to 2 W m^{-2} for Laszlo and Pinker (1993), and - 1.7 W m^{-2} for Rossow (1993). In the LW, although some methods have been proposed for retrieving the LW radiation at the surface (Darnell et al., 1983; Frouin et al., 1988; Zhi and Harshvardhan, 1993; Stephens et al., 1994), usually from a combination of satellite data and temperature and humidity profiles from other sources, there is presently no available climatology of the LW radiation at the surface other than the climatologies built up from bulk parameterizations applied on near surface ship observations over the oceans (Esbensen and Kushnir, 1981; Oberhuber, 1988).

4. Sensitivity of climate to variations of solar output: A radiative transfer perspective

The connection between the sunspot number and the variation of the total solar irradiance seems well established on an 11-year time scale (Foukal and Lean, 1988). The amplitude of the change (about 0.5 W m^{-2} over the last 11-year cycle) and the time-scale for oceanic processes (several decades) are such that a change in climate induced by this change in radiative forcing is very unlikely (Hoffert et al., 1980, 1988; Wigley and Raper, 1990; Reid, 1991). However, the "Maunder Minimum", a period with very few sunspots that lasted over the second half of the 17 th century, coincided with the Little Ice Age, a long and severe cold period over Western Europe and the North Atlantic area. Recently Ribes (1993) and Nesme-Ribes et al. (1993), using a combination of surrogate data and simplified solar dynamics theory, have estimated the relative deficit in the total solar irradiance to 0.4 %, i.e., 5 to 6 W m^{-2} over a period of at least 50 years. This corresponds to a decrease by 1 W m^{-2} of the solar radiation absorbed by the Earth's atmosphere-surface system after accounting for the geometry and reflection by the surface, clouds and the atmosphere. The LMD GCM is then used by Nesme-Ribes et al. (1993) to study the impact of a variation of the solar output on the model simulated climate. They apply a radiative forcing of $\pm$ 4 W m^{-2} on the net downward solar radiation, mimicking the $\mp$ 4 W m^{-2} change in net upwelling longwave radiation computed for doubling or halving the atmospheric CO_2 concentration. In their analysis, the response of a climate variable x to the Maunder Minimum conditions is computed as

$$\Delta x = \Delta S_0' \frac{\partial x}{\partial S_0}$$

with

$$\frac{\partial x}{\partial S_0} = \frac{x(S_0+\Delta S_0') - x(S_0-\Delta S_0')}{2\Delta S_0'}$$

for $\Delta S_0' = 1$ W m^{-2}.

The main findings are that the simulated cooling appears to be compatible with the temperature data from the Little Ice Age, and that the model response to this change in incoming solar flux is very similar to the response to a change in CO_2 concentration, another globally homogeneous external forcing parameter. Using the same methodology, and prescribing the change in sea surface temperature from the results of Nesme-Ribes et al. (1993) with the LMD GCM, Royer et al. (1993) have repeated the simulation with the climate version of the Arpege model of Meteo-France.

Even without addressing the validity of the assumption made by Nesme-Ribes et al. (1993) and Royer et al. (1993) of the linear behaviour of most of the model parameters when the forcing used in the model to estimate the derivatives is set to a value 8 times bigger than the "observed" forcing (see Palmer, 1993; Holton, 1993, for comments on the "linear thinking" prevalent in most of today's climatic studies), the overall methodology can also be questioned from the point of view of the radiative processes involved in a variation of solar output, and of the radiation transfer parametrizations to be used for representing them.

Both studies have assumed that the change in total solar irradiance is distributed uniformly over the solar spectrum, so that the same percentage decrease is applied in all spectral intervals of the shortwave radiation schemes used in the model simulations. This assumption can be questioned from the studies of the variations of the solar output over solar cycle 21. Using measurements from the Solar Backscatter UV experiment on the Nimbus-7 satellite, and spectrometer measurements from the Solar Mesosphere Explorer, Lean (1989) points out the large contribution of the UV radiation to the changes in total irradiance: "although only 1 percent of the Sun's energy is emitted at UV wavelengths between 200 and 300 nm, the decrease in this radiation from 1 July 1987 to 30 June 1985 accounted for 19 percent of the decrease in the total irradiance over the same period".

Essentially all of the incident solar radiation at $\lambda < 0.295$ µm is absorbed by the atmospheric gases (O, N, O_2, O_3, and N_2) and does not reach the Earth's surface (Coulson, 1975). Very little of it reaches the tropopause level. All radiation at wavelength λ shorter than 0.08 µm is absorbed between 120 km and 160 km depending on wavelength and ionization properties of the gases (Friedman, 1960). If h_e is the height at which the incoming radiation is reduced to 1/e of its extraterrestrial value, the absorption of longer wavelengths occurs in the 0.104 - 0.130 µm region ($h_e \approx 80$ to 100 km), in the Schumann continuum of oxygen (0.145 - 0.176 µm and $h_e \approx 100$ km), in the Schumann-Runge bands (between the continuum and 0.2026 µm, with h_e between 40 and 80 km), and in the Hartley band of ozone (0.220 - 0.295 µm and $h_e \approx 40$ km). Radiation in the near-UV (0.295 - 0.40 µm) is partly absorbed by the Huggins bands of ozone ($\lambda < 0.32$ µm), partly backscattered to space, partly transmitted to the surface.

Any variations in the shorter UV radiation is known to strongly affect the ozone photochemistry in the upper stratosphere, thus the UV flux reaching the lower layers, and in turn the ozone production rate due to molecular oxygen dissociation as the partitioning between nitrogen species is very sensitive to the UV flux available in the lower stratosphere (Cariolle and Déqué, 1986) . Therefore the possibility of "top-down" interactions affecting the UV radiation, the ozone distribution, and eventually the troposphere is one possible scenario that cannot be ruled out.

The other "boundary" is the ocean surface. There again, the absorption of solar radiation by the mixed layer is not spectrally flat. Solar irradiance measurements in the 400 - 1000 nm spectral region by Paulson and Simpson (1977) show that the assumption of an exponential decay with depth of the solar radiation in the ocean is a poor approximation in the upper 5 m, because of preferential absorption by the hydrosols at the short wavelengths and by water at the long wavelengths (Sellers, 1965). The assumption of an exponential decay, as used in the oceanic mixed layer of the LMD GCM (Le Treut et al., 1985; Simonot and Le Treut, 1986) becomes a good approximation only below 10 m depth for the remaining blue-green light..

The results of Nesme-Ribes et al. (1993) indicate that the sensitivity of the climate to the variations of the solar output seems dominated by "bottom-up" interactions, where a change in sea surface temperature, induced by the change in solar radiative forcing at the tropopause, drives the response of the model. This may only be the result of calculations with a model neglecting the possible interactions between the UV radiation and the ozone photochemistry, and with a shortwave radiation scheme including only two spectral bands (0.25 - 0.68 μm, and 0.68 - 4.00 μm). The combination of both the UV and visible ozone absorption in the first spectral interval might lead to an improper vertical distribution of the change in solar forcing. All these factors, plus the specified spectrally flat change in SW flux, are l ikely to overemphasize the response of the SST, and to force a $2xCO_2$ type response. In principle, the Météo-France Climate model should be better suited to the problem as it may be run including a prognostic ozone equation with provision for the interaction between the UV radiation and ozone photochemistry. However, the model used in Royer et al. (1993)'s study includes an even coarser SW radiation scheme (with only one spectral interval to represent the whole solar spectrum) and uses the changes in SST obtained by the LMD GCM, which might force it to respond following "bottom-up" interactions.

With respect to a better representation of the radiative transfer interactions with the ozone photochemistry and the oceanic boundary, it would be interesting to repeat Nesme-Ribes et al. (1993) experiment with a model including the proper tools, i.e., a radiation scheme able to cope with differential spectral variations in the different parts of the UV, visible, and near infrared spectrum, interactive ozone photochemistry. and a more accurate description of the radiative transfer in the first layer of the oceanic mixed layer model, to cross-check that the "bottom-up" response found by Nesme-Ribes et al. (1993) and Royer et al. (1993) is not an artefact of an inadequate representation of the radiation transfer.

References

Betts, A.K., and W. Ridgway, 1988: Coupling of the radiative, convective, and surface fluxes over the Equatorial Pacific. J. Atmos. Sci., 45, 522-536.

Cariolle, D., and M. Deque, 1986: Southern hemisphere medium-scale waves and total ozone disturbances in a spectral general circulation model. J. Geophys. Res., 91D, 10825-10846.

Chou, M.-D., 1982: Atmospheric solar heating rate in the water vapor bands. J. Climate Appl. Meteor., 25, 1532-1542.

Coulson, K.L., 1975: Solar and Terrestrial Radiation, Academic Press, New York, 322 pp.

Darnell, W.L., S.K. Gupta, and W.F. Staylor, 1983: Downward longwave radiation at the surface from satellite measurements. J. CLimate Appl. Meteor., 22, 1956-1960.

Darnell, W.L., W.F. Staylor, S.K. Gupta, N.A. Ritchey, and A.C. Wilber, 1992: Seasonal variation of surface radiation budget derived from ISCCP-C1 data. J. Geophys. Res., 97D, 15741-15760.

Ebert, E.E., and J.A. Curry, 1992: A parametrisation of ice cloud optical properties for climate models. J. Geophys. Res., 97D, 3831-3836.

Ellingson, R.E., J. Ellis, and S.B. Fels, 1991: The Intercomparison of Radiation Codes used in Climate Models: Longwave results. J. Geophys. Res., 96D, 8929-8954.

Ellingson, R.E., and Y. Fouquart, 1991: The Intercomparison of Radiation Codes for Climate Models: An overview. J. Geophys. Res., 96D, 8925-8927.

Esbensen, S.K., and Y. Kushnir, 1981: "The Heat Budget of the Global Ocean. An Atlas based on Estimates from Surface Marine Observations". Rep. No. 29, Climatic Research Institute, Oregon State University, Corvallis, OR.

Fouquart, Y., 1987: Radiative transfer in climate models. NATO Advanced Study Institute on Physically-Based Modelling and Simulation of Climate and Climatic Changes. Erice, Sicily, 11-23 May 1986. M.E. Schlesinger, Ed., Kluwer Academic Publishers, 223-284.

Fouquart, Y., B. Bonnel, and V. Ramaswamy, 1991: Intercomparing shortwave radiation codes for climate studies. J. Geophys. Res., 96D, 8955-8968.

Freidenreich, S.M., and V. Ramaswamy, 1993: Solar radiation absorption by CO_2, overlap with H_2O, and a parameterization for general circulation models. J. Geophys. Res., 98D, 7255-7264.

Friedman, H., 1960: The Sun's ionizing radiations. In "Physics of the Upper Atmosphere" (J.A. Ratcliffe, ed.). Academic Press, New York.

Frouin, R., C. Gautier, and J.-J. Morcrette, 1988: Downward longwave irradiance at the ocean surface from satellite data: Methodology and in situ validation. J. Geophys. Res., 93 C, 597-619.

Geleyn, J.-F., and A. Hollingsworth, 1979: An economical analytical method for the computation of the interaction between scattering and line absorption of radiation. Beitr. Phys. Atmosph., 52, 1-16.

Gupta, S.K., W.F. Staylor, W.L. Darnell, A.C. Wilber, and N.A. Ritchey, 1993: Seasonal variation of surface and atmospheric cloud radiative forcing obver the globe from satellite data. J. Geophys. Res., 98D, 20761-20778.

Hoffert, M.I., A.J. Callegari, and C.-S. Hsieh, 1980: The role of deep-sea heat storage in the secular response to climate forcing. J. Geophys. Res., 85, 6667-6679.

Hoffert, M.I., A. Frei, and V.K. Narayanan, 1988: Application of Solar Max ACRIM data to analysis of solar-driven climatic variability on earth. Clim. Change, 13, 267-285.

Holton, J.R., 1993: The quasi-biennal oscillation in the earth's atmosphere and its links to longer period variability. in The Solar Engine and its Influence on Terrestrial Atmosphere and Climate, NATO ARW 920946, Paris, 25-29 October 1993.

Laszlo, I., and R.T. Pinker, 1993: Shortwave cloud-radiative forcing at the top of the atmosphere, at the surface and of the atmospheric column as determined from ISCCP C1 data. J. Geophys.Res., 98D, 2703-2718.

Lean, J., 1989: Contribution of ultraviolet irradiance variations to changes in the Sun's total irradiance. Science, 244, 197-200.

Le Treut, H., J.-Y. Simonot, and M. Crépon, 1985: A model for the sea-surface temperature and heat content in the North Atlantic ocean, in Coupled Ocean-Atmosphere Models, J. Nihoul, ed., Elsevier Oceanogr. Ser. Vol. 40, Elsevier, New York, 439-445.

Li, Zhanqing, and H.G. Leighton, 1993: Global climatologies of solar radiation budgets at the surface and in the atmosphere from 5 years of ERBE data. J. Geophys. Res., 98D, 4919-4930.

Livingston, W., 1993: Surrogates for total solar irradiance. in The Solar Engine and its Influence on Terrestrial Atmosphere and Climate, NATO ARW 920946, Paris, 25-29 October 1993.

McClatchey, R.A., R.W. Fenn, J.E.A. Selby, F.E. Volz, and J.S. Garing, 1971: Optical properties of the atmosphere. AFCRL 71-0279, Environ. Res. Paper No. 354, Bedford, Mass., 91 pp. (also 1972, AFCRL-72-0497, ERP 411, 3rd ed., 108pp)

McMillin, L.M., and H.E. Fleming, 1976: Atmospheric transmittance of an absorbing gas: A computationally fast and accurate transmittance model for absorbing gases with constant mixing ratios in inhomogeneous atmospheres. Appl. Opt., 15, 358-363.

McMillin, L.M., H.E. Fleming, and M.L. Hill, 1979: Atmospheric transmittance of an absorbing gas: A computationally fast and accurate transmittance model for absorbing gases with variable mixing ratios. Appl. Opt., 18, 1600-1606.

Morcrette, J.-J., 1990: Impact of changes to the radiation transfer parameterizations plus cloud optical properties in the ECMWF model. Mon. Wea. Rev., 118, 847-873.

Morcrette, J.-J., 1991: Radiation and cloud radiative properties in the ECMWF operational weather forecast model. J. Geophys. Res., 96D, 9121-9132.

Morcrette, J.-J., and Y. Fouquart, 1986: The overlapping of cloud layers in shortwave radiation parameterizations. J. Atmos. Sci., 43, 321-328.

Nakajima, T., M.D. King, J.D. Spinhirne, and L.F. Radke, 1991: Determination of the optical thickness and effective particle radius of clouds from reflected solar radiation measurements. Part II: Stratocumulus observations. J. Atmos. Sci., 48, 728-750.

Nesme-Ribes, E., E.N. Ferreira, R. Sadourny, H. Le Treut, and Z.-X. Li, 1993: Solar dynamics and its impact on solar irradiance and the terrestrial climate. J. Geophys. Res., in press.

Oberhuber, J.M., 1988: An atlas based on the COADS dataset: The budget of heat, buoyancy and turbulent kinetic energy at the surface of the global ocean. Max-Planck-Institute for Meteorology Report No.15, Hamburg 13, F.R.G., 196 pp.

Palmer, T.N., 1993: A nonlinear dynamical perspective on climate change. Weather, 48, 314-325.

Palmer, T.N., 1993: Extended-range atmospheric prediction and the Lorenz model. Bull. Amer. Meteorol. Soc., 74, 49-65.

Paulson, C.A., and J.J. Simpson, 1977: Irradiance measurements in the upper ocean. J. Phys. Oceanogr., 7, 952-956.

Ramanathan, V., 1987: Atmospheric general circulation and its low frequency variance: Radiative influences. in: Short and Medium Range Weather Prediction. Collected papers presented at WMO/IUGG NWP Symposium, Tokyo, Japan, 4-8 Aug. 1986, ed. by T. Matsuno, Special Volume of the J. Meteor. Soc., 151-175.

Ramanathan, V., R.D. Cess, E.F. Harrison, P. Minnis, B.R. Barkstrom, E. Ahmad and D. Hartmann, 1989: Cloud-radiative forcing and climate: Results from the Earth Radiation Budget Experiment. Science, 243, 57-63.

Ramanathan, V., and W. Collins, 1991: Thermodynamic regulation of ocean warming by cirrus clouds deduced from observations of the 1987 El Nino. Nature, 351, 27-32.

Ramaswamy, V., and S.M. Freidenreich, 1992: A study of broadband parameterizations of the solar radiative interactions with water vapor and water drops. J. Geophys. Res., 97D, 11487-11512.

Randall, D.A., Harshvardhan, D.A. Dazlich, and T.G. Corsetti, 1989: Interactions among radiation, convection, and large-scale dynamics in a general circulation model. J. Atmos. Sci., 46, 1943-1970.

Reid, G.C., 1991: Solar total irradiance variations and the global sea surface temperature record. J. Geophys. Res., 96D, 2835-2844.

Rodgers, C.D., and C.D. Walshaw, 1966: The computation of the infrared cooling rate in planetary atmospheres. Quart. J. Roy. Meteor. Soc., 92, 67-92.

Rossow, W.B., 1993: Satellite observations of radiation and clouds to diagnose energy exchanges in the climate. in Energy and Water Cycles in the Climate System. E. Raschke, D. Jacob, Eds., NATA ASI Vol. 15, Springer-Verlag Berlin Heidelberg, 123-141 (Part I); 143-164 (Part II).

Royer, J.-F., F. Chauvin, and G. Pélissier, 1993: Sensitivity of the atmospheric circulation to solar irradiance variations. in The Solar Engine and its Influence on Terrestrial Atmosphere and Climate, NATO ARW 920946, Paris, 25-29 October 1993.

Ribes, J.C., and Nesme-Ribes, E., 1993: The solar sunspot cycle in the Maunderminimum AD 1645 to AD 1715. Astron. Astrophys., 276, 549-563.

Scott. N.A., and A. Chédin, 1981: A fast line-by-line method for atmospheric absorption computations: The automatized atmospheric absorption atlas. J. Appl. Meteor., 20, 801-812.

Sellers, W.D., 1965: Physical Climatology, University of Chicago Press, 272 pp.

Simonot, J.-Y., and H. Le Treut, 1986: A climatological field of mean optical properties of the world ocean. J. Geophys. Res., 91C, 6642-6646.

Slingo, A., and J.M. Slingo, 1988: The response of a general circulation model to cloud longwave radiative forcing: I. Introduction and initial experiments. Quart. J. Roy. Meteor. Soc., 114, 1027-1062.

Smith, E. A., and Lei Shi, 1992: Surface forcing of the infrared cooling profile over the Tibetan plateau. Part I: Influence of relative longwave radiative heating at high altitude. J. Atmos. Sci., 49, 805-822.

Stephens, G.L., G.G. Campbell, and T.H. Vonder Haar, 1981: Earth Radiation budgets. J. Geophys. Res., 86, 9739-9760.

Stephens, G.L., A. Slingo, and M. Webb, 1993: On measuring the greenhouse on Earth. in High Spectral Resolution Infrared Remote Sensing for Earth's Weather and Climate Studies. A. Chedin, M.T. Chahine, and N.A. Scott, Eds., NATO ASI Series, Vol. 19, Springer-Verlag, Berlin, Heidelberg, 395-417.

Stephens, G.L., and S.-C. Tsay, 1990: On the cloud absorption anomaly. Quart. J. Roy. Meteor. Soc., 116, 671-704.

Taylor, J.P., 1992: Sensitivity of remotely sensed effective radius of cloud droplets to changes in LOWTRAN version. J. Atmos.Sci., 49, 2564-2569.

Tian, L., and J. Curry, 1989: Cloud overlap statistics. J. Geophys. Res., 94D, 9925-9936.

Zhi, H., and Harshvardhan, 1993: A hybrid technique for computing the monthly mean net longwave surface radiation over oceanic areas. Theoret.Appl.Climatol., 47, 65-80.

Table 1: Transmission through atmospheres with climatological distributions of water vapour. Results are given for a vertical and a slant path (with $\theta = 60\ °$)

The last column is for computations with a narrow-band model with absorption parameters derived as in Rodgers and Walshaw (1966). The two other computations correspond to a narrow-band model whose absorption parameters are defined separately for self- and foreign-broadened lines (of either Lorentz or sub-Lorentz shape) and account for the actual distribution of the line intensity over more than one 10 cm^{-1} spectral interval.

	Lorentz lines self / foreign	Sub-Lorentz lines self / foreign	Rodgers-Walshaw (Lorentz foreign)
Tropical	0.8022 0.7656	0.8024 0.7658	0.8314 0.7983
Mid-Lat. Summer	0.8217 0.7872	0.8219 0.7873	0.8465 0.8152
Mid-Lat. Winter	0.8804 0.8536	0.8805 0.8538	0.8935 0.8689
Sub-Arctic Summer	0.8408 0.8085	0.8409 0.8086	0.8611 0.8318
Sub-Arctic Winter	0.9069 0.8843	0.9070 0.8844	0.9162 0.8950

Table 2: Radiative fluxes at the top of the atmosphere and surface computed for different cloud overlap assumptions. Results are for a Mid-Latitude Summer atmosphere including a typical convective cloud (see text).

	TCC	TTR	TSR	CLFT	CSFT	STR	SSR	CLFS	CSFS
MAX	0.400	-258.4	923.3	33.3	-132.5	-70.2	648.9	12.2	-183.9
MRN	0.400	-258.4	923.1	33.3	-132.7	-70.2	648.5	12.2	-184.3
RAN	0.774	-221.4	905.0	70.3	-149.2	-48.5	624.4	34.0	-206.6

TCC is the total cloud cover corresponding to the cloud overlap assumption
TTR and TSR are the net terrestrial and solar radiation fluxes at the top of the atmosphere, respectively. STR and SSR are the net terrestrial and solar radiation fluxes at the surface, respectively. CLFT and CSFT are the cloud longwave and shortwave forcings seen from the top of the atmosphere; CLFS and CSFS are the cloud longwave and shortwave forcings seen from the surface. All quantities except TCC are in W m^{-2}.

Table 3: As in Table 2, but for a MLS atmosphere including three overlapping high-, mid-, and low-level clouds (see text).

	TCC	TTR	TSR	CLFT	CSFT	STR	SSR	CLFS	CSFS
MAX	0.300	-259.8	823.8	31.9	-232.0	-60.9	558.1	21.5	-274.7
MRN	0.657	-251.2	823.0	40.5	-232.8	-48.0	555.8	34.4	-277.0
RAN	0.657	-251.2	820.5	40.5	-233.8	-48.0	553.2	34.4	-277.8

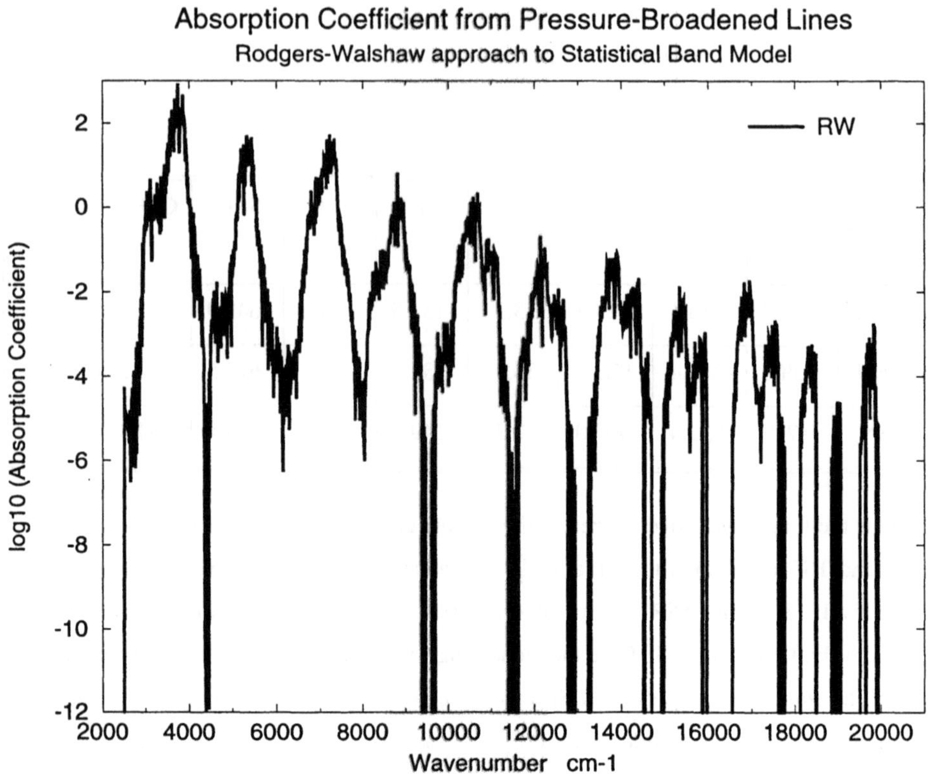

Figure 1: Water vapour absorption coefficients over the solar spectrum (2500-21000 cm^{-1}) in 10 cm^{-1} wide intervals.

Panel 1: The absorption coefficient ($k\pi\alpha_o$ / δ^2) is computed from line spectroscopic parameters following Rodgers and Walshaw (1966).

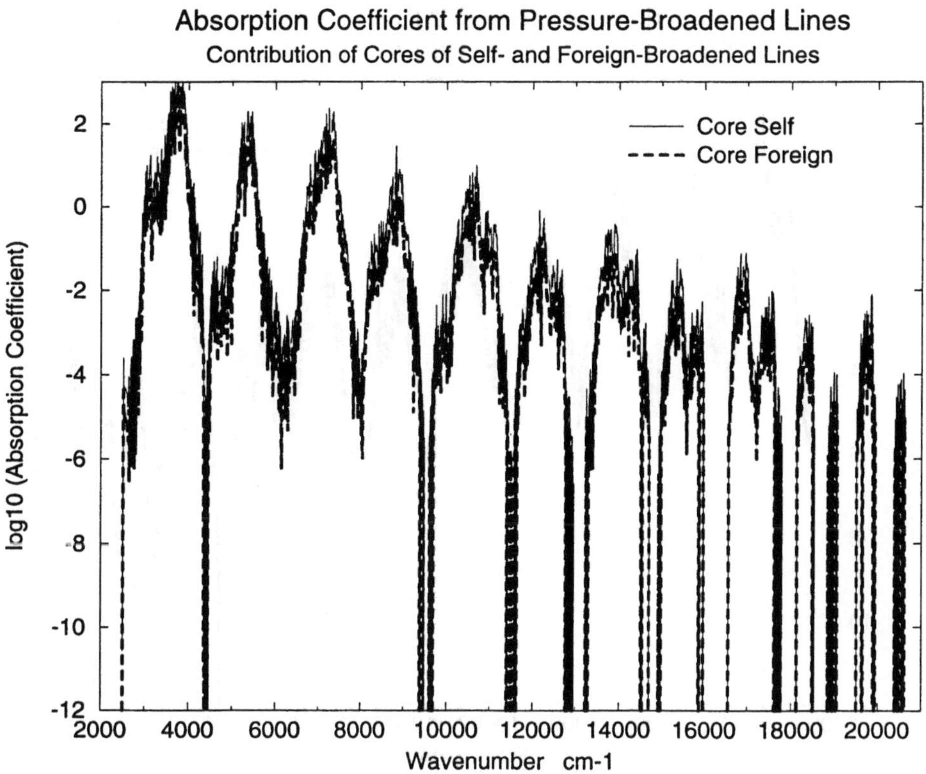

Figure 1 following

Panel 2: only the fraction of the line intensity pertaining to the given 10 cm^{-1} wide spectral interval containing the line centre is considered when computing the absorption coefficient.

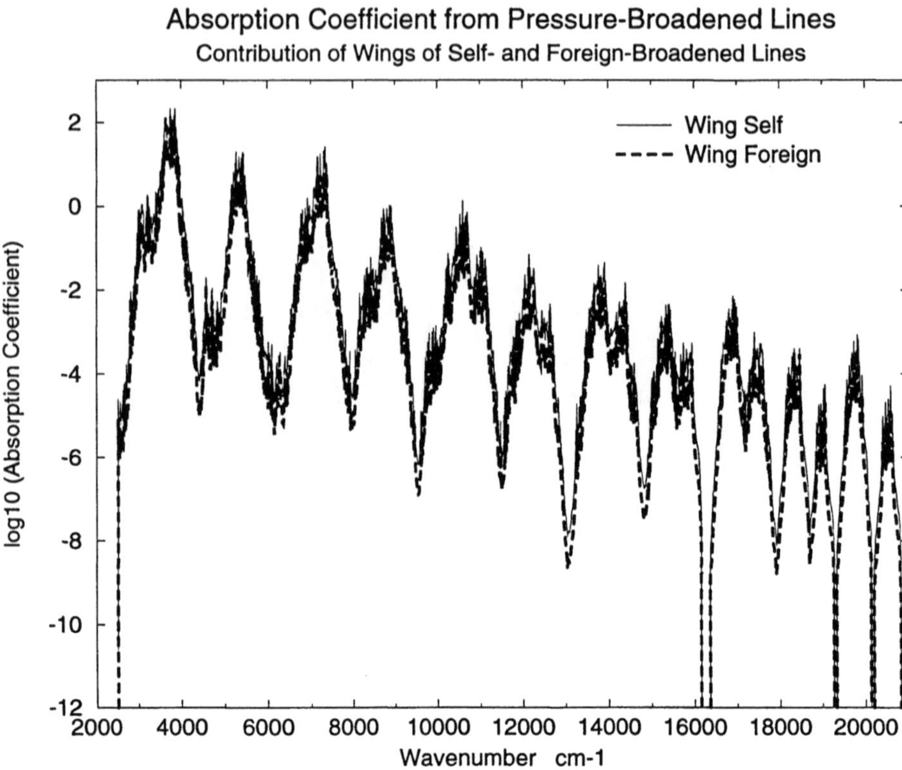

Figure 1 following

Panel 3: Continuum-like absorption coefficient corresponding to the sum of the wing effect of all lines outside the 10 cm^{-1} wide spectral interval. In panels 2 and 3, both the self- and foreign-broadened half-widths of each line is accounted for in the computations.

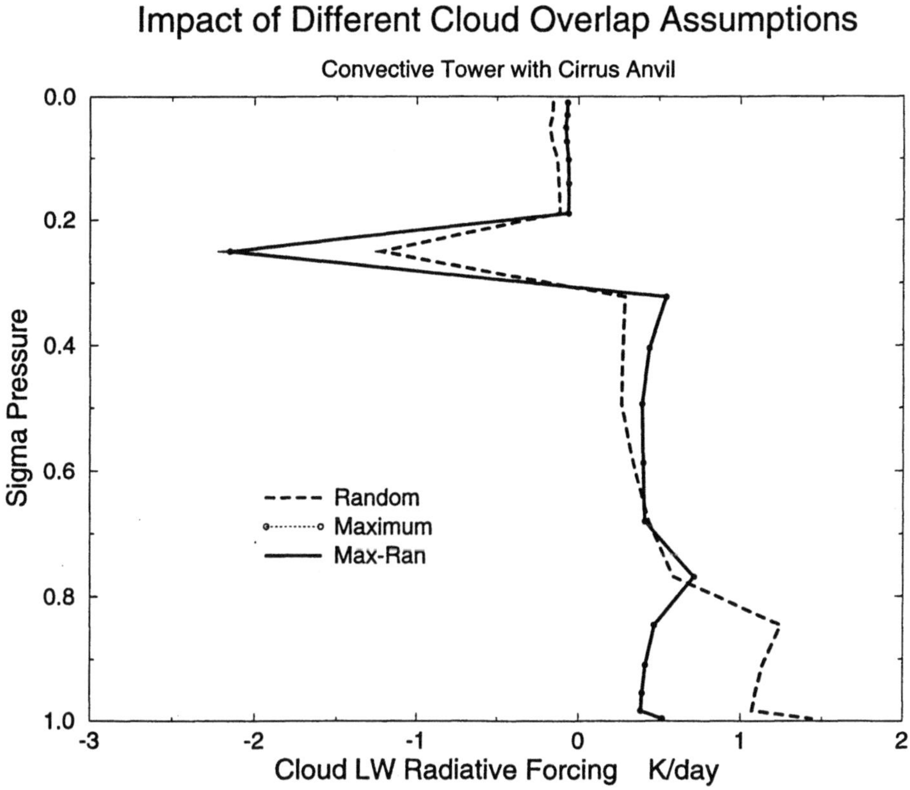

Figure 2:

2a- The impact of the random, maximum, and maximum-random cloud overlap assumption on the profiles of the cloud longwave radiative forcing

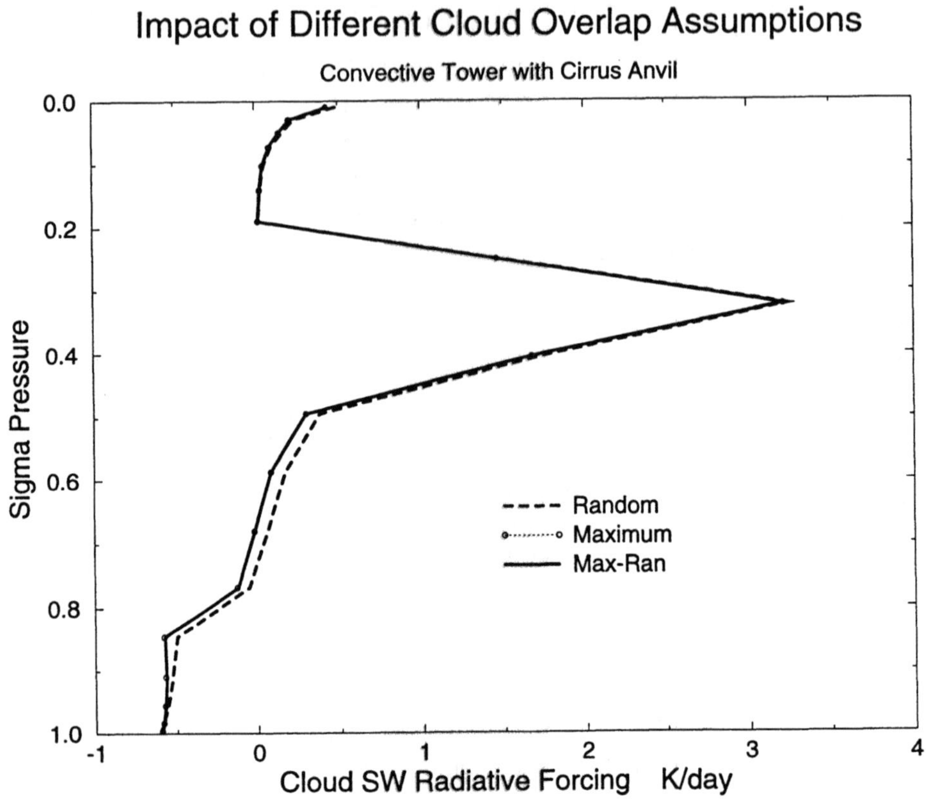

Figure 2 following:

2b- cloud shortwave radiative forcing

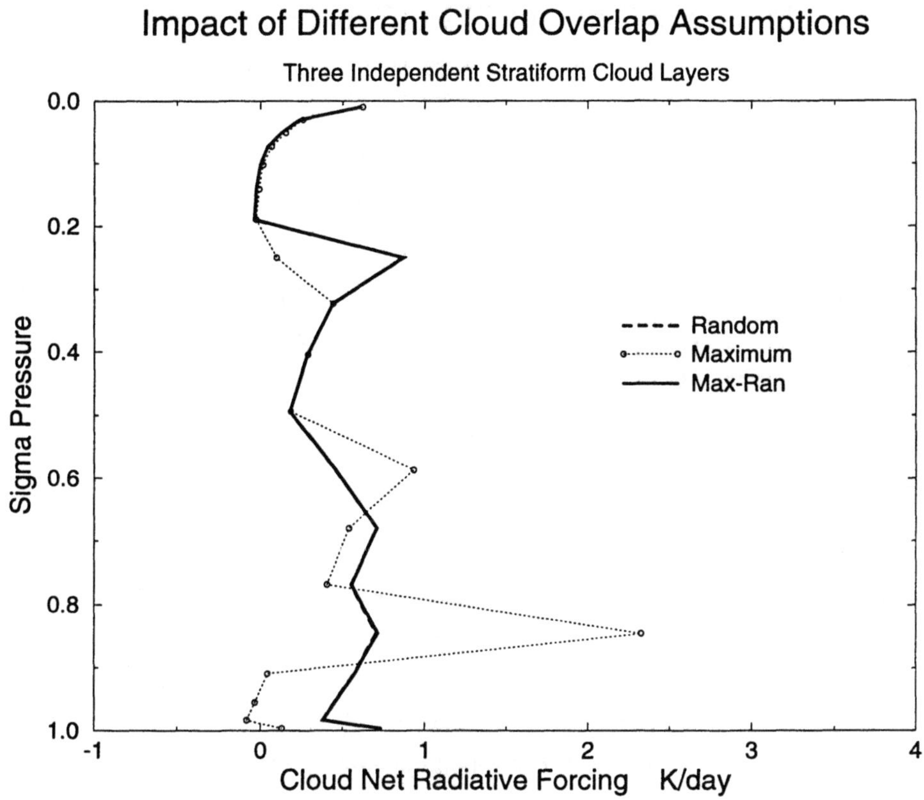

Figure 2 following:

2c- cloud net radiative forcing. The climatological mid-latitude summer atmospheric profile includes a convective tower topped by a cirrus anvil (ee text).

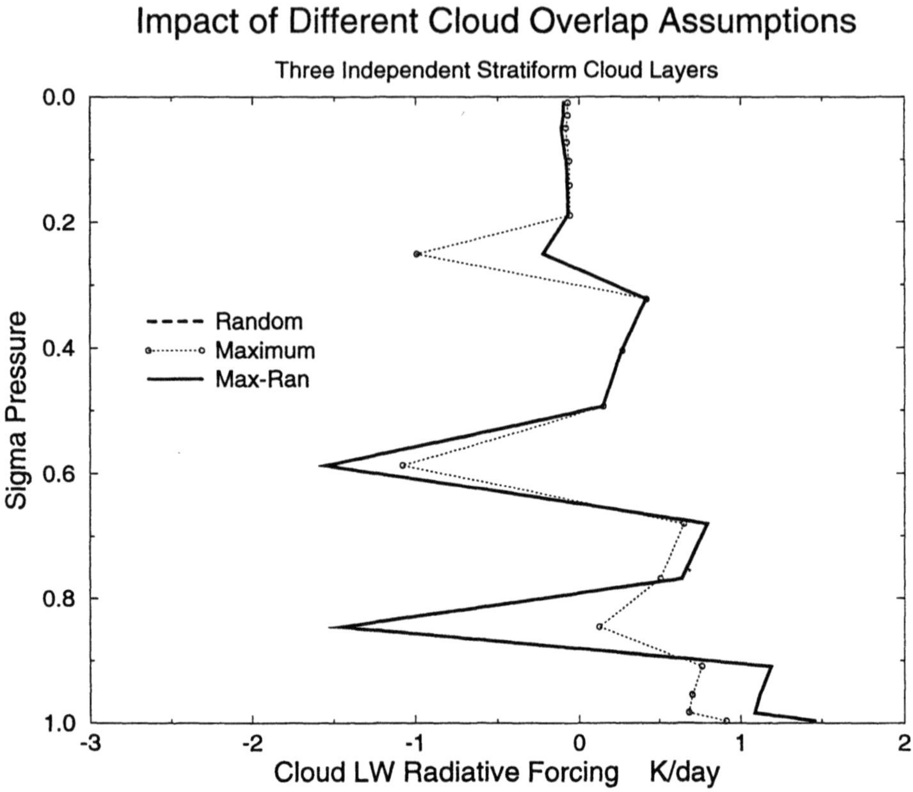

Figure 3:
3a- As in Figure 2, but for an atmosphere including three independent layers of high stratiform clouds.

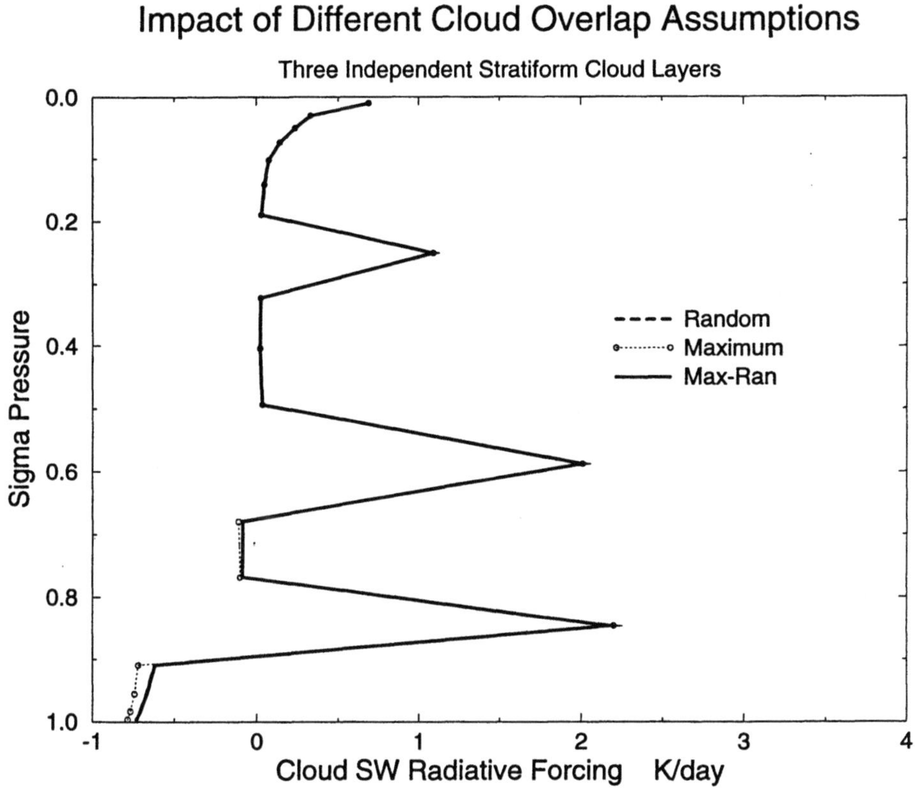

Figure 3b following 3b- middle stratiform clouds

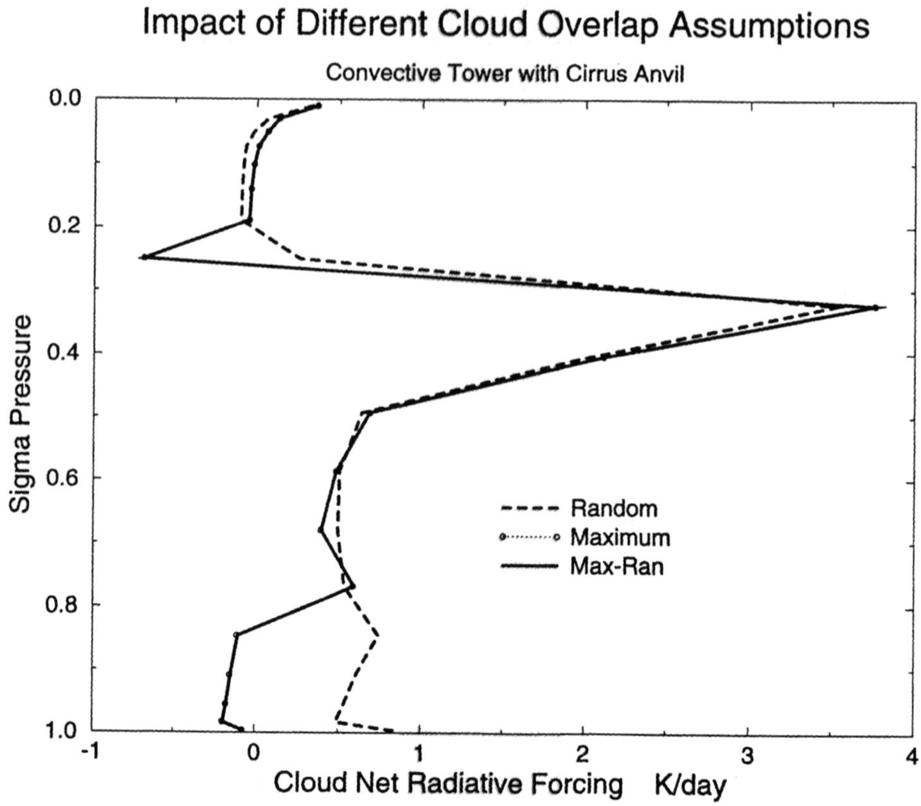

Figure 3c following 3c- low-level stratiform clouds

CLIMATE SENSITIVITY:
CLOUD AND WATER FEEDBACKS AND THEIR ASSESSMENT

Hervé Le Treut, Zhao-Xin Li and Sandrine Bony

Laboratoire de Météorologie Dynamique du CNRS
Ecole Normale Supérieure, 24 rue Lhomond
75231 Paris cedex 05, France

ABSTRACT:
The climate sensitivity may be defined as the rate of change in the radiative budget at the top of the atmosphere. This definition is applied to the LMD GCM results in order to evaluate the main processes which account for the model response to an external forcing, such as an increase in the greenhouse gases concentration, or a modification of the solar constant. The sensitivity associated with these long term changes of the climate system is compared with that corresponding to shorter time-scales, such as the response to the seasonal forcing or the interannual variability. This constitutes a first step toward the validation of the model, since the model behaviour at short time scales may be evaluated by comparison with observed data.

1 Introduction

Climate sensitivity may be defined as the rate of change of the climate system in response to an external perturbation. Due to the complexity of the climate system, and to the fact that no equilibrium other than statistical is ever attained, this notion is dificult to translate in a simple

NATO ASI Series, Vol. I 25
The Solar Engine and Its Influence
on Terrestrial Atmosphere and Climate
Edited by Elizabeth Nesme-Ribes

parameter. The radiation at the top of the atmosphere has very often been used to characterize climate changes. One reason is of practical nature: the radiation at the top of the atmosphere is measured from satellite-borne radiometers, such as those of the Earth Radiation Budget Experiment (ERBE). The other reason is the recognition that these radiation fluxes constitute the only mechanism by which the Earth-Atmosphere system exchanges energy with the outer space and therefore sums up the behaviour of the whole system. If we describe the climate system through its surface temperature T, an external parameter λ, and some internal variables x_i, with the associated radiative budget noted as $B(\lambda, T, x_i)$, one may view a climate change as the transition from one equilibrium state to another, then

$$dB = \frac{\partial B}{\partial \lambda}\Delta\lambda + \frac{\partial B}{\partial T}\Delta T + \Sigma\frac{\partial B}{\partial x_i}\Delta x_i = 0 \qquad (1)$$

And solving this equation to get ΔT, one may define a forcing term $\Delta B_0 = \frac{\partial B}{\partial \lambda}\Delta\lambda$, a sensitivity parameter $\beta = -1/(\frac{\partial B}{\partial T})$ and a feedback parameter $s_i = -\frac{\partial B}{\partial x_i}\frac{\Delta x_i}{\Delta T}/\frac{\partial B}{\partial T}$, so that

$$\Delta T = \beta\frac{\Delta B_0}{(1 - \Sigma\, s_i)} \qquad (2)$$

This approach of the sensitivity problem is not original, and many authors have used it previously (Cess, 1976; Hansen et al, 1984; Schlesinger, 1988; Wetherald and Manabe,1988), but it is useful here to define precisely those concepts.

Determining the climate sensitivity then means finding a regression between the radiative budget at the top of the atmosphere and the surface temperature. The aim of this short paper is to show how this simple definition may be applied to the LMD (Laboratoire de Météorologie Dynamique) GCM. For that purpose we will examine two different climate scenarios: warming due to the greenhouse effect increase, modification of the solar constant. Then we try to apply the same diagnostics to shorter climate fluctuations, simulated by the model but also observed by satellites. The comparison of those different elements will help us to evaluate the ability of the present models to predict future climate changes, and the possibilities of validation offered by past climate changes such as the little ice age.

2 Model description

The general design of the LMD GCM was first described by Sadourny and Laval (1984). The model dynamics is formulated in finite differences. The physical package used in the present experiments includes the radiation scheme developed at the Laboratoire d'Optique Atmosphrique (LOA), which is presently used in the operational model of the European Centre for Medium Range Weather Forecast (ECMWF). The model includes a full hydrological cycle. The water vapour condensation is simulated differently for stratiform and convective clouds. In the case of stratiform clouds, a simple statistical scheme is used to produce condensation on a fraction of the grid only, and therefore before a complete saturation of the grid area. Convective clouds are produced using both a Kuo scheme and a moist adiabatic adjustment scheme. The condensed water is retained as cloud water before it is precipitated to the ground: it can be advected by the large scale circulation, and reevaporated in drier air conditions. This is done through a cloud water budget equation and cloud water is one of the model prognostic variables. The cloud optical properties are determined using the predicted cloud water content. This cloud parameterization is described in Le Treut and Li (1991). The ground physics is very simple in this version of the model: a simple bucket model with a maximum capacity of 150 mm is used to compute soil moisture. The ground temperature is also computed using a single soil layer. This simple surface physics is consistent with the fact that this version of the model does not include the diurnal cycle: this simplicity is technically very useful when making the sensitivity and feedback analysis, because off-line radiation calculations are then easier to carry out.

3 Climate sensitivity experiments

3.1 General design

The LMD GCM was used at low resolution (48 points in longitude, 36 in latitude and 11 vertical levels) to make various sensitivity experiments to an increase in greenhouse gases (Le Treut et al., 1994) and to modifications of the solar constant (Nesmes-Ribes et al., 1993). The details of these experiments are described in the above papers. We recall here briefly the basic design of the experiments.

First the GCM is integrated for 10 years using observed Sea Surface Temperature (SST) conditions. This yields an estimation of the ocean heat transport, which we then use to integrate the atmospheric model coupled to an homogenous ocean layer of 50-meter depth. This procedure is a standard procedure: changes in the ocean circulation are therefore not among the mechanisms which can affect the model sensitivity. But the accumulation of heat by the ocean is considered by the model, and this offers a simple closure to the atmospheric feedbacks. As the heat capacity of the atmosphere is smaller than that of the ocean, we can note also that the heat being stored in the Earth/Atmosphere system is very accurately described by the temperature of this homogeneous surface ocean layer. This tends to reinforce the definition of the climate sensitivity made above, which uses relation between the radiation at the top of the atmosphere and the variations of the SST.

Three experiments were performed: (1) a control experiment corresponding to the present climate. (2) a 2×CO_2 experiment (the CO_2 amount increasing from 320ppm to 640ppm). (3) an experiment where the solar constant is increased by 1.66% in order to get an increase in the surface temperature similar to that of the 2×CO_2 experiment. A fourth experiment with a corresponding decrease of the solar constant is not analyzed here.

3.2 Main results

The global surface temperature increase of the model compared to the control experiment is 3.9°C for experiment (2) and 4.1°C for experiment (3). We use the diagnostics of the $\frac{\partial B}{\partial x_i}\Delta x_i$ terms, where x_i is the water vapour, the cloud distribution and cloud optical thickness, or the surface albedo, to determine which are the main processes which account for this temperature increase. This is done by doing off-line radiative computations where the model parameters are changed one by one from their values in the control experiment to their values in the perturbed climate. The fact that the model does not present any diurnal cycle allows to do these computations with a reasonable accuracy while using time-averaged values for the atmospheric parameters. The results are shown on Figure 1.a for the 2×CO_2 experiment and in Figure 1.b for the solar constant experiment. We can note the large contribution of the water vapour feedback, the generally positive cloud feedback and the surface albedo feedback limited to high latitudes. We may note that the effect of changes in the temperature lapse-rate are not distinguished from the effect of vertically uniform changes of temperature in this diagram. They will be shown later when we describe the model greenhouse effect. We may also note that the water vapour contribution here is different from the clear-sky greenhouse effect observed through ERBE measurements, because it does not include the changes in the vertical temperature lapse-rate.

In Li and Le Treut (1992), or Le Treut et al. (1994) the dependence of the model sensitivity on the parameterization of cloudiness was emphasized. The response of the model was shown to be very dependant on the details of the parameterizations. Here we compare the feedback terms in response to two different external forcings for a given set of parameterizations. These results are displayed in Figures 1.a and 1.b. The similarity between the two is extremely striking, as one forcing concerns the solar radiation, and the other affects primarily the terrestrial radiation. The fact that the sensitivity depends so little on the forcing pleads for using past climate changes to validate the models. The two external forcings considered here, however, are similar in two aspects: they are global, and they are imposed continuously to the climate system, on very long time scales (several decades), so that a quasi-equilibrium is reached. Results may be very different if we consider shorter fluctuations, as discussed in the next section.

4 The validation of individual feedback processes: the clear-sky greenhouse sensitivity

4.1 Position of the problem

The analysis of climate fluctuations at short time scales (for example seasonal or interannual variations) makes it possible to study with a very good accuracy some of the processes that are involved in climate sensitivity, and to provide some elements of validation to climate models. One is however immediately faced to the fact that those short-term climate variations are associated with very specific forcings. The seasonal cycle corresponds to a back and forth motion of the sun in the North/South direction. The

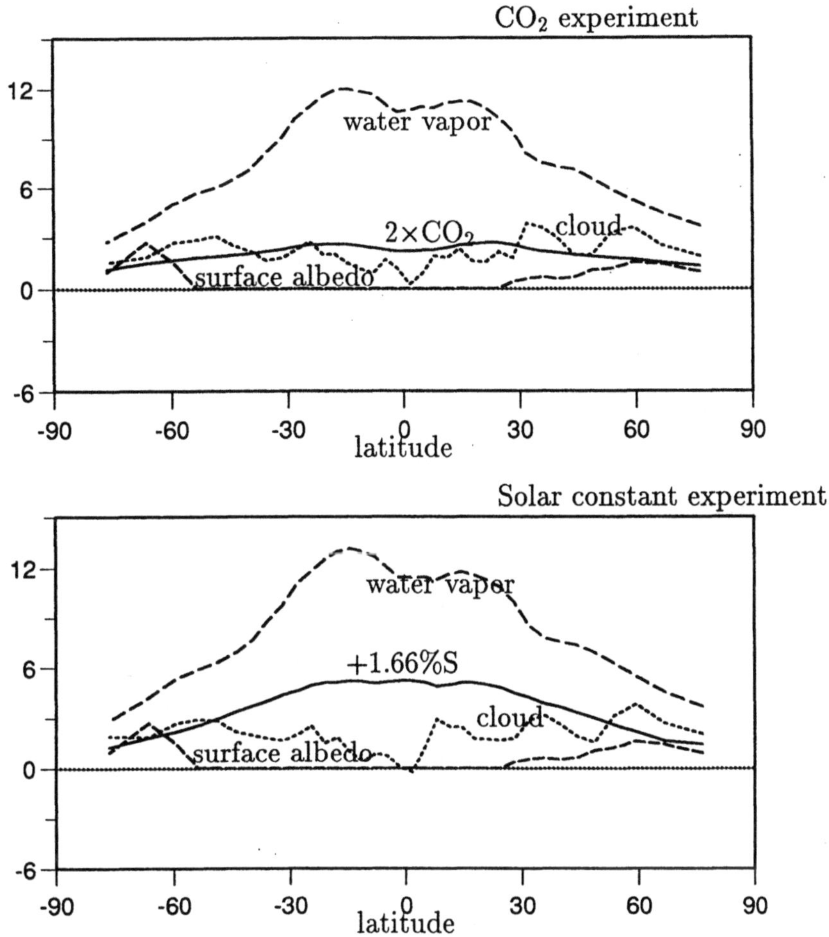

Figure 1: Zonal means (W m^{-2}) of the direct forcing and feedbacks at the top of the atmosphere for the doubled CO_2 experiment (top) and the increased solar constant experiment (bottom).

interannual climate fluctuations mainly result from SST changes due to changes in the ocean circulation. This is very different from the greenhouse forcing which is primarily characterized by its distribution along the vertical axis, causing a warming in the troposphere and a cooling in the stratosphere. This is illustrated here by focusing on a single effect, which is the role of the water vapour feedback. This term appears as the strongest feedback effect in our decomposition and is strongly linked with the cloud feedbacks. Its climatic importance is very large, and its validation through comparison with modern observed data is extremely important. It is also relatively simple to examine because water vapour may be described by less parameters than clouds, which must be defined through their shape, water content, repartition between water and ice, droplet size, etc. The ERBE experiment (Barkstrom, 1984), which provides measures of the clear-sky longwave radiation at the top of the atmosphere, the Sea Surface Temperature (SST) data from Reynolds analyses (Reynolds, 1988), the precipitable water W retrieved from SSM/I (Special Sensor Microwave Imager) observations, give a unique opportunity to examine the link between all those parameters.

4.2 Methodology

The greenhouse effect corresponds to the absorption (and subsequent reemission) of infrared radiation emitted by the Earth surface by atmospheric trace gases or clouds. A simple manner to characterize this effect is to compare the outgoing longwave radiation LW at the top of the atmosphere with the surface longwave emission $\epsilon\sigma T^4$, with ϵ being the surface emissivity (close to unity over ocean regions) and σ the Stefan-Boltsmann constant. We may then define, following Raval et Ramanathan (1989), a normalized greenhouse factor:

$$g = 1 - \frac{\mathrm{LW}}{\sigma \mathrm{SST}^4} \tag{3}$$

which provides a gross measure of the greenhouse effect. The derivative of this term with respect to the observed surface temperature constitutes one component of the climate sensitivity, as described in the first paragraph.

Raval and Ramanathan (1989), Stephens and Greenwald (1991) or Duvel and Bréon (1991) have used the geographical inhomogeneities of the climatic system, its seasonal variations, or its interannual variability to obtain regression coefficients between g (in clear-sky conditions), SST and W. The thermodynamic law of Clausius-Clapeyron explains a very clear mean relation between W and SST, the dominant effect of water vapour absorption in the infrared explains a clear mean relation between g and W. On the whole there is a mean positive regression between g and SST, which represents the positive feedback of water vapour.

This mean relation between g, W and SST nevertheless is marked by a very large dispersion, and it is not granted that a climate change would occur along the line of the mean regression. Indeed the studies of Duvel et al. (1993) or Bony and Duvel (1994) have shown that a decrease of the clear-sky greenhouse effect at high latitudes could be associated with an increase of the precipitable water, due to simultaneous changes in the vertical temperature lapse-rate.

It is therefore important to compare the sensitivity associated with different climate fluctuations in order to know what are the hopes for a validation of individual feedback processes.

4.3 The greenhouse sensitivity associated with different climate variations

It is possible to use the results of a GCM to diagnose a posteriori the normalized greenhouse factor g for clear-sky conditions, and study its linear regression with surface temperature. This is done for four cases:

- the seasonal cycle
- the interannual variability (difference between the years 87 and 88)
- the model response to a CO_2 increase
- the model response to an increase of the solar constant (these two experiments are the ones described above)

The sensitivity parameter is computed for various ocean regions of the globe referenced by their mean surface temperature.
As shown by Figure 2, the LMD GCM agrees rather well with the observations in terms of the seasonal and interannual variability, with however some discrepancies in tropical regions and cold SST regions (SST $< 10^oC$). Figure 2 also points out the fact that for a given SST change, the observed clear-sky greenhouse effect sensitivity is different when considering seasonal or interannual variations, and when considering high latitudes or tropical regions. This suggests that the sensitivity of the vertical atmospheric structure strongly depends on the mean climatic state, and also on the nature of the forcing that is applied to the climate system. In view of these results, implications of model validation at short time scales for the purpose of the validation of model's sensitivity to a long-term forcing (such as an anthropogenic increase of greenhouse gases for example) depends on whether short term and long term sensitivities are comparable.

Figure 3 compares the linear regression coefficient between the clear-sky greenhouse effect and the SST computed for the different climate scenarios,

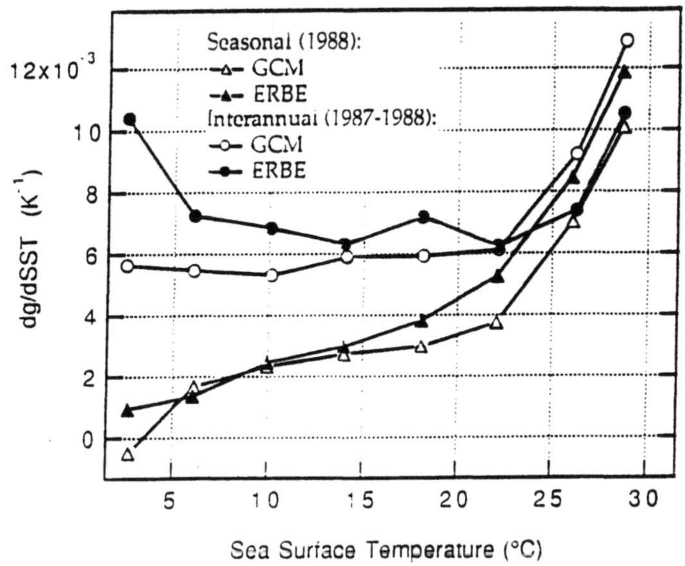

Figure 2: Linear regression coefficient between clear-sky greenhouse effect and SST variations (seasonal or interannual), computed for various ocean regions of the globe referenced by their mean surface temperature, on the basis of LMD GCM simulations and of satellite data.

and shows that for a given region of the globe and a given SST change, the short term and long term sensitivities of the greenhouse effect are quite different. In consequence, the meaning of model validation at short time scales for the purpose of model's sensitivity validation depends on whether we can interprete the differences between the sensitivities associated with the various climate scenarios.

This can be done by diagnosing separately the contribution of a given parameter to the change of the greenhouse factor. These diagnostics (not shown here) indicate that the vertical structure of the atmosphere is mainly responsible for the discrepancies between the different scenarios wih regard to the clear-sky greenhouse effect sensitivity. If we diagnose g by letting the relative humidity of the air column and its temperature lapse-rate unchanged under the climate fluctuation which is studied, we find that the sensitivities of the four scenarios are almost identical, with very little

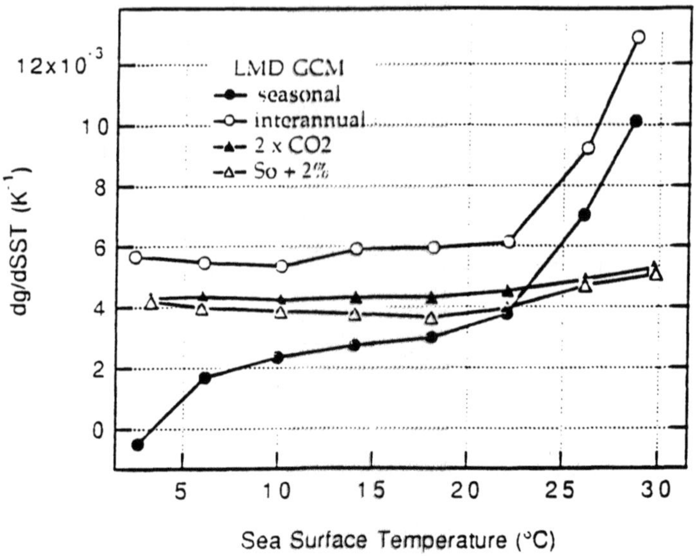

Figure 3: Same as figure 2 but for the LMD GCM only, when considering seasonal variations, interannual variations, a doubled CO_2 experiment and a solar constant increase experiment.

dependence in latitude.

These changes in the vertical structure of the atmosphere are more important for short term climate fluctuations, which are associated with large migrations of the dynamical structures of the atmosphere (the Hadley cells for example), and depend very much on latitude: this explains the general shape of Figure 3. The warmer climates associated with increased CO_2 or solar constant appear to depend much less on the vertical structure of the atmosphere, perhaps because the atmospheric circulation is less affected by global and long-term forcings (Nesme-Ribes et al., 1993) than by short term fluctuations: this appears as a favourable circumstance for model validation, because it indicates that validating the short term variability of the modeled vertical structure of the atmosphere poses strong constraints on the model's performance.

5 Conclusions

The response of the LMD GCM to two different global external forcings, exerted over long time scales (several decades) is explained qualitatively and quantitavely by very similar feedback effects, and leads to virtually identical climate change scenarios. The situation is very different for shorter fluctuations, which are more easily observable, but which do not allow a global equilibrium of the climate system. The peculiarities of the local or regional forcing, the associated changes in the vertical structure of the atmosphere, makes it essential to go back to the study of individual processes, and do not allow the simple derivation of a single "sensitivity" parameter. These short time-scale validation studies, however, put very serious constraints on the model, and may increase our confidence in their results.

References

[1] Barkstrom, B. R. (1984). The Earth Radiation Budget Experiment (ERBE). Bull. Am. Meteorol. Soc., 65, 1170–1185.

[2] Bony, S. and Duvel, J.-P. (1994). Influence of the vertical structure of the atmosphere on the seasonal variation of precipitable water and greenhouse effect. J. Geophys. Res. (in press).

[3] Cess, R. D. (1976). Climatic change: An appraisal of atmospheric feedback processes by employing zonal climatology. J. Atmos. Sci., 33, 1831–1843.

[4] Duvel, J. P. and Bréon, F. M. (1991). The clear-sky greenhouse effect sensitivity to a sea surface temperature change. J. Clim., 4, 1162–1169.

[5] Duvel, J.-P., Kandel, R. S., Bony, S., and Le Treut, H. (1993). Relation between the precipitable water and the greenhouse effect: Lapse rate influence. In Deepack, A., editor, IRS'92: Current Problems in Atmospheric Radiation, pp. 158–162, Tallin, Estonia.

[6] Hansen, J., Lacis, A., Rind, D., Russell, G., Stone, P., Fung, I., Ruedy, R., and Lerner, J. (1984). Climate sensitivity: Analysis of feedback mechanisms. In Hansen, J. E. and Takahashi, T., editors, Climate Process and Climate Sensitivity, volume 29 of Geophys. Monogr. Ser., pp. 130–163. AGU, Washington, D. C.

[7] Le Treut, H. and Li, Z. X. (1991). Sensitivity of an atmospheric general circulation model to prescribed SST changes: Feedback effects associated with the simulation of cloud optical properties. Clim. Dyn., 5, 175–187.

[8] Le Treut, H., Li, Z. X., Forichon, M., and Butel, R. (1994). Sensitivity study of the lmd gcm to greenhouse forcing associated with two different cloud water parameterizations. submitted to J. Clim.

[9] Li, Z. X. and Le Treut, H. (1992). Cloud-radiation feedbacks in a general circulation model and its dependence on clouds modeling assumptions. Climate Dynamics, 7, 133–139.

[10] Nesme-Ribes, E., Ferreira, E. N., Sadourny, R., Le Treut, H., and Li, Z. X. (1993). Solar dynamics and its impact on solar irradiance and the terrestrial climate. J. Geophys. Res., 98(A11), 18.923–18.935.

[11] Raval, A. and Ramanathan, V. (1989). Observational determination of the greenhouse effect. Nature, 342, 758–761.

[12] Reynolds, R. W. (1988). A real-time global Sea Surface Temperature analysis. J. Clim., 1, 75–86.

[13] Sadourny, R. and Laval, K. (1984). January and July performance of the LMD general circulation model., pp. 173–197. New Perspectives in Climate Modelling, edited by A. Berger and C. Nicolis, Elsevier, New-York.

[14] Schlesinger, M. E. (1988). Quantitative analysis of feedbacks in climate model simulations of CO2-induced warming. In Schlesinger, M., editor, Physically-based modelling and simulation of climate and climatic change, volume C243 of NATO ASI Series, pp. 653–735.

[15] Stephens, G. L. and Greenwald, T. J. (1991). Observations of the Earth's radiation budget in relation to atmospheric hydrology, 1, Clear-sky greenhouse effect and water vapor feedback. J. Geophys. Res., 96, 15,311–15,324.

[16] Wetherald, R. T. and Manabe, S. (1988). Cloud feedback processes in a general circulation model. J. Atmos. Sci., 45(8), 1397–1415.

Panel Discussion on PUTTING SOLAR VARIABILITY IN PERSPECTIVE

K.P. Shine
Department of Meteorology
University of Reading
2 Earley Gate
Reading RG6 2AU
United Kingdom

Introduction

Assessment of the potential importance of solar variability on climate requires an understanding of other mechanisms which may lead to climate change. Here an attempt is made to provide a perspective on the timescale of the past century or so.

Two related ways in which an assessment can be made are discussed. One is to use the concept of radiative forcing which is defined as the change in net (solar and thermal infrared) irradiance (in W/sq.m) at the tropopause following, for example, a change in solar output or greenhouse gas concentration ; a positive forcing is defined as one which tends to warm the climate system. The rationale for this definition, along with some subtleties, is discussed in the Intergovernmental Panel on Climate Change Working Group 1 reports (IPCC 1990 ; IPCC 1992) and will not be dwelt on here. It is assumed that there is a one-to-one correspondence between the radiative forcing and the consequent climate change (here crudely taken to be the change in global mean surface temperature) ; again there are a number of issues involved in making this assumption which will not be dwelt on here. The best-guess values of the radiative forcing due to each climate change mechanism are estimated independently of each other ; their sizes can then be compared to assess their relative importance.

NATO ASI Series, Vol. I 25
The Solar Engine and Its Influence
on Terrestrial Atmosphere and Climate
Edited by Elizabeth Nesme-Ribes

The radiative forcing can be related to a global mean temperature change by a climate sensitivity parameter, which can be calculated using a general circulation model (IPCC 1990). Its value is determined by feedbacks involving clouds, water vapour and surface albedo. Currently, model estimates of its value vary by at least a factor of three, probably due to uncertainties in the way clouds respond to climate change. Hence, it is more straightforward to compare radiative forcing than relate each forcing mechanism to some possible temperature change.

The radiative forcing due to many possible climate change mechanisms is itself rather uncertain. A second technique for estimating the relative importance of climate change mechanisms has been in use for many years (Hansen *et al.*, (1981) is one example) and has become popular again recently. This attempts to constrain the uncertainties using a simple climate model together with the observed temperature record by obtaining a best fit between model and observations. I call this approach the "fitting technique". It is my contention that the fitting technique is so ill-conditioned that, whilst of interest, it adds little or no extra information to our knowledge of radiative forcing mechanisms. Indeed, it can be misleading.

In this discussion I will first present some estimates of radiative forcing from the recent literature. I will then outline my reservations about the fitting technique. It is first important to recognise that the complex coupled ocean atmosphere system on our planet can undergo variations in its own climate due, for example, to changes in partitioning of heat between the ocean and the atmosphere. Thus climate does not need to be "forced" to undergo a change. Indeed, this point is vital to my discussion of the fitting technique. Coupled ocean-atmosphere general circulation models (e.g. Stouffer *et al.*, 1994) have been integrated for periods of many centuries without any imposed radiative forcing. They show changes in global mean temperature on decadal time scales, purely due to the model's own internal variability, of a size similar to those seen in observations during the past century ; observed changes on the century time scale do, however, seem larger than those found in the model. These models are, of course, much simplified compared to the real ocean-atmosphere system, but they are at least suggestive of the great care needed in interpreting observed temperature variations and relating them to supposed causes. The signal due to climate forcing mechanisms may not (yet) be outside the internal "noise" of the climate system.

Estimates Of Radiative Forcing

GREENHOUSE FORCING : One of the most discussed forcing mechanisms in recent times is that due to increased concentrations of greenhouse gases. There is a natural greenhouse effect, mainly due to water vapour, but with important contributions from carbon dioxide and ozone, which is important for maintaining the observed surface temperature. These gases absorb and re-emit infrared radiation ; the wavelengths of most importance are between 4 and 500 microns. The re-emission of radiation towards the surface helps maintain the surface temperature ; a simplistic calculation indicates that without these gases the Earth's surface would be some 30 deg C cooler than it is now.

The enhanced greenhouse effect, or the greenhouse forcing, results from an increase in atmospheric concentrations of a number of gases. Well-documented changes in the concentrations of gases such as carbon dioxide, methane, nitrous oxide and the chlorofluorocarbons have been used as input to radiative transfer models. The effect of these gases is termed direct forcing, as human activity has led to increased emissions, and hence concentrations, of these gases. The indirect forcing is when emissions of a gas, which needs not itself be a greenhouse gas, lead to chemical reactions which lead to a change in the concentration of other greenhouse gases. For example, increased emissions of carbon monoxide and hydrocarbons are believed to have increased tropospheric ozone concentrations and the chlorofluorocarbons are believed to have led to stratospheric ozone depletion (IPCC 1990 ; IPCC 1992).

Estimates of the direct forcing due to greenhouse gas changes are reasonably consistent. Although various authors choose slightly different periods and different trace gas concentration changes, estimates for the change since the pre-industrial era (i.e. since about 1800) are 2.3 W/sq.m (IPCC 1990), 2.1 W/sq.m (Hansen *et al.*, 1993), 2.3 W/sq.m (Shi and Fan 1992) and 2.6 W/sq.m (Hauglustaine *et al.*, 1994). About 50% of this forcing is due to changes in carbon dioxide.

For the indirect forcing, uncertainties are larger. Hansen *et al.*, (1993) estimate -0.2 W/sq.m for stratospheric ozone change. Hauglustaine *et al.*, (1994) estimate a total indirect forcing of +0.7 W/sq.m mostly due to a model-calculated increase in tropospheric ozone. It seems likely that over century timescales the indirect greenhouse forcing due to ozone changes has been positive, although the balance between tropospheric increases and stratospheric decreases over the past decade or so is much less clear.

TROPOSPHERIC AEROSOL FORCING : Human activity is also believed to have led to increased concentrations of sub-micron aerosol particles in the atmosphere. These aerosols may play two roles in radiative forcing. The first, called the direct effect, is that the aerosols scatter and absorb solar radiation ; the net effect is most easily represented by the changes in planetary albedo. Depending on the aerosol properties, the aerosols can lead to either an increase or a decrease in the albedo ; an increase in albedo would be a negative forcing (i.e. a tendency to cool the climate system) and a decrease, would lead to a positive forcing (i.e. a tendency to warm the climate system). Of most current interest are increased aerosol emissions due to sulphur emissions, mostly associated with fossil fuel burning, and due to biomass burning in the tropics.

An indirect aerosol forcing may also occur and is more subtle. Some of the aerosols act as cloud condensation nuclei (CCN) which allow water vapour to more easily condense to form cloud droplets. There are normally an abundance (several hundred per cubic cm) of natural CCN. However, the increased concentrations of aerosols may have led to an increase in the number of CCN. This means that when a given amount of water condenses to form a cloud it leads to more, but smaller, cloud water droplets. This leads to an increase in the available surface area per unit mass of droplets which can interact with solar radiation ; there is a consequent increase in cloud albedo which would cause a negative radiative forcing. This mechanism is much more speculative, on a global scale, as there are many uncertainties in relating aerosol emissions to cloud albedo. An added complication is that if these aerosols are sufficiently absorbing (such as due to the presence of elemental carbon) the cloud albedo could decrease.

Estimates for the global mean forcing since pre-industrial times due to the direct effect of tropospheric aerosols range from -0.25 W/sq.m (Hansen *et al.*, 1993) to -1 W/sq.m (Charlson *et al.*, 1992), although all authors acknowledge large uncertainties. The direct effect of biomass burning is similarly uncertain, and estimates range from -0.25 W/sq.m (Hansen *et al.*, 1993) to -1 W/sq.m (Penner *et al.*, 1992).

The indirect effect of aerosols, due to cloud albedo changes, is much more uncertain. Kaufman and Chou (1993) regard their value of -0.45 W/sq.m since pre-industrial times to be a conservative estimate ; Penner *et al.*, (1992) conclude that the indirect effect of biomass burning must be comparable with the sulphate effect. Hansen *et al.*, (1993) estimate a combined (i.e. sulphate and biomass) effect of -0.5 W/sq.m, with at least a factor of two uncertainty. However, even the sign of the indirect effect is not yet certain. Kaufman and Nakajima (1993) have analysed satellite imagery over the Amazon during biomass burning episodes and found cloud albedo to decrease. Measurements in the former Soviet Union around cities have also been interpreted as indicating a decreased cloud albedo in polluted circumstances (e.g. Binenko and Harshvardhan 1992).

An overall estimate of the effect of aerosols is impossible to make with any confidence. On the basis of the above, a range between zero and -4 W/sq.m since pre-industrial times is plausible but it is difficult to rule out the possibility that the true value lies outside this range. Hence tropospheric aerosols could either partially offset the greenhouse forcing or, at the other extreme, completely overwhelm it. However, it must be noted that even if the aerosol and greenhouse forcing combine to give a value of zero averaged over the globe, there still could be a climate change due to regional variations in the forcing ; this is the subject of much current research.

VOLCANIC AEROSOLS : Volcanic eruptions can lead to massive injections of sulphur gases into the stratosphere ; these quickly oxidize to sulfuric acid which can then readily form sub-micron aerosols. These scatter radiation back to space and, for all plausible size ranges, such volcanic aerosols are believed to cause a negative forcing. The effect of volcanic aerosols is transient as they are removed from the stratosphere, and subsequently washed out, on a time-scale of a couple of years. The eruption of Mount Pinatubo in the Philippines in June 1991 is estimated to have caused peak forcings of up to -2 W/sq.m (e.g. Hansen *et al.*, 1993, Minnis *et al.*, 1993). A possible complication in the estimation of the climatic effects of volcanic eruptions concerns the possibility that the aerosols falling out of the stratosphere affect cirrus cloud formation ; Sassen (1992) presents some observational evidence for such an effect, although there is enormous difficulty in trying to estimate whether it is important on a global scale.

It is unlikely that changes in volcanic activity have caused a trend on the century time scale ; however, variations in volcanic activity during the century may have led to important decadal changes in stratospheric aerosol load (e.g. Robock 1991 ; Sato *et al.*, 1993) ; for example, the period 1925-1960 was notably lacking in large eruptions and this period coincided with a period of global mean surface warming. The main problem with understanding the effect of volcanoes is that the emissions are only known to reasonable accuracy for eruptions over the past 15 years or so.

SOLAR VARIABILITY : Estimates of change in solar output are, of course, the topic of many other contributions to this volume. In terms of radiative forcing, a change in solar constant must be multiplied by a factor of (1-alpha)/4 where alpha is the planetary albedo. The value of 4 in the denominator accounts for the fact that the surface area of the planet is one-quarter the cross-sectional area of the Earth as seen by the Sun ; the (1-alpha) factor accounts for the fraction of solar energy that is scattered back to space without warming the planet. For the Earth, alpha is about 0.3, so that (1-alpha)/4 is about 0.175. In terms of radiative forcing (i.e. 0.175 times the variation in total solar irradiance) over the past century, Hansen *et al.*, (1993) estimate +0.3 W/sq.m. Changes since the Maunder Minimum may be double (e.g. Lean *et al.*, 1992) or even treble this (Nesme-Ribes *et al.*, 1993) although these estimates are subject to large uncertainties. These are the overall increases in this period ; superimposed on this increase, shorter time period variations have been proposed. Such changes may well be of significance for the earlier part of the century when greenhouse and aerosol forcings were much smaller ; their significance since then depends to a large extent on the degree of compensation between the greenhouse and aerosol forcing.

However, as discussed elsewhere in this volume, the variability in solar irradiance is not, at least for the 11-year cycle, constant with wavelength ; the actual change at the tropopause, which is of more relevance to climate change, after accounting for the absorption in the stratosphere and possible consequent changes in ozone, is, as yet, unclear, and the subject of current research.

ESTIMATING FORCING USING THE FITTING TECHNIQUE

An alternative method for assessing past forcings is to use statistically based comparisons with observed changes in surface temperature, either directly, or else by using a climate model with some representation of oceanic thermal inertia to calculate the evolution of the temperature. This modelled temperature series can then be compared with the observed and some measure of explained variance can be computed. Three recent examples of this technique using simple global or hemispheric energy balance models are Kelly and Wigley (1992), Schlesinger and Ramankutty (1992) and Shi *et al.*,

(1993). The methodology varies, but the results do not show good consistency.

Kelly and Wigley (1992) use their best estimate of greenhouse gas and sulphate aerosol forcing and impose a speculative representation of the effect of solar cycle length variations on solar constant with an undetermined amplitude B. They then find the climate sensitivity and B which yield the largest value of explained variance. The results show that the explained variance is highest when the solar cycle effects are included on their own ; however, it is only slightly lower when greenhouse gases and tropospheric aerosols are included. I do not suggest that the solar-only case is plausible. However, the fact that it gives such a high explained variance indicates that the quality of the fit is a very poor guide to the contribution of different forcing mechanisms.

Schlesinger and Ramankutty (1992) adopt a similar approach and, further, explore the sensitivity to the size of the sulphate aerosol forcing. They obtain the highest explained variance when including greenhouse gases, solar cycle effects and tropospheric aerosols, for a pre-industrial to 1978 contribution of aerosols of -1.5 W/sq.m ; this includes indirect effects and can be considered, to a limited extent, to include the contribution of non-sulphate aerosols.

Neither of these two studies included the effects of volcanic eruptions which earlier work (e.g. Hansen *et al.*, 1981) indicated may be important. Shi *et al.*, (1993) achieve a higher explained variance than the other studies for a combination of greenhouse gases and volcanic eruptions and require no need for either tropospheric aerosols or solar forcing. The conclusions of this study are seriously at variance with those of Kelly and Wigley (1992) and Schlesinger and Ramankutty (1992). This further illustrates that agreement with observational records is a very poor constraint on forcing estimates.

Further problems with this fitting approach are :

(a) It ignores the effect of unforced natural variability ; this variability might require that a large fraction of the variance does not need to be explained in terms of forcing mechanisms. As mentioned in the Introduction, this unforced variability may be of a similar size to observed temperature variations. Crudely, the question must be posed "how much of the variance needs to be explained ?"

(b) It assumes that the surface temperature record is of uniform reliability since the mid 19th century so that the modelled temperatures are required to fit the observations equally well for all periods. It is extremely difficult to place error bars on the global mean temperature series, but IPCC (1990) and IPCC (1992) state that the observed global mean warming between 1865 and the present day to be between 0.3 and 0.6 deg C, indicating substantial uncertainty ; Hansen and Wilson (1993) indicate that differences between individual years of less than 0.1 deg C cannot be regarded as significant. Attempting to fit this temperature series without considering error bars is reminiscent of the problems encountered in attempting to retrieve atmospheric temperature profiles from satellite observations (see e.g. Houghton *et al.*, 1984, Chapter 7) ; ignorance of the error bars leads to an ill-conditioned problem. Crudely, the question must be posed "how much of the variance is real and what are the observational error bars ?"

(c) It either assumes values for climate sensitivity and effective thermal inertia, or uses them as floating variables in the maximisation of explained variance. If the fitting technique allows too many floating variables then a good fit is almost guaranteed, particularly since the temperature series shows a good deal of autocorrelation. The question is then "what guarantee is there that the best fit is of any more geophysical significance than other possible solutions ?"

I contend that these fitting methods ultimately give us no extra information beyond that which can be obtained from a direct estimate of the radiative forcing ; indeed the fitting technique explicitly relies on these to verify the plausibility of the solution. In my opinion they can neither be used to confirm nor deny the existence of a solar influence on climate

Acknowledgements

This Panel Discussion was written whilst I was in the process of writing early drafts of the IPCC Working Group 1's 1994 Scientific Assessment. I must thank my co-authors in this task, Y. Fouquart, S. Solomon, J. Srinivasan and V. Ramaswamy whose contributions helped me in the preparation of this discussion ; however, the strongly expressed views herein are entirely my own responsibility.

References

Binenko, V., Harshvardhan, I. (1992). Aerosol effect in radiation transfer, in : "Aerosol Effects In Climate", ed. S.G. Jennings. University of Arizona Press, Tucson, 190-232

Charlson, R.J., Schwarz, S.E., Hales, J.M., Cess, R.D., Coakley, J.A., Hansen, J.E., Hoffman, D.J. (1992). Climate forcing by anthropogenic aerosols. Science, 255: 423-430

Hansen, J., Johnson. P., Lacis, A., Lebedeff, S., Lee, P., Rind, D., Russell, G. (1981). Climate impacts of increasing atmospheric carbon dioxide. Science, 213: 957-966

Hansen, J., Wilson, H. (1993). Commentary on the significance of global temperature records. Climatic Change, 25: 185-191

Hansen, J., Lacis, A., Ruedy, R., Sato, M., Wilson, H. (1993). How sensitive is the world's climate ?. National Geographic Research and Exploration, 9: 142-158

Hauglustaine, D.A., Granier, C., Brasseur, G.P., Megie, G. (1993). The importance of atmospheric chemistry in the calculation of radiative forcing on the climate system. J. Geophys. Res., 99: 1173-1186

Houghton, J.T., Taylor, F.W., Rodgers, C.D. (1984). Remote sounding of atmospheres. Cambridge University Press

IPCC (1990). Climate Change - The IPCC Scientific Assessment. (Houghton, J.T., Jenkins, G.J., Ephraums, J.J. (Eds)). Cambridge University Press

IPCC (1992). Climate Change 1992 - The Supplementary Report to the IPCC Scientific Assessment, eds. J.T. Houghton, B.A. Callendar, S.K. Varney. Cambridge University Press

Kaufman, Y.J., Chou, M.D. (1993). Model simulations of the competing climatic effects of SO2 and CO2. J. Climate, 6: 1241-1252

Kaufman, Y.J., Nakajima, T. (1993). Effects of Amazon smoke on cloud microphysics and albedo - analysis from satellite images. J. Appl. Meteorol., 32: 729-744

Kelly, P.M., Wigley, T.M.L. (1992). Solar cycle length, greenhouse forcing, and global climate. Nature, 360: 328-330

Lean, J., Skumanich, A., White, O. (1992). Estimating the Sun's radiative output during the Maunder minimum. Geophys. Res. Lett., 19: 1591-1594

Minnis, P., Harrison, E.F., Gibson, G.G., Denn, F.M., Doelling, D.R., Smith, Jr ; W.L. (1993). Radiative forcing by the eruption of Mt Pinatubo deduced from NASA's Earth Radiation Budget Experiment data. Science, 259: 1411-1415

Nesme-Ribes, E., Ferreira, E.N., Sadourny, R., Le Treut, L., Li, Z.X. (1993). Solar dynamics and its impact on solar irradiance and the terrestrial climate. J. Geophys. Res., 98: 18923-18935

Penner, J.E., Dickinson, R.E., O'Neill, C.A. (1992). Effects of aerosol from biomass burning on the global radiation budget. Science, 256: 1432-1434

Robock, A. (1991). The volcanic contribution to climate change of the past century, in : "Greenhouse-Gas-Induced Climate Change ; A critical appraisal of simulations and observations", ed. M.E. Schlesinger, Elsevier: 429-443

Sassen, K. (1992). Evidence for liquid-phase cirrus cloud formation from volcanic aerosols, climatic implications. Science, 257: 516-519

Sato, M., Hansen, J.E., McCormick, M.P., Pollack, J.B. (1993). Stratospheric aerosol optical depths, 1850-1990. J. Geophys. Res., 98: 22987-22994

Schlesinger, M.E., Ramankutty, N. (1992). Implications for global warming of intercycle solar irradiance variations. Nature, 360: 330-333

Shi, G., Fan, X. (1992). Past, present and future climatic forcing due to greenhouse gases. Adv. Atmos. Sci., 9: 297-286

Shi, G.Y., Guo, J.D., Fan, X.B., Wang, B., Wang, L.X. (1993). Climate Change and its Causes, in Proceedings of the International Conference on Regional Environment and Climate Changes in East Asia, Taipei, Taiwan

Stouffer, R.J., Manabe, S., Vinnikov, K.Y. (1994). Model assessment of the role of natural variability in recent global warming. Nature, 367: 634-636

ASPECTS OF A DECADAL SUN-ATMOSPHERE CONNECTION

K. Labitzke and H. van Loon[1]
Meteorologisches Institut, Freie Universität Berlin
Carl-Heinrich-Becker-Weg 6-10
12165 Berlin
Germany

ABSTRACT

This paper brings up-to-date our correlations between the 10.7cm solar flux and 30mb heights, and our composites of temperatures and geopotential heights grouped according to the extremes of the 11-year solar cycle. It shows that our earlier results are robust. Furthermore, we demonstrate that the large correlation coefficients between the solar flux and the 30mb heights to a great extent are associated with temperature variations on a decadal scale in the middle and upper troposphere in the outer tropics-subtropics. Comparison between the vertical structure of the temperature difference between the extremes of the solar cycle and the vertical structure of the temperature difference between years of strong and weak Hadley circulations show that the two are similar, which suggests that a solar influence on the upper troposphere and lower stratosphere may include changes in the Hadley circulation.

INTRODUCTION

The growing interest in the variability of climate has sharpened the realization that our knowledge of the changes in and forcing of the atmosphere is imperfect. As a result of our limited insight into the causes of climate variations interest has revived in

[1]Permanent affiliation: National Center for Atmospheric Research, Boulder, CO, USA. NCAR is sponsored by the National Science Foundation.

NATO ASI Series, Vol. I 25
The Solar Engine and Its Influence
on Terrestrial Atmosphere and Climate
Edited by Elizabeth Nesme-Ribes

possible solar influences on the atmosphere. The popular belief of the past century was - to paraphrase Godson (1991) - that the variable solar activity had no influence on climate, or that if it did, the effects should always be the same with regard to sign and size. The scepticism is due to the fact that the statistical investigations, which dominate the research, although they often have suggestive results, have been limited by their failure to explain how the sun could produce the observed variations in the atmosphere.

Our papers referred to below are no exception in this respect, but the results they present are of sufficient statistical significance that they should be taken into account - even if their underlying cause should turn out not to be solar.

EFFECTS OF THE 11-YEAR SUNSPOT CYCLE

SUMMER

If the 11-year solar cycle has a noticeable effect on the earth's atmosphere, there is no self-evident reason why the outcome should be the same everywhere, or that it should appear as a purely radiative effect. The statistics presented below (based on the historical daily analyses at the Freie Universität Berlin) suggest that dynamics play a major role in the spatial distribution of the atmosphere's response to the 11-year solar variability. This is evident in Fig. 1, which shows the correlation between a 36-year series of 30-mb height in summer and the 10.7cm solar flux. This flux is a good index of the 11-year sunspot cycle. We have computed such correlations every year during the past five years, starting with a sample of 31, and they have remained stable with each additional year. The correlations over more than three quarters of the area are locally significant above the 1% level; and over one third of the area the local significance is above the 0.1% level. The map as a whole is field significant at better than the 1% level (see Livezey and Chen, 1983, for an explanation of field significance). The shape of the correlation pattern, which is repeated in each month of the year, implies that the probable influence of the 11-year solar cycle is areally concentrated by atmospheric motions.

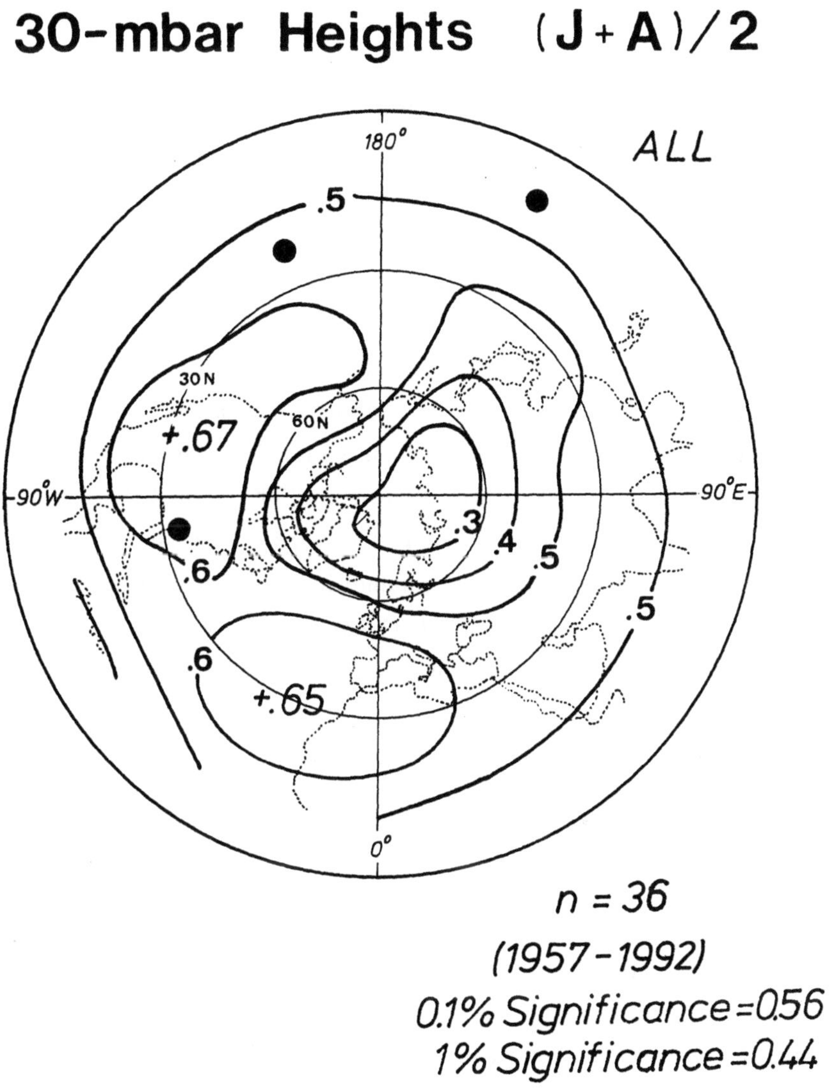

Figure 1. Correlations between 30mb height in July-August and the 10.7cm solar flux, n = 36, and the 0.1% local significance level is 0.56. Dots mark the positions of stations in Fig. 2. Updated from van Loon and Labitzke (1990).

In order for the 30-mb geopotential heights to contain a periodicity similar to the 11-year sunspot cycle, as the correlations in Fig. 1 indicate that they do over large areas, the temperature of the air column below or the surface pressure, or both, must contain the same periodicity. As is evident in Figs. 2a and b, this periodicity does not exist in the surface pressure but it is obvious in the temperature: the figure shows the difference in the temperature between solar maxima and minima from the ground to 30 mb (about 24 km) at stations in areas of Fig. 1 with statistically significant correlations. They have in common that the troposphere is warmer in solar maximum and, in Figs. 2a and b, that the difference is negative in the levels about the tropopause (100 mb), and furthermore that on average the air in the stratosphere is warmer in the solar maxima. At the station nearest the axis of highest correlation, Charleston, the average temperature difference increases with height above 700 mb to 1.6°C just below the tropopause; at Truk the largest average difference is 0.8°C. The association between the 11-year solar cycle and the global temperature in the upper troposphere and lower stratosphere was studied by Sellers and Liu (1988). In agreement with Fig. 2, they found that the low frequency variation of the first EOF in the temperature responds "primarily and strongly" to decadal variations in the 10.7cm solar flux.

The association between tropospheric temperature and stratospheric height is illustrated in Fig. 3 by the time series of 30mb and 100mb height and of the 200mb temperature at Charleston, South Carolina: a decadal oscillation, in-phase with the 11-year solar cycle, is evident in the heights, and a corresponding oscillation in the 200mb temperature has existed in the temperature of the upper troposphere since data became available in 1951.

The pattern of correlation in Fig. 1 is common to all months of the year, and it is therefore strongly reflected in the correlation between the 35-year record of the annual mean values of the 30mb height and the 10.7cm solar flux. The strongest association between the two quantities is along 30N, especially over the western hemisphere.

The shape of the two-monthly mean vertical profiles of the temperature difference between the extremes of the solar cycle at Hawaii is the same throughout the year in Fig. 2c: the summer values differ little from the annual mean, the largest deviatons are in winter and spring, and the smallest in autumn. There is not much difference between solar maximum and minimum near the tropopause, and in the stratosphere the sign of the difference is not consistent, so it is obvious that the large positive correlation

a)

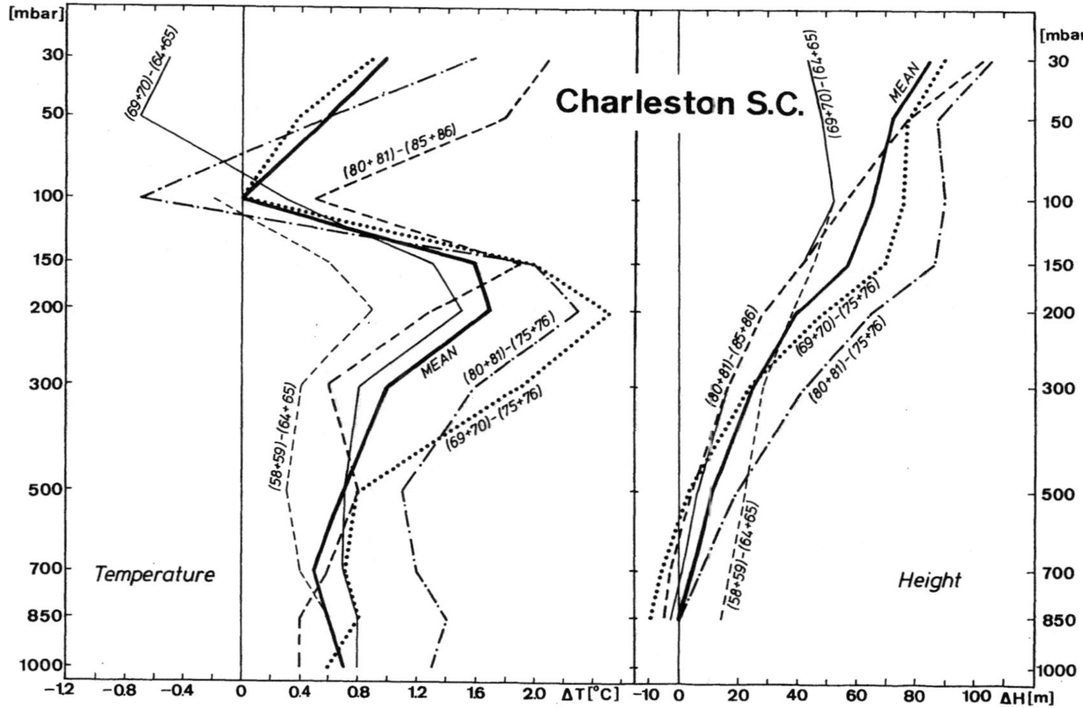

Figure 2. (a) and (b) show the change of temperature and geopotential height in July-August from minimum to maximum in the 11-year solar cycle at Charleston, South Carolina, and Truk (7.5°N, 152°E); the heavy line is the average (from Labitzke and van Loon (1992), (c) the average temperature change from minimum to maximum in the solar cycle at Lihue (22°N, 159°W) through the year and for the annual mean (heavy solid line).

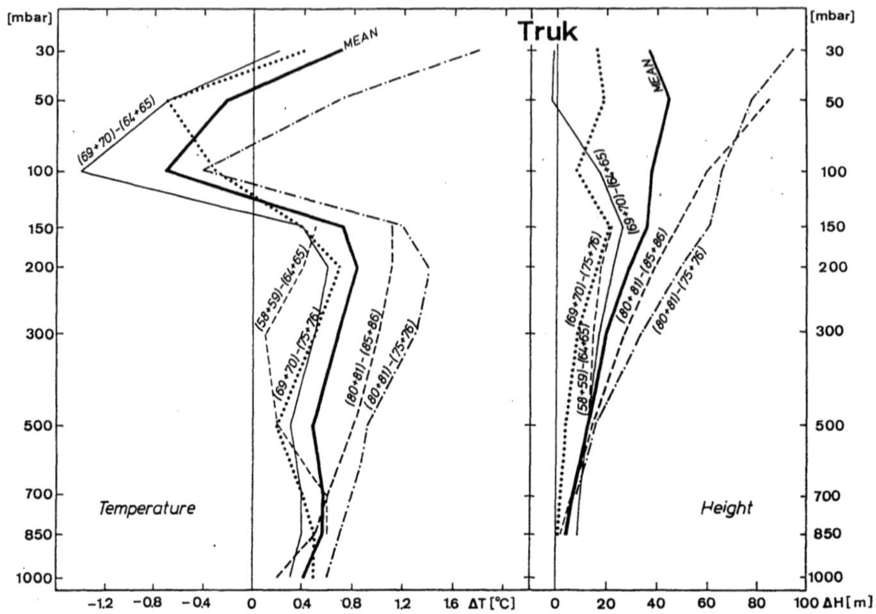
Truk
[mbar]
30
50
100
150
200
300
500
700
850
1000
MEAN
(69+70)-(64+65)
(58+59)-(64+65)
(69+70)-(75+76)
(80+81)-(85+86)
(80+81)-(75+76)
Temperature
Height
-1.2 -0.8 -0.4 0 0.4 0.8 1.2 1.6 ΔT [°C]
0 20 40 60 80 100 ΔH [m]

2b)

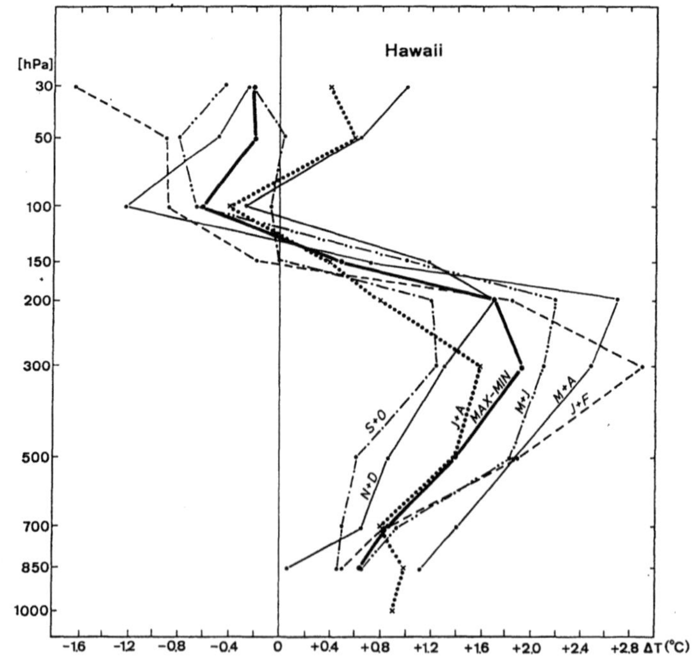
Hawaii
[hPa]
30
50
100
150
200
300
500
700
850
1000
S+O
N+D
J+A
MAX-MIN
M+J
M+A
J+F
-1.6 -1.2 -0.8 -0.4 0 +0.4 +0.8 +1.2 +1.6 +2.0 +2.4 +2.8 ΔT (°C)

2c)

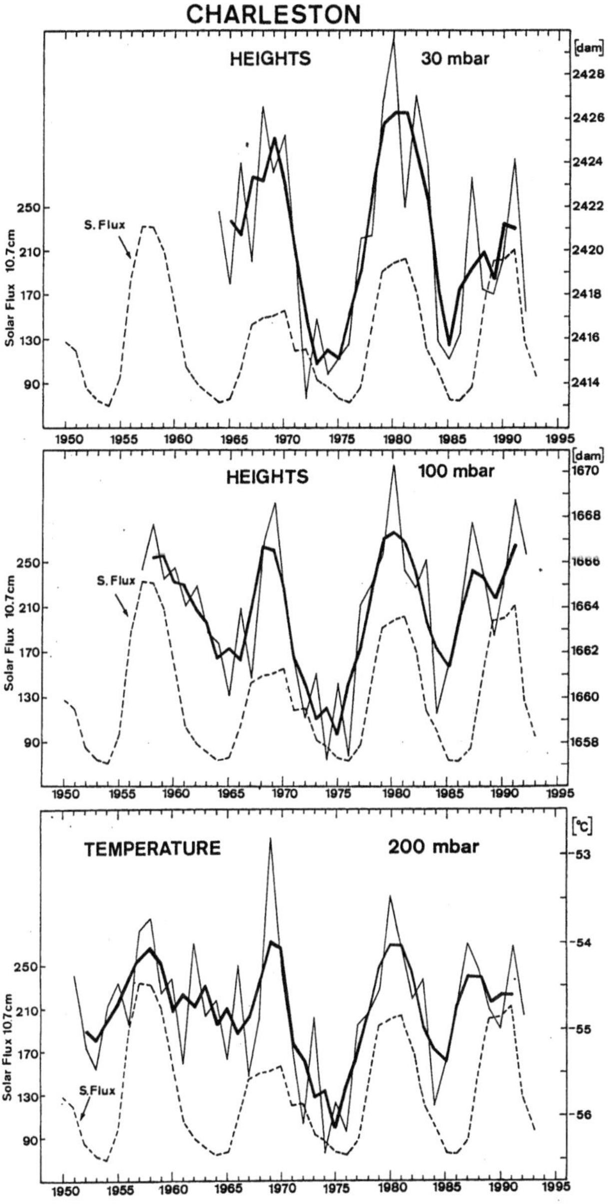

Figure 3. Time series of 30mb and 100mb height and of 200mb temperature at Charleston, South Carolina. The heavy line is the three-year running mean, updated from van Loon and Labitzke, 1990.

between the annual-mean 30mb height and the 11-year sunspot cycle in the tropical-subtropical North Pacific Ocean in Fig. 4 is principally due to decadal variations in the temperature of the middle and upper troposphere.

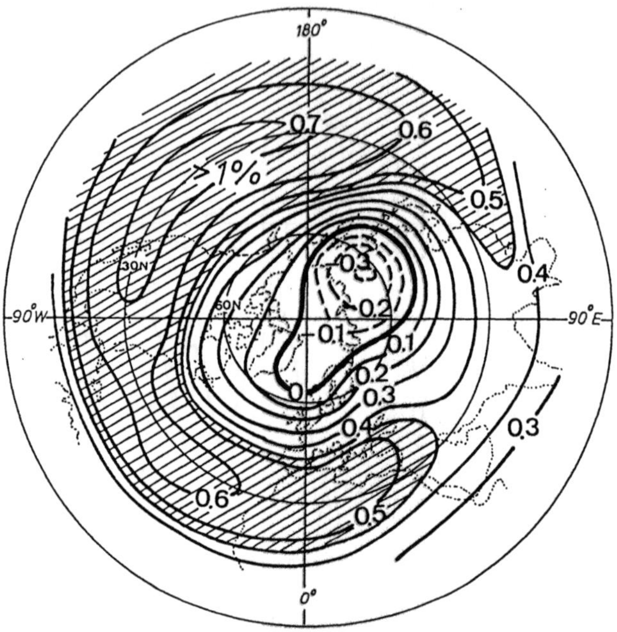

Figure 4. Correlations between the annual mean 30mb height, n = 35, and the 10.7cm solar flux. In the hatched areas the local significance exceeds the 1% level; the 0.1% local significance level = 0.57. Updated from van Loon and Labitzke, 1990.

A positive temperature difference between solar extremes in the troposphere at Truk (Fig. 2b) is also present through the year; but here the tropopause temperature is markedly lower at solar maximum than in the minimum, and the stratosphere is warmer in the maxima. The shape of the vertical profiles in Fig. 2, and the positions of the stations in the vicinity of the Intertropical Convergence Zone (Truk) and in the outer-tropics/subtropics (Hawaii, Charleston) suggest that the temperature difference between the extremes of the sunspot cycle is associated with a difference in the strength of the Hadley circulation. If that is so, the variations of the Hadley circulation must contain a decadal component which at its peaks would produce a warmer troposphere at Truk than in its valleys, presumably through greater latent-heat release in the ITCZ; and produce a warmer troposphere to the north in the descending branch of the Hadley circulation through stronger subsidence at the peaks of the cycle. The lower *tropopause* temperature at Truk in the decadal peaks than in the valleys (negative difference in Fig 2b) could then be ascribed to stronger adiabatic cooling of the tropopause levels when lifted in the more vigorous Hadley circulation.

The idea of the Hadley circulation being stronger at solar maxima than minima is tested in a preliminary way below. For this purpose we have chosen a period with intense tropical convection in the Pacific Ocean: January-February 1983, which was the mature stage of an extraordinary warm event in the Southern Oscillation, and January-February 1985, when the low-latitude convection in the tropical Pacific Ocean was weaker. In a warm event the outer-tropical North Pacific is prone to drought in winter because of stronger than normal subsidence on the poleward side of the Hadley circulation (Ropelewski and Halpern, 1987, their Figs. 2 and 3).

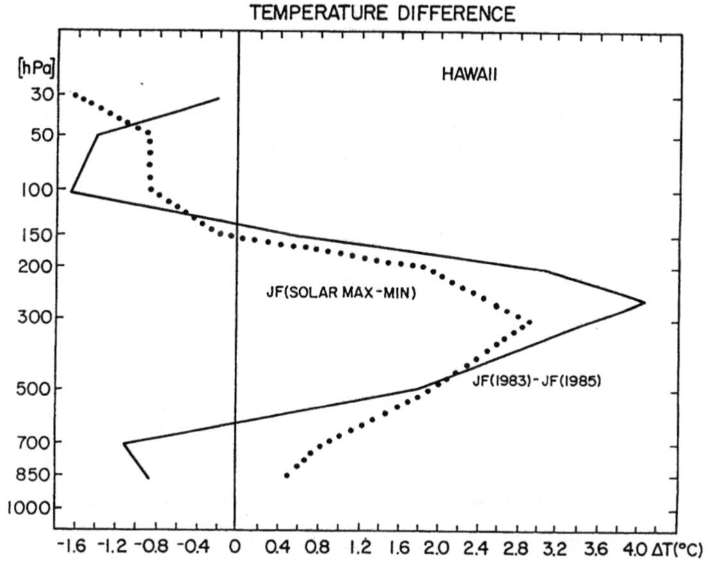

Figure 5. The temperature difference for January-February, 1983 minus 1985, for an area about Hawaii, and the temperature difference between the extremes of the 11-year solar cycle at Lihue, Hawaii.

Figure 5 shows the vertical profiles of the temperature difference between January-February 1983 and 1985 from a mean of four grid points about the Hawaiian Islands, and the average difference between sunspot maximum and minimum in the temperature at Lihue in January-February. The two profiles have the same shape, and we therefore propose, as a working hypothesis, that the mechanism behind the influence of the 11-year sunspot cycle on the upper troposphere/lower stratosphere includes a variation of the Hadley circulation on a decadal scale. This hypothesis will be the basis of further diagnosis and numerical experiment.

MIDWINTER

The basic shape of the correlation pattern between 30-mb height and the 11-year solar cycle, a crescent of high correlations along 30°N on the water-dominated side of the hemisphere (Figs. 1 and 4), is also present in January-February (Fig. 6a) but the correlations are lower than in the ten other months of the year, reaching the 5% significance level only between 20°N and 45°N over the Pacific Ocean. It was found, however, that the stratosphere responds differently to solar changes in either extreme of the Quasi-Biennial Oscillation (QBO) in the wind of the equatorial stratosphere (Labitzke, 1987). In the east years (Fig. 6b) the correlation pattern is the same as in the ten other months of the year, and over more than half of the map the correlations are locally significant above the 5% level. If the solar effect expresses itself mainly where the crescent of high correlations lies, then - owing to the marked teleconnections in winter (Shea et al., 1992) - the changes at high latitudes must be opposite to those in the tropics and subtropics and thus account for the large negative correlations in the Arctic (Fig. 6b). In the east years of the QBO, the polar vortex tends, according to the figure, to be stronger at high and weaker at low solar activity, whereas theory (Holton and Tan, 1980, 1982) and model experiment (e.g., Dameris and Ebel, 1990) indicate that the vortex should generally be weak in east years.

Conditions are different in the west years of the QBO (Fig. 6c): in these years there are no statistically significant correlations over most of the map, the only noteworthy ones (positive) being over the arctic, and there is no sign of the strong teleconnections between higher and lower latitudes which otherwise dominate the winter climate of the stratosphere (Shea et al., 1992).

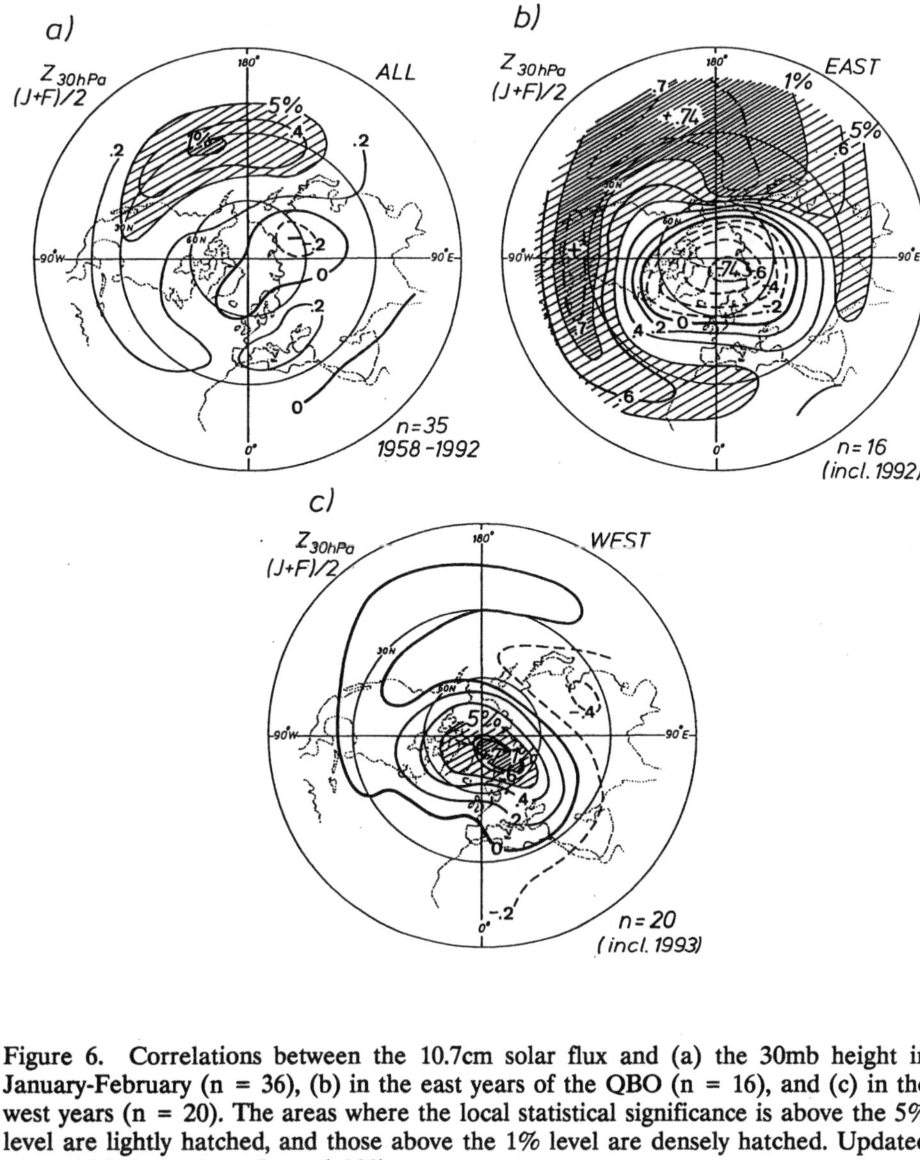

Figure 6. Correlations between the 10.7cm solar flux and (a) the 30mb height in January-February (n = 36), (b) in the east years of the QBO (n = 16), and (c) in the west years (n = 20). The areas where the local statistical significance is above the 5% level are lightly hatched, and those above the 1% level are densely hatched. Updated from Labitzke and van Loon (1988).

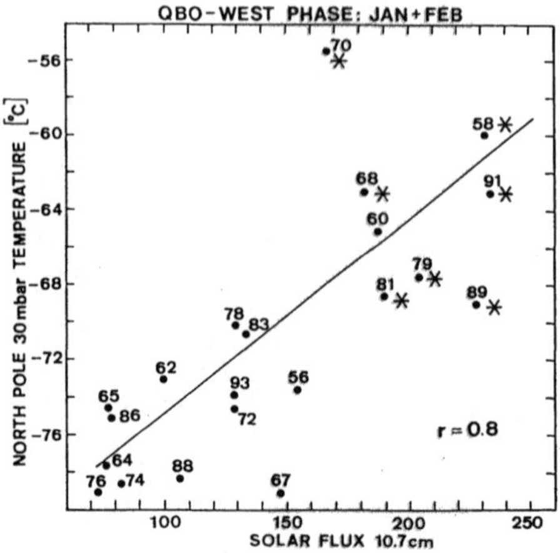

Figure 7. 30mb temperature at the North Pole in January-February plotted against the 10.7cm solar flux in west years of the QBO. Asterisks denote years with major midwinter warmings. Updated from van Loon and Labitzke (1990).

In the west years the arctic stratosphere is positively correlated with the solar cycle because the major midwinter warmings in these years take place only at solar maximum (Fig. 7). Our correlations suggest that high solar activity brings on a disturbed state in winter: at solar minimum the vortex is intense in the west and weak in the east years which, as mentioned above, is predicted by theory and confirmed by experiment; at solar maximum the vortex weakens in the west and strengthens in the east years, contrary to theory.

CONCLUSION

The correlations of Labitzke and van Loon (1988, 1992) and van Loon and Labitzke (1990) have remained stable and increasingly statistically significant with the addition of 5-7 years of data. The basic pattern of our correlations with the sunspot cycle is shaped like a crescent with the largest correlations between 20°N and 45°N in the Pacific-Atlantic area, and it is present in every month. The correlations are statistically significant through most of the year, but the significance is especially high in summer and for the annual average. The statistically insignificant correlations in January-

February change when the data are grouped according to the wind near 50 mb above the equator (the QBO). The east-wind years then show a statistically significant pattern, the same as that in other months, whereas the west-wind years have no significant correlations outside the Arctic; there they are positive in contrast to the east years.

The temperature difference at tropical and subtropical locations between the extremes of the 11-year solar cycle suggests that the Hadley circulation on an average is stronger at solar maximum than minimum.

ACKNOWLEDGEMENT.

We thank the members of the Stratospheric Research Group, Freie Universität Berlin, for technical and professional assistance. Part of the research was funded by the Bundesministerium für Forschung und Technologie, Projekt 07KF2012.

REFERENCES

Dameris, M., and A. Ebel, 1990: The quasi-biennial oscillation and major stratospheric warmings: A three-dimensional model study, Ann. Geophys, **8**, 79-86.

Godson, W. L., 1991: The recognition and assessment of decadal-scale variation of the effects of solar variability on climate. J. Geomagn. Geoelectr., 43, 785-793.

Holton, J. R., and H.-Ch. Tan, 1980: The influence of the equatorial QBO in the global circulation at 50mb. J. Atmos. Sci., 37, 2200-2208.

Holton, J. R., and H.-Ch. Tan, 1982: The quasi-biennial oscillation in the northern hemisphere lower stratosphere. J. Met. Soc. Japan, 60,140-148.

Labitzke, K., 1987: Sunspots, the QBO, and the stratospheric temperature in the North Polar region. Geophys. Res. Lett., 14, 535-537.

Labitzke, K., and H. van Loon, 1988 : Association between the 11-year solar cycle, the QBO and the atmosphere. Part I: The troposphere and stratosphere in the Northern Hemisphere in winter. J. Atmos. Terr. Phys., 50, 197-206.

Labitzke, K., and H. van Loon, 1992: Associations between the 11-year solar cycle and the atmosphere. Part V: Summer. J. Clim., 5, 240-251.

Livezey, R. E., and W. Y. Chen, 1983: Statistical significance and its determination by Monte Carlo techniques, Mon. Weath. Rev., 111, 46-58.

Ropelewski, C.F. and M.S. Halpern, 1987: Global and regional scale precipitation patterns associated with the El Nino/Southern Oscillation, Mon. Weath. Rev., **115**, 1606-1626.

Sellers, W. D., and W. Liu, 1988: Temperature patterns and trends in the upper troposphere and lower stratosphere. J. Clim., 1, 573-581.

Shea, D.J., H. van Loon, and K. Labitzke, 1992: Point correlations of geopotential height and temperature at 30mb and between 500mb and 30mb. NCAR/TN-368.

van Loon, H., and K. Labitzke, 1990: Association between the 11-year solar cycle, the QBO and the atmosphere. Part IV: The Stratosphere, not grouped by the phase of the QBO. J. Clim., 3, 827-837.

Viewpoints concerning solar activity effects on the atmospheric circulation and climate

C.J.E. Schuurmans
KNMI
P.O. Box 201
3730 AE DE BILT

Summary

Eleven year and other long-term associations between solar activity and terrestrial climate are commonly ascribed to gradual changes in solar luminosity related to·the activity of the sun. In this contribution it is suggested that such long-term associations also might be the result of an accumulation of atmospheric responses to individual solar events. Statistical evidence is presented that solar flare and proton events are followed by or associated with certain characteristic changes in the lower earth's atmosphere. Confirmation of the atmospheric effects is obtained for independent key day samples. The causing agent is as yet not identified. It might be enhanced UV, energetic protons or a temporal increase of total solar irradiance. From a physical point of view the latter might be the most realistic option.

Introduction

This short contribution is intended to draw attention to a possible solar influence on terrestrial climate which is not often considered. The most common approach is to think in terms of a long-term statistical relation between sunspot number and some atmospheric quantity, being the result of a long-term, cyclic, or at least gradual change of solar luminosity (this meeting).
A second point which I wish to emphasize if that changes in local or regional climate parameters are the result of changes

NATO ASI Series, Vol. I 25
The Solar Engine and Its Influence
on Terrestrial Atmosphere and Climate
Edited by Elizabeth Nesme-Ribes

in the frequency of occurrence of atmospheric circulation types.

This means that in the commonly used approach we try to understand, e.g. by climate model simulations, how small changes of the solar constant (of order 1 W/m^2) may systematically change the circulation regime, gradually or abruptly from one state into another.

In the alternative which I like to present solar activity is characterized by individual short term events. These events may force or favour some specific type of circulation and the cumulative effect of these events will lead to the observed long-term changes. The question is: how realistic is this second possibility?

Solar events

Nearly 25 years ago Schuurmans and Oort (1969) studied the response of the atmospheric circulation to solar flares. In their statistical study they found that systematic changes occurred of the height of the 500 mb level. The average reponse pattern shows positive changes in the latitude belt 40-70°N, with maxima over Europe, Eastern Asia, Gulf of Alaska and the New Foundland-Greenland area. Negative changes were found equatorward of 40°N and over the polar region. In general the pattern at mid-latitude shows an enhanced tendency to more meridional or blocking circulations. The solar flare sample used consisted of 81 carefully selected flares, which occurred during July 1957 through 1959, the socalled IGY-period and its extension in 1959.

For one of the areas of strong positive height changes a more detailed analysis was made, on the basis of reliable radiosonde observations. Aerological station Ship B (located at 56°30'N, 51°oo'W) was chosen for this purpose. It was found that the positive height changes during the occurrence of flares increased in the troposphere up to 300 mb and then gradually declined into the lower stratosphere. These height changes are hydro-statically related to temperature changes, which on the basis of the same radiosonde measurements turned out to be positive in the troposphere and strongly negative at or just above the tropopause level. Fig. 1, taken from Schuurmans and Oort (1969) shows the corresponding profile. For comparison the profile of the standard error of the mean of 81 random key day changes is also shown.

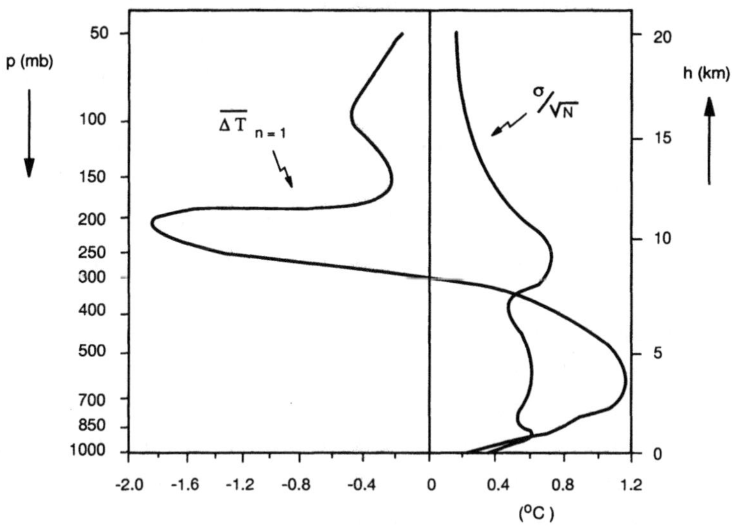

Fig. 1. Average temperature change in 24 hours at Ship B after the occurrence of a solar flare, compared to the standard error of the mean. The number of flare cases is N = 81.

The authors concluded that the large cooling at tropopause level is very unlikely due to pure chance. Unfortunately a convincing or even speculative explanation of a solar flare-induced cooling could not be offered. An attempt was made in Schuurmans (1969), where among a further analysis of the atmospheric response to the above mentioned IGY-flare sample, the response to an individual strong flare on June 1, 1960 (not belonging to the sample of 81 cases) was investigated. This flare case was discussed in Hartman (1963) as being the possible cause of an abnormal tropospheric warming of the Northern Hemisphere. The vertical profile of temperature change after the June 1, 1960 flare at the aerological station

Ship B, turned out to be exactly equal to the one in Fig. 1. The magnitude of the tropospheric warming for this individual case was 9°C at 500 mb, while the cooling at 200 mb amounted to -12 °C.

QBO-relation

The subject was taken up again after the publications of van Loon and Labizke on the associations between the 11-year solar cycle, the QBO and the atmosphere (van Loon and Labitzke, 1988) and Labitzke and van Loon, 1988, 1989).
Again a sample of solar events was used, this time consisting of 72 socalled proton events selected by Tinsley (1989). These events are related to solar flares. They occurred over the period 1955-1984 and are to a very large extent independent of the earlier used IGY-flare sample.
Aerological observations of Ship B however could no longer be used since this ocean weather ship had stopped observations in the course of the period considered. Instead data of aerological station De Bilt (52°06'N, 5°11'E) were used. This station has a homogeneous series of observations over the period since 1955 and is like Ship B located in an area where the original analysis of Schuurmans and Oort (1969) found positive height changes at 500 mb after solar flares. In Fig. 2 the mean temperature changes at 200 and 500 mb are shown, as a result of a superposed epoch analysis from 10 days before to 21 days after the solare proton events.
It may be concluded that the result confirms the earlier found warming of the troposhere and cooling of the lower stratosphere. Fig. 2 is taken from Schuurmans (1991).

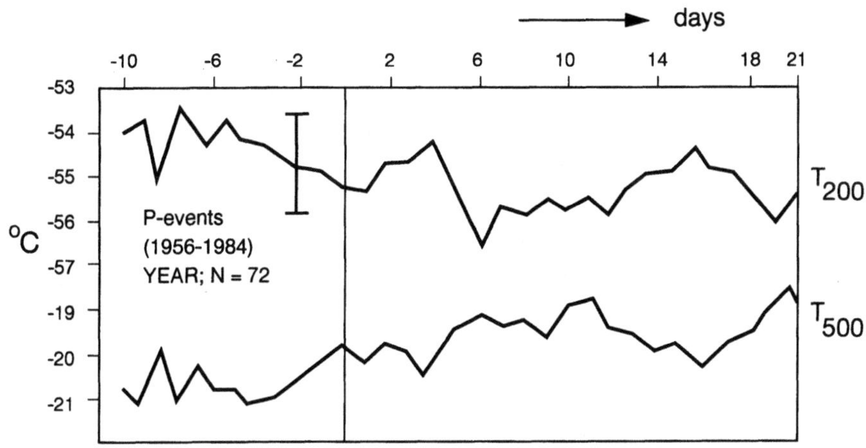

Fig. 2. Average temperature at 200 and 500 mb at De Bilt, from 10 days before till 21 days after a solar proton event. The error bar gives the standard error of the mean. Sample size N = 72.

From the same publication table I is reproduced. It shows that the cooling at 200 mb is most pronounced in relation to solar events occurring during the East phase of the QBO. The subsamples however are relatively small, which means that the QBO-East preference could be just accidental.

Conclusions

How realistic is the option that the cumulative effect of atmospheric reponses to individual solar events is the cause of the statistical relations between long-term solar activity changes and climate?
In the above only one part of the answer has been touched upon. From the statistical studies discussed it might be concluded that apparently solar flare events are associated with systematic changes in the earth's lower atmosphere. Since

as yet we heave no explanation of these associations it is impossible to answer the above question fully. If at least we should know the real forcing mechanism, it might be possible that by model simulation experiments the net long-term response to the intermittent forcings could be established. At present however we have to deal with the highly hypothetical mechanism of upper atmosphere proton forcing, the less hypothetical UV-forcing and the physically and meteorologically realistic forcing of increases in total solar irradiance.

References

Hartmann, W. (1963). Einfluss einer Sonneneruption auf die flächen mitteltemperatur der Nordhalbkugel in 500 mb. Met. Rundschau, 16: 150 - 152.

Labitzke, K. and van Loon, H. (1988) Associations between the 11-year solar cycle, the QBO and the atmosphere. Part I: the troposphere and stratosphere in the northern hemisphere in winter. J. Atm. Terr. Phys. 50: 197-206.

Labitzke, K. and van Loon, H. (1989). Associations between the 11-year solar cycle, the QBO and the atmosphere. Part III: Aspects of the association J. of Clim. 2: 554-565.

Schuurmans, C.J.E. and Oort, A.H. (1969). A statistical study of pressure changes in the troposphere and lower stratosphere after strong solar flares. Pure and Appl. Geophys. 75: 233-246.

Schuurmans, C.J.E. (1969). The influence of solar flares on the tropospheric circulation. Med. en Verh. KNMI No. 92, 122 pp. Staatsuitgeverij, 's Gravenhage.

Schuurmans, C.J.E. (1991). Changes of the coupled troposphere and lower stratosphere after solar activity events. J. Geomag. Geoelectr. 43 (supplement): 767-773.

Tinsley B.A. (1989). Personal communication.

Van Loon, H. and Labitzke K. (1988). Association between the 11-year solar cycle, the QBO and the atmosphere. Part II: Surface and 700 mb in the northern hemisphere in winter. J. of Clim. 1: 905-920.

Table 1 Average temperature response at 200 mb to proton events: day +1 to +10 minus day -10 to -1.

cases	N	°C
all	72	-1.4
winter	25	-1.9
summer	47	-1.2
QBO-west	39	-0.3
QBO-east	33	-2.4
winter, QBO-east	12	-1.7
summer, QBO-east	21	-3.1

CONCERNING VARIOUS TYPES OF SOLAR FORCING AND THEIR EFFECTS.

J.C. Pecker
Collège de France
11 place Marcelin Berthelot
75005 PARIS

Variations in solar irradiance (not luminosity !) affect the whole Earth in a way that is controlled by the inclination of solar radiation with respect to the Earth's surface. But, as J. Lean has said, the spectral distribution also changes and is sensitive to the atmospheric structure, different at high and low latitudes. Moreover, the ionized particles from the Sun are really entering the terrestrial realm from the polar regions, first producing auroral and geomagnetic phenomena.

Therefore, certain effects of solar activity cycle can be detected at high latitudes only. Phenomena at lower latitudes are affected by local structures (orogenic structure, oceanic distribution and so forth). Planetary or global climatic indices might therefore not properly reflect the solar influence, in view of the low weight of the polar regions in these indices.

The precipitation rate is an example. In a pioneering paper, J. Xanthakis (1973) had shown the excellent correlation of the precipitation index (refer to his publication for the definition) in the polar regions 70°-90°. However, in less northerly regions 60°-70°, the correlation becomes an anticorrelation, and finally disappears at moderate latitudes $<60°$.

In a previous paper (Pecker, 1982), I have studied precipitation data from eight equatorial stations, and in Uppernivik (Greenland), from the original sources, and I have performed the power analysis of these data. Interestingly enough, one could find quite different typical periods from place to place, sometimes coinciding roughly with the roughly 11-year period of the solar data. But clearly, this had no special meaning, and the precipitations did not

NATO ASI Series, Vol. I 25
The Solar Engine and Its Influence
on Terrestrial Atmosphere and Climate
Edited by Elizabeth Nesme-Ribes

appear to be linked at all with solar activity, except again in the polar regions.

References :

Xanthakis, J. (1973). Solar activity and precipitations in Solar activity and related interplanetary and terrestrial phenomena, ed. J. Xanthakis, Springer-Verlag: 20-48

Pecker, J.C. (1982). Caractères météorologiques du climat, et activité solaire, in Compendium in Astronomy, eds., E.G. Mariopoulos *et al.*, Reidel, Dordrecht: 151-160

Panel Discussion on the SOLAR AND CLIMATIC RELATIONS

E. Friis-Christensen
Solar-Terrestrial Physics Division
Danish Meteorological Institute
Lyngbyvej 100
DK-2100 Copenhagen Ø
DENMARK

Introduction

On a long time scale, over thousands of years, changes in the orbit of the Earth around the Sun have been associated with drastic changes in the Earth's climate. The reason for these changes is thought to be changes in the received solar energy and in particular its distribution on the Earth's surface. Although small, these changes seem sufficient to have initiated large oscillations in the Earth's climate in a manner still poorly understood. Similarly, variations in the activity of the Sun itself may have an influence on the climate of the Earth, and a number of correlations have been presented that indicate the possibility of a direct link. However, no convincing physical mechanism has yet been presented that explains the observations.

Due to the lack of a plausible physical mechanism the increasing number of correlations between solar activity changes and climate parameters of different kinds have been accompanied by significant skepticism in the scientific community. This is also the reason why a consensus in this field, both regarding the significance of the published correlations and the proposed physical mechanisms, has been impossible to reach. It is obvious that correlations of this

NATO ASI Series, Vol. I 25
The Solar Engine and Its Influence
on Terrestrial Atmosphere and Climate
Edited by Elizabeth Nesme-Ribes

kind can never prove a physical connection between solar activity variations and climate. But the correlations, in particular when refined to reveal spatial differences, may provide clues to the understanding of the physical mechanisms.

One of the aims of the workshop is to evaluate the climatic effects of possible solar forcings. Two principal different approaches may be used for this purpose. The first is to examine past climate changes and compare them with known or inferred variations in solar output. The second is to use climate models to calculate the climatic effects of given (and realistic) solar radiative forcings.

Solar activity changes are known to take place with different periodicities. The most conspicuous periodicity is the so-called 11-year sunspot cycle. In fact this cycle is not exactly 11 years long, neither is it constant. The period is varying between approximately 8 and 14 years. A longer, well-known period is the Gleissberg period which is 80 to 90 years long.

Decadal time scale variations

As the clearest cycle of solar activity is in the 11-year variation it is with this periodicity we would expect the largest corresponding solar induced climatic variation. However, there is no reason why the climatic response should be the same everywhere on the globe. Labitzke and van Loon (1993) have reviewed recent investigations of possible effects of solar variations on the Earth's temperature and on the height of constant pressure levels in the stratosphere. They emphasize the fact that the correlations between solar activity and climate parameters for the stratosphere on a decadal scale have a specific spatial distribution over the globe with highest values between 20°N and 45°N in the Pacific - Atlantic area during most of the year but especially high and statistically significant during summer. Only during winter (January and February) the correlations are statistically insignificant but they improve considerably when the data are grouped according to the wind near 50 mb above the equator (the QBO). The east-wind years show a highly significant pattern corresponding to the other months whereas the west-wind years show no correlations outside the Arctic.

These results illustrate that a possible effect of the varying solar activity does not appear as a purely radiative effect. The dynamics of the general circulation pattern play a major role in the spatial distribution of the atmosphere's response to solar variability.

Another specific property regarding the response of the climate to the 11-year solar activity cycle is the vertical distribution in the atmosphere. At a station close to the peak of the correlation between the 30 mb height and the solar flux in the 10.7 cm wavelength region, the average temperature difference between solar maximum years and solar minimum years is largest just below the tropopause.

Long-term variations

On a longer time scale corresponding to the Gleissberg period, the global temperature, particularly the Northern Hemisphere land air temperature, has been found to be nearly perfectly correlated with the long-term variation of solar activity during the entire interval of systematic temperature measurements from 1860 to 1990 (Friis-Christensen and Lassen, 1991). They also demonstrated that the length of the solar cycle is superior to the sunspot number as an index of long-term variability of the Sun, and they concluded that this parameter appears to be a possible indicator of long-term changes in the total energy output of the Sun.

A subsequent comparison with the temperature series reconstructed from proxy data prior to 1860 by Groveman and Landsberg (1979) revealed that there is a good association between the variations in this temperature series and in the solar cycle length record since the beginning of regular solar observations in the middle of the eighteenth century (Friis-Christensen and Lassen, 1992). This independent result for a separate time interval indicates that the association between solar cycle length and hemispheric average temperature found for the last 130 years has probably been present for at least 240 years.

Physical mechanisms

The simplest physical mechanism for the found correlations is the possibility of variations in the total solar irradiance that are sufficiently large to have an observed temperature effect on climate. Satellite measurements over approximately one solar cycle have shown that the irradiance is not constant, but model calculations show that it varies too little (less than 0.1%) during a solar cycle to be of major importance for climate. However, no measurements yet exist that do exclude the possibility of larger variations in total irradiance over a longer period of time. On the contrary, Lean *et al.*, (1992) from correlations between total solar irradiance measurements and Ca II solar emissions estimate that in the absence of surface solar magnetism (as was probably the case during the Maunder sunspot minimum period around 1700) the total solar irradiance may have been reduced by about a quarter of a per cent.

Both the vertical distribution and the spatial distribution of the atmospheric response to solar activity variations may constitute important clues to the physical mechanism by which solar activity variations could influence the atmosphere. Regarding the 11-year solar cycle it is known that the UV variation has a larger amplitude than the total solar irradiance. However, it is not possible based on the correlations alone to uniquely determine whether the possible effect is through the troposphere or through warming of the stratosphere due to absorption of UV radiation by ozone. The latter would change the vertical distribution of temperature and thereby the height and temperature of the tropopause. Labitzke and van Loon (1993), however, regard it more plausible that during high solar activity the ocean surface at low latitudes warm more during solar maximum than during solar minimum. This would intensify the Hadley circulation during solar maxima thereby creating warming in the subtropics by increased subsidence which would effect both the tropopause and the lower stratosphere.

Schlesinger objected to this hypothesis and argued that the small observed change in total solar radiation of approximately 0.1% during the last solar cycle could only have a minor effect on the ocean temperature. He further argued that an effect on the ocean temperature, if real, should be observable in a corresponding distribution of the measured sea surface temperatures. This had apparently not been reported. Labitzke argued that the relative warming of the tropical water in solar maximum years need not be large in order to create changes in the global circulation. The drastic change in the global atmospheric circulation in the years after 1977 was accompanied by an average change in the sea surface temperature of only 0.3°C which is not far from observational accuracy. Another problem with this hypothesis, according to Schlesinger, is the thermal inertia of the ocean which would considerably reduce any decadal solar forcing effect through the warming of the ocean whereas solar forcing with longer time scales might have a measurable effect.

Schlesinger emphasized the difference in the apparent response to the solar variations of the stratosphere and the troposphere and found it important to distinguish between a "top-down" (stratosphere-to-troposphere) response versus a "bottom-up" (troposphere-to-stratosphere) response to putative solar-irradiance variations. He recommended that we test for these two possible response structures using general circulation model (GCM) experiments.

The hypothesis of an increased Hadley circulation in response to an increased total solar irradiance may also be tested by means of GCMs. An experiment with this objective was performed by Sadourny (1994), this volume. In evaluating the result of this experiment particular emphasis was placed on the response of the Hadley circulation. The result of the experiment did not show an increase of the Hadley circulation as a response to increased solar luminosity. On the contrary, the GCM experiment showed a small weakening accompanied by a slight meridional extension. The present capability of GCM's may, however, not yet be sufficient to determine such issues definitively.

There is still a major uncertainty regarding the role of some of the important processes, including the treatment of water vapour feedback, effects of clouds, and realistic atmosphere-ocean coupling processes. One of the crucial results of these models is the climate sensitivity, which is usually expressed in terms of the equilibrium response of the global temperature corresponding to a radiative forcing equivalent to that for a doubling of the CO_2 concentration in the atmosphere. The range of climate sensitivity that results from different models is still relatively large, varying by a factor of three from 1.5 to 4.5 °C for a doubled CO_2.

A different modelling approach is to use more simple energy-balance climate models. In these models a specific global or hemispheric climate sensitivity has to be assumed. Such models were used in the investigations by Kelly and Wigley (1992) and Schlesinger and Ramankutty (1992) who explore the relative role of possible solar forcing versus greenhouse forcing. Schlesinger and Ramankutty find strong circumstantial evidence that there have been intercycle variations in solar irradiance which have contributed to the observed temperature changes since 1856, but their model calculations indicate, also in agreement with Kelly and Wigley, that since the nineteenth century greenhouse gases have been the dominant contributor to the observed temperature changes. An important result regards the best estimate of the climate sensitivity, which is reduced by approximately 50% when the Sun's past influence is included.

This result demonstrates the importance of establishing a firm basis for the possible solar forcing.

The merits of the energy-balance models are that they are simple and inexpensive in terms of computer resources. Friis-Christensen questioned the relevance of these models because they assume that the global climate can be calculated simply by means of global averages of radiative forcings, some of which are assumed to be relatively well described. As a first-order approximation these models may constitute an appropriate tool but caution must be exercised in the interpretation of the results. It must be realized that average global values may not constitute an adequate description of the very complex atmospheric processes that determine the average global temperature. Even if energy-balance models only intend to consider global-mean response to different forcings, a global-mean climate sensitivity is not necessarily appropriate, taking into account the multitude of non-linear effects that is characteristic of the Earth's climate. The different radiative forcings, including solar irradiance, greenhouse gases, and anthropogenic sulphate aerosols, all have different spatial distributions and, in case of the latter, a particularly inhomogeneous distribution that makes it rather dubious if a one-dimensional climate model is at all appropriate in a quantitative distinction of the effects of these different forcings.

Schlesinger objected to this criticism and referred to the results of the existing GCM experiments, which indicate a similar climate sensitivity, regardless of the nature of the radiative forcing, be it solar luminosity or long-wave radiative forcing.

A reason for this is that in the climate models both forcings act through a warming of the surface and that the major effect on the temperature is caused by the feed-back mechanisms primarily due to water vapour and clouds. The feed-back mechanisms are assumed to be independent on the nature of the primary warming.

Schlesinger also drew attention to the fact that the observed global temperature record contains a 65-79 year oscillation (Schlesinger and Ramankutty, 1994) which is dominated by the variation in the North Atlantic and North American region. From this he argued that this oscillation could not be due to solar forcing which would generate a global response. He therefore concluded that a large part of the oscillation which includes much of the warming in this century is due to an internal oscillation of the atmosphere-ocean system. Arguing against this interpretation Friis-Christensen referred to the presentation by Labitzke (this volume) who showed that the apparent temperature response to solar variations is characterized by being very inhomogeneous over the globe, because the changes in the temperature are brought about by changes in the general circulation.

Schuurmans emphasized that the hypothesis of variations in total solar irradiance as cause of global temperature changes leads to inconsistent or unrealistic results, provided the climate models are basically correct. If the variability of the solar constant is of the order 0.1%, as in fact observed during the last solar cycle, the climate sensitivity has to be very large. This would necessarily lead to a considerable, and not observed, effect of the monotonicly rising greenhouse gas concentrations. On the other hand, if a reasonable climate sensitivity is assumed, the variation of the solar constant over long-term (century) time scales should be significantly larger than actually observed to be compatible with the found correlation. Therefore he argued that the apparent solar effect on climate might not only act through a gradual change in the total solar radiation but also through other types of solar activity variations. A possible cumulative effect of a number of individual solar flare effects could be imagined based on reported analyses of troposphere and lower stratosphere effects of individual solar proton events (Schuurmans, 1991). Schuurmans therefore argued that part of both the 11-year cycle and the long-term variation in climate could be due to such effects thereby reducing the climate sensitivity to radiative forcing.

References

Friis-Christensen, E., and Lassen, K. (1991). Length of the solar cycle : An indicator of solar activity closely associated with climate. Science, 254: 698-700

Friis-Christensen, E., and Lassen, K. (1992). Global temperature variations and a possible association with solar activity variations, in Scientific Report 92-3, Danish Meteorological Institute

Groveman, B.S., and Landsberg, H.E. (1979). Simulated Northern Hemisphere temperature departures 1579-1880. Geophys. Res. Lett., 6: 767-769

Kelly, P.M., and Wigley, T.M.L. (1992). Solar cycle length, greenhouse forcing and global climate. Nature, 360: 328-330

Lean, J., Skumanich, A., and White, O. (1992). Estimating the Sun's radiative output during the Maunder minimum. Geophys. Res. Lett., 19: 1591-1594

Labitzke, K., and van Loon, H. (1993). Some recent studies of probable connections between solar and atmospheric variability. Ann. Geophysicae, 11: 1084-1094

Sadourny, R. (1994). Sensivity of Climate to Long-term variations of the Solar output. This volume

Schlesinger, M.E., and Ramankutty, N. (1992). Implications for global warming of intercycle solar irradiance variations. Nature, 360: 330-333

Schlesinger, M.E., and Ramankutty, N. (1994). 65-70 Year oscillations in the global climate system. Nature, 367: 723-726

Schuurmans, C.J.E. (1991). Changes of coupled troposphere and lower stratosphere after solar activity events. J. Geomag. Geoelectr., 43 Suppl.: 767-773

GRASPING AT SHADOWS? A SELECTIVE REVIEW OF THE SEARCH FOR SUNSPOT-RELATED VARIABILITY IN TREE RINGS.

K.R. Briffa
Climatic Research Unit
University of East Anglia
Norwich NR4 7TJ, U.K.

INTRODUCTION

Tree-ring data, and climate timeseries reconstructed from them, possess a number of attributes which recommend them for studies of the possible influence of solar variability on climate. They are continuous. They have annual resolution and precise calendrical dating, and they frequently display a strong affinity with climate variability that can be quantified and verified. Tree-rings therefore facilitate the extension of our knowledge of natural climate variability by centuries and sometimes even millenia.

This short paper is, of necessity, a very selective review of tree-ring-related research relevant to the issue of how, or indeed whether, solar variability impacts upon climate and so affects the growth of trees. As requested, I also briefly review the evidence for a solar influence in a recently produced timeseries representing Fennoscandian summer temperatures since A.D. 500. Due to the space limitation, I have chosen to focus on the evidence for circa 11-year (solar sunspot cycle); 22-year (solar magnetic cycle); and 80-90 year (sunspot envelope modulation) effects, though brief mention is also made of the possible influence of longer-timescale solar variation.

Finding (or not finding) periodic behaviour in tree-ring or other climate-related timeseries that might correspond to solar-related variability (e.g. 11, 22 or 80-90 year cycles) obviously means very little in itself. Random climate variability will produce such behaviour some of the time. To compound the problem, we do

NATO ASI Series, Vol. I 25
The Solar Engine and Its Influence
on Terrestrial Atmosphere and Climate
Edited by Elizabeth Nesme-Ribes

not know enough about the nature of the true 'solar signal' that we would expect to find in climate series. Even if there were 'real' solar-induced variations, they could be masked by natural climate noise (i.e. non solar-forced variability). Without some *a priori* appreciation of what variables to explore; what strength of effect to expect; what regional or seasonal manifestation and what response times to consider, it would seem that we are left to indulge in what are no more than mere 'fishing' exercises. Some would consider that this is just what many analyses represent. In this review I discuss several papers which together hopefully present something of an overview of the subject to allow the general reader to judge the evidence.

THE BEGINNINGS

Investigations into possible solar-related variability in tree rings date back to the start of what can be termed the 'modern era' of the science of dendroclimatology itself. A.E. Douglass, the founder of the Laboratory of Tree-Ring Research at the University of Arizona was, in fact, an astronomer. In 1919 he published "Climatic Cycles and Tree-Growth: A study of the annual rings of trees in relation to climate and solar activity". This summarized much of his research into the development of tree-ring chronologies and their comparison with sunspot data mainly in America but also in several European countries. Douglass was responsible for the development of much of the methodological base that underlies the study of climate and tree growth today, but his motivation was entirely "*from the standpoint of astronomy and a desire to understand the variations of the sun*" (Douglass, 1919, page 9). He assumed that climate was driven by the sun and that within climate records lay information on past solar variability. He rightly considered that meteorological records were comparatively short, and so he sought to extend them using tree-ring records. Douglass considered that solar (and consequently tree-ring) cycles were likely to be transitory phenomena; making them unsuitable for study based on harmonic analysis. He coined the term 'cyclics'

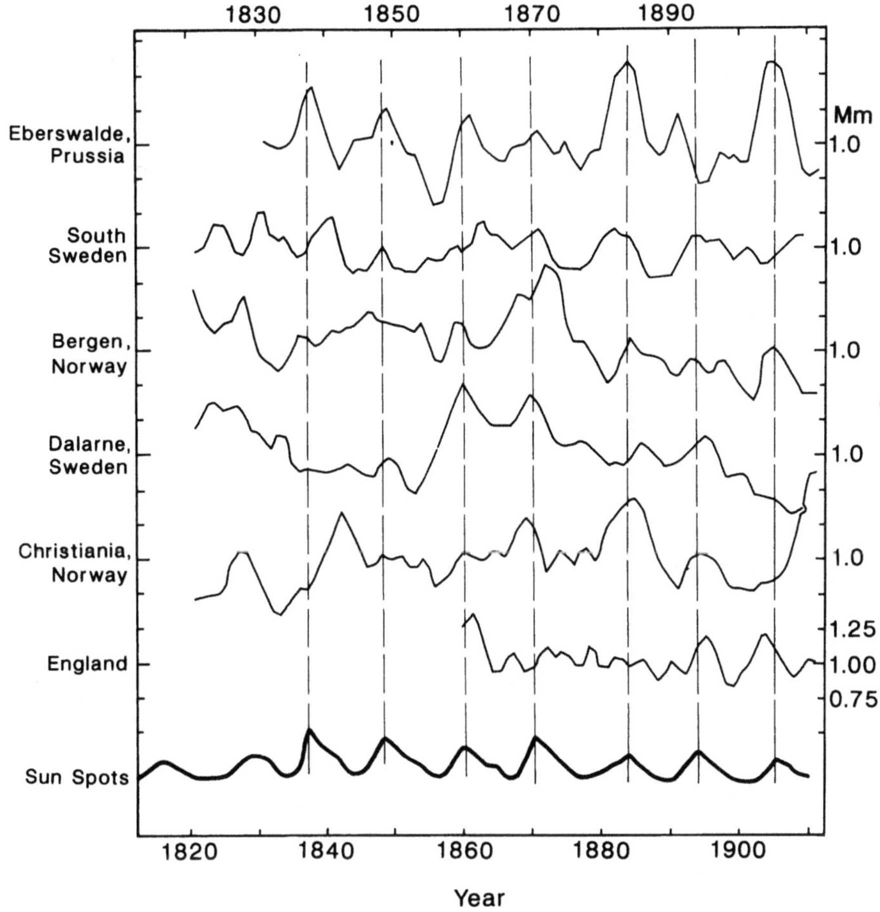

Figure 1. Smoothed ring-width chronologies from different locations compared with a record of sunspot numbers (after Douglass, 1928).

to distinguish such variations from true cycles. He even invented a machine, the "cycloscope" to analyze the phenomena in tree-ring records (see Douglass, 1943).

He appears never to have doubted the reality of solar forcing on tree-ring widths and was content to smooth (and sometimes lag by several years) his tree ring data series to illustrate this. He later summarized his conviction that "*the mass of cyclics ... showed a preference for values which were simple factions of 1, 2 or 3 times the length of the sunspot cycle, taking it as near 11.3 years*" (Douglass, 1943, page 24: see also Schulman, 1936). The reader can judge from Figure 1 (taken from Douglass, 1928) something of the extent to which his conviction was justified.

SUBSEQUENT DEVELOPMENTS

Stimulated to explore the reality, or otherwise, of Douglass' conviction that the influence of solar variability was widely evident in tree-ring records, LaMarche and Fritts (1972) described an extensive investigation into the spectral characteristics of more than 50 tree-ring chronologies originating from locations in western north America, from Canada to Mexico. They also examined the power spectra of regional tree-growth principal component amplitudes. They found no convincing evidence of any significant power corresponding to 11-year or 22-year cycles.

They also smoothed their tree-ring series to emphasise the variability at about 11 years and used harmonic dial analysis to test the association between peaks in the sunspot record and the tree-ring data. Again this yielded no real evidence of an association. They concluded that their investigations produced no convincing evidence of consistent relationships between ring width and solar variation. They then went on to qualify the statement by saying that their results did "... *not rule out the possibility of solar-related oscillations ... which undergo frequent phase shifts or reversals*". However, they further stated that they believed "*that a continued search for empirical associations between terrestrial time series ... and ... sunspot numbers is likely to prove unrewarding."*

Cook and Jacoby (1979) reconstructed drought indices in the Hudson Valley region of New York State back to 1694. The variance spectrum of the reconstructions contained "*statistically significant quasi-periodicities of 11.4 and 26 years*". However, the correspondence in the phase and amplitude between these cycles and variations in sunspot records were not (and appear not to have been subsequently) explored.

Any review of tree-rings and solar variability must draw attention to the work of Mitchell et al. (1979). This remains the most convincing evidence of a possible link yet published. Using up to 65 tree-ring sites, these authors reconstructed Palmer Drought Severity Indices, from 1700 onwards, for each of 40 regions of the west and central United states. These data were then transformed into a series of indices representing annual values of the area experiencing drought of different degrees of severity. Variance spectra of the various severity indices consistently showed peaks corresponding to a 22-year periodicity. Realistic assumptions used to calculate the significance of these peaks showed them to be unlikely to have arisen by chance. What is more, using harmonic dial analyses of the smoothed drought indices, it was shown that droughts tended to occur consistently during the two to three years following Hale minima (i.e. after every alternate 11-year cycle). They also showed that the filtering procedures they employed were not themselves causing this apparent relationship. The authors then went on to examine the longer-term relationships between the sunspot record and their drought area indices by comparing smoothed records of each (Figure 2). While this again shows some intriguing similarity between drought and sunspot activity, it has to be noted that the long-term drought record leads the sunspot record by a decade or so, as was clearly pointed out by Mitchell et al.

This work stimulated much discussion and engendered a number of critical comments, including questions on the implications of the non-normality of drought area indices and the influence of a possible 18.6 lunar forcing effect. These and other aspects of

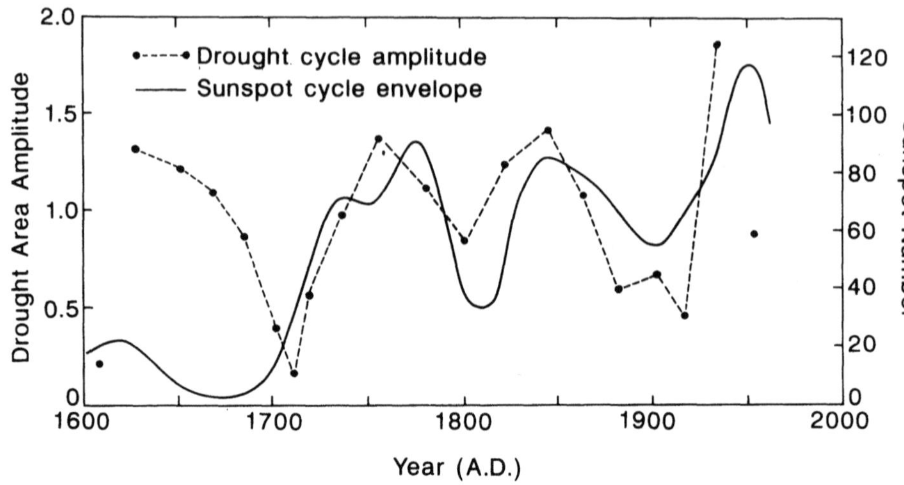

Figure 2. Changing amplitude of the 22-year drought cycle in western North America, as reconstructed from tree- ring data, compared to changes in the Hale sunspot envelope (after Mitchell et al., 1979).

the original work were specifically addressed in a later paper (Stockton et al., 1983). This concluded that the 22-year rhythm was a robust and persistent feature in the drought area record after 1600, and that there was evidence of an 18.6-year lunar component, though this was physically decoupled from the solar-related variation.

The Mitchell et al. (1979) results are even more intriguing when one considers an important later analysis of reconstructed annual regional precipitation for four separate climate divisions centred in Iowa, Oklahoma, Montana and Wyoming (Stockton and Meko, 1983). While the variance spectrum of a series representing the mean reconstructed precipitation in all regions (1750-1977) showed a (poorly resolved) concentration of variance corresponding to periods around 22 years, similar analyses of the separate regional records showed that this possible sunspot-related peak was not ubiquitous within the regions (being mostly evident in Iowa). Furthermore, when the mean regional series was analyzed over different periods the peak was found to be totally absent or weak in some periods of the record (e.g. 1714-1801). Stockton and Meko (1983) concluded that while their results indicated "... *difficulties with drought prediction at specific locations based on a 22-year recurrence period, they do not refute a possible solar drought connection* ...".

Later work based on tree-ring data from the U.S. 'corn belt' (mostly in Iowa and Illinois) indicated no periodicity in the eastern corn belt but, between 1680 and 1980, evidence of a circa 18.3-year cycle in the western part, that has apparently increased in length from near 18 to near 21 years over this period (Meko et al., 1985). The authors conclude that "*the periodic component of drought near 20 years is too weak and irregular to be of use in drought forecasting.*"

RECENT EVIDENCE

Taylor and Rose (1992) undertook a form of evolutionary spectral analysis of 12 1000-year-plus ring-width chronologies from the Colorado Plateau region of the United States. Moisture availability is the dominant growth-limiting factor affecting the variability of these series. Various 'oscillations' were identified within the range 20-to-80 years but none was consistent over most of the record and none could be considered coincident with possible solar-variability. The authors do, however, suggest that "*a noteworthy relationship exists between the times of low sunspot activity and abrupt changes in the power spectrum.*" They base this conclusion on the fact that a relatively high number of the chronologies display statistically significant oscillations (of various wavelengths) apparently only during periods of normal or high sunspot activity and these general manifestations of oscillatory behaviour seem to break down when sunspot minima occur. The oscillations themselves are not at solar-related frequencies but Taylor and Rose (1992) argue that the changes in solar activity may induce rearrangements of the climate system and hence disturb the 'normal' cyclic behaviour of the ring-width timeseries. Their results are certainly interesting and the authors are correctly cautious

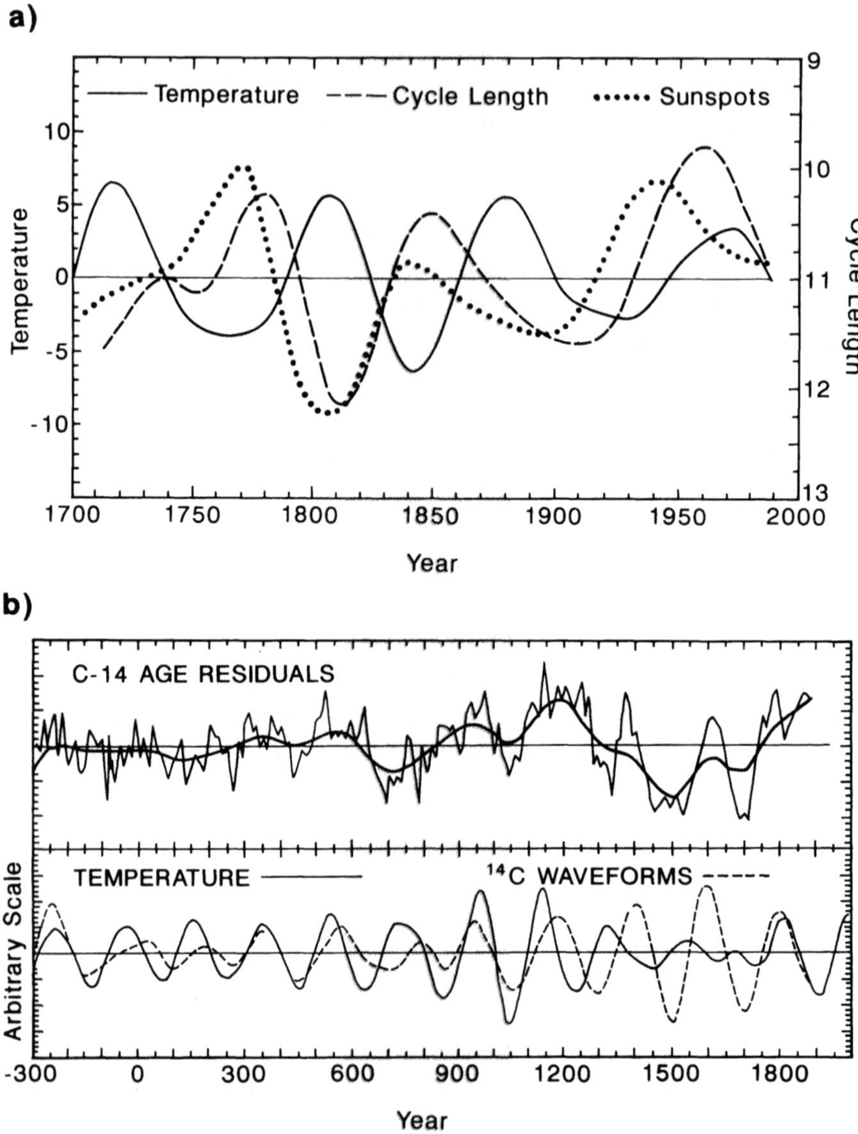

Figure 3a. The near-80-year waveform identified by Singular Spectrum Analysis of a reconstructed summer temperature series for Tasmania compared with the sunspot number envelope and a record of sunspot cycle length. **b.** Comparison of the 200-year temperature waveform with an equivalent waveform extracted from a series of ^{14}C anomaly data, after removal of low-frequency variance. The ^{14}C residual series has also been smoothed to highlight century-timescale variations (after Cook et al., 1994).

about this conclusion. It must, nevertheless, be said that any objective assessment of these results would have to consider them as equivocal.

Recently developed reconstructions of summer temperatures (Nov.-April) in Tasmania, at present spanning more than 2000 years (Cook et al.,1991,1992,1994), provide one opportunity to explore possible solar-related variability, over a long period, in the Southern Hemisphere. Cook et al. (1994), describe four principal modes of oscillatory behaviour in their data, corresponding to periods of about 31, 56, 79 and 204 years. They suggest that the latter two might be related to long-term solar variation. Figure 3a shows the circa 80-year temperature oscillation in conjunction with both the envelope of sunspot number and a series representing sunspot cycle length. On the basis of this Figure, the authors suggest a "*reasonably consistent lag relationship of 30-40 years between ... temperatures and solar-cycle length since the early 1700s*". Though, to be fair to them, they also state that this coincidence hardly represents conclusive proof.

A sceptical (or perhaps dispassionate) assessment of the association of what amounts to three oscillations in either of the cycle length or sunspot curves and four oscillations in the reconstructed temperature data might conclude with equal justification that there is little or no evidence of any association between the reconstructions and solar-related curves, above what might be expected by chance. However, a comparison of the 200-year waveform against a high-pass filtered ^{14}C series (a proxy for sunspot numbers, e.g. see Stuiver and Quay, 1980 and this volume) is perhaps more interesting (Figure 3b).

These two series appear to be in phase for much of the period between 300 BC and the present, except for a notable mismatch between AD 1300 and 1700. While this is inconclusive as regards a statistically significant association it would seem to offer more circumstantial evidence in support of one than does the 80-year waveform.

EVIDENCE OF SOLAR-VARIABILITY IN FENNOSCANDIAN TREE RINGS

Early in the history of Fennoscandanian dendrochronology several authors investigated the periodic character of pine ring-width chronologies. A short review is given in Erlandsson (1936). He himself investigated pine growth at four sites in northern Fennoscandia (Torneträsk, Kiruna and Karesuando: all relatively close to the limit of pine growth in northern Sweden; and Sodankylä in northern Finland). He was well aware of the early work of Douglass described above. Erlandsson reviewed earlier Fennoscandian work and he explored the use of alternative methods to calculate periodicities in his pine data, principally during the interval 1700-1930. He demonstrated apparent periodicities of about 2.8, 5.8, 11, 23 and 33 years. The first of these had the highest amplitude but the circa 23 periodicity (between 21 and 23 years at the different sites) was also shown to be significant. Erlandsson (1936) considered that the maximums in the ~22-year period corresponded well with alternate 11-year sunspot maxima, whereas the 11-year periodicity in his tree rings and the 11-year sunspot period comparison "*shows no great congruence*". He also suggested that the 33-year period maxima (as had been shown by Douglass, 1919) equated to every third sunspot maximum.

His harmonic analyses of these tree-ring data suggested a periodicity closer to 35 years, which he pointed out was the same length as the 'global' terrestrial climate cycle proposed by Brückner (1890) of (34.8 0.7 years; see also Clough, 1905). He then went on to review the evidence for a correspondence between this periodicity in Fennoscandian trees and evidence of warm periods that Brückner had calculated from data (including that of Rykatschew, 1887; and see Hann, 1932) on lake levels, precipitation and temperature and the formation and disappearance of ice on Russian rivers.

Erlandsson was aware that controversy surrounded the reality of Brückner's periods (a brief review is given in his paper) but he came to the conclusion that his tree-ring evidence supported the validity of periodic warm and cold periods corresponding to Brückner's maxima and minima. On the basis of his data from

Torneträsk, which extended from the mid 15th century, he even listed 7 (warm) maxima and six (cold) minima prior to those listed by Brückner (1890).

Later Ording (1941), investigating pine data from Sørfold in northwest Norway, confirmed that his data were also characterized by major wavelengths of 2.75, 8.5, 11, 23, 35 and 57 years, which "*made themselves felt ... with considerable strength and regularity*".

Sirén (1961) constructed a ring-width series for Scots pine growing in northern Finland extending from 1181. Sirén and Hari (1971) showed that this series contained significant spectral peaks corresponding to about 3.3, 3.6, 7.4, 22.8, 32 and 80-96 years. A spectrum calculated with the same data but from 1463-1960, showed major peaks at about 18, 23, 30 and 90 years.

RECENT TREE-RING RECONSTRUCTIONS IN FENNOSCANDIA

In their preliminary work reconstructing mean summer temperatures representing a large region of northern Fennoscandia, Briffa and Schweingruber, (1992) calculated a single power spectrum of a reconstructed timeseries spanning the years from 1580 to 1978, and found a number of notable peaks at around 2, 3 and 4 years but also at 33.25, 38.0 and 88.7 years. They pointed out the coincidence between peaks in these data and the Sirén and Hari (1971) Finnish tree-ring data at about 33 and 80-96 years. However, unlike the Finnish tree-rings, the reconstructed temperatures did not appear to provide evidence for a near-22-year periodicity.

Subsequent work incorporating ring-width and tree-ring density data from sites around Lake Torneträsk, resulted in the production of an equivalent 'summer' (April to August) reconstruction spanning nearly 1500 years after A.D. 500 (Briffa et al. 1990, 1992). This provided an opportunity to examine 'periodic' variability over a much longer period. Simple spectral analyses of the complete series and of separate sub-sections revealed evidence of what seem to be some reasonably time-stable quasi-periodic variability along with that of more transitory 'periodicities'.

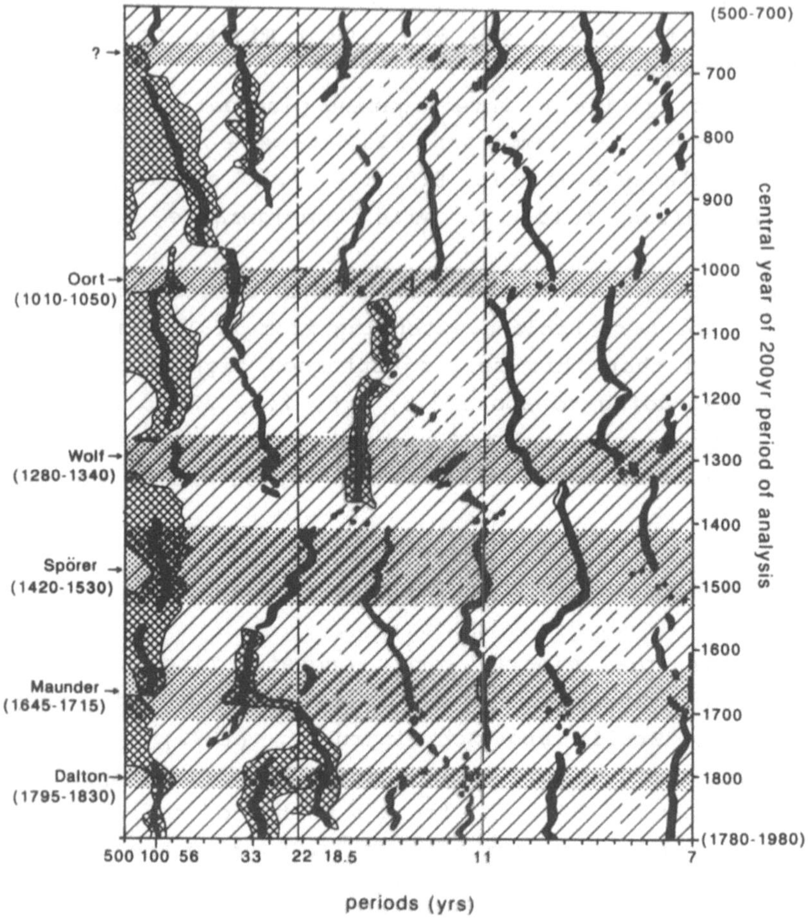

Figure 4. Summary of an evolutionary spectral analysis of reconstructed summer temperatures in Fennoscandia (Briffa et al., 1992). Maximum Entropy Method spectra are calculated for a 200-year window (with a prediction error filter length of 50 with 350 spectral estimates) successively shifted by five years (i.e. 195-year overlaps). Only the major peaks are shown, with relative variance concentration indicated according to the shading: from solid (highest), through crosshatching, single hatching to blank. Horizontal stippled bands show the periods of sunspot minima (with the question mark indicating an unnamed period). The position of the 11-year and 22-year periodicities are shown, for reference, as vertical dashed lines.

The time-stable peaks were 2.1, 2.5, 3.1, 3.6, 4.8 and ~32-33 years, as well as for a range of poorly-resolved periods between about 55 and 100 years. Again, what is notable in these analyses is the lack of any evidence for a circa 22-year periodicity. The only evidence for an 11-year cycle was restricted to the last third of the data (after 1500). A variance concentration corresponding to a time range of about 85-100 years was significant in all but the early third (i.e. not before 1000) of the data. Briffa et al. (1992) considered that this was not inconsistent with the possibility of a Gleisberg-type cycle but they chose to lay stress on the apparent significance of the ~32-year periodicity. Given the resolution of their spectra, this is not meaningfully distinguishable from the cycle proposed by Brückner (1890). However, Briffa et al. (1992) did not investigate this possibility or compare the earlier data with the maxima and minima proposed by Erlandsson (1936). Instead they concluded that these data "*offer no strong evidence for solar-related forcing of summer temperatures in these regions.*"

Figure 4 summarizes the results of an evolutionary spectral analysis of the Briffa et al. (1992) Fennoscandian summer temperature reconstruction. It is based on a 200-year sliding window with 195-year overlaps. This illustrates how the major peaks in the spectra change with time. Only frequencies between 0.002 and 0.143 (periods between 100 and 7 years) years are indicated. For reference, vertical dashed lines show the positions of 11-year and 22-year periods.

What is striking in this Figure is the concentration of variance at about 80-100 years and its apparent stability through much of the last 1500 years. The circa 30-40 year variance highlighted in the earlier analyses also appears to be an important feature of the series for much of this time, with the notable exception of the 15th century. A near 11-year periodicity is apparent during most of 1300-1800 but it is absent at other times. There is clearly little evidence for a near 22-year cycle in almost the whole record, except perhaps for a period around 1400 and 1700.

Following Taylor and Rose (1992), the periods of supposedly

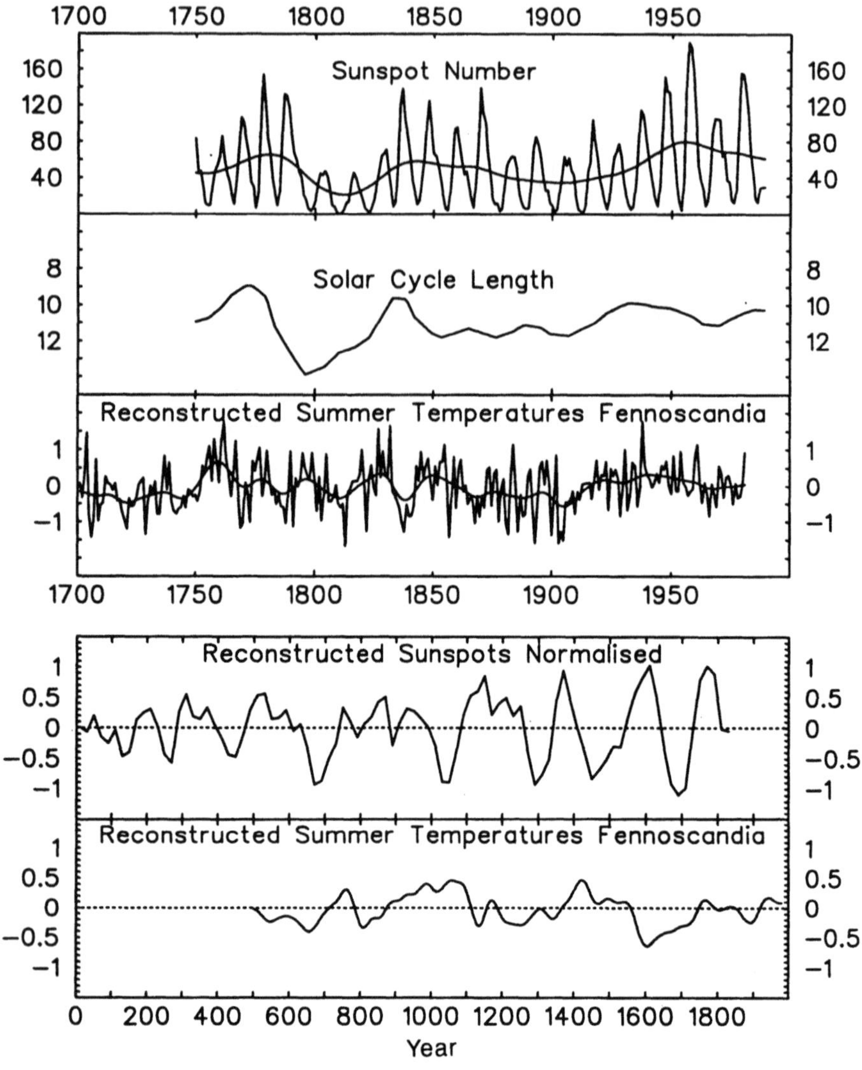

Figure 5. **The upper box** shows series of annual sunspot number, sunspot cycle length (note inverted scale) and reconstructed summer temperatures (°C anomalies) after A.D. 1700. The temperature data have been smoothed with a 10-year filter and are overlain with a curve showing 25-year smoothed data. **The lower box** shows normalized reconstructed sunspot numbers (after Stuiver et al., 1991) compared with 100-year smoothed Fennoscandian temperatures from A.D. 500.

reduced sunspot activity are also indicated on Figure 4. My subjective assessment of times when the variance pattern appears to be disrupted yields the following periods: the early 700s; around 1000; early-to-mid 1300s; mid-to-late 1500s and the mid-to-late 1700s. The second and third of these would seem to be most clear. Subsequent comparison with the periods of sunspot minima show an interesting coincidence between the Oort and the period around 1000, but the other minima appear to be randomly distributed with regard to the highlighted periods.

Briffa (1994) examined the specific question of whether periods of low sunspot activity were also, on average, periods of significantly cool summers in the Fennoscandian reconstruction. Average 'summer' temperatures during each of 5 sunspot minima were compared with the overall mean reconstructed over nearly 1500 years. While the Maunder Minimum was undoubtedly relatively cool, 50-year periods before and following the 1645-1715 period were also notably cool, while other low-sunspot periods were either 'near normal' or even warm. He concluded for Fennoscandia that "*these observations argue against any simple conclusions linking periods of low sunspot activity with generally cool conditions.*" The truth of this can be seen in Figure 5.

SUMMARY AND CONCLUSIONS

A number of other papers would have been discussed here, given more space. These include Mori (1981), Sonett and Suess (1984), Guiot (1987)and others. No reference has been made of the work of Curry (e.g. 1991) who finds evidence for a near 11-year cycle in many tree-ring chronologies in the United States. Neither has any reference been made to the large body of early Russian (more strictly ex Soviet Union) work on cyclic behaviour in tree-ring time series and its relationship with solar variability: work which formed the foundation for predicting ecosystem or agricultural changes in the former Soviet Union. A starting point for anyone interested in this work is Kairiukstis et al. (1990).

Despite these shortcomings I have reviewed a reasonable

sample of the possible evidence of solar-related variability in tree rings. Clearly any time series containing variability within the range of several years up to several decades would be expected to show some fortuitous similarity to the sunspot record, at least over part of its length. Applying appropriate band-pass filtering and allowing some shifting of series (to take account of possible lag effects) mean that it would be difficult not to find periods when tree-ring records seem to match the sunspot record. Having seen some of the tree-ring-related evidence shown here, the sceptic might consider that it amounts to nothing more than fortuitous coincidence. A rigid (some would say correct) application of statistical methods (see for example reviews of climate/solar relationships by Pittock, 1978, 1979; Goldberg, 1982) would seem to justify this attitude.

However neither does the evidence allow us to reject a solar influence on climate on timescales of decades to centuries. Given that the climate system is dynamical and non-linear we might not expect evidence of various external forcings to be directly recognizable in simple (often local) climate records. Longer, well-resolved proxies of climate and solar variability are still required to explore this question further, but we also require modelling studies of the possible non-linear response of the climate signal to changes in solar output to help identify the expression of the solar-signal for which we are searching.

What does seem clear, from the studies cited here and in many reviews of the possible influence of solar variability on climate, is that changes in solar forcing, at least on the timescales on which we have focussed, are not of sufficient magnitude to be of use in any prediction sense. However the search for evidence to the contrary will no doubt continue. To what extent this will represent a continued 'grasping at shadows' remains to be seen.

ACKNOWLEDGEMENTS

I thank Doug Hoyt and Lennart Bengsston for comments and the U.K. Dept. Environment for support under contract PECD/7/12/78.
Ed Cook provided the data used to plot Figure 4.

REFERENCES

Briffa, K.R. 1994: Tree-ring evidence of northern Fennoscandian summer temperatures during the Maunder Minimum. (In) Climate Trends and Anomalies in Europe during the Maunder Minimum Period (1675-1715). *Paläoklimaforschung* (In press).

Briffa, K.R. and Schweingruber, F.H. 1992: Recent dendroclimatic evidence of northern and central European summer temperatures. (In) *Climate Since A.D. 1500*, (R.S. Bradley and P.D. Jones, Eds.) 366-392. Routledge, London.

Briffa, K.R. Bartholin, T., Eckstein, D., Jones, P.D. Karlen, W., Schweingruber, F.H. and Zetterberg, P. 1990: A 1,400-year tree-ring record of summer temperatures in Fennoscandia. *Nature 346*, 434-439.

Briffa, K.R., Jones, P.D., Bartholin, T.S., Eckstein, D., Schweingruber, F.H., Karlen. W., Zetterberg, P. and Eronen, M. 1992: Fennoscandian summers from A.D. 500: temperature changes on short and long timescales. *Climate Dynamics 7*, 111-119

Brückner, E. 1890: Klimaschwankungen seit 1700 nebst Bemerkungen über die Klimaschwankungen der Diluvialzeit *Geograph. Abhandl. herausgeb. v A. Penck. 4*, Wien.

Clough H.W. 1905: Synchronous variations in solar and terrestrial phenomena. *Astrophysical Journal 10*, 42-75.

Cook, E.R. and Jacoby, G.C., Jr. 1979: Evidence for quasi-periodic July drought in the Hudson Valley, New York. *Nature 282*, 390-392.

Cook, E.R., Bird, T., Peterson, M., Barbetti, M., Buckley, B., D'Arrigo, R., Francey, R. and Tans, P. 1991: Climate change in Tasmania inferred from a 1089-year tree-ring chronology of subalpine huon pine. *Science 253*, 1266-1268.

Cook, E.R., Bird, M., Peterson, M., Barbetti, M., Buckley, B., D'Arrigo, R. and Francey, F. 1992: Climatic change over the last millennium in Tasmania reconstructed from tree rings. *The Holocene 2*, 205-217.

Cook, E.R., Buckley, B. and D'Arrigo, R. 1994: Decadal-Scale Oscillatory modes in a millennia-long temperature reconstruction from Tasmania. Paper prepared for the National Academy of Sciences Workshop in "*The Natural Variability of the Climate System on 10-100 Year Timescales*" Irvine, California Sept. 21-24, 1992.

Currie, R.G. 1991: Deterministic signals in tree-rings from North America. *International Journal of Climatology 11*, 861-876.

Douglass, A.E. 1919: Climate cycles and Tree Growth. Carnegie Institution of Washington Publication 289. Volume 1. Washington.

Douglass, A.E. 1928: *Climatic Cycles and Tree-Growth. A Study of the Annual Rings of Trees in Relation to Climate and Solar Activity*. Carnegie Institution of Washington. Pub. 289, Gibson Brothers, Washington.

Douglass, A.E. 1943: Advances in dendrochronology. 1943. *Tree Ring Bulletin 9*, 18-24.

Erlandsson, S. 1936: Dendro-Chronological studies. Data, 23. *Frn Stockholms Högskolas Geokronologiska Institut*, Stockholm. 116 p.

Goldberg, R.A. 1982: A review of reported relationships linking

solar variability to weather and climate (In) *Solar Variability, Weather, and Climate.* National Research Council. Geophysics Study Committee. 19-32. National Academy Press, Washington.
Guiot, J. 1987: Reconstruction of seasonal temperatures in central Canada since A.D. 1700 and detection of the 18.6- and 22-year signals. *Climate Change 10,* 249-268.
Hann, J.v. 1932: Handbuch der Klimatologies. Bd.1. Allgemeine Klimalchre. 4:te Aufl. ausgegeb. von K. Knoch. Stuttgart.
Kairiukstis, L. and Shiyatov, S.G. (with Kocharov, G.E., Mazepa, V., Dubinskaite, J., Vaganov, E., Bitvinskas, T. and Jones, P.D.) 1990: Tree Rings in the study of future change (In) *Methods of Dendrochronology. Applications in the Environmental Sciences* (E.R. Cook and L.A. Kairiukstis, Eds.) 289-340. IIASA/Kluwer, Dordrecht.
LaMarche,V.C.,Jr. and Fritts, H.C. 1972: Tree rings and sunspot numbers. *Tree Ring Bulletin 32,* 19-33.
Meko, D.M., Stockton, C.W. and Blasing, T.J. 1985: Periodicity in tree rings from the Corn Belt. *Science 229,* 381-384.
Mitchell, J.M., Jr., Stockton, C.W. and Meko, D.M. 1979: Evidence of a 22-year rhythm of drought in the western United States related to the Hale solar cycle since the 17th century. (In) *Solar-Terrestrial Influences on Weather and Climate* (B.M. McCormac and T.A. Seliga, Eds.) 125-143, Reidel, Dordrecht.
Mori, Y. 1981: Evidence of an 11-year periodicity in tree-ring series from Formosa related to the sunspot cycle. *Journal of Climatology 1,* 345-353.
Ording, A. 1941: Årringanalyser pa gran og furu. *Meddelelser fra Det Norske Skogforsøksvesen 25,* 354 pp.
Pittock, A.B. 1978: A critical look at long-term sun-weather relationships. *Reviews of Geophysics and Space Physics 16,* 400-420.
Pittock, A.B. 1979: Solar cycles and the weather: successful experiments in autosuggestion? (In) *Solar-Terrestrial Influences on Weather and Climate* (B.M. McCormac and T.A. Seliga, Eds.) 181-191, Reidel, Dordrecht.
Rykatschew, M. 1887: Über den Auf- und Zugang der Gewässer des Russischen Reiches. Zweit. Suppl.-Bd. z. Repert. f. Meteorol. St. Petersburg.
Schulman, E. 1936: Tree-rings and cycle analysis. *Tree-Ring Bulletin 11,* 19-22.
Sirén, G. 1961: Skogsgranstallen som Indikator för klimafluktuationerna i norra Fennoskandien under historisk tid. *Communicationes Instituti Forestalis Fenniae 54,* 66pp.
Sirén, G. and Hari, P. 1971: Coinciding periodicity in recent tree rings and glacial clay sediments. *Report of the Kevo Subarctic Research Station 8,* 155-157.
Sonett, C.P. and Suess, H.E. 1984: Correlation of bristlecone pine ring widths with atmospheric ^{14}C variations: a climate-sun relation. *Nature 307* 141-143.
Stockton, C.W. and Meko, D.M. 1983: Drought recurrence in the Great Plains as reconstructed from long-term tree-ring records. *Journal of Climate and Applied Meteorology 22,* 17-29.
Stockton, C.W., Mitchell, J.M., Jr. and Meko, D.M. 1983: A

reappraisal of the 22-year drought cycle. (In) *Weather and Climate responses to Solar Variations* (B.M. McCormac, Ed.) 507-515, Colorado Associated University Press, Boulder, Colorado.
Stuiver, M. and Quay, P.D. 1980: Changes in atmospheric carbon-14 attributed to a variable sun. *Science 207*, 11-19.
Stuiver, M., Braziunas, T.F., Becker, B. and Kromer, B. 1991: Climatic, solar, oceanic, and geomagnetic influences on late-glacial and Holocene atmospheric $^{14}C/^{12}C$ change. *Quaternary Research 35*, 1-24.
Taylor, K. and Rose, M. 1992: Relationship of solar activity and climate oscillations on the Colordao Plateau. *Journal of Geophysical Research 97*, 15,803-15,811.

PERSPECTIVES ON THE CLIMATE OF THE LAST 500 YEARS

R.S. Bradley,
University of Massachusetts,
Amherst, MA 01003-5820, U.S.A.

Abstract

High resolution proxy records of past climate can be used to extend our perspective on regional and hemispheric changes of climate back in time by several hundred years. Using historical, tree-ring and ice core data, we examine climatic variations during the period commonly called the "Little Ice Age". A reconstruction of summer temperatures reveals that, for the northern hemisphere as a whole, the coldest summer temperatures of the last 560 years were between 1570 and 1730, and in the early to mid-19th century. At other times, temperatures were relatively high, in some regions equalling or exceeding levels typical of the early to mid 20th century. Temperatures during the Maunder Minimum (1655-1714) correspond to the coldest 60 year interval of the last few centuries at many locations, but the period of the Spörer Minimum was not particularly unusual. In the context of the last 500-600 years, unusually warm conditions have prevailed since the 1920s ; this may be related to a relative absence of major explosive volcanic eruptions and higher levels of greenhouse gases.

1. Introduction

For most parts of the world, instrumentally-recorded records of climate span a period of less than a century, providing a very limited perspective on climate variability and its relationship to potentially important forcing factors. To obtain a longer perspective requires reliance on climate-dependent natural phenomena which have preserved, in some way, a climate signal (Bradley, 1985 ; Bradley and Jones, 1992). To utilize such material in paleoclimatic reconstruction, two fundamental criteria must be met : (1) the material must

NATO ASI Series, Vol. I 25
The Solar Engine and Its Influence
on Terrestrial Atmosphere and Climate
Edited by Elizabeth Nesme-Ribes

be capable of being accurately dated, and (2) the climatic signal must be clearly separable from non-climatic noise. Given that such records are difficult to find, a further attribute should be that the record represents climate variations of more than purely local significance. This encounters the difficulty of determining the "spatial representativeness" of a record, and whether this "spatial field" has remained constant over time. It is likely that the spatial field is frequency-dependent, such that lower frequency components of a record represent a wider geographic area than higher frequency components. Thus, a few widely separated records may be sufficient to represent century-scale variations (10^2-10^3 year) over a large region whereas decadal scale variations (10^1-10^2 year) may require a larger network of data. Thus, in determining the relationship between forcing factors and climate response (whether low or high frequency) the network density required to adequately detect a signal must be considered.

2. Types of Record Applicable to Paleoclimatic Reconstruction of the Last 500 Years

The period before the beginning of the 20th century is often referred to as the "Little Ice Age", largely because of the abundant evidence that alpine glaciers throughout the world were significantly larger in preceding centuries. Unfortunately, records of glacier fluctuations are insufficient to address the relationship between forcing and climate response. Geological evidence of glacier position cannot provide a continuous time series ; moraines can delimit forward ice positions but the extent of recession between advances is not determinable by geomorphological evidence. Moraines cannot be accurately dated ; usually ice advances are 'bracketed' by radiocarbon dates on organic material over-ridden by ice advance, and material subsequently growing on the stable moraine. Even if this interval was in reality very short, the uncertainties in radiocarbon dating material with a time age of 0-500 years B.P. (because of changes in ^{14}C production rate) mean that assigning a unique date range to individual samples is often impossible. The problem is only compounded when using lichonometrical dating in which the lichen growth curve is defined by ^{14}C-dated material. Some of these difficulties are overcome when historical records provide precise documentation of glacier position (e.g. Messerli *et al.*, 1978) but such cases are rare outside of Alpine Europe and the vast majority of glaciers studied rely on geological methods to define glacier positions in the past. Even if dating of glacier position was not a problem, the value of such evidence for paleoclimatic purposes is far from clear. Glaciers may lose considerable mass by down-wasting before the terminal position changes significantly. Individual glaciers respond quite differently to climatic changes depending on their topographic situation, size and 'activity ratio' (that is the effective rate at which snowfall moves through a glacier from accumulation area to ablation zone). Finally, it is an obvious fact that glaciers are limited in their geographic extent and, hence, cannot provide widely distributed evidence for past climatic conditions. All of these factors contribute to limiting the value of glacial evidence in determining *quantitatively* former climatic conditions over a wide area. Far preferable are continuous, accurately dated, proxy records which have been calibrated with modern, instrumentally-recorded climatic data. Such records (discussed in somewhat more detail in Duplessy *et al.*, this volume) include :

a) historical data
b) tree ring data
c) ice core data
d) coral data
e) varved sediment data

Historical records can, potentially, provide seasonally-resolved quantitative estimates of past climate over wide geographic regions, though at present only European and East Asian sources have been adequately utilized.

Tree ring data includes both ring width and ring density variations. Records are available from all continental areas (except Antarctica) though most series are from extra-tropical regions. High latitude and high altitude trees generally provide paleotemperature estimates ; trees in semi-arid regions generally provide paleo-precipitation estimates though even in wetter areas, records of precipitation changes can sometimes be obtained.

Ice cores provide multi-proxy records of paleoclimate but changes in ^{18}O accumulation rate and (summer) melt conditions are of primary interest in examining recent centuries. ^{18}O in polar regions is generally considered to be an indicator of annual temperature (Dansgaard *et al.*, 1973) and though there are innumerable problems in applying this simple interpretation to regions beyond the interior of Antarctica (Bradley, 1985 ; Jones *et al.*, 1993) it continues to be equated with mean annual temperature. Other useful climate indicators include the fraction of a core containing 'melt features' (produced by the re-freezing of percolating surface melt water) which provides a useful index of summer temperature conditions, and accumulation rate changes which indicate past precipitation amounts (assuming no net loss during the year).

Corals provide a unique high resolution record of sea-surface temperatures from changes in the (temperature dependent) oxygen isotopes in the carbonate skeleton of the corals. In some cases, salinity variation is the most important factor influencing isotopic content, in which case the changes reflect precipitation and runoff from adjacent continental regions (Dunbar and Cole, 1993).

Varved sediments, from both lacustrine and marine environments, are annually laminated records of past environmental conditions in the lake or oceanic region. There are few ocean areas where varved sediments are known to occur (generally upwelling coastal regions where anoxic bottom waters are found) but varved lake sediments are quite widespread across all continents. Providing the records are clearly annual and a strong climatic signal can be demonstrated, these records can provide useful data from many regions of the world.

In the following section, we discuss the evidence provided by such records in reconstructing the climate of the last few centuries. It should be noted, however, that most proxies are quite season-specific, and provide information only about the time of year when the proxy is in some way registering the climate signal. In comparing proxies, it is necessary to ensure that the same seasonal signals are being examined. This often limits large scale comparisons to the summer season.

3. The Climatic Record of the Last 500 years

Using a variety of proxy temperature records, an estimate of the northern hemisphere summer paleotemperatures over the last 500 years has been made (Bradley and Jones, 1993). This estimate is based on records from three regions (North America, Europe and East Asia) which are known to "track" temperature variations of the hemisphere as a whole during the period of instrumental records (i.e. they explain $> 70\%$ of the northern hemisphere temperature variation on a decadal time-scale). The limited number of similar records from the southern hemisphere precludes a similar analysis being made for this region at present.
In the northern hemisphere, temperatures were generally cooler than in the 20th century for most of the preceding 500 years, but not continuously so (Figure 1). Overall, the coldest periods were between 1570 and 1730, and in the early to mid-19th century. In some regions (e.g. Europe in the 18th century) temperatures were as high as in the early 20th century. Thus, the notion of a prolonged and uninterrupted cool interval, implied by the term "Little Ice Age", is not supported by the limited evidence available. Rather, the last 500 years experienced both cool and warm intervals, with temperatures generally increasing over the last 400 years.

It was noted earlier that most proxies are related to summer temperature and therefore the northern hemisphere reconstruction described above is based on data for this season only. In a few regions, temperature reconstructions have been made for all seasons of the year, based largely on historical data (e.g. Switzerland ; Pfister, 1992 ; eastern and northern China, Wang, 1991). In comparing these different seasonal records, it is apparent that summer temperatures are not always in phase with the other seasons. Therefore, it must be recognized that our perspective on past climate from the available proxies may be somewhat biased by the preponderance of summer temperature indicators. However, for at least the period of instrumental records (1901-1990) inter-annual summer temperature variations over the northern hemisphere are highly correlated with annual temperature variations ($r \geq 0.88$, for $n=90$; Jones and Briffa, 1992).

4. Solar Activity Variations and the Record of Past Climate over the Last 500 years

It is often argued that climatic variations of the last few centuries are largely the result of either solar or volcanic forcing (e.g. Wigley, 1991 ; Porter, 1986). Here we examine the relationship between estimates of long-term summer temperature variations and solar irradiance changes. Recent studies have shown a strong correlation between temperature changes over the last 100 years and solar cycle length, which is considered to be an index of solar irradiance variations (Friis-Christensen and Lassen, 1991). However, the change in solar cycle length has occurred at a time when greenhouse gas concentrations have also been rising, so separating the influence of each forcing factor is problematical. Theoretical considerations suggest that the magnitude of irradiance change (indicated by the change in solar cycle length) is too small to have been of great climatic significance, and that forcing by greenhouse gases is of more importance (Kelly and Wigley, 1992 ; Schlesinger and Ramankutty, 1992). Another way of examining this issue is to compare irradiance variations with temperature changes prior to the 20th century, since the influence of anthropogenic greenhouse gases was small before 1900. Figure **2** shows estimated solar irradiance (based on several measures of solar variability calibrated against recent satellite-based irradiance measurements ; Hoyt and Schatten, 1993) compared with the reconstructed summer temperature record for the northern hemisphere, discussed above. The correspondence is poor ($r = 0.49$, for decadal averages, $n=29$) especially between 1770 and 1880. However, ^{10}Be measurements in polar ice cores, which serve as an index of short-term solar variability, do show a very good correlation with the long-term record of temperature in central England (Beer *et al.*, this volume). Given the uncertainties in both the reconstruction of past solar irradiance, and of paleotemperatures, there is a real need to undertake further analysis of these data, especially using ^{10}Be from many sites as a proxy of solar irradiance changes.

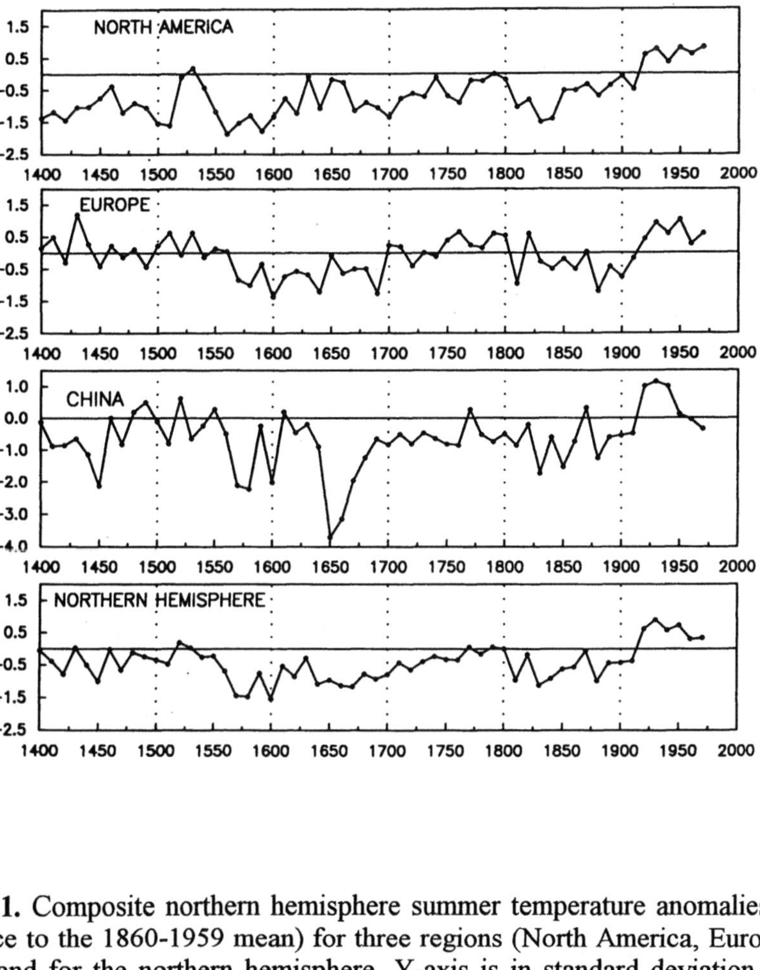

Figure 1. Composite northern hemisphere summer temperature anomalies (with reference to the 1860-1959 mean) for three regions (North America, Europe and China) and for the northern hemisphere. Y-axis is in standard deviation units ; each series is the average of a set of normalized, decadal anomalies, derived from different proxy records. The largest decadal temperature anomaly (1600-1609) is estimated to be 0.43°C below the 1950-79 average for the northern hemisphere [from Bradley and Jones, 1993].

5. The Maunder Minimum

The period of reduced sunspot activity from A.D.1655-1714, known as the Maunder Minimum (Eddy, 1976) is often cited as a period of reduced solar

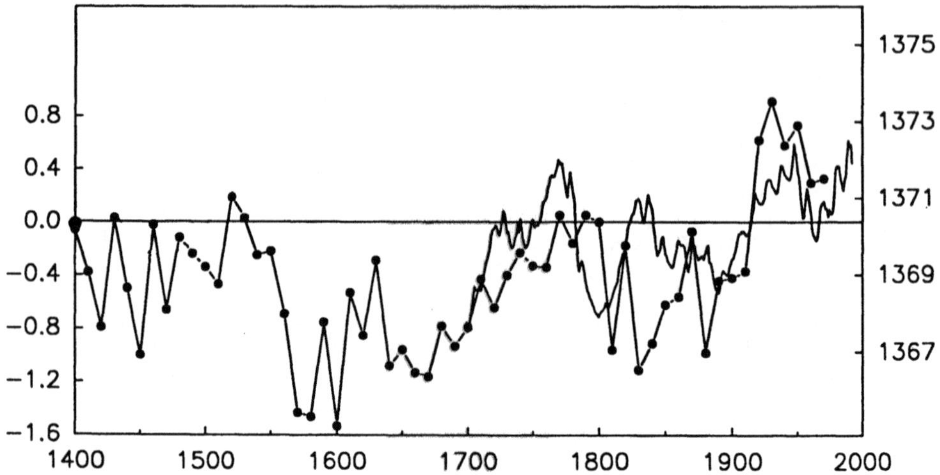

Figure 2. Mean normalised northern hemisphere summer temperature anomalies (with reference to the 1860-1959 mean) [same as lowest panel of Figure 1, shown as the line with circles ; from Bradley and Jones, 1993] and solar irradiance changes (Watts per square meter) (from Hoyt and Schatten, 1993).

irradiance. Recent studies indicate that other indices of solar activity may not have changed at that time, so it is not clear if solar irradiance was, in fact, lower (cf. Silverman, 1993). Without entering that debate, we simply note here that many proxy records (especially those from high latitudes) did record the most extreme conditions of the last few centuries at the time of the Maunder Minimum (i.e. in comparing 50-60 year means for periods before, during and after the Maunder Minimum, the Maunder Minimum is frequently the coldest interval of record).

For example, records of melting conditions on polar ice caps in Spitsbergen, Ellesmere Island and Greenland all show the lowest levels of melting over the past 600-700 years (indicating low summer temperatures and/or low net radiation) occurred during the Maunder Minimum. Oxygen isotope values

(^{18}O) were also lowest at this time, suggesting low temperatures during periods of precipitation on the ice caps. Similarly, many historical records from Korea, and from northern and eastern China indicate that the coldest summers of the last 500 years were during this interval ; winters were also exceptionally cold, in many regions. Tree ring records from northern Patagonia also indicate low summer temperatures at the time of the Maunder Minimum. Of course, there are also proxy records which show nothing unusual about the Maunder Minimum, and to obtain a complete perspective, as dense a network as possible must be examined, and the anomalies mapped, to determine the spatial distribution of the extremes. These maps may then be compared with general circulation model simulations of circulation changes associated with a reduction in solar irradiance (e.g. Lean and Rind, 1994). By combining proxy records and model simulations, it should eventually be possible to understand the significance of small changes in solar forcing on the general atmospheric circulation.

6. Summary and Conclusions

A set of proxy records are available which provide a representative estimate of northern hemisphere summer temperature changes, on a decadal scale for the last 500 years. Although generally lower temperatures prevailed before the 20th century, there were relatively warm intervals in the 18th and 16th centuries, and in some regions these approached 20th century levels. Estimates of solar irradiance changes do not correspond well with the paleotemperature record before 1900, though there are clearly limitations in both reconstructions. The Maunder Minimum period was one of the coldest 60 year periods of the last few centuries at many, but not all, sites around the world. A comprehensive study of long-term temperature proxies, focusing on this interval, is needed to establish the pattern of large-scale circulation anomalies.

References

Bradley, R.S. (1985). Quaternary Paleoclimatology, Chapman and Hall, London 472

Bradley, R.S., and Jones, P.D. (1992). Climate since A.D. 1500 : Introduction, in Climate Since A.D. 1500, eds. R.S. Bradley and P.D. Jones, Routledge, 1-16

Bradley, R.S., and Jones, P.D. (1993). Little Ice Age summer temperature variations : their nature and relevance to recent global warming trends. The Holocene, 3: 367-376.

Dansgaard, W., Johnsen, S.J., Clausen, H.B., and Gundestrup, N. (1973). Stable isotope glaciology. Meddelelser øm Grønland, 197: 1-53

Dunbar, R., and Cole, J. (1993). Coral records of ocean-atmosphere variability. Special Report No. 10, NOAA Climate and Global Change Program. Washington D.C., 38

Eddy, J. (1976). The Maunder Minimum. Science, 192: 1189-1202

Friis-Christensen, E., and Lassen, K. (1991). Length of the solar cycle : an indicator of solar activity closely associated with climate. Science, 254: 698-700

Hoyt, D.V., and Schatten, K.H. (1993). A discussion of plausible solar irradiance variations, 1700-1992. J. Geophysical Research, A11: 18895-18906

Jones, P.D., and Briffa, K.R. (1992). Twentieth century global surface air temperature variations, Part 1. The Holocene, 2: 165-179

Jones, P.D., Marsh, R., Wigley, T.M.L., and Peel, D.A. (1993). Decadal timescale links between Antarctic Peninsula ice core oxygen-18, deuterium and temperature. The Holocene, 3: 14-26

Kelly, P.M., and Wigley, T.M.L. (1992). Solar cycle length, greenhouse forcing and global climate. Nature, 360: 328-330

Lean, J., and Rind, D. (1994). Solar variability : implications for global change, in Eos Transactions of the American Geophysical Union, 75: 1

Messerli, B., Messerli, P., Pfister, C., and Zumbuhl, H.J. (1978). Fluctuations of climate and glaciers in the Bernese Oberland, Switzerland, and their geoecological significance. Arctic and Alpine Research, 10: 247-260

Pfister, C. (1992). Monthly temperature and precipitation in central Europe 1525-1979, quantifying documentary evidence on weather and its effects. In Climate Since A.D. 1500, eds. R.S. Bradley and P.D. Jones. Routledge, 118-142

Porter, S. (1986). Pattern and forcing of northern hemisphere glacier hemisphere temperature variations during the last millennium. Quaternary Research, 26: 27-48

Schlesinger, M.E., and Ramankutty, N. (1992). Implications for global warming of intercycle solar irradiance variations. Nature, 360: 330-333

Silverman, S. (1993). Is the Maunder Minimum real ?, in Eos Transactions of the American Geophysical Union, 74: 450

Wang, Shao-Wu (1991). Reconstruction of paleo-temperature series in China from the 1380s to the 1980s. Wurzburger Geographische Arbeiten, 80: 1-19

Wigley, T.M.L. (1991). Climate variability on the 10-100 year time scale, observations and possible causes. In Global Changes of the Past, ed. R.S. Bradley, Office of Inter-Disciplinary Earth Studies, UCAR, Boulder: 83-101

THE 10.8 YEARS SOLAR CYCLE RECORDED IN SEA SEDIMENTS DURING THE LAST 4000 YEARS

G.Cini Castagnoli, G.Bonino, C.Taricco
Istituto di Cosmogeofisica del CNR and
Istituto di Fisica Generale dell'Universita'
Corso Fiume 4
10133 Torino
ITALY

Summary

The thermoluminescence (TL) and carbonate profiles are measured in shallow water Ionian Sea cores, dated precisely by radiometric and tephroanalysis methods. Their time series, covering the last 4000 years, are obtained at time intervals of 3.096 and 3.87 years. We show that the 10.8 yr cyclicity is always present in each of time series analysed. A study of the amplitudes and phases of the wave is performed by the method of superposition of epochs. The phase shift of the TL and of the $CaCO_3$ waves is $\sim\pi$. The persistency of the wave is assured in the last 3000 years by the cyclogram analysis. The power involved in this component of the signals is however small. Since the time series which we have extracted from the Ionian Sea cores bear some relation to the climatic factors forcing the deposition of the coastal carbonatic mud, these results may be interpreted as a proof that the Schwabe cycle of the Sun, associated with the

NATO ASI Series, Vol. I 25
The Solar Engine and Its Influence
on Terrestrial Atmosphere and Climate
Edited by Elizabeth Nesme-Ribes

irradiance variability, is directly involved in the climatic changes of the terrestrial archives.

Introduction

In previous papers we have analysed the variations of the thermoluminescence (TL) signal recorded (at spatial intervals of 2.5 mm) in the fine polymineralic crystals of the carbonatic material of the shallow water Ionian Sea core GT14, spanning the last 1800 yr (Cini Castagnoli and Bonino, 1988a,b,c,d; 1990a,b; 1991). An accurate measurement of the deposition rate of the sediment in this area by radiometric and tephroanalysis methods allows to transform the depth profiles into time series, with a precision of about 1%, at least in the last two millennia, where the dates of the major volcanic eruptions of the Campanian area are historically known. Phenomenological similarities of the TL profile and of the radiocarbon record in tree-rings or the variations of the mean annual sunspot number R_Z suggested a partial solar control of the TL signal. In particular, several time series analysis methods revealed the presence in the TL signal of the periodicities of 10.8 and 22 years, corresponding to the well known Schwabe and Hale solar cycles.

In this paper, we focus our attention on the 10.8 yr periodicity. In particular:

a) we confirm the presence of a 10.8 yr periodicity in a new TL time series spanning the last 3100 yr, measured in a longer core (GT89-3) taken in the same area, sampled at spatial intervals of 2 mm, corresponding to 3.096 yr,

b) we show that this periodicity is present also in the total carbonate time series of the GT14 and GT89-3 cores, measured with the sampling interval of 2.5 mm, corresponding to 3.87 yr,

c) we confirm the presence of a 10.8 yr periodicity in a new $CaCO_3$ time series, measured in the GT91-1 core taken 30 km apart from the previously mentioned cores.

These results show that the solar variability (in this case the Schwabe cycle of sunspot number) is imprinted in the records. This imprinting may be generated through climatic factors and suggests the influence of solar variability on climate.

Cores dating

The main characteristics of the three Ionian Sea cores GT14, GT89-3 and GT91-1, considered in this paper, are reported in Tab.1.

CORE (Ionian Sea)	COORDINATES	WATER DEPTH (m)	CORE LENGTH (m)	TIME INTERVAL (yr) (before 1979 AD)
GT14	39°45'53"N 17°53'30"E	166	1.17	1807
GT89/3	39°45'43"N 17°53'55"E	178	2.81	4350
GT91/1	39°59'23"N 17°45'52"E	250	2.93	4540

Tab.1: Main characteristics of the three Ionian Sea cores GT14, GT89-3 and GT91-1.

To transform the depth profiles of the cores into well defined time series it is necessary to carefully determine the sedimentation rate and to verify whether it remained constant over the time interval spanned by the cores.
A detailed evaluation of the dating of the cores is reported elsewhere (see for example Bonino et al., 1993). We recall here that the sedimentation rate of these cores has been conducted by considering: a) the measurement of the ^{210}Pb ($T_{1/2}$ = 22.3 yr) activity in "excess" with respect to the ^{226}Ra isotope ($T_{1/2}$ = 1600 yr) in the natural ^{238}U radioactive chain, b) the measurement of the activity of ^{137}Cs in the uppermost layers of the sediment, mainly produced by nuclear-bomb testing in the atmosphere, c) a careful tephroanalysis of event markers in the sediments.
The measurement of the amount of "excess" ^{210}Pb, present in a sediment as a function of depth, allows for a direct dating of the core until approximately 150 yr BP. A maximum concentration of ^{137}Cs has been attained in the years 1963-1964 AD, as a result of the intense activity of the nuclear weapon-testing in the atmosphere. The presence of this maximum warranties that the top of the core has not been disturbed.
A careful tephrochronology allows for verifying whether the sedimentation rate determined by the ^{210}Pb method in the upper part of the sediment has been constant for the entire length of the core. In the Ionian Sea cores which we have analysed, the eruptions in the Campanian volcanic area have prooven to be particularly significant for the cores dating (well-known examples of dated plinian and subplinian volcanic events are the Ischia (1301 AD), Pollena (472 AD) and Pompei (79 AD) eruptions). The results of the dating procedures allow for the accurate evaluation of the sedimentation rate,which is $s = (0.0646 \pm 0.0007)$ cm/yr. The results of this analysis demonstrate that the sedimentation rate is constant, at least in the last 2000 years. Furthermore it is uniform for the whole area considered. In fact, we find the same volcanic ash layers at the same depths for different cores. The total $CaCO_3$ profile measured in the different cores are identical for the overlapping parts. This is apparent in Fig.1, where the $CaCO_3$

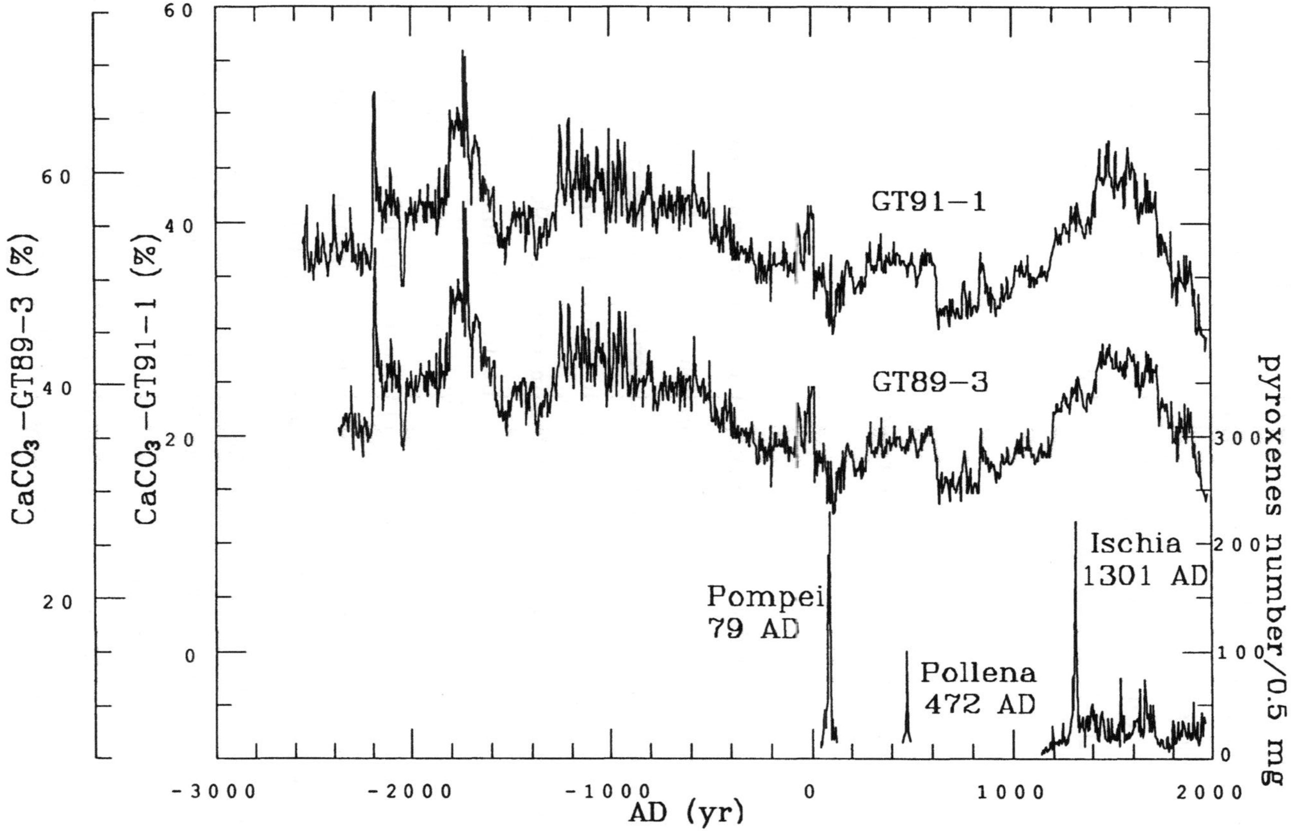

Fig.1: Carbonate profiles of GT89-3 and GT91-1 sea cores, showing the extreme correlation between the carbonate records. The historic records of pyroxene from volcanic eruptions shown at the base of the figure provide exact time 'benchmarks'.

profiles of the cores GT89-3 and GT91-1 are reported as a function of time, together with the clinopyroxene volcanic record. The reference (top) level is 1979 AD for all the cores and the samples were numbered progressively backward in time from this level.

The uniformity of the sediment deposition and the test of the volcanic markers along the cores allow for the reliability of the time scale to about 1%.

The TL analysis of the GT89-3 Ionian Sea core

To measure the TL profile, 1 g of material was taken from adjacent layers equispaced at a distance of 2 mm. The preparation of the samples was done in red light treating the wet material by successive dispersion, washing in NaOH, water and acetone, using the centrifuge on each step to preserve the original composition of polyminerals. After drying in oven at 40°C overnight, the sediment powder was gently sieved and the fraction $\leq 44\ \mu m$ was weighted in samples of 15 mg for the TL measurements. Glow curves were then obtained by using a TL analyser, described by Miono and Otha, (1969). The TL signals were recorded in a neutral atmosphere of ultrapure nitrogen at a heating rate of 5 °C/s; the readings of the glow curves were taken at different temperatures between 300°C and 380 °C and then averaged.

The TL signal, released by heating in the TL analyser, was stored in the sea sediments by means of the following mechanisms. When crystalline minerals are exposed to ionizing radiation, free electrons and holes are produced and a small fraction of them becomes trapped in crystal defects. If the crystals are heated, these trapped charges are released and are able to recombine, generating the TL signal by a radiative process. The captured charges in the crystal defects may also be released by light: for this reason, the potential TL is dictated by an equilibrium environmental level of light and background particles. This predepositional TL is augmented by the TL acquired *in situ* (after deposition) from the local radioactivity concentration in the core. In our very recent sediments this effect is negligible.

We obtained the TL profile of GT89-3 core by measuring the TL signal in 1000 samples of the upper 200 cm of the core. In order to test the aliasing effects in the analysis of the time series, while the previously analysed GT14 core was sampled at equispaced intervals $\Delta d = 2.5$ mm, corresponding to time intervals $\Delta t = 3.87$ yr, the GT89-3 core has been sampled at $\Delta d = 2$ mm, corresponding to $\Delta t = 3.096$ yr. The time series of the GT14 core spans ~1800 yr, while the GT89-3 TL time series spans ~3000 yr.

To study the principal periodic components of the TL record, we have analysed the data by the method of superposition of epochs (S.E.). Following this method, we divided the TL series into segments of fixed length T which are then superposed. This gives the possibility to explore the presence of a T-year periodicity. In this way, all the periodicities in the time series are averaged to about zero, except those with period T/n, where n is an integer number. The T-year basic interval has then been divided into subdivisions, and the mean and the standard deviation of the experimental data falling into each subdivision has been computed. It is thus possible to obtain the shape, the period, the amplitude and the phase of each periodic component of the signal. The phase information is particularly important for finding the possible connections between the TL profile and the solar activity and it is hardly obtainable from power spectral analysis.

The least-square fit of the superposed data to a sinusoidal wave has been subsequently evaluated and the amplitude and the phase of the waves have been determined, together with the correlation coefficient of the superposed data to the sinusoid.

By this method, we found several periodicities. Here we present the results obtained for the 10.8 yr cycle, due to its possible relation with the fundamental Schwabe cycle of solar activity. In Fig.2 we show the results of the S.E.. The basic interval has been divided into 5 subdivisions. To better visualize the wave, both the superposed data and the least-square fit sinusoid have been repeated twice. The amplitude, phase and correlation coefficient for the GT89-3 core are reported in Tab.2, together with the values of the same parameters calculated for the core GT14 and reported in Tab. 1 of the paper Cini Castagnoli et al. (1988d). We note that the lower r^2 for GT14 is only a consequence of the fact that the basic interval was there divided into 20 subdivisions.

	A	ϕ	r^2
GT14	1.0	0.94	0.64
GT89-3	1.2	0.79	0.95

Tab.2: Parameters of the S.E. for the GT14 and GT89-3 cores (T = 10.8 yr).

We note the good agreement between the two set of data. The wave amplitude A = 1.2 corresponds to a peak to peak variation of ~3% of the average (80 arbitrary units) TL signal. We notice that the time interval covered by the new TL time series is nearby doubled with respect to the old one. Clearly the sampling interval and the time series length do not introduce spurious effects on the determination of this periodicity and on its phase.

The $CaCO_3$ analysis of the Ionian Sea cores

For the carbonate measurements, the cores GT89-3 and GT91-1 have been divided into 1124 and 1173 samples respectively, with equal spatial thickness $\Delta d = 2.5$ mm, corresponding to $\Delta t = 3.87$ yr.

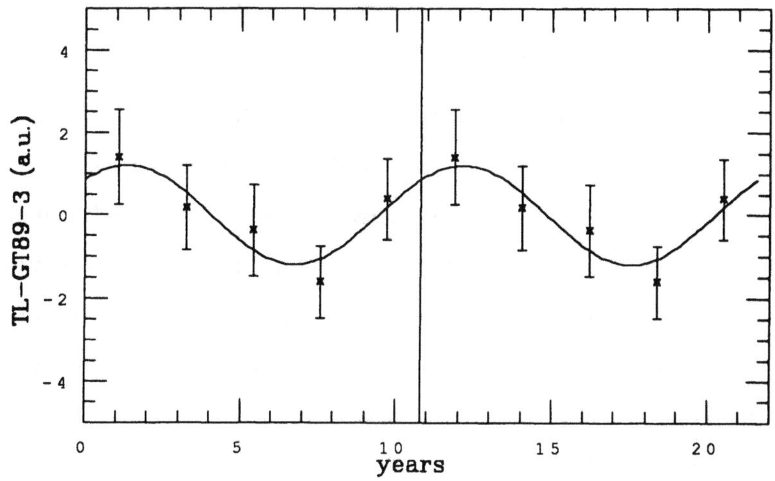

Fig. 2: Superposed data and least-square-fit sinusoid at T=10.8 yr for the TL time series of the GT89-3 sea core, as given by the method of superposition of epochs. A=1.2, ϕ=0.79, r^2=0.95. The phases are referred to 247.4 AD.

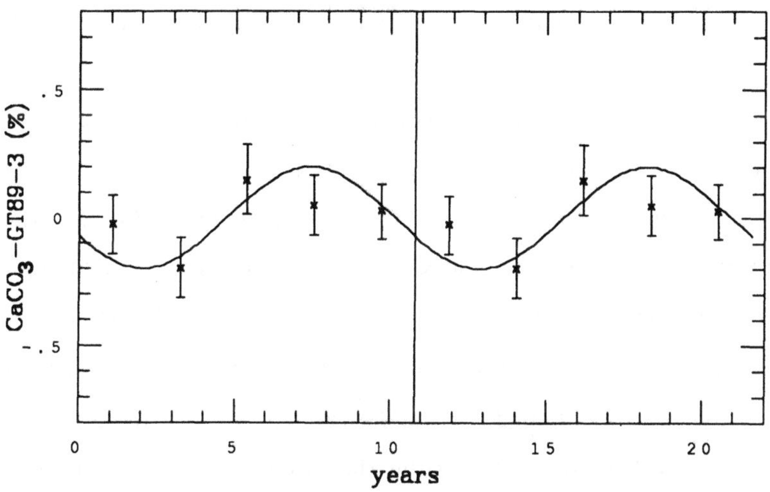

Fig. 3: Superposed data and least-square-fit sinusoid at T=10.8 yr for the $CaCO_3$ time series of the GT89-3 sea core, as given by the method of superposition of epochs. A=0.2, ϕ=3.52, r^2=0.82. The phases are referred to 247.4 AD.

The carbonates of the core are mostly $CaCO_3$, containing low concentrations (less than 0.7%) of magnesium. The integrated $CaCO_3$ content of each sample was determined by titration with EDTA, using the procedure given in Barnes (1959). The material of each sample was dried at 110^oC overnight and the $CaCO_3$ from the powder (typically 100 mg) was leached by boiling 2 or 3 times with 50 ml of 2% acetic acid. The solution was made up to a known volume and, after addition of a buffer solution, it was titrated with a 0.01 M solution of sodium salt of EDTA. Titration was performed on three aliquots of the solution and frequent titration of standard solutions was carried out to ensure reproducibility. The triplicate analysis agreed within 0.3%.

The carbonate record should be primarily climatic, e.g., temperature-precipitation-related and/or biogenic. Purely oceanic forcing is difficult to explain; invoking atmospheric exchange of CO_2 involves the complex chemistry of the upper ocean, with respect to $(HCO_3)^-$ and $(CO_3)^{2-}$ (Broecker and Peng, 1970). Consideration of the temperature-solubility problem is beyond the scope of this paper and, in any event, may be of greater complexity than for the open global ocean.

The $CaCO_3$ time series are pratically coincident within the sampling interval (see Fig.1), even if the cores are ~30 km apart from each other. To study the principal periodic components, we have analysed tha data by the method of S.E., described in the previous chapter. The $CaCO_3$ time series show several significant periodicities at about 1100, 690, 500, 340, 250, 200 and 90 yr (Cini Castagnoli et al., 1990c, 1992a,b). Here we search for the existence of the Schwabe solar cycle; in analogy to the results obtained for the TL records, we use the S.E. method at T = 10.8 yr. In Fig.3 we show the results: the carbonate wave amplitude A = 0.2 corresponds to a peak to peak variation of ~1% of the average (38%) of the total $CaCO_3$ concentration. By comparing the Fig.2 and the Fig.3, we see that the $CaCO_3$

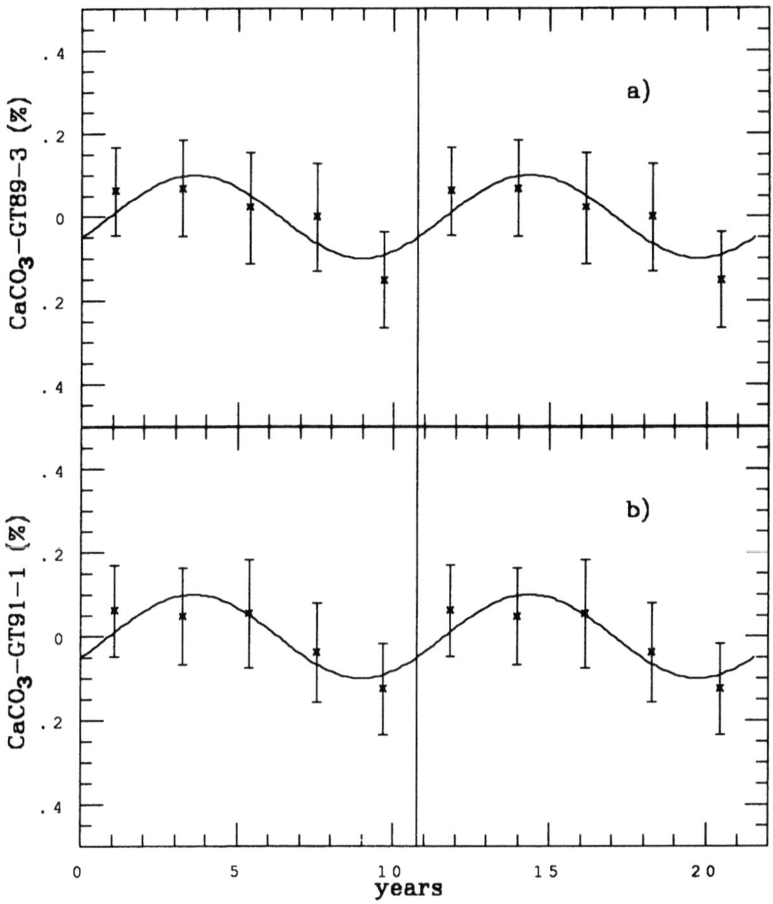

Fig. 4a,b: Superposed data and least-square-fit sinusoid at T=10.76 yr for the $CaCO_3$ time series a) of the GT89-3 sea core: A=0.1, ϕ=5.75, r^2=0.87 and b) of the GT91-1 sea core: A=0.1, ϕ=5.75, r^2=0 92, as given by the method of superposition of epochs. The phases are referred to 247.4 AD.

and TL waves are about in antiphase. However the best correlation coefficient between $CaCO_3$ experimental data and the least-square sinusoidal fit wave is given at 10.76 yr period. A comparison for this period of the GT89-3 and GT91-1 $CaCO_3$ records is shown in Fig.4a,b. We note that the phases of the two waves are identical, although the two cores were taken ~30 km apart.

In order to investigate the persistency of this wave in the last 4500 yr covered by the time series of the core GT91-1, we have applied the cyclogram method to the carbonate record. This method allows a time series to be analysed interval-wise by plotting running averages of the complex Fourier amplitudes for a given periodicity (Galli,1988). A window with length T is moved along the time series, continuously testing for a chosen periodicity τ by least-square fitting a sinusoid with fixed period τ to the data points in the window. For each window the output of the method is the amplitude R and the phase ϕ of the test sinusoid. The cyclogram is obtained by plotting the complex Fourier amplitudes $\rho=a+ib$ of the test sinusoid (where $a = R \cos\phi$ and $b = R \sin\phi$) in sequence. When the period of the test sinusoid τ is not close to any well-defined periodicity present in the data, the cyclogram resembles a random walk on the complex Fourier plane. When the chosen periodicity is present and correctly determined, the cyclogram tends to a straight line. For test periods longer than a "true" period present in the data, the cyclogram tends to turn clockwise. Thus computing several cyclograms for different test periods around a periodicity of interest (whose existence has already been indicated by the results for example of the method of S.E.) allows for testing its stability along the time series. We note that the phase cyclogram obtained by putting always R = 1 represents a useful tool for qualitatively assessing the persistency of periodic components.

In Fig.5.a we show the phase cyclogram for the test period τ = 10.74 yr, with a window T = 100 yr (equivalent to 25 data points). The cyclogram is clearly straight and shows that 30 independent vectors (whose extrema are indicated by squares) are aligned with approximately the same phase. The cyclogram turns

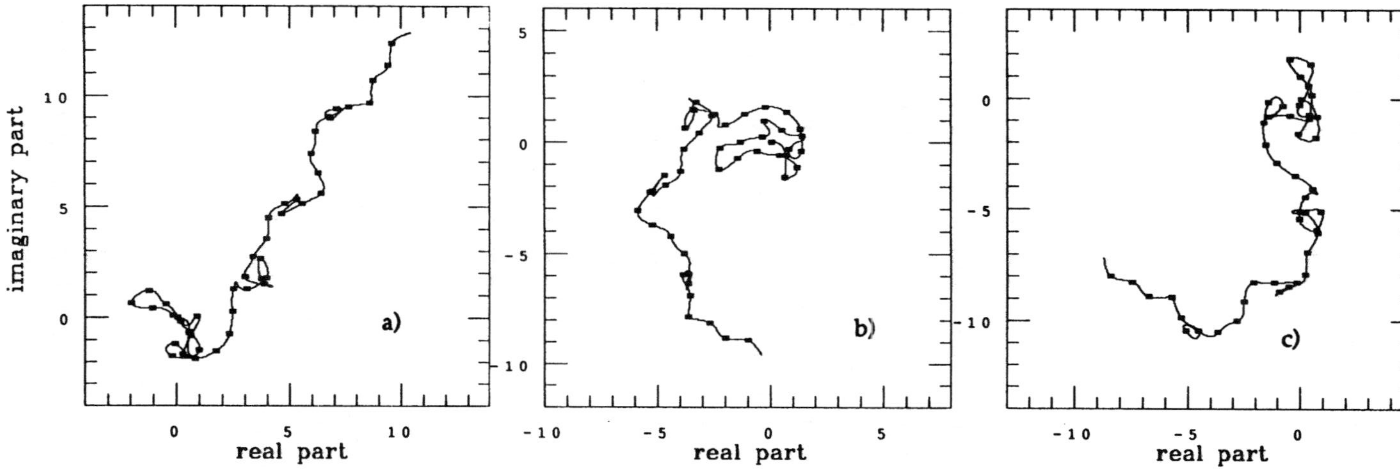

Fig. 5a,b,c: Cyclograms of the $CaCO_3$ time series of the GT91-1 sea core for a window T=100 yr and a test period: a) τ=10.74 yr, b) τ=10.72 yr, c) τ=10.76 yr.

anticlockwise when performed on a test period $\tau = 10.72$ yr (Fig.5.b) and it turns clockwise for $\tau = 10.76$ yr (Fig.5.c), confirming the non-random character of this component of the Schwabe cycle.

Conclusions

The 10.8 yr cyclicity, previously found in the TL record of the Ionian Sea core GT14, is confirmed in a longer TL time series of the GT89-3 core, sampled at different spatial intervals. A similar periodicity of ~10.75 yr is found in the carbonates profiles of the same cores and also in a core taken in the same area, ~30 km apart. The persistency of the wave with the same phase in the last 3000 yr is assured by the cyclogram analysis.

The 10.8 yr cycle is one of the components of the TL record (Cini Castagnoli et al., 1989) found in the range of the Schwabe solar cycle. Further studies are in progress in the new records in order to explore the existence of close-by decennial periodicities.

Acknowledgments. We are grateful to Prof. C.Castagnoli for helpful discussions and continuous support and to Mr. A.Romero for the attendance in taking the cores and in TL measurements.

References

Barnes HA (1959) *Apparatus and methods of Oceanography,* edited by Allen and Unwin, London

Bonino G, Cini Castagnoli G, Callegari E, Zhu GM (1993) Radiometric and tephroanalysis dating of recent Ionian Sea cores. *Nuovo Cimento* C 16: 155-162

Broecker WS, Peng TH (1970) *Tracers in the Sea.* Palisades, Eldigio Press, New York, p.690

Cini Castagnoli G, Bonino G (1988a) Solar imprint in sea sediments: the thermoluminescence profile as a new proxy record, in *Secular Solar and Geomagnetic Variation in the Last 10,000 Years,* edited by F.R.Stephenson and A.W.Wolfendale, Kluwer Academic Publ.

Cini Castagnoli G, Bonino G (1988b) Thermoluminescence in sediments: a tool for environmental investigations, in *Solar-Terrestrial Relationships and the Earth Environment in the Last Millennia,* Proc. of the Int. School of Phys. "E. Fermi", Course XCV (1985), edited by G. Cini Castagnoli, Elsevier, Amsterdam: 317-328

Cini Castagnoli G, Bonino G, Provenzale A (1988c) On the Thermoluminescence Profile of an Ionian Sea Sediment: Evidence of 137, 118, 12.1 and 10.8 yr Cycles in the Last Two Millennia. *Nuovo Cimento* C 11: 1-12

Cini Castagnoli G, Bonino G, Provenzale A (1988d) The thermoluminescence profile of a recent sea sediment core and the solar variability. *Sol. Phys.* 117: 187-197

Cini Castagnoli G, Bonino G, Provenzale A, Serio M (1990a) On the Solar origin of the thermoluminescence profile of the GT14 core. *Sol. Phys.* 127: 357-377

Cini Castagnoli G, Bonino G, Provenzale A, Serio M (1990b) On the presence of regular periodicities in the thermoluminescence profile of a recent sea sediment core. *Phil. Trans. R. Soc. Lond.* A 330: 481-486

Cini Castagnoli G, Bonino G, Caprioglio F, Serio M, Provenzale A, Zhu GM (1990c) The carbonate profile of two recent Ionian Sea cores: evidence that the sedimentation rate is constant over the last millennia. *Geoph. Res. Lett.* 17: 1937-1940

Cini Castagnoli G, Bonino G, Provenzale A (1991) Solar-Terresrial relationships in recent sea sediments, in *The Sun in Time,* edited by C.P.Sonett, M.S.Giampapa, M.S.Matthews, University of Arizona Press.: 562-586

Cini Castagnoli G, Bonino G, Provenzale A, Serio M, Callegari E (1992a) The $CaCO_3$ profiles of deep and shallow Mediterranean Sea cores as indicators of past Solar-Terrestrial relationships. *Nuovo Cimento* C 15: 547-563

Cini Castagnoli G, Bonino G, Serio M (1992b) Common spectral features in the 5500-year record of total carbonate in sea sediments and radiocarbon in tree rings. *Radioc.* 34: 798-805

Galli M (1988) Time series analysis with power spectrum and cyclograms, in *Solar-Terrestrial Relationships and the Earth Environment in the Last Millennia,* Proc. of the Int. School of Phys. "E. Fermi", Course XCV (1985), edited by G. Cini Castagnoli, Elsevier, Amsterdam: 246-273

Miono S, Otha M (1969) The possible variation of ^{14}C predicted from thermoluminescence dating of the Japanese ancient pottery, in *Proc. 16th Int. Cosmic Ray Conf.,* vol.2: 263-268.

Panel Discussion on PALEOCLIMATIC DATA

J.C. Duplessy[1], R.S. Bradley[2] and K. Briffa[3]

[1] Centre des Faibles Radioactivités, Laboratoire mixte CNRS - CEA, Gif-sur-Yvette, France
[2] University of Massachusetts, Amherst, U.S.A.
[3] University of East Anglia, U.K.

1. Introduction

Analysis of the possible influence of variations of the energy emitted by the Sun in dictating the climate of the last past millennia is important because of the possible effect of solar variations in enhancing or reducing the effects of global warming by greenhouse gases released in the atmosphere since the beginning of the industrial era. However, instrumental records of climate are generally limited to the last two centuries or less and the reconstruction of climatic conditions prevailing during the few centuries which preceded the development of industry requires an extensive use of paleodata.

A meaningful reconstruction of the "Little Ice Age", or similar cold spells of short duration, is made particularly difficult because it critically depends on both the accuracy of the estimates of climatic parameters derived from various transfer functions and the resolution of the paleoclimatic records. The panel therefore discussed the accuracy of the paleodata for reconstructing the recent past (i.e. about the last two millennia), the reliability of evidence for a "Little Ice Age", and the uniqueness of this event.

NATO ASI Series, Vol. I 25
The Solar Engine and Its Influence
on Terrestrial Atmosphere and Climate
Edited by Elizabeth Nesme-Ribes

2. What is the potential accuracy of the proxy data ?

In order to reconstruct past climate, quantitative estimates of physical parameters such as mean monthly or seasonal temperature and precipitation are required. However the historical or geological data are different in nature. They often rely on plants and animals, which are sensitive to environmental changes, but are not passive recorders (Eddy and Oeschger, 1993). In addition, in the same way as a written manuscript can be altered after its deposition in a library, geological deposits can be altered by natural hazards or the activity of microorganisms living in the sediments or rocks. For each paleorecorder, one must determine what forcing factors are primarily responsible for the variability in the data and what non-climatic factors have possibly disturbed the record. In this section, we address the reliability and accuracy of climate estimates derived from historical documents, ice cores, tree rings, continental deposits (in lake and peat bogs), corals and ocean sediments and focus on the problems associated with the study of a period as recent as the last millennium.

2.1 Historical records

Diaries, annals and chronicles can provide very high frequency information on any event of dramatic character (flood, drought, storm, exceptionally cold or warm spells). However, it is difficult to quantify, in terms of modern climate, what are often qualitative descriptions of meteorological conditions in the past. The wealth of material is such that any anomalous season since about AD 1400 is probably already known in Europe (Le Roy Ladurie, 1971 ; Lamb, 1982 ; Pfister, 1985). Numerous data are available in China and Japan and there is much potential for further analysis in other countries such as Korea and India (Bradley, 1993). Unfortunately, there have been very few systematic studies of historical records in other regions, particularly in the Americas, Africa and Southeast Asia. Colonial era records from such regions almost certainly contain useful climatic information (e.g. drought frequency in Indonesia as related to extreme ENSO events) but such resources have yet to be tapped.

2.2 Ice records

Glaciers and ice sheets consist of layers of snow compacted under their own weight to form ice. An ice core therefore provides a detailed (and possibly complete) sequence of precipitation samples from the past. The $^{18}O/^{16}O$ ratio of the ice is an index of the air temperature at the time snow formed in the atmosphere prior to its deposition and eventual transformation to ice. Because water molecules containing heavy isotopes (^{18}O or D) evaporate less rapidly from surface water and condense more readily in clouds than $H_2{}^{16}O$ molecules, lower concentrations of heavy isotopes are found in precipitation condensed at lower temperature (Dansgaard, 1964). Empirical linear relationships have been found between $^{18}O/^{16}O$ ratio of recent ice and mean annual air temperature. They are valid for modern conditions and will be valid for earlier climatic regimes only if the source of water and cooling mechanisms of water vapour along its trajectory to the glacier have not changed significantly and if the distribution of snowfall events throughout the year has remained approximately the same.

Long-term isotopic records from Greenland or Antarctica are so similar that there is no doubt that they reflect both global trends and regional climatic changes (Lorius *et al.*, 1985 ; Dansgaard *et al.*, 1993). However, a study of the noise of samples representing a ten-year average in East Antarctica was made for drilling sites 10 km apart. It shows that a decade with a mean temperature 1 °C cooler or warmer than the long time average cannot be reconstructed based on one single core despite the great accuracy of the isotopic measurements (Petit *et al.*, 1982). Records with a greater resolution are thus necessary. Greenland is a particularly favourable place since the accumulation of snow is generally an order of magnitude larger than in Antarctica and provides the possibility of reconstructing seasonal cycles for more than one century. However, with depth the progressive thinning of the ice and diffusion artificially reduce the amplitude of these cycles in the isotope record.

2.3 Tree rings

In many areas of the world, especially in mid- to high-latitude regions, tree-growth is strongly affected by the annual cycle of climate. This seasonal limitation produces clear annual rings which can be dated and measured to give quantitative estimates of varying tree-growth from year to year (Fritts, 1976 ; Hughes *et al.*, 1982 ; Schweingruber, 1988). By comparing and combining the ring-patterns of many trees within a site or region and by amalgamating the overlapping records from modern, historical and even sub-fossil material, very long records of mean tree-growth can be assembled (Baillie, 1982). These records can be compared with measured instrumental data to establish statistical transfer functions which are then used to give estimates of past climate. Though empirically-based, tree-ring reconstructions are routinely compared with independent instrumental data to establish and quantify their veracity (Fritts *et al.*, 1990).

The potential for reconstructing various climate parameters depends critically on the specific nature of the climate forcing at particular tree-growth sites. High-altitude and high-latitude trees respond principally to temperature change. In semi-arid regions the limitation is available soil moisture, and information on precipitation-related variables is more feasibly extracted. In less extreme environments the strength of the climate 'signal' in tree-ring data will be reduced, relative to that in harsher, more marginal, regions. The recovery of climate information is more difficult in these less extreme areas, as it is more likely to be 'contaminated' with non-climatic effects such as disturbance or competition.

When developing mean site chronologies, the original measured time series for individual trees are standardized to remove the variance associated with tree aging. This imposes a restriction on the long-timescale variance in the resulting indices and limits the potential for reconstructing long-timescale climate change (Cook and Briffa, 1990). Typically, variations on the order of several centuries and longer are not represented in most dendroclimatic reconstructions. In exceptional cases, however, where trees are very old (perhaps thousands of years) or 'conservative' methods of standardization are employed, this limit is considerably extended (e.g. Dupouey *et al.*, 1992 ; Briffa *et al.*, 1992).

Extensive networks of tree-ring data, comprising densitometric as well as ring-width parameters, are under construction in many areas of the Northern and Southern hemispheres and several projects around the world are constructing continuous chronologies over thousands of years (e.g. see papers in Bradley and Jones, 1992). Tree-rings will therefore increasingly provide absolutely-dated and annually-resolved reconstructions of a range of temperature, precipitation- and even circulation-related patterns, with spatial representation ranging from individual sites, through regional-mean up to continental-scale patterns. The potential use of other extractable parameters, such as carbon, hydrogen and oxygen isotopes (e.g. White, 1983 ; Leavitt and Long, 1984 ; Luckman *et al.*, 1985) will ensure that tree-rings take on an increasingly important role, bridging the gap between instrumental climate records and other, longer-timescale but less-well-resolved, information such as that derived from pollen.

2.4 Continental deposits

Lake sediments and peat bogs collect pollen grains which are indicative of the local/regional vegetation, which is itself linked to the regional climate. However the short-term stability of the vegetation may prevent accurate detection of decadal climatic variations of small amplitude.

Dating of continental deposits is generally difficult and the time scale of most continental records has a resolution that is no better than a few decades. Annually laminated sediments (varves) offer a unique opportunity to date continental deposits with an annual resolution. Still, even in these cases, the time scale of the paleoclimatic record is often hampered by missing varves or processes which result in several laminations within a single year. The inertia of the vegetation is such that pollen analysis does not allow full benefits to be derived from the high resolution offered by varves. By contrast, the $^{18}O/^{16}O$ ratio of carbonates precipitated locally in a lake offers an efficient paleoclimatic tool to reconstruct variations of water temperature ; the uncertainties of the estimates are similar to those derived from the isotopic analysis of ice cores.

2.5 Ocean sediments

Cores from the ocean floor contain clay minerals originating from the erosion of the continents, fossils of animals (foraminifera, radiolarians) and plants (coccoliths, diatoms). With reference to modern conditions, transfer functions are empirically derived equations for calculating quantitative estimates of past sea surface temperatures from paleontological data. Ideally, this problem requires the formulation of a model that explains the biological responses in terms of the multitude of factors comprising the observed ecological system. Techniques, which use multivariate statistical procedures, were first applied to marine plankton data by Imbrie and Kipp (1971) and provide typically sea surface temperature estimates with an accuracy of 1.5°C. The $^{18}O/^{16}O$ ratio of foraminiferal calcite or diatom silica is a function of both the temperature and the isotopic composition (i.e. the salinity) of the ocean water when and where the microorganisms lived (Emiliani, 1955). If the salinity would remain constant, the accuracy of temperature estimates deduced from $^{18}O/^{16}O$ measurements would be close to 0.3°C. If salinity is varying with time, the simultaneous use of micropaleontologic transfer functions and $^{18}O/^{16}O$ ratio analysis enables both the temperature and salinity of the past ocean to be reconstructed (Duplessy *et al.*, 1991). Unfortunately, the resolution of the oceanic record is generally poor for recent paleoclimates. Material is deposited at a fairly constant though slow sedimentation rate, of the order of a few centimeters per thousand years. Only in exceptional cases (such as the mouth of a river, or oceanic zone where deep currents deposit their sedimentary load) the sedimentation rates may be higher than 20cm per thousand years, allowing detection of changes occurring in less than one century. However, even in these favourable cases, the activity of benthic organisms mixes the sedimentary record, blurring the paleoclimatic record and reducing the amplitude of the climatic variations registered in the sediments. The great potential accuracy of the oceanic paleodata can be fully used only when varved sediments are deposited in rare deep trenches where the oxygen content of the water is zero and no benthic life can survive. For instance, such conditions are found in some areas of the Gulf of California ; the isotope analysis of diatoms showed that El Niños were more intense than today at the beginning of the 20th century (Juillet- Leclerc and Schrader, 1987).

2.6 Corals

Corals are warm water animals which build their skeleton in aragonite. They are present in the tropical basins of the three major oceans. The $^{18}O/^{16}O$ ratio of their skeleton, like that of the shells of foraminifera, is a simple function of both sea surface temperature and salinity. In this latter case, the full potential of oxygen isotope geochemistry can be used, because corals grow annual layers which may be sampled to reconstruct detailed seasonal information from the marine environment. Such detail offers the rare opportunity to reconstruct short-lived extremes as well as a high-resolution view of the natural variability in space and time of the monsoon or the El Niño Southern Oscillation.

The use of coral records is a relatively new approach and their potential is not fully explored (Dunbar and Cole, 1993). Continuous coral records may span several centuries and floating chronologies offer high-resolution windows into much older periods. It is expected that over the last 300 years, stable isotope and geochemical measurements will permit the reconstruction of past hydrological conditions (temperature, salinity, runoff, etc) with monthly to seasonal resolution. The same method could be used to generate mean annual values for the last millennium and mean decadal values for the last 18,000 years.

3. How reliable is the evidence for a "Little Ice Age" ?

The term "Little Ice Age" is widely used to describe the period of a few centuries between the Middle Ages and the present century (Grove, 1988). It is considered as a cold period, because mountain glaciers in many parts of the world expanded. In Europe, historical chronicles testify to the occurrence of more frequent cold weather spells, significantly cooler than present-day conditions (Pfister, 1985). Unfortunately, few areas of the world have the wealth of historical and high resolution paleoclimatic proxy records which are available for Europe, so it is difficult to assess, in detail, the global or even hemispheric extent of climatic fluctuations over the last 500-1000 years. Based on the analysis of different types of proxy climate records from many regions of the northern hemisphere, it appears that temperatures were not uniformly cold during the last 500 years, though overall temperatures were lower than 20th century averages.

In general, the coldest periods were in the late 16th and early 17th centuries, and in the early to mid-19th century, though the coldest individual decade of summer temperatures in the last 500 years in the northern hemisphere was 1600 - 1609 (0.43°C below 1950-79 averages) (Bradley and Jones, 1993). Warm intervals (sometimes exceeding 20th century levels) were also experienced. No such estimates are yet available for other seasons, or for the southern hemisphere. However, it is worth noting that these conclusions are based entirely on mid- to high latitude continental data (30-70°N) ; virtually no comparable records exist for the Tropics and sub-Tropics, which make up 50% of the hemisphere. In these regions, climatic changes on this timescale are often manifested most strongly in hydrological changes, which are especially apparent where the levels of closed basin lakes can be reconstructed (Street-Perrott and Harrison, 1985). Nevertheless, there are still few records with the necessary resolution from such regions.

Similarly, there is a dearth of records from the southern hemisphere. High resolution records from the southern hemisphere are especially needed to understand the important forcing mechanisms. If volcanic aerosols were a primary factor modulating climate over the last millennium, the paleotemperature record of the northern and southern hemispheres is likely to have been different, since aerosols from explosive eruptions tend to remain largely in the hemisphere of the eruption (except for near-equatorial events) and thus affect solar radiation receipts in each hemisphere differently (Lamb, 1970). On the other hand, if solar forcing has been of more significance, temperatures in the two hemispheres should have responded more in phase. Unfortunately, the number of southern hemisphere records available to answer these questions are too few at present, and the extent to which temperature changes in the two regions co-varied on even a decade to century scale remains unknown. In short, the spatial and temporal pattern of climate variations in the "Little Ice Age" period are still poorly understood, which makes it difficult to determine the most likely causative (forcing) factors for the observed environmental changes. In particular, there is currently insufficient evidence to unequivocally identify solar activity variations as an important factor in the changes. Certainly, there is no evidence to support the notion that the Maunder Minimum (approximately A.D. 1655-1715) was a uniquely cold period in the last 500 years, and thus the term "Little Ice Age" should not be considered as synonymous with the Maunder Minimum.

4. Was the "Little Ice Age" unique ?

The "Little Ice Age" had such a short duration that the insolation forcing which is responsible for major long-term climatic changes during the Quaternary (the so-called Milankovitch orbital variations) has remained essentially constant during the whole climatic event. Hence, other causes need to be considered to explain this climatic oscillation. It is interesting to note that the climatic record of the Holocene shows events of a similar duration and magnitude (or even stronger) also occurred in the past, particularly in the late Holocene (after 5,000B.P.).

Isotope analyses of the GRIP ice core from Central Greenland provides evidence for general instability of past climate and show that Holocene conditions, at least in Greenland, were in fact exceptionally stable as compared to the climatic evolution of the last 150,000 years (GRIP Project Members, 1993). In particular, the GRIP record suggests that climate in Greenland during the last interglaciation, which has been warmer than the present one, was cut by a series of severe cold periods, which began abruptly and lasted from decades to centuries. However, this information is not confirmed by the GISP core, which was drilled about 30 km from the GRIP core (Grootes *et al.*, 1993). The climate records of both cores agree for the upper 2700 meters (representing about 87,000 years), but exhibit significant differences for the older periods. The differences between the records must result from flow deformation at one or both of the cored localities. Inclined layering present in the deep parts of both cores may indicate such flow deformation and the abrupt temperature changes of the Eemian interglacial must be confirmed by the study of high sedimentation rate cores from the North Atlantic Ocean or new drillings in northern Greenland.

In conclusion, the "Little Ice Age" is not a unique feature of recent climate but rather one of a series of multi-decade-to-century scale oscillations in climate. The causes of such oscillations must be understood before it will be possible to reliably anticipate future anthropogenic impacts on climate.

5. References.

Baillie, M.G.L. (1982), in Tree Ring Dating and Archaeology, Croom Held, London: 357 pages

Bradley, R.S. (1993), in High Resolution Records of Past Climate from Monsoon Asia : the last 2000 years and beyond, in PAGES Core Project Office Report 93-1, Bern, Switzerland: 46 pages

Bradley, R.S., and Jones, P.D. (eds.) (1992), in Climate Since A.D. 1500, Routledge, London: 679 pages

Bradley, R.S., and Jones, P.D. (1993). Little Ice Age summer temperature variations : their nature and relevance to recent global warming trends. The Holocene, 3: 367-376

Briffa, K.R., Jones, P.D., Bartholin, T.S., Eckstein, D., Schweingruber, F.H., Karlin, W., Zetterberg, P., and Eronen, M. (1992). Fennoscandian summers from A.D. 500 : temperature changes on short and long timescales. Climate Dynamics, 7: 111-119

Cook, E.R., and Briffa, K.R. (1990). A comparison of some tree-ring standardization methods, in Methods of Dendrochronology, eds. E.R. Cook and L.A. Kariukstis. Kluwer/IIASA, Dordrecht: 153- 162

Dansgaard, W. (1964). Stable isotopes in precipitation. Tellus, 16: 436-468

Dansgaard, W., *et al.* (1993). Evidence for general instability of past climate from a 250-kyr ice-core record. Nature, 364: 218-220

Dunbar, R.B., and Cole, J.E. (1993), in Coral Record of Ocean-Atmosphere Variability, eds. R.B. Dunbar and J.E. Cole, NOAA Special Report, 10: 38 pages

Duplessy, J.C., Labeyrie, L.D., Juillet-Leclerc, A., Maitre, F., Duprat, J., and Sarnthein, M. (1991). Surface salinity reconstruction of the North Atlantic Ocean during the last glacial maximum. Oceanologica Acta, 14: 311-324

Dupouey, J.L., Denis, J.B., and Becker, M. (1992). A new method of standardization for examining long-term trends in tree-ring chronologies, in : Tree Rings and Environments, eds. T.S. Bartholin, B.E. Berglund, D. Eckstein and F.H. Schweingruber, LUNDQUA Report 34, Dept. of Quaternary Geology, Lund University, Sweden, 85-88

Emiliani, C. (1955). Pleistocene temperatures. J. Geology, 63 : 538-578

Eddy J.A., and Oeschger, H. (1993). Global Changes in the Perspective of the Past. J. Wiley and sons, New York: 383 pages

Fritts, H.C. (1976), in Tree Rings and Climate. Academic Press, London: 567 pages

Fritts, H.C., Guiot, J., Gordon, G.A., and Schweingruber, F.H. (1990). Methods of calibration, verification and reconstruction, in Methods of

Dendrochronology, eds. E.R. Cook and L.A. Kariukstis, Kluwer/IIASA, Dordrecht, 163-217

GRIP Project Members (1993), in Climate instability during the last interglacial period recorded in the GRIP ice core. Nature, 364: 203-207

Grootes, P.M., Stuiver, M., White, J.W.C., Johnsen, S., and Jouzel, J. (1993). Comparison of oxygen isotope climate record of the greenland summit area. Nature, 366: 552-554

Grove, J.M. (1988), in The Little Ice Age, Methuen, London: 498 pages

Hughes, M.K., Kelly, P.M., Pilcher, J.R., and LaMarche, V.C., Jr. (1982), in Climate from Tree Rings, eds. M.K. Hughes, P.M. Kelly, J.R. Pilcher, and V.C. LaMarche, Cambridge University Press, Cambridge: 457 pages

Imbrie, J., and Kipp, N.G. (1971). A new micropaleontological method for quantitative paleoclimatology : Application to a late pleistocene Caribbean core. The Late Cenozoic Glacial Ages, ed. K. K. Turekian: 71-181

Juillet-Leclerc, A., and Schrader, H. (1987). Variations of upwelling intensity recorded in varves sediment. Nature, 329: 146-149

Lamb, H.H. (1982), in Climate History and the Modern World. Methuen, London: 387 pages

Lamb, H.H. (1970). Volcanic dust in the atmosphere, with a chronology and assessment of its meteorological significance, in Philosophical Transactions of the Royal Society of London, A266: 425-533

Le Roy Ladurie, E. (1971). Times of Feast, Times of Famine : A History of Climate since the Year 1000 translated bu B. Bray, New York, Doubleday: 426 pages

Leavitt, S.W., and Long, A. (1984). Sampling strategy for stable carbon isotope analysis for tree rings in pine. Nature, 311: 145-147

Lorius, C., Jouzel, J., Ritz, C., Merlivat, L., Barkov, N.I., Korotkevich, Y.S., and Kotlyakov, V.M. (1985). A 150000 Year climatic record from Antarctic ice. Nature, 316: 591-596

Luckman, B.H., Hamilton, J.P., Jozsa, L.A., and Gray, J. (1985). Proxy climatic data from tree rings at Lake Louise, Alberta : a preliminary report. Géographie et Physique du Quaternaire, 39: 127-140

Petit, J.R., Jouzel, J., Pourchet, M., and Merlivat, L. (1982). A detailed study of snow accumulation and stable isotope content in Dome C (Antarctica). J. Geophys. Res., 87: 4301-4308

Pfister, C. (1985), in Das Klima der Schweiz von 1525-1860 und seine Bedeutung in der Geschichte von Bevölkerung und Landwirtschaft. 2 vols. Academica Helvetica, Bern: 746 pages

Schweingruber, F.H. (1988), in Basics and Applications of Dendrochronology. Kluwer, Dordrecht: 329 pages

Street-Perrott, A., and Harrison, S.P. (1985). Lake levels and climate reconstruction, in Paleoclimate Analysis and Modelling eds. A.Hecht, J.Wiley, New York, 291-340

White, J.W.C. (1983). The climatic significance of D/H ratios in white pine in the northeastern United States., in PHD dissertation, Columbia University, New York

SENSITIVITY OF CLIMATE TO LONG-TERM VARIATIONS OF THE SOLAR OUTPUT

R. Sadourny
Laboratoire de Météorologie Dynamique du CNRS
École Normale Supérieure, Paris

1. INTRODUCTION

Fluctuations of solar irradiance can have a significant impact on terrestrial climate only if they are strong enough and last long enough to induce significant warming of the upper ocean. This is hardly true for the eleven year cycle, whose amplitude in terms of mean absorbed solar flux is less than 0.1 Wm^{-2}. The Maunder Minimum case is somewhat different, as the solar luminosity in the second half of the XVIIth Century may have been less than its present value by as much as 0.4%, i.e. about 1 Wm^{-2}. The Maunder Minimum is therefore a possible candidate for explaining the so-called Little Ice Age episode. It is, however, very difficult to really assess the relevance of such an hypothesis, as the chronology of the Little Ice Age itself is somewhat blurred, and we lack quantitative data on various possible other perturbation mechanisms that could have induced cooling, such as volcanic aerosols or a significant change in the deep ocean circulation. Concerning the Little Ice Age itself, the present data situation is not so good : XVIIth Century paleotemperature data are scarce, and the most abundant data concern glacier extensions, which cannot be directly used for climate model validation ; and we cannot expect fast improvement of this situation.

Meanwhile, it is relatively easy to perform equilibrium climate simulations to assess the response of terrestrial climate to long-term variations of solar irradiance, and calculate the amplitude and geographical structure of the

NATO ASI Series, Vol. I 25
The Solar Engine and Its Influence
on Terrestrial Atmosphere and Climate
Edited by Elizabeth Nesme-Ribes

response for a perturbation of the order of magnitude of the Maunder Minimum. This has been done by Nesme-Ribes *et al.*, (1993), and gives temperature variations not too far from observations, in the very scarce areas where we do have observations. Still, the problem here is our confidence in the model. We know that our current models are insufficiently validated, as far as climate sensitivity is concerned : climate sensitivity depends crucially on complicated processes like cloud radiative forcing, which are difficult to quantitatively assess and parametrise in general circulation models (Cess *et al.*, 1990). The model used in Nesme-Ribes *et al.*, (1993) is one of the most sensitive at hand, and may therefore have overestimated the impact of the Maunder Minimum perturbation.

Here I shall discuss some aspects of the climate sensitivity to solar output variations. The discussion will be based on the paper by Nesme-Ribes *et al.*, (1993). I shall concentrate mainly on qualitative aspects and give some interpretations of the dynamics of climate response, which I believe are not too dependent on the particular characteristics of the model wich was used in the experiments.

2. SOLAR WARMING *vs.* GREENHOUSE WARMING

Figure **1** compares the zonally average meridional structure of warming as obtained from the LMD general circulation model coupled with a slab ocean, for an 1.67 % increase in solar irradiation and a doubling of the CO_2 content (the two cases inducing a radiative perturbation of $4 Wm^{-2}$ at the top of the atmosphere). The most striking features are the strong similarity of the tropospheric warming distributions, and the weak stratospheric cooling which occurs even in the solar perturbation case. The explanation given in Nesme-Ribes *et al.*, (1993) relies on the additional greenhouse effect of water vapour : tropospheric warming induces tropospheric moistening and decreases the upwelling long-wave flux reaching the lower stratosphere. Note however that the atmospheric model used in these experiments has a too coarse vertical resolution for giving a precise description of the stratospheric response ; the reader is referred to the work of Royer *et al.*, (1993) for a more detailed description.

The close similarity of the tropospheric warming structures displayed in Figure 1 reflects in fact a similarity of the whole three-dimensional structure of tropospheric climate response, including its geographical distribution. The structure of the response appears mainly driven by the internal feedback processes, in absence of specific features of the forcing perturbation. Both a solar luminosity variation and a change in atmospheric CO_2 are deprived of geographical and seasonal signature. This would not be the case, for example, in the variations of solar forcing driven by changes in orbital parametres : changes in orbital forcing produce specific geographical effects.

The main characteristics of climate warming in the troposphere can be described as follows. The pole-to-equator temperature gradient in the lower troposphere decreases as a result of the snow-albedo feedback and the associated stronger polar warming. This also means a decrease of the energy deficit at higher latitudes. On the other hand, the pole-to-equator moisture gradient increases, due to the very strong dependency of saturation mixing ratio on temperature, as expressed by the Clausius-Clapeyron law : the increase of $\partial q_s/\partial T$ with temperature is so large (about 15 times larger in the tropics than in polar regions) that it easily overcomes the stronger polar warming. In the tropics, the warming is strongest in the upper troposphere, due to enhancement of convection by strong moistening of lower layers ; and this strong upper tropospheric warming in the tropics is in turn associated with a strengthening of the subtropical jets. The mass fluxes of the Hadley cells decrease, but their meridional extent increase, so that they penetrate more into the subtropics ; and they carry as well less energy. The subtropical dry regions extend poleward, while precipitation and soil moisture increase in midlatitude winter.

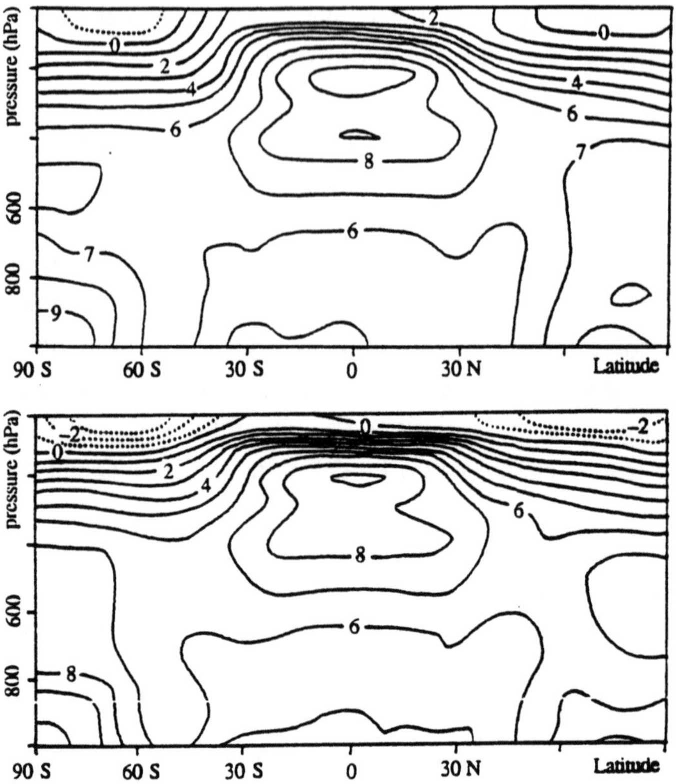

Figure 1. Annually and zonally average temperature changes for a 1.67% increase of the solar constant (above) and a doubling of the atmospheric CO_2 (below) in K, as simulated by the LMD general circulation model.

3. WARMING DYNAMICS AND ENERGETICS IN THE TROPICS

3.1. Energetic efficiency of the Hadley-Walker circulation

Tropical circulation is driven by the need for the atmosphere to transport energy from its main source region (the Intertropical Convergence Zone, especially its continental extensions where solar heating cannot be absorbed by the surface to a significant amount) to its subtropical sink regions (most obviously the desert belt of the winter hemisphere, but also the colder, cloud free eastern sides of the tropical oceans). In the tropics, the absence of significant horizontal temperature gradients makes energy transport by horizontal transient mixing quite inefficient, and the only structure that can transport energy is then the large-scale rolls of the Hadley and Walker circulations. Energy transport by the Hadley-Walker rolls occurs as a differential between enthalpy and latent heat import, driven by the lower branches, and potential energy export, driven by the upper branches. The efficiency of such a thermal engine to extract energy from a given source area can be defined as the ratio of the net energy extracted to the enthalpy and latent heat inflow needed to extract it ; in other words,

$$\eta = - \frac{S}{F(c_p\ T) + F(Lq)} \tag{1}$$

where S is the time-average integral of the energy input over the source area, and F denotes a time-average energy flux across a closed cylinder encircling the source area, counted positively outwards. An alternative form equivalent to **(1)** is

$$\eta = - \frac{F(gz) + F(c_p\ T) + F(Lq)}{F(c_p\ T) + F(Lq)} \tag{2}$$

As the lower branch mass flux equilibrates the upper branch flux, **(2)** can be approximately written

$$\eta = - \frac{g\Delta z + c_p \Delta T + L\Delta q}{c_p \Delta T + L\Delta q} \tag{3}$$

where Δ refers to differences between upper branch and lower branch values. The efficiency η is thus a gross measure of the mean stability of the atmosphere in intermediate areas separating energy sources from energy sinks. Atmospheric data for the Hadley cell (*e.g.*, Oort and Peixoto 1992) show that it is of the order of 10%. One basic question then is whether this efficiency undergoes significant change in a climate warming process (a related question being to what extent it depends on details of the parametrisation of moist convection in atmospheric models).

Denoting by primes the derivatives with respect to the warming parameter (i.e., the solar flux or the atmospheric content in greenhouse gases), the efficiency tendency is given by

$$\frac{\eta'}{1+\eta} = \frac{\Delta z'}{\Delta z} + (1+\eta)\frac{c_p \Delta T' + L\Delta q'}{g\Delta z} \tag{4}$$

Separating the lapse rate effect from the tropopause lift effect in the temperature term :

$$\Delta T' = \frac{\partial T}{\partial z}\Delta z' + \Delta z \frac{\partial T'}{\partial z} \tag{5}$$

where vertical derivatives refer to mean lapse rates, we get the equivalent form

$$\frac{\eta'}{1+\eta} = \left[1 + (1+\eta)\frac{c_p}{g}\frac{\partial T}{\partial z}\right]\frac{\Delta z'}{\Delta z} + (1+\eta)\frac{c_p}{g}\frac{\partial T'}{\partial z} + (1+\eta)\frac{L\Delta q'}{g\Delta z} \tag{6}$$

One sees from (6) that the net effect of tropopause lift is to enhance efficiency (the square bracket is approximetely 0.3) ; efficiency is also enhanced by the decrease of lapse rate ($\partial T'/\partial z > 0$), but is strongly decreased by moistening of the lower layers ($\Delta q' < 0$). In the LMD model the moistening effect balances almost exactly the two other effects and η does not vary significantly in the warming process. The fact that η is relatively insensitive to changes in the thermodynamic structure of the atmosphere looks rather reasonable ; it would be of interest to compare several models or parametrisations of moist convection on this specific point.

3.2. Weakening and poleward expansion of the Hadley cell

Just like the Hadley-Walker circulation, the Hadley cell itself is basically driven by the need of the atmosphere to transport the excess of solar energy from the meteorological equator to the energy sinks of the subtropical desert belts. (Of course the ocean also takes part in the process, which makes things somewhat more complex if we look at the whole system). The deficit of the subtropical belt with respect to the meteorological equator will indeed govern the magnitude of energy transport in the tropics. In the version of the LMD model used here, an increase of the solar constant or an increase of CO_2 loading induces a stronger increase of radiative forcing in the subtropics than in the equatorial belt (see Li and Le Treut, 1992, for the CO_2 case) : in particular, the water vapour and the cloud feedbacks appear significantly stronger in the subtropics (probably because the secondary effects of moistening are felt more strongly in a warm, relatively dry atmosphere) than in the already moist equatorial belt. Then, because the subtropical energy deficit weakens, the Hadley circulation must reorganise itself to transport less energy towards the subtropics, especially on the winter hemisphere side. This is indeed observed in equilibrium simulations, where the ocean circulation is held fixed ; it is unlikely, though, that things would change very much if the ocean response was taken into account, as the energy sinks of the Hadley circulation are basically over continental areas. Now, how can a Hadley cell reduce its energy transport in a warmer climate ? If, as we have seen above, the efficiency η does not vary significantly, then the mass flux of the Hadley cell must decrease. If η does not vary, then the meridional energy flux, approximated as

$$F = \frac{\eta}{1 + \eta} Mg\Delta z = -\eta M \, (c_p \, \Delta T + L\Delta q) \qquad (7)$$

where M is the Hadley cell mass flux, must vary according to

$$\frac{F'}{F} = \frac{M'}{M} + \frac{\Delta z'}{\Delta z} = \frac{M'}{M} + \frac{\Delta(c_p \, T' + Lq')}{\Delta(c_p \, T + Lq)} \qquad (8)$$

For an increase of 0.4% in solar irradiance, the LMD model approximately gives –3%, –6% and +3% for these three terms. Figure 2 illustrates the mass flux decrease by comparing the Hadley cells of warm and cold climates obtained by increasing or decreasing the solar irradiance by 1.67% (4 Wm^{-2} in mean absorbed solar energy). Another approach to explaining the weakening of Hadley cell mass flux has been given in Sadourny (1994). It is based on an analysis of the longitudinal component of the baroclinic source term of the vorticity equation in the tropical belt. The magnitude of the longitudinal vorticity source is proportional to the equilibrium Hadley cell mass flux, as the baroclinic vorticity source is equilibrated by diffusion. In a global warming situation, the variation of the baroclinic source term is controlled by the warming itself (T') and the change of temperature lapse rate ($\partial T'/\partial z$), which both tend to decrease the source term, and by the change of meridional temperature gradient ($\partial T'/\partial \varphi$), which is rather small in the tropics due to the homogeneization of angular momentum and temperature by Hadley motion (Held and Hou 1980).

The meridional extension of the Hadley cell is another important matter, as the position of the descending branch determines the location of desert and semi-desert areas. The latitude of Hadley cell subsidence depends on the strength of the zonal wind and therefore on the magnitude of the pole-to-equator temperature gradient (Held and Hou 1980). The strength of the jet increases as the Hadley cell extends closer to the poles, due to the eastward deviation of poleward moving air masses in the upper troposphere. In turn, geostrophic balance imposes that stronger jets be associated with stronger meridional temperature gradients. This is the case in the upper troposphere of a warm climate, due to strong convective warming of upper tropical layers. To produce stronger tropospheric jets in equilibrium with such thermal gradients, the Hadley cell will have to expand further poleward. The poleward expansion of the winter hemisphere Hadley cell in a warmer climate is also clearly seen in Figure **2**.

4. EFFECTS OF WARMING IN MIDDLE LATITUDES

The response of stationary waves and baroclinic transients to climate warming has been discussed by Held (1993). In particular, the decrease of the lower tropospheric pole-to-equator temperature gradient may yield weaker baroclinic transients in the lower troposphere (although moistening might to some point counteract that effect). The simulations of Nesme-Ribes *et al.*, (1993) show a decrease of the poleward enthalpy fluxes and an increase of the poleward latent heat fluxes. Both fluxes are dominated by lower tropospheric contributions. The poleward enthalpy flux due to transient eddies is more or less proportional to the square of the lower tropospheric meridional temperature gradient : during climate warming, it decreases rather strongly, due to the mixing of weaker contrasts by weaker transients. The most interesting feature is the increase of the poleward moisture flux in the Southern Hemisphere and Boreal winter. The strengthening of the meridional moisture gradient is powerful enough to induce enhanced mixing of moisture in spite of the weaker transients. Even more importantly, in spite of the increase of evaporation due to warming, the wintertime soil moisture increases in the midlatitudes of the Northern Hemisphere (similarly, the precipitation-minus-evaporation budget increases over midlatitude oceans). The increase of precipitation is a direct result of the strengthening of meridional mixing of moisture : it is thus remotely driven by the amplitude of tropical moistening which forces the increase of the meridional moisture gradient. Evaporation, on the other side, is a local effect driven by the Clausius-Clapeyron equation at local temperatures : hence the predominance of precipitation increase over evaporation increase in the surface moisture budget.

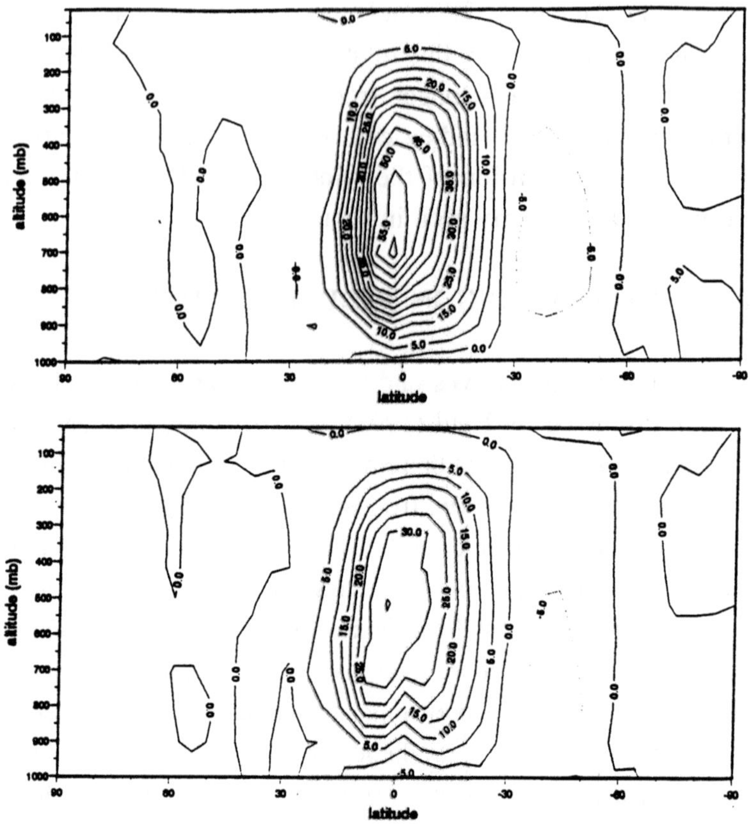

Figure 2. Hadley circulation during Boreal summer, illustrated by the zonaly average meridional streamfunction, for values of solar luminosity decreased (above) and increased (below) by 1.67%, as simulated by the LMD general circulation model (in units of 10^{10} kg s^{-1}).

There are other strong signals in the midlatitude response. One of them is the pronounced summer drying of continental areas ; in Boreal summer the meridional mixing of water vapour is no longer important, local effects become predominant and drying follows the evaporation increase due to local warming. Drying, on the other hand, is also observed in winter in the surroundings of the Mediterranean sea ; this effect could be related to the northward expansion of the Boreal winter Hadley cell (Section 3.2), which produces a northward shift of the Saharan subsidence branch and desertification of Mediterranean countries.

5. CONCLUSION

An interesting feature of the response of tropospheric climate to a long-term increase in the solar output is its close similarity to the global warming induced by an increase of the atmospheric loading in greenhouse gases. The reasons for this are not totally clear. The fact that the troposphere is thermally forced mainly from below is probably important for the similarity of responses.

The main features of tropical response to global warming, be it due to CO_2 or to solar constant increase, are strong low level moistening, high level warming, and weakening of the Hadley cell, probably due to the fact that the energy deficit of subtropical deserts, relative to the tropical belt, decreases in the warming process. This seems to occur both for a change in solar luminosity and a change in greenhouse gas loading. The explanations, however, are somewhat different, which underlines the difficulty of explaining the similarity of responses.

The most prominent response in middle latitudes is the strong increase of precipitation during winter, remotely driven by tropical moistening which strongly enhances the meridional moisture gradient and therefore, meridional moisture mixing by transients and planetary waves. As evaporation is locally driven at relatively low temperatures, the wintertime soil moisture (or precipitation-minus-evaporation balance) increases strongly. During summer, on the opposite, soil humidity decreases because the water balance is then mainly driven by local evaporation, which increases with local temperature. A final point of interest is the fact that Southern Europe and the Mediterranean area tend to be drier than present also during winter, a behaviour possibly related to the poleward expansion of the Hadley cell. The mechanism which drives this expansion is the following. In the tropical belt, warming induces strong low level moistening ; through the moist convection process, this moistening induces strong upper level warming ; the resulting upper-tropospheric increase of the meridional temperature gradient strengthens the upper level jets ; and, through the angular momentum balance, this strengthening forces a poleward expansion of the Hadley cell.

REFERENCES

Cess, R.D., Potter, G.L., Blanchet, J.P., Boer, G.J., Del Genio, A.D., Déqué, M., Dymnikov, V., Galin, V., Gates, W.L., Ghan, S.J., Kiehl, J.T., Lacis, A.A., Le Treut, H., Li, Z.X., Liang, X.Z., McAvaney, B.J., Meleshko, V.P., Mitchell, J.F.B., Morcrette, J.J., Randall, D.A., Rikus, L., Roeckner, E., Royer, J.F., Schlese, U., Sheinin, D.A., Slingo, A., Sokolov, A.P., Taylor, K.E., Washington, W.M., Wetherald, R.T., Yagai, I., and Zhang, M.H. (1990). Intercomparison and interpretation of climate feedback processes in nineteen atmospheric general circulation models. J. Geophys. Res., 95: 16601-16615

Held, I.M. (1993). Large-scale dynamics and global warming. Bull. Amer. Meteor. Soc., 74: 228-241

Held, I.M., and Hou, A.Y. (1980). Nonlinear axially symmetric circulations in a nearly inviscid atmosphere. J. Atmos. Sci., 37: 515-533

Li, Z.X., and Le Treut, H. (1992). Cloud-radiation feedbacks in a general circulation model and their dependence on cloud modelling assumptions. Climate Dynamics, 7: 133-139

Nesme-Ribes, E., Ferreira, E., Sadourny, R., Le Treut, H., and Li, Z.X. (1993). Solar dynamics and its impact on solar irradiance and the terrestrial climate. J. Geophys. Res., 98: 18923-18935

Oort, A., and Peixoto, J. (1992), in Physics of climate, ed. American Institute of Physics

Royer, J.F., Chauvin, F., and Pélissier, G. (1994). Sensitivity of the atmospheric circulation to solar irradiance variations. This volume

Sadourny, R. (1994). Maunder Minimum and the Little Ice Age : Impact of a long-term variation of the solar flux on the energy and water cycle. NATO Advanced Study Institute on Long-Term Climate Variations, in ed. J.C. Duplessy, Springer-Verlag, in press

HAVE SOLAR-IRRADIANCE VARIATIONS INFLUENCED CLIMATE?

Michael E. Schlesinger and Navin Ramankutty
Department of Atmospheric Sciences
University of Illinois at Urbana-Champaign
105 South Gregory Avenue
Urbana, IL 61801
USA

1. INTRODUCTION

In the past two years we published two papers in *Nature* which examined the issue of whether or not putative solar-irradiance variations have influenced the earth's observed surface temperatures during the past 130 years (Schlesinger and Ramankutty, 1992, 1994). Here we summarize the approach and findings of these papers.

2. IMPLICATIONS FOR GLOBAL WARMING OF INTERCYCLE SOLAR-IRRADIANCE VARIATIONS

In this first study, we used our hemispherically resolved energy-balance climate/upwelling-diffusion ocean model to simulate the temperature change from 1638 contributed by greenhouse gases (GHGs), anthropogenic sulphate aerosols (ASAs), and solar-irradiance variations of the form posited by Friis-Christensen and Lassen (1991; hereafter FL). We did this by prescribing the GHG concentration and ASA emission histories, the former based on the Intergovernmental Panel on Climate Change (IPCC) 1990 report (Houghton *et al.,* 1990), and the latter based on Schlesinger *et al.* (1992). From these histories we calculated the corresponding change in the net downward radiative flux at the tropopause – the radiative forcing. The GHG radiative forcing was calculated from the relationship given by Shine *et al.* (1990), and was taken to be the same for both Northern and Southern Hemispheres. The ASA radiative forcing was calculated in terms of a prescribed parameter, $\Delta F_{SO4}(1978)$ – the ASA radiative forcing in 1978 – with 90% of the global radiative forcing taken to occur in the Northern Hemisphere and 10% in the

NATO ASI Series, Vol. I 25
The Solar Engine and Its Influence
on Terrestrial Atmosphere and Climate
Edited by Elizabeth Nesme-Ribes

Southern Hemisphere. The radiative forcing due to putative solar-irradiance variations was taken in both hemispheres as $\Delta F_{Sun}(t) = \beta \cdot [P_o - P(t)]$, where P(t) is the length of the '11-year' sunspot cycle given by Gleissberg (1944) for June 1638–December 1744 and FL for January 1745–June 1985, P_o = 11.018 years is the average cycle-period over June 1638–June 1985, and β is a parameter. The sensitivity of the climate system, ΔT_{2x} – the change in the equilibrium global-mean surface temperature in response to a doubling of the CO_2 concentration – was also prescribed in the simple climate/ocean model.

For a prescribed value of $\Delta F_{SO4}(1978)$, the model was run from 1638 through 1985 – the latest year for which P(t) was estimated by FL – for different values of ΔT_{2x} and β. For each run, a constant was added to the model-simulated temperature changes relative to 1638 to convert them into model-simulated temperature anomalies from the 30-year (1951-80) average temperature so that they could be compared with the observed global-mean surface temperature anomalies from the 30-year (1951-80) average temperature (Folland *et al.*, 1992). The constant was determined analytically for each model simulation to minimize the root-mean-square error (RMSE) between the observed and model-simulated temperature anomalies (Schlesinger *et al.*, 1992). For each value of $\Delta F_{SO4}(1978)$, the model was run for many values of ΔT_{2x} and β. This rerunning was done in a computationally efficient manner by an optimization code to determine the best value of ΔT_{2x} and β which minimized the RMSE.

The results we present below differ somewhat from those presented in our *Nature* paper. Here we use: (1) a modified version of the simple climate model which explicitly includes the land in each hemisphere (Ramankutty, 1994); (2) solar-cycle lengths given by Friis-Christensen and Lassen (1992), which extend only through 1983, rather than the solar-cycle lengths we digitized from FL and Gleissberg (1944); and (3) the observed surface temperatures of the IPCC 1992 report (Houghton *et al.*, 1992) rather than the observed data from the United Kingdom Meteorological Office.

The resulting optimal model-simulated global-mean surface temperature anomalies, contributed jointly by the GHGs, ASAs for either $\Delta F_{SO4}(1978) = 0$ or -1.1 Wm^{-2}, and the Sun are shown in Fig. 1, panels A and C, superimposed on the observed temperature anomalies. The values of ΔT_{2x} and β obtained by the optimization are also presented in Fig. 1. For both $\Delta F_{SO4}(1978)$ values the model-simulated temperature anomaly increases from 1700 to 1770, decreases until about 1810, increases until 1850, remains nearly constant until 1890, increases until 1945, remains nearly constant until 1960, and then increases to 1983.

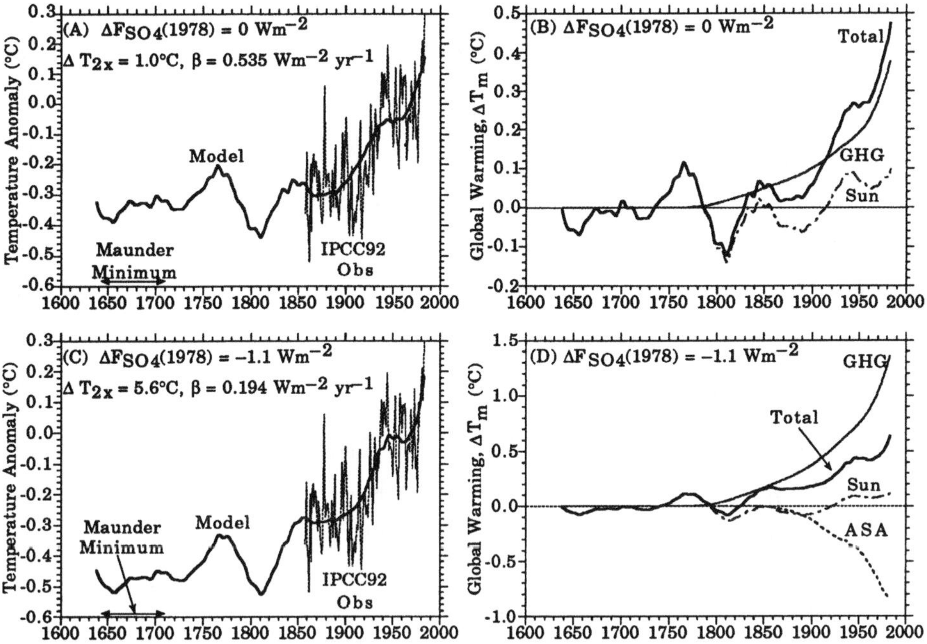

Fig. 1. Optimal model-simulated global-mean surface temperature anomalies, contributed jointly by GHGs, ASAs with $\Delta F_{SO4}(1978) = 0$ or -1.1 Wm^{-2}, and FL solar-irradiance variations, superimposed on observed temperature anomalies (panels A, C), together with their individual contributions (panels B, D).

The contribution of each forcing component is presented in Fig. 1, panels B and D, for $\Delta F_{SO4}(1978) = 0$ and -1.1 Wm^{-2}. For $\Delta F_{SO4}(1978) = 0$ (-1.1) Wm^{-2} the temperature change due to GHGs permanently exceeds the magnitude of the temperature change due to the Sun after 1829 (1825), and their difference increases to a maximum of 0.28°C (1.24°C) in 1982 (1983). These results do not support the conclusion of the Marshall Institute report (Jastrow *et al.*, 1992) "that the sun has been the controlling influence on climate in the last 100 years, with the greenhouse effect playing a smaller role."

The intercycle solar-irradiance anomaly, $\Delta S = \Delta F_{Sun} \cdot [4/(1 - \alpha_p)]$, with α_p = 0.3 the planetary albedo, is presented in Fig. 2 for the two cases of Fig. 1, both in Wm^{-2} and as a percent of the mean solar irradiance measured from 1980–1989 by the Solar Maximum Satellite (SMS), S_o = 1367.5 Wm^{-2} (Willson and Hudson, 1991; R. C. Willson, personal communication, 1992). The intra-cycle solar-

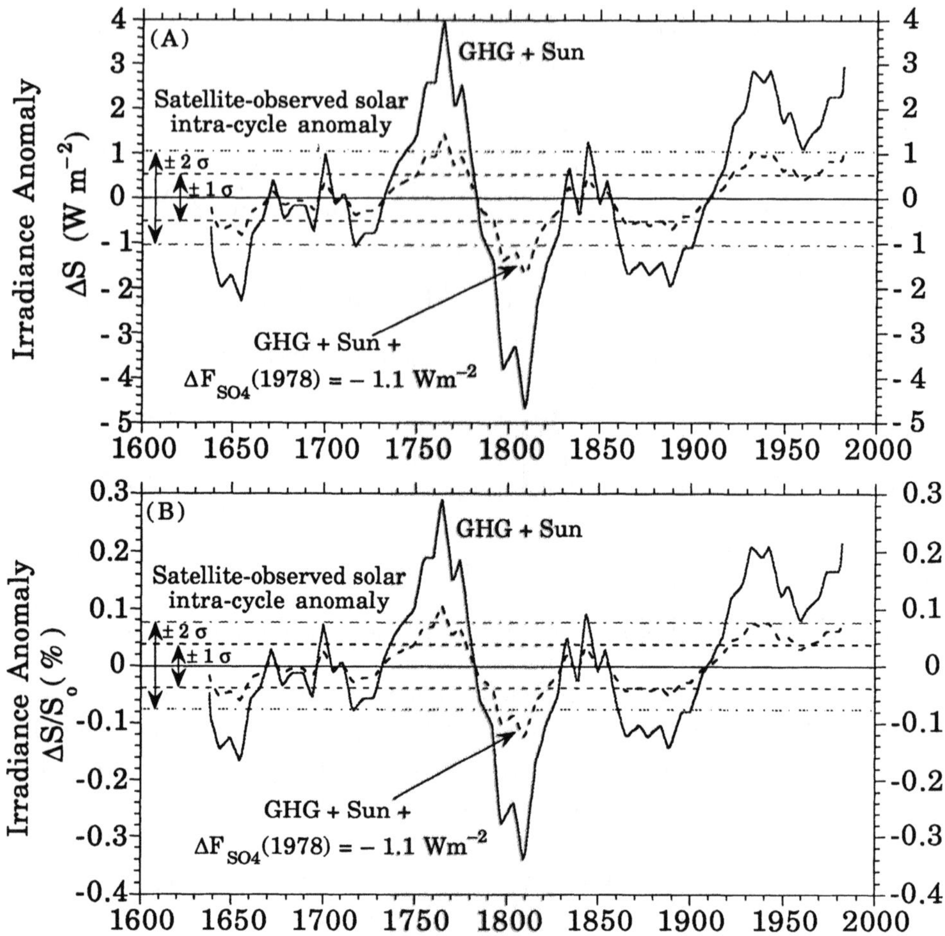

Fig. 2. Intercycle solar-irradiance variation for the two cases of Fig. 1, ΔS and $\Delta S/S_o$, where S_o = 1367.5 Wm^{-2} is the mean solar irradiance measured by the Solar Maximum Satellite (SMS) from 1980–1989. The values of $\pm\sigma$ and $\pm 2\sigma$ for the SMS observations, σ(intra) = 0.62 Wm^{-2} (0.046% of S_o), are also shown.

irradiance variation given by the standard deviation of the SMS observations during 1980 to 1989 (obtained from the National Geophysical Data Center, Boulder, CO) σ(intra) = 0.52 Wm^{-2} (0.038% of S_o, is also shown for comparison. For ΔF_{SO4}(1978) = –1.1 Wm^{-2}, ΔS ranges from –1.7 Wm^{-2} to 1.4 Wm^{-2} (–0.12% to 0.10%), with a standard deviation σ(inter) = 0.62 Wm^{-2} (0.046%). This intercycle variation is within the "several tenths of a per cent" estimated by

Baliunas and Jastrow (1990) based upon observations of the visual-brightness and magnetic-activity variations of 74 solar-type stars. For $\Delta F_{SO4}(1978) = 0$, ΔS ranges from –4.7 Wm^{-2} to 4.0 Wm^{-2} (–0.34% to 0.29%), with σ(inter) = 1.72 Wm^{-2} (0.126%). This intercycle variation is within the 0.1–3% variation found by Lockwood *et al.* (1992) for 25 solar-type stars. Consequently, we concluded that these "results constitute strong evidence, albeit circumstantial, that there have been variations in solar irradiance which have contributed to the observed temperature changes since 1856."

3. 65-70 YEAR OSCILLATION IN THE GLOBAL CLIMATE SYSTEM

In this second study, we used our hemispherically resolved energy-balance climate/upwelling-diffusion ocean model to remove from each of the four observed global and hemispheric-mean, annual-mean surface temperature data sets[4] (Fig. 3), the temperature change from 1765 contributed by GHGs and ASAs. We did this as described above, but without any contribution from putative solar-irradiance variations.

The resulting optimal model-simulated global-mean surface temperature anomalies, contributed jointly by the GHGs and ASAs, are shown in Fig. 3, panels A, C, E and G, superimposed on the observed temperature anomalies – a different model-simulated curve for each observational data set. The values of ΔT_{2x} and $\Delta F_{SO4}(1978)$, obtained by the optimization are also presented in Fig. 3. The ΔT_{2x} values extend over $1.9^{o}C \leq \Delta T_{2x} \leq 2.6^{o}C$, which is in the lower half of the IPCC range, $1.5^{o}C \leq \Delta T_{2x} \leq 4.5^{o}C$ (Mitchell *et al.*, 1990); the $\Delta F_{SO4}(1978)$ values extend over $-0.37\ Wm^{-2} \leq \Delta F_{SO4}(1978) \leq 0\ Wm^{-2}$, which is at the bottom of the range estimated by Schlesinger *et al.* (1992), $-2.0\ Wm^{-2} \leq \Delta F_{SO4}(1978) \leq -0.2\ Wm^{-2}$.

For each observed temperature-anomaly data set we subtracted the corresponding optimal model-simulated temperature anomaly to obtain the residual, detrended temperature anomaly (Fig. 1, panels B, D, F and H). Each of these reveals what, in the terminology of the World Meteorological Organization (Mitchell *et al.*, 1966), is a climatic oscillation.

Because the timescale of the oscillation is about half the length of the observed record, the standard analysis tool – power spectrum analysis (PSA) – is not sufficiently powerful to yield a statistically significant spectral analysis. Accordingly, we employed a comparatively new spectral analysis tool – Singular Spectrum Analysis (SSA) (Broomhead *et al.*, 1986; Fraedrich, 1986; Vautard and Ghil, 1989; Vautard *et al.*, 1992). SSA differs from PSA only in that the functions used in the finite mathematical series to represent the time series are not sines and cosines, but instead are determined by the structure of the time series itself. SSA determines the eigenvalues and eigenvectors of the autocovariance (Toeplitz) matrix constructed from the time series data. The eigenvectors – the so-called empirical orthogonal functions (EOFs) of the time series – are to SSA what the sines and cosines are to PSA. The eigenvalues, when divided by the total variance of the time series, give the percent thereof explained by each EOF; this allows the EOFs to be descendingly ordered in terms of the percentage of total variance they explain. SSA also determines the time varying amplitude of each EOF – the so-called principal components, PC(t).

When SSA is applied to each detrended observed surface temperature time series, the first two eigenvalues have nearly identical explained variances – about 14% each – which are well separated from, and above, the explained variances of all the other eigenvalues. Such a pair of eigenvalues is indicative of an oscillation, in this case, one which explains about 28% of the variance of each detrended time series. Moreover, the separation between these two eigenvalues and all the other eigenvalues is many times larger than the 95% confidence band surrounding each eigenvalue. Thus the pair of eigenvalues is statistically significant. The corresponding eigenvectors, EOF_1 and EOF_2, have an oscillatory structure, as do $PC_1(t)*EOF_1$ and $PC_2(t)*EOF_2$. The sum $PC_1(t)*EOF_1 + PC_2(t)*EOF_2$ for each detrended time series is shown in Fig. 3, panels B, D, F and H, superimposed thereon. This sum, which is *de facto* a low-pass filtering of the time series with a 'data-adaptive' filter, clearly reveals an oscillation. It is seen that the 'amplitude' and 'period' of the oscillation vary over the length of the observational record, hence the oscillation is not regular.

Having discovered an oscillation in the global-mean surface temperature record, we next examined whether the oscillation occurred everywhere over the globe or only in certain regions. To do this we divided the globe between 60°S and 75°N latitude into 11 latitude-longitude regions. For each region we used the 5°x5° latitude-longitude gridded data of Jones *et al.* (1991) to calculate the average annual-mean temperature anomaly for every year there were any observations within the region. From each regional time series we then subtracted the optimal model-simulated regional temperature anomaly contributed by GHGs and ASAs, taken equal to the simulated hemispheric-mean temperature anomaly for land, times the fraction of land within the region, plus the simulated hemispheric-mean temperature anomaly for ocean, times the fraction of ocean within the region. We then performed SSA on each of the detrended regional time series. The results showed that the oscillation

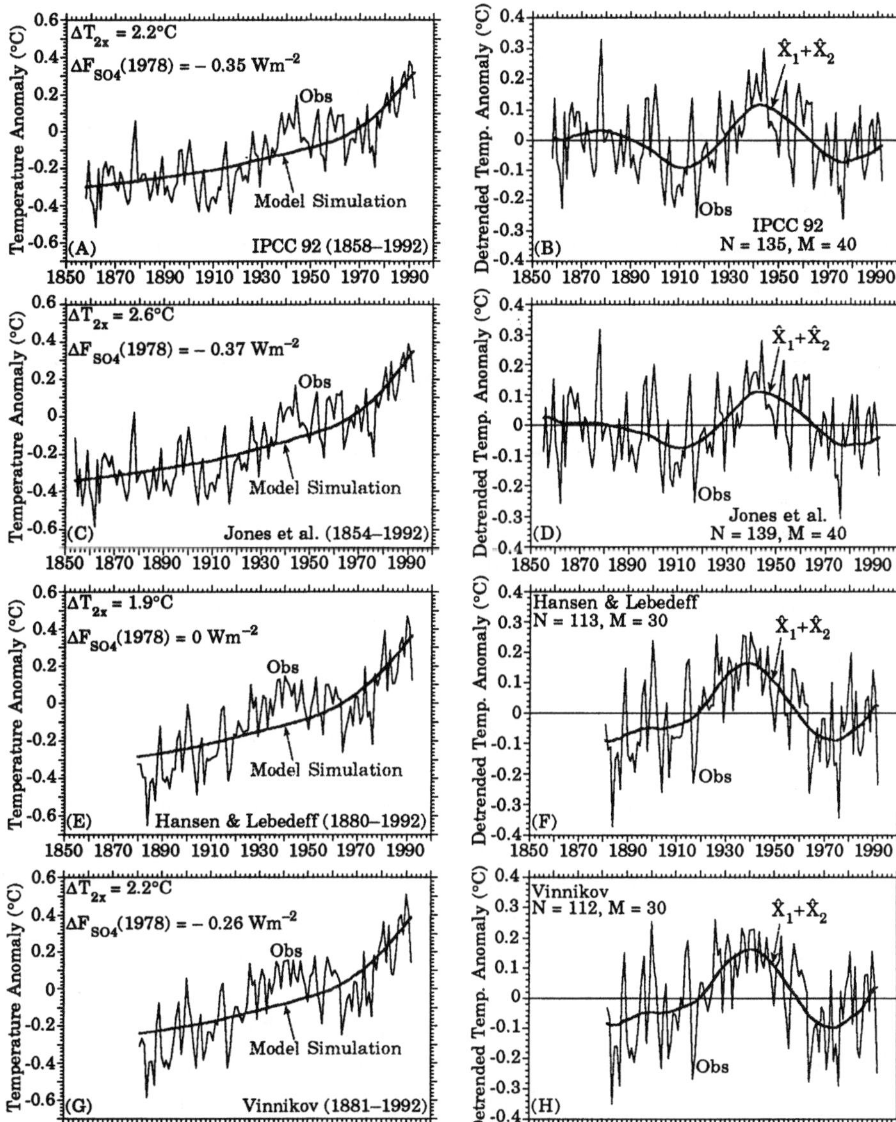

Fig. 3. Optimal model-simulated global-mean surface temperature anomalies, contributed jointly by GHGs and ASAs, superimposed on the observed temperature anomalies (panels A, C, E, G); together with their difference (Obs) and $PC_1(t)*EOF_1 + PC_2(t)*EOF_2$ (= $\hat{X}_1 + \hat{X}_2$) from SSA (panels B, D, F, H). N and M are the lengths of the times series and data window in years.

does not exist in the Southern Hemisphere, nor everywhere in the Northern Hemisphere, but instead only over the North Atlantic Ocean, North America, Europe and northern Africa. Thus the oscillation in the global-mean surface temperature is the statistical result of the regional long-timescale oscillations which occur in the Northern Hemisphere. These regional oscillations are not regular. Table 1 presents an estimate of the average 'amplitudes' and 'periods' for these regional oscillations, together with those for the oscillation in global-mean temperature.

Table 1. Estimated average 'amplitude' and 'period' of the oscillations

Region	'Amplitude' (°C)	'Period' (Years)
North Atlantic Ocean	0.33	76
North America	0.28	88
Europe	0.14	84
Northern Africa	0.11	50
Global Mean	0.09	67

The geographical structure of the oscillation is revealed in Fig. 4 by the product of the differences of 20-year-averaged temperature anomalies, centered at the extrema of global-mean oscillation, 1900–1920 minus 1935–1955 times 1935–1955 minus 1970–1990. The negative (positive) product shown by dark (light) shading is where the temperature anomaly changed (did not change) from negative during 1900–1920 to positive during 1935–1955 and back to negative during 1970–1990. In the nonshaded regions there is no product because data are absent during at least one of the 20-year averaging periods. Figure 4 indicates that the oscillation exists over the North Atlantic Ocean, North America, western Eurasia and northern Africa, and does not exist over most of the Southern Hemisphere, in the regions where data exist back to 1900.

4. DISCUSSION AND CONCLUSION

There are three possible sources for the 65-70 year "global" oscillation: (1) random forcing of the ocean by the atmosphere, such as by white noise; (2) external oscillatory forcing of the climate system, such as by a variation in the solar irradiance; and (3) an internal oscillation of the atmosphere-ocean system.

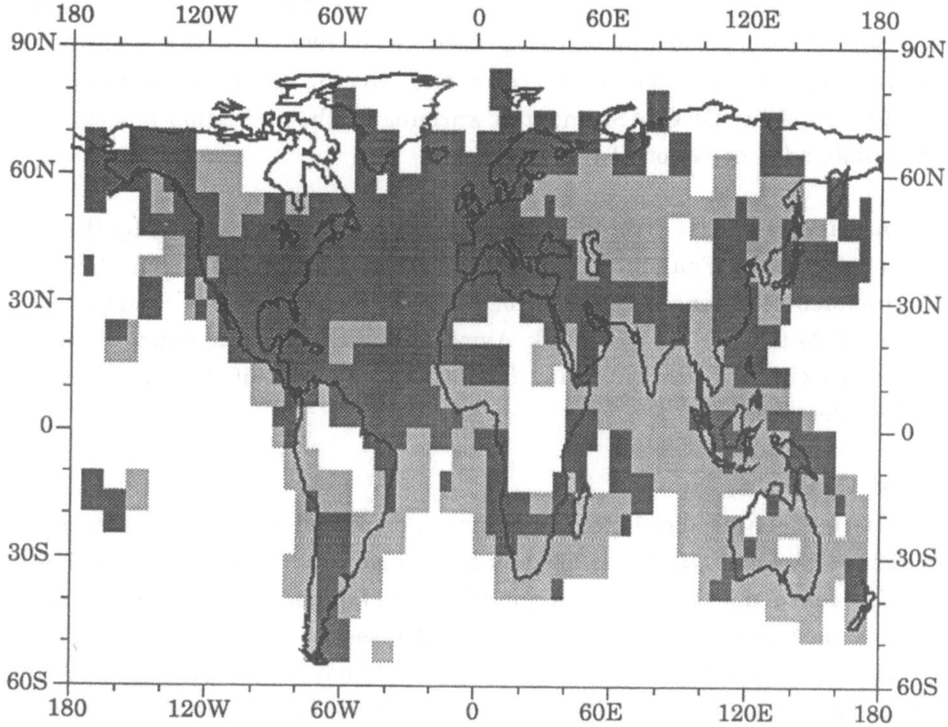

Fig. 4. Product of the differences of 20-year-averaged temperature anomalies, centered at the extrema of global-mean oscillation. Negative (positive) product shown by dark (light) shading.

It is unlikely that white-noise forcing is the source of the oscillation because such forcing should generate an ocean-wide response, but the oscillation is not pan-oceanic. It is unlikely that putative variations in solar irradiance (e.g., Friis-Christensen and Lassen, 1991) are the source of the oscillation because solar forcing should generate a global response, but the oscillation is not global. Consequently, the most probable cause of the oscillation is an internal oscillation of the atmosphere-ocean system. This conclusion is supported by a growing body of observational evidence (Bjerknes, 1964; Levitus, 1989; Kushnir, 1994; Taylor *et al.*, 1993; Greenland Ice-core Project (GRIP) Members, 1993; Dansgaard *et al.*, 1993) and coupled atmosphere/ocean general circulation model simulation results (Delworth *et al.*, 1993).

Comparison of the regional and global-mean temperature changes caused by the oscillation with those induced by GHG+ASA forcing shows that the rapid rise in global-mean temperature between about 1908 and 1946, and the subsequent reversal of this warming until about 1965 were the result of the oscillation. In the North Atlantic and North American regions, the domination of the GHG+ASA-induced warming by the oscillation has obscured and confounded detection of this warming.

Acknowledgements: The research reported here was supported by the U.S. National Science Foundation and the Carbon Dioxide Research Program, Environmental Sciences Division, of the U.S. Department of Energy under Grant ATM-9001310. We thank Ayman Ghanem for assistance with the figures herein, and Wanqiu Wang for making us aware of the Mitchell *et al.* (1966) reference.

ENDNOTES

1. This model was developed by Schlesinger, based on the formulation by Hoffert *et al.* (1980), and was used by Schlesinger and colleagues in studies of the influence on the observed global-mean temperature record of ASAs (Schlesinger *et al.*, 1992), possible greenhouse-gas induced future climate change (Bretherton *et al.*, 1990), and greenhouse policy (Schlesinger and Jiang, 1991; Hammitt *et al.*, 1992; Schlesinger, 1993; Lempert *et al.*, 1994).

2. It may appear surprising that the temperature observations are of the departures from an unspecified 30-year average, rather than simply the temperatures themselves. The reason for this is that surface air temperatures are measured at the differing elevations of the observing stations. If these station temperatures were area-weighted and then averaged to form hemispheric and global means, one would have averaged the cold temperatures of high-altitude stations with the warm temperatures of low-altitude stations. Clearly, such an average temperature would give a result distorted by the orography of the earth's surface. This distortion is avoided by using the departure of temperature from the local long-term mean temperature for each month, and then averaging the temperature departures in time and space.

3. This standard deviation is for annual-mean values, while that in Schlesinger and Ramankutty (1992) is for daily values.

4. Two of these observed data sets are based only on the distribution of surface

air temperature over land (Hansen and Lebedeff, 1988; Vinnikov *et al.*, 1990), and two are based on these data plus the distribution of sea surface temperature (Jones *et al.*, 1991; Folland *et al.*, 1992).

5. "Climatic oscillation: A fluctuation in which the variable tends to move gradually and smoothly between successive maxima and minima. (Contrast vacillation.) Climatic vacillation: A fluctuation in which the climatic variable tends to dwell alternately around two (or more) average values, and to drift from one average value to the other at regular or irregular intervals. (Contrast oscillation.) Climatic fluctuation: A climatic inconstancy that consists of any form of systematic change, whether regular or irregular, except trend and discontinuity. Characterized by at least two maxima (or minima) and one minimum (or maximum) including those at the end points of the record."

6. This detrending is undoubtedly imperfect, as the model simulation lacks regional specificity. However, it is preferable to apply SSA to a somewhat imperfectly detrended time series than to the nondetrended time series. Eventually, improvements in detrending might be achieved by using the geographical distribution of temperature change simulated by a calibrated atmosphere-ocean general circulation model for the time-dependent GHG and ASA forcing since 1850.

REFERENCES

Baliunas, S., and R. Jastrow, 1990: Evidence for long-term brightness changes of solar-type stars. Nature, 348, 520-523.

Bjerknes, J., 1964: Atlantic air-sea interaction. In Advances in Geophysics 10. Landsberg, H. E. ,and F. Van Mieghem (Eds.), Academic Press, New York, pp. 1-82.

Bretherton F. P. et al., 1990: Time-dependent greenhouse-gas-induced climate change. In Climatic Change: The IPCC Scientific Assessment, J. T. Houghton, G. J. Jenkins and J. J. Ephraums (Eds.), Cambridge University Press, 173-193.

Broomhead, D. S., and G. P. King, 1986: Extracting qualitative dynamics from experimental data. Physica, 20D, 217-236.

Dansgaard, W. et al., 1993: Evidence for general instability of past climate from a 250-kyr ice-core record. Nature, 364, 218-220.

Delworth, T., S. Manabe and R. J. Stouffer, 1993: Interdecadal variations of the thermohaline circulation in a coupled ocean-atmosphere model. J. Climate,

6, 1993-2011.

Folland, C. K., T. R. Karl, N. Nicholls, B. S. Nyenzi, D. E. Parker and K. Ya. Vinnikov, 1992: Observed climate variability and change. In Climate Change 1992 – The Supplementary Report to the IPCC Scientific Assessment. Houghton, J. T., B. A. Callander and S. K. Varney (Eds.), Cambridge University Press, pp. 135-170.

Fraedrich, K. J., 1986: Estimating the dimensions of weather and climate attractors. J. Atmos. Sci., 43, 419-432.

Friis-Christensen, E., and K. Lassen, 1991: Length of the solar cycles: An indicator of solar activity closely associate with climate. Science 254, 698-700.

Friis-Christensen, E., and K. Lassen, 1992: Solar activity parameters used in geophysical studies at DMI. Scientific Report 92-8, Danish Meteorological Institute, 11 pp.

Gleissberg, W., 1944: A table of secular variations of the solar cycle. Terrest. Magnet. Atmos. Electr. 49, 243-244.

Greenland Ice-core Project (GRIP) Members, 1993: Climate instability during the last interglacial period recorded in the GRIP ice core. Nature, 364, 203-207.

Hammitt, J. K., R. J. Lempert and M. E. Schlesinger, 1992: A sequential-decision strategy for abating climate change. Nature, 357, 315-318.

Hansen, J., and S. Lebedeff, 1988: Global surface air temperatures: Update through 1987. Geophys. Res. Lett. 15, 323-326.

Hoffert, M. I., A. J. Callegari and C.-T. Hsieh, 1980: The role of deep sea heat storage in the secular response to climatic forcing. J. Geophys. Res., 85, 6667-6679.

Houghton, J. T., G. J. Jenkins and J. J. Ephraums (Eds.), 1990: Climate Change: The IPCC Scientific Assessment, Cambridge University Press, 365 pp.

Houghton, J. T., B. A. Callander and S. K. Varney (Eds.), 1992: Climate Change 1992 – The Supplementary Report to the IPCC Scientific Assessment, 200 pp, Cambridge University Press.

Jastrow, R., W. Nierenberg and F. Seitz, 1992: Global Warming Update: Recent Scientific Findings. George C. Marshall Institute, Washington, DC, 32 pp.

Jones, P. D., T. M. L. Wigley and G. Farmer, 1991: Marine and land temperature data sets: a comparison and a look at recent trends. In Greenhouse-Gas-Induced Climatic Change: A Critical Appraisal of

Simulations and Observations. Schlesinger, M. E. (Ed.), Elsevier, pp. 153-172.

Kushnir, Y., 1994: Interdecadal variations in North Atlantic Sea surface temperature and associated atmospheric conditions. J. Clim., 7, 141-157.

Lempert, R. J., M. E. Schlesinger and J. K. Hammitt, 1994: The impact of potential abrupt climate changes on near-term policy choices. Climatic Change (in press).

Levitus, S., 1989: Interpentadal variability of temperature and salinity at intermediate depths of the North Atlantic Ocean, 1970-1974 versus 1955-1959. J. Geophys. Res., 94, 6091-6131.

Lockwood, G. W., B. A. Skiff, S. L. Baliunas and R. R. Radick, 1992: Long-term solar brightness changes estimated from a survey of Sun-like stars. Nature, 360, 653-655.

Mitchell, J. F. B., S. Manabe, T. Tokioka and V. Meleshko, 1990: Equilibrium Climate Change. In Climate Change: The IPCC Scientific Assessment. Houghton; J. T., G. J. Jenkins and J. J. Ephraums (Eds.), Cambridge University Press, pp. 131-172.

Mitchell, Jr., J. M., B. Dzerdzeevskii, H. Flohn, W. L. Hofmeyr, H. H. Lamb, K. N. Rao and C. C. Wallen, 1966: Climatic Change. World Meteorological Organization Technical Note No. 79, 79 pp.

Ramankutty, N., 1994: An empirical estimate of climate sensitivity. M. S. Thesis, Department of Atmospheric Sciences, University of Illinois at Urbana-Champaign, 167 pp.

Schlesinger, M. E., 1993: Greenhouse Policy. National Geographic Research & Exploration, 9(2), 159-172.

Schlesinger, M. E., and X. Jiang, 1991: Revised projection of future greenhouse warming. Nature, 350, 219-221.

Schlesinger, M. E., and N. Ramankutty, 1992: Implications for Global Warming of Intercycle Solar-Irradiance Variations. Nature, 360, 330-333.

Schlesinger, M. E., and N. Ramankutty, 1994: An Oscillation in the Global Climate System of Period 65-70 Years. Nature, 367, 723-726.

Schlesinger, M. E., X. Jiang and R. J. Charlson, 1992: Implication of Anthropogenic Atmospheric Sulphate for the Sensitivity of the Climate System. In Climate Change and Energy Policy: Proceedings of the International Conference on Global Climate Change: Its Mitigation Through Improved Production and Use of Energy, Rosen, L, and R. Glasser (Eds.), American Institute of Physics, New York, 75-108.

Shine, K. P., R. G. Derwent, D. J. Wuebbles and J.-J. Morcrette, 1990: Radiative forcing of climate. In Climate Change: The IPCC Scientific

Assessment, Houghton, J. T., Jenkins, G. J. and Ephraums, J. J. (Eds.), Cambridge University Press, 41-68.

Taylor, K. C. et al., 1993: The flickering switch of late pleistocene climate change. Nature, 361, 432-436.

Vautard, R., and M. Ghil, 1989: Singular spectrum analysis in nonlinear dynamics, with applications to paleoclimatic time series. Physica, 35D, 395-424.

Vautard, R., P. Yiou and M. Ghil, 1992: Singular spectrum analysis: A toolkit for short, noisy chaotic signals. Physica, 58D, 95-126.

Vinnikov, K. Ya., P. Ya. Groisman and K. M. Lugina, 1990: Empirical data on contemporary global climate changes (temperature and precipitation). J. Clim., 3, 662-677.

Willson, R. C., and H. S. Hudson, The Sun's luminosity over a complete solar cycle, 1991: Nature, 351, 42-44.

SENSITIVITY OF THE ATMOSPHERIC CIRCULATION TO SOLAR IRRADIANCE VARIATIONS

J.F. Royer, F. Chauvin and G. Pélissier
Météo-France, Centre National de Recherches Météorologiques,
42 av. Coriolis, 31057 Toulouse Cédex, France.

INTRODUCTION

The question of the possible influence of variations in the solar irradiance is of great interest in view of the recent findings concerning solar variability. Atmospheric general circulation models are the most appropriate tool to investigate this problem. Since the pioneering study of Wetherald and Manabe (1975), a number of such experiments have been reported in the literature (Sadourny, 1993). The influence of solar irradiance variation on the atmosphere is twofold: a direct radiative forcing through the absorption of short-wave solar radiation by the atmospheric constituents, and an indirect effect by modification of the ocean surface temperature. These two effects have rather different time-scales. The computation of solar irradiance variation over the ocean require the use of coupled ocean-atmosphere models which need at least ten years to reach equilibrium because of the thermal inertia of the ocean. On the other hand, the direct atmospheric effects are felt on monthly time scales. Recently, a sensitivity experiment to solar irradiance variations has been performed by Nesme-Ribes et al (1993) using the LMD GCM coupled to a mixed-layer ocean. This experiment has revealed a systematic response of the atmospheric circulation bearing a striking similarity to the response patterns obtained with doubling CO_2 experiments. It therefore seems important to confirm this response using another GCM. In order to isolate the direct atmospheric effect from that due to changes in ocean temperatures it was decided to run the experiment using the sea surface temperature response computed from the LMD coupled run, and use the same framework of experiment with another GCM.

The model used for rerunning the experiment is the new climate version of the Arpege model developed at Meteo-France. Although the code was completely rewritten, this new model has very similar characteristics to those of the previous Emeraude climate model, whose climatology has been documented in several validation studies and sensitivity experiments (Planton et al, 1991;

NATO ASI Series, Vol. I 25
The Solar Engine and Its Influence
on Terrestrial Atmosphere and Climate
Edited by Elizabeth Nesme-Ribes

Royer et al, 1990; Déqué and Royer, 1991). One of the most interesting and original features of this model is the use of a large number of vertical layers (30 levels) allowing a more detailed resolution of the stratospheric circulation than most other GCMs. Moreother, the model includes a prognostic equation for ozone including its transport and a simplified representation of its photochemistry (Cariolle and Déqué, 1986). The numerical technique used to represent the horizontal variations of atmospheric fields involve a spectral expansion in spherical harmonics truncated at wavenumber 42. The nonlinear terms of the dynamical equations and the physical parameterizations are computed on a quasi-regular grid of about 2.8° resolution. Numerous experiments with spectral models have shown that this horizontal resolution adequately represents the large-scale dynamical processes. Detailed description of the physical processes are provided in technical reports of the Arpège model, as well as the basic climatology of the model in 10 year integrations (Déqué et al, 1993).

DESCRIPTION OF THE SENSITIVITY EXPERIMENT

The sensitivity of the atmospheric circulation to solar irradiance variation has been studied by using the same variations in solar forcing and sea surface temperature as in the experiment of Nesme-Ribes et al (1993). The reference climate of the model has been computed from a control integration corresponding to the current conditions with a solar irradiance of 1365 W/m2, and using as boundary conditions over the ocean the seasonally varying monthly mean sea surface temperature (SST) specified from analysed observations averaged over the 10 year period 1979-1988. Then two 14 month simulations were performed with increased (SOP) and decreased (SOM) solar irradiance. The solar irradiance variations were of +/- 1/60 (1.67 %), which corresponds to variations of about +/- 4 Wm2 in the mean short wave radiative input to the Earth-atmosphere system. In each simulation the monthly mean SST anomalies were specified by using the anomalies averaged over the last 10 year of the coupled simulation of the LMD model, which were kindly provided by the authors. The first 2 months of the simulations have been discarded from the analysis, in order to let the atmospheric circulation adjust to its new boundary conditions.

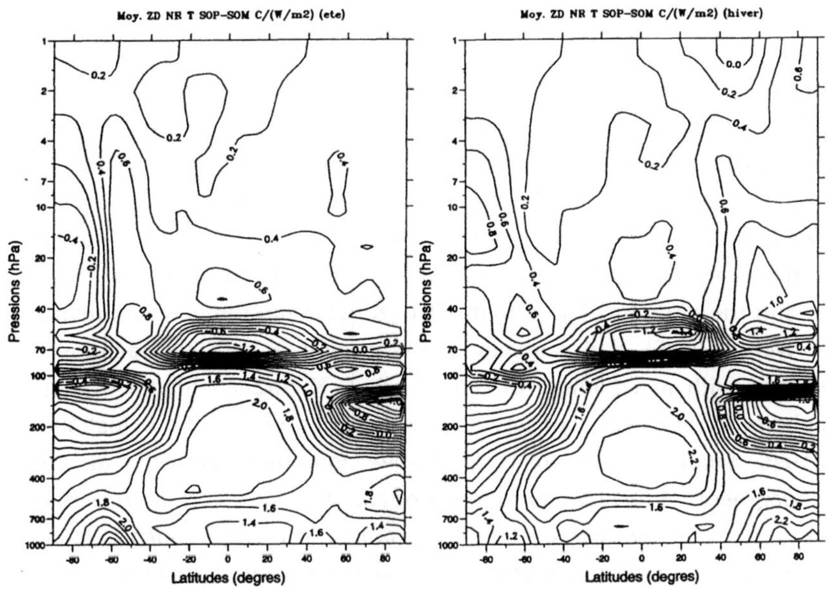

Fig. 1. Latitude height distribution of the zonal average of the normalized temperature response in K /(W/m2) for June-July-August (left) and December-January-February (right)

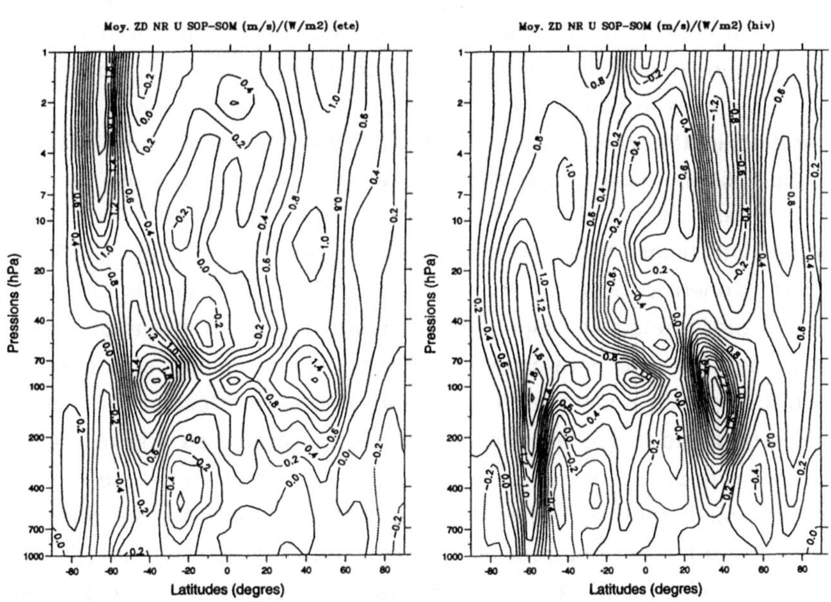

Fig. 2. Same as figure 1 for the zonal (positive eastwards) wind component in m/s (/W/m2).

ZONAL MEAN RESPONSE

For comparison with the LMD results, we will display the normalized response defined as the difference between the increased (SOP) and decreased (SOM) solar luminosity, divided by the mean difference in solar input (8 W/m2). This normalized response, which gives the sensitivity to a variation of 1 W/m2 in the short wave radiative input (assuming a linear response), has been interpreted by Nesme-Ribes et al. (1993) as the possible difference between the present climate and the Little Ice Age climate. The zonal mean temperature response (fig 1) for winter (DJF) and summer (JJA) shows a positive response in the troposphere with a maximum of more than 2 K (m2/W) in the tropical upper troposphere. A cooling is observed in the lower stratosphere just above the tropopause. Such features are in qualitative agreement with the results of the LMD experiment. However, the upper tropospheric warming in our simulation is slightly larger, and the low stratosphere cooling in high latitudes is located at a lower altitude, so that the increase of the temperature gradient between the pole and equator in the upper troposphere is larger than in the LMD simulation. A new feature which is displayed by our simulation is that this stratospheric cooling is limited to a rather shallow layer just above the tropopause, and is replaced by a very moderate warming in the middle stratophere. This warming can be caused in part by ozone absorption of the increased solar flux, yet dynamical changes in stratospheric circulation may also play a part in extending the warming in the winter hemisphere. This middle stratospheric response to solar irradiance increase is completely different from the response to a carbon dioxide increase, which produces, by an inverted greenhouse effect, a cooling increasing with height throughout the stratosphere (Mahfouf et al, 1993). The lower stratosphere cooling observed in the present simulation is caused by an increase in the tropopause height. As the warming due to insolation increase is rather modest in the stratosphere it has only a minor effect on ozone concentration which undergoes a very slight decrease. Most of the zonal mean responses obtained in our experiment can be interpreted as due to the vertical shift of the tropopause caused by the surface temperature response.

The zonal wind response (fig 2) indicates a small increase and upward displacement of the subtropical westerly jet particularly in the winter hemisphere. This response is similar to that observed in CO2 increase experiments (Mahfouf et al., 1993), and can be interpreted as caused by the

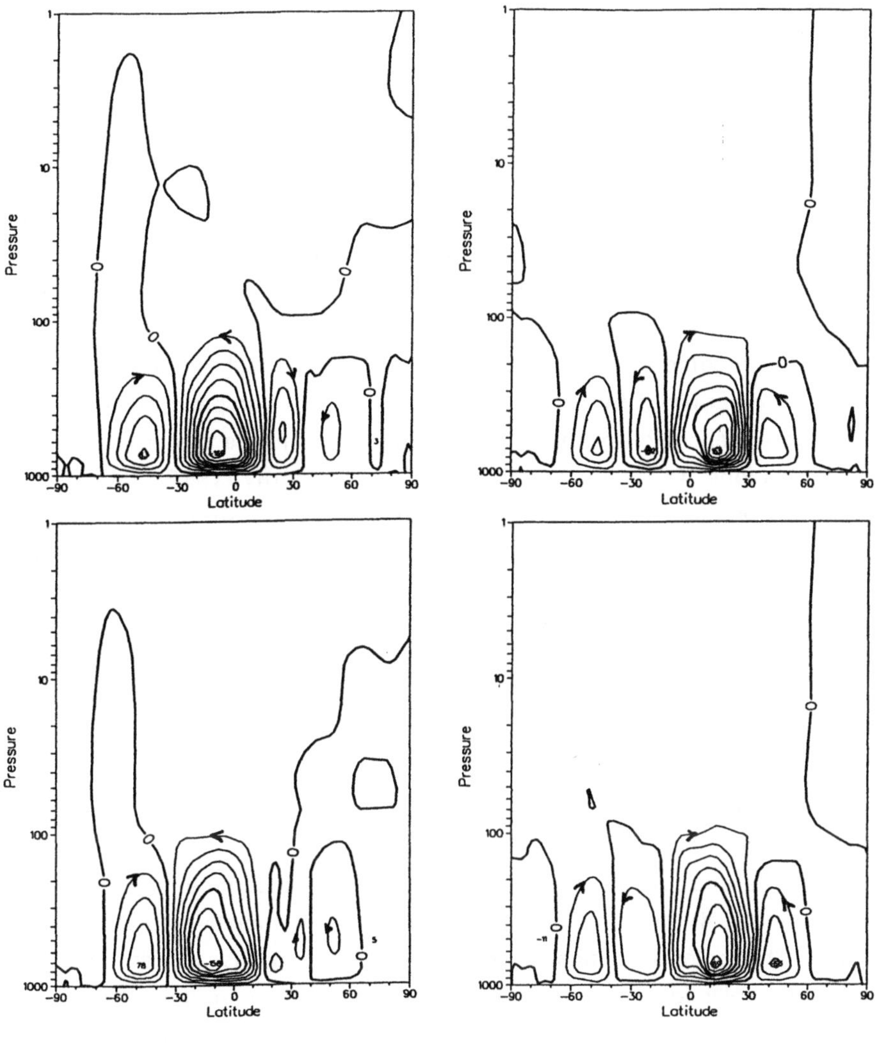

Fig. 3. Stream functions of the mean meridional circulation, respectively for boreal summer (JJA: left) and winter (DJF: right) in the decreased (SOM: top) and increased (SOP: bottom) solar constant experiments

increased pole-to-equator temperature gradient in the upper troposphere, and by the upward shift of the tropopause. This is also consistent with an increase of the height of the transport of westerly momentum by the poleward branch of the Hadley circulation which is shifted upwards in the winter hemisphere . The intensity of the Hadley circulation (Fig 3) is not significantly changed in the winter hemisphere, while it is strongly reduced in the summer hemisphere. The response of the Hadley cell in the winter hemisphere is weaker than what is obtained in the LMD simulation, and this may be due to the different repartition of tropospheric warming. As pointed out by Held (1993), the reduction of temperature gradient near the surface and its increase in the upper troposphere have opposing effects on the baroclinic activity in midlatitudes, so that the response is determined by the balance of these two competing effects, which can be very sensitive to model formulation and to details in the physical parameterizations. According to the analysis of Sadourny (1993), the meridional temperature gradient effect in upper layers can act to strengthen the Hadley cell, while the vertical temperature lapse rate effect and the warming effect tend to weaken the Hadley cell. We can thus interpret the small response of the winter Hadley cell in our simulation, as a near cancellation of the different effects, whereas for the summer Hadley cell, for which the solar forcing is maximum, the lapse rate and warming effects tend to dominate the response.

SURFACE RESPONSE

Surface temperature response maxima, found over the continents in the summer hemisphere (fig 4), are the direct response to the radiative forcing by solar variation. The polar amplification of the warming in the winter hemisphere, where direct solar forcing is minimal, can be only due to an indirect greenhouse effect of an increase in the water vapour content of the atmosphere. In spite of the different nature, and the different seasonally varying geographical signature of the solar forcing, this pattern is qualitatively similar to the response to CO_2 increase.

A consequence of the temperature increase in summer over the midlatitude continents is a decrease of soil moisture (fig 5), which is caused mainly by an increase of evaporation. The sea level pressure response (fig 6) indicates an increase of anticyclonic conditions over the continents, whereas cyclonic conditions are intensified in high latitudes over the Atlantic ocean. The

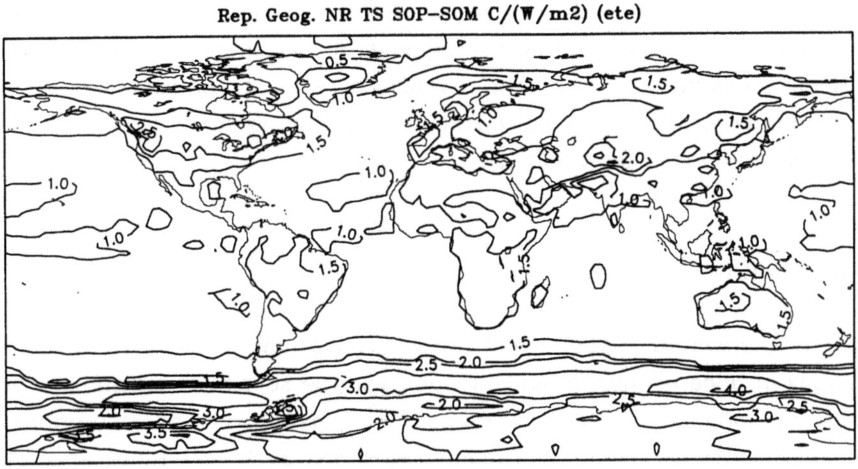

Fig. 4. Horizontal distribution of the normalized surface temperature response in K/(W/m2) for June-July-August

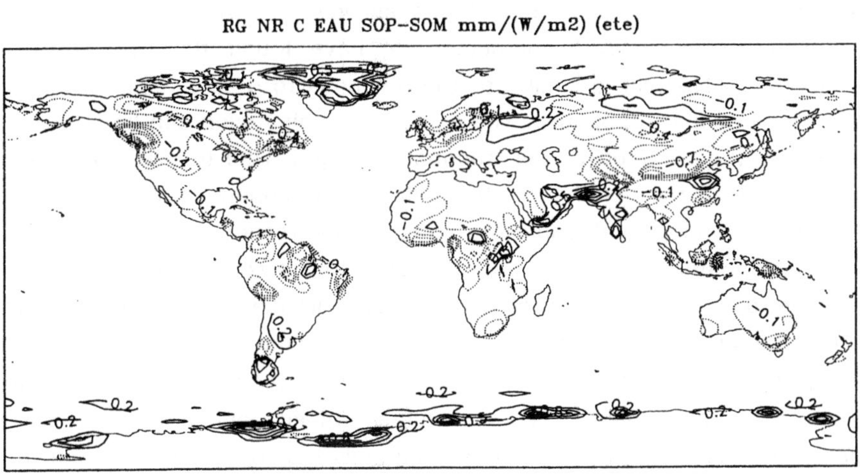

Fig. 5. same as figure 4 for the soil moisture in mm (/W/m2)

reduction of surface pressure over the circum Antarctic depression belt is an indication of increased baroclinic activity in this area. Consistent with the surface pressure response, the overall increase in precipitation is mainly concentrated in the equatorial regions, and at high latitudes over the oceans in the regions of increased cyclonicity.

CONCLUSION

Intercomparison of model simulations is important in order to identify the sources of uncertainties in model responses. The sensitivity experiment to solar constant variations with the Arpège model has confirmed the tropospheric response obtained in the LMD model. As found by Nesme-Ribes et al (1993), and Sadourny (1993) the tropospheric response to solar irradiance variations is qualitatively similar to the response due to changes in carbon dioxide concentration. However, due to the different horizontal resolution and physical parameterizations of the two models, some dynamical aspects of their response, which are more sensitive to details of the repartition of the tropospheric thermal response, are slightly different. Thanks to its higher vertical resolution in the stratosphere, the new simulation has been able to provide new details on the stratospheric response. In particular we have found that the stratospheric temperature response is similar to the response to greenhouse forcing only in the vicinity of the tropopause. The response in this region can be interpreted as caused by the variation of the tropopause height. Due to ozone absorption of the solar flux, the forcing in the middle stratosphere is different from the greenhouse forcing, and leads to a stratospheric response of opposite sign and smaller magnitude. It can be expected that the stratospheric response will be sensitive to the spectral variations of the solar irradiance, particularly in the UV part of the spectrum. However, the radiation scheme used in this version of the Arpège model does not provide enough spectral resolution to investigate this effect. New experiments using a more refined radiation scheme will be necessary in order to compute more accurately the stratospheric response.

RG NR PS SOP–SOM hPa/(W/m2) (ete)

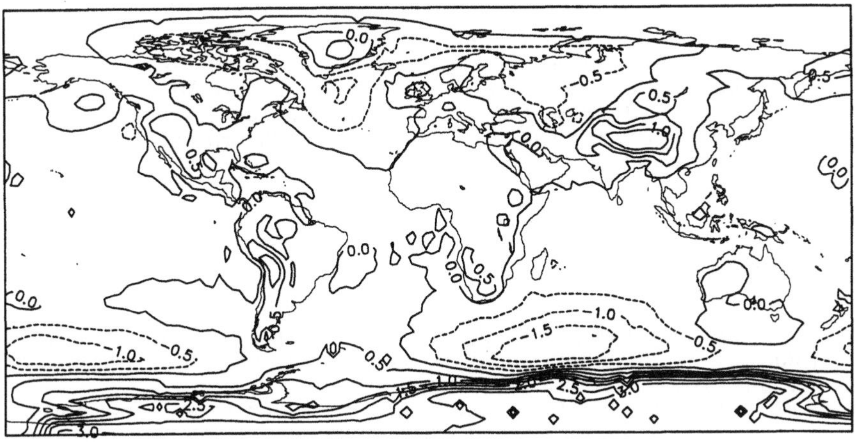

RG NR PS SOP–SOM hPa/(W/m2) (hiver)

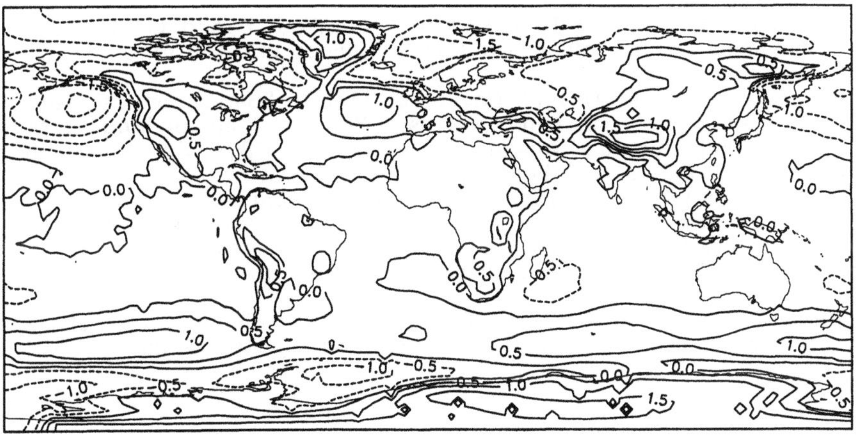

Fig. 6. same as figure 5 for the sea-level pressure in hPa (/W/m2) for summer (JJA: top) and winter (DJF: bottom)

ACKNOWLEDGEMENTS

We thank the team of Météo-France who have developed the Arpège model, in particular J.F. Geleyn and M. Déqué. We thank E Nesme-Ribes and R Sadourny for encouraging us to perform this experiment, M Forichon for preparing the SST files of the LMD experiment, and D Stephenson for useful suggestions. Financial support from PNEDC and Commission of the European Communities is gratefully acknowledged.

REFERENCES

Cariolle D and Déqué M (1986) Southern hemisphere medium-scale waves and total ozone disturbances in a spectral General Circulation Model. J Geophys Res 91 D10: 10 825-10 846

Déqué M and Royer JF (1991) GCM response of the mean zonal surface heat and water budgets to a global sea surface temperature anomaly. Dynamics of Atmospheres and Oceans 16: 133-146

Déqué M, Dreveton C, Braun A, Cariolle D (1993) Arpège/IFS the climate version n°O. Note de Travail du Groupe de Modélisation de Grande Echelle et Climat n°23, CNRM, 42 av. Coriolis 31057, Toulouse Cédex.

Held I (1993) Large-scale dynamics and global warming. Bull Amer Meteor Soc 74: 228-241

Nesme-Ribes E, Fereira EN, Sadourny R, Le Treut H and Li ZX (1993) Solar dynamics and its impact on solar irradiance and the terrestrial climate. J Geophys Res (in press)

Mahfouf JF, Cariolle D and Royer JF (1993) "Simulation de l'effet d'un doublement du gaz carbonique sur l'ozone stratosphérique". Note aux Comptes Rendus de l'Académie des Sciences, t.316, Série II: 61-68

Planton S, Déqué M and Bellevaux C (1991) Validation of an annual cycle simulation with a T42-L20 GCM. Climate Dynamics 5: 189-200

Royer JF, Planton S and Déqué M (1990) A sensitivity experiment to removal of Arctic Sea-Ice with the French spectral GCM. Climate Dynamics 5: 1-17

Sadourny R (1993) Climate sensitivity, Maunder minimum and the Little Ice Age. (This volume)

Wetherald RT and Manabe S (1975) The effects of changing the solar constant on the climate of a general circulation model. J Atmos Sci 32: 2044-2059

COMMENT ON THE OCEAN ALBEDO AND ABSORPTION

P. Braconnot, Laboratoire de Modélisation du Climat et de L'Environnement, CEA-DSM, Orme des Merisiers, CE-SACLAY, 91191 GIF SUR YVETTE CEDEX.

According to W. Levingston a change of the solar energy distribution (with total flux conserved) could have an impact on climate through changes of the ocean albedo or absorption. His idea is based on the water absorption curve given by Smith and Baker (1978), showing a sharp increase in absorption at wavelength longer than 500nm. In the following we comment on this hypothesis.

THE SURFACE ALBEDO OF THE OCEAN.

The sea surface albedo results from three components: the reflection of the direct and diffuse solar radiation, which define the reflectance of the sea surface, and the upward back scattering from below the sea surface. The reflectance, which is independant of the absorption and scattering properties of the water, can be calculated from Frenel's formula, and is a function of the zenith angle and the surface roughness (Ivanoff, 1977). The back scattering is largely determined by the optical properties of the first meters of the ocean column (see bellow) and hence has a spectral dependance. However, in the open ocean, integrated over the visible part of the spectrum, it represents only a small percentage (about 1%) of the impinging flux at the surface (Morel 1988). Therefore, if the sun changed its color the surface albedo of the ocean would remain the same.

THE ABSORPTION OF THE OCEAN.

Below the surface of the ocean the solar radiation is selectively absorbed with depth. Infrared radiation is absorbed within the first centimeters of the water column, whereas radiation in the visible part of the spectrum penetrates more deeply. The absorption of sea water is a combination of the absorption of pure water and of the absorption of the dissolved and particulate matter (mainly phytoplankton and their derivatives). For example (Morel, 1978), the blue waters of the Sargasso Sea (Fig. 1a) contain a low amount of plant matter, and the available energy concentrates with increasing depth in the 450-480nm spectral band, which corresponds to the minimum absorption band of water. In the green waters of the Mauritanian upwelling (Fig. 1b) where phytoplankton is more abundant the residual light is centered around 560 nm, wavelength of minimal absorption by the algal pigments. Thus, due to the varying optical properties of oceanic waters, there is a broad range of possible absorption around the world. However, most of the oceanic waters belong to the case I water of the classification of Jarlov (1976), for which there is a close relationship between the phytoplankton concentration and the optical properties (Morel 1988)

NATO ASI Series, Vol. I 25
The Solar Engine and Its Influence
on Terrestrial Atmosphere and Climate
Edited by Elizabeth Nesme-Ribes

A spectral shift of the impinging solar irradiance at the sea surface may induce a modification of the depth distribution of heating in the water column. Roughly speaking, more blue implies more heating at depth, and more red implies more heating at the surface. The climatic impact can be inferred from simulations with an oceanic general circulation model by changing the depth at which the solar radiation is allowed to penetrate. This has been done by Blanke and Delecluse (1993) in two simulations of the equatorial Atlantic. In the first simulation the solar radiation penetrates in the first meters of the ocean, so that about half of the visible radiation is absorbed within the first top decimeters of the ocean, while the remaining part is absorbed within the first ten meters. This is representative of the real conditions. In the second simulation the solar radiation penetrates only in the first layer of the ocean (10m). This case is representative of a color shift towards redder wavelength (with the same total energy), so that there is nearly no radiation into the visible part of the spectrum. Note that this change in the model is equivalent to a dramatic change in the solar spectrum which is unlikely to happen. Fig. 2 shows that, for such a change, the impact on the depth of the mixed layer (depth at which the temperature is that of the surface minus one degree) in the equatorial Atlantic is confined in the east of the basin in regions where the mixed layer is shallow, with a maximum of about 10 meters. Everywhere else, the depth of the mixed layer is driven by turbulent mixing. The impact is thus limited to locations where the penetration of the solar radiation determines the mixed layer depth. In these simulations, the ocean was forced by observed windstress, so that the feedbacks with the atmosphere that can amplify a small initial change of the sea surface temperature field are not incorporated. Hower, in a more realistic scenario this impact would be still smaller than the results presented here, and the optical properties of sea waters would modulate the ocean response.

In conclusion, it is not obvious that variations in depth distribution of ocean heating can link climate changes to shifts of the solar spectrum during the solar cycle.

Acknowledgments. I would like to thank Olivier Marchal and Sylvie Joussaume for helpful discussions.

REFERENCES

Blanke B. and P. Delecluse, 1993 : Variability of the Tropical Atlantic Ocean Simulated by a General Circulation Model with two different mixed-layer physics, J. Phys. oceanogr., 23, 1364-1388.

Ivanoff, A., 1977 : Oceanic absorption of solar energy. In modelling and prediction of the upper layer of the ocean, E. B. Kraus, ed. Pergamon press, Oxford, pp 47-71.

Jerlov N.G., 1976 : Marine optics, Elsevier oceanography series, Vol. 14, Elsevier, 231pp.

Morel, A., 1978 : Available, usable, and stored radiant energy in relation to marine photosynthesis, Deep Sea Res., 25, 673-688.

Morel, A., 1988 : Optical modeling of the upper ocean in relation to its biogenous matter content (case I Waters), J. Geophys. Res., 93, 10749-10768

Smith, R.C. and K.S. Baker, 1978 : Optical classification of natural waters, Limnol. and Oceanogr., 23, 260-267.

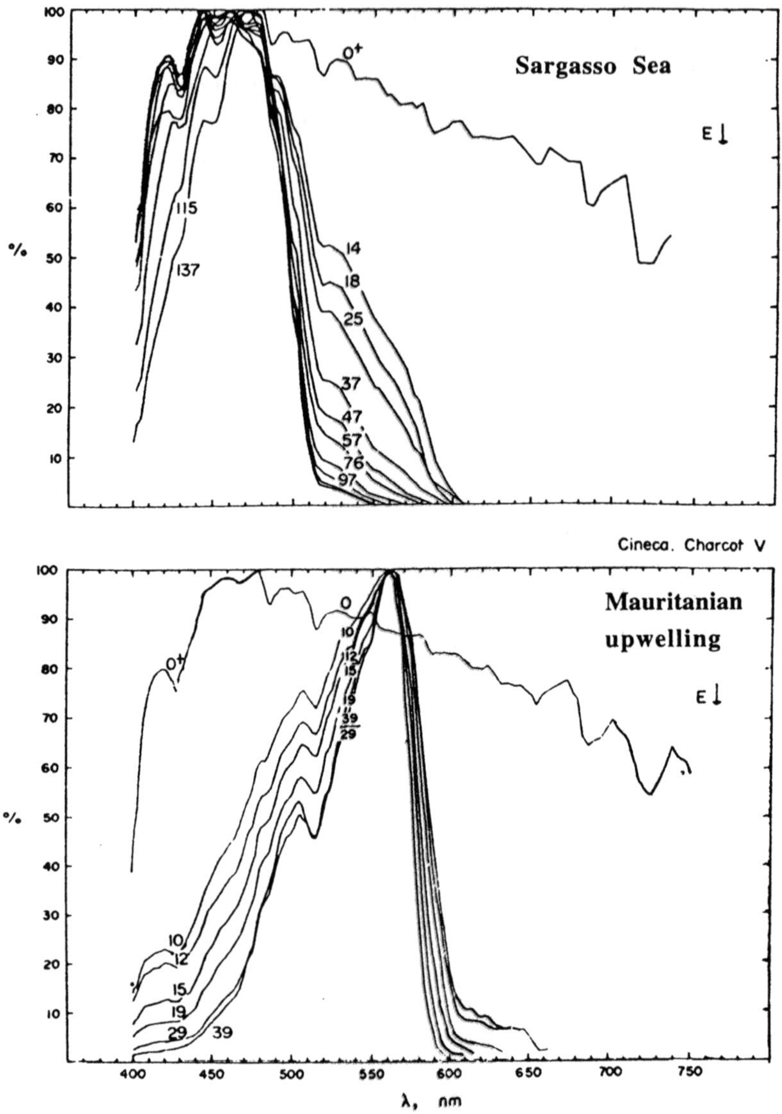

Fig. 1 : Spectral downwelling irradiance in two typical oceanic situations. For each depth the curve is normalized by its maximum. The notation O+ stands for the daylight spectrum obtained above the surface. Adapted from Morel (1978)

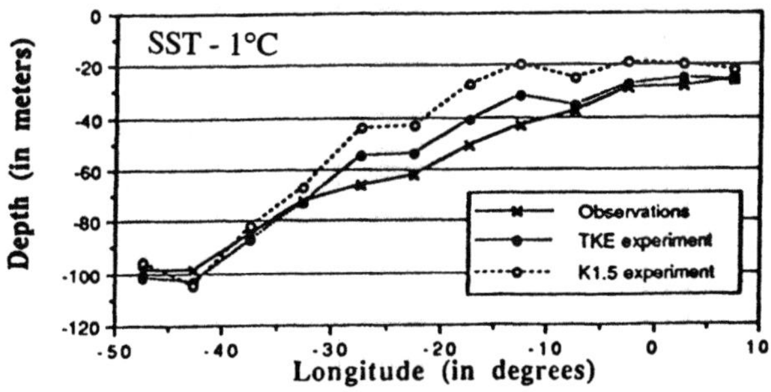

Fig. 2 : Equatorial section of the SST-1°C depth in August 1983 for the experiment with penetrative solar radiation (TKE) and the experiment with no penetration of solar radiation (K1.5). From Blanke and Delecluse (1993).

Panel discussion on CLIMATIC RESPONSE TO VARIATION IN THE SOLAR FORCING

Leader : L. Bengtsson
Max Planck Institute for Meteorology
Bundesstr. 55, D-20146 Hamburg 13

In order to explore the sensitivity of the climate to changes in solar output, comprehensive numerical models are needed that can reproduce the characteristic response of the climate system to changes in external forcing. Due to the inherent dynamic instability of the ocean and atmospheric circulation, a large number of internal processes do vary in the climate system, on timescales ranging from a few days to several hundred years. The fluctuations are not necessarily periodic, but often show up with distinct spectral amplitudes at a number of specific frequencies.

It is, for this reason, extremely difficult to allocate a specific observed variation in a given quantity in a given frequency domain to a specific external forcing, unless it describes a very substantial part of the total variance.

Consequently, not only is it difficult to detect the effect of a given solar cycle in the observations, but also to exclude the possibility that the observed variations are not rather due to internal fluctuations in the climatic system.

At first approximation, the response of climatic numerical models to changes in solar forcing is more or less equivalent to changes in greenhouse gas forcing, in spite of the fact that different parts of the energy spectrum are affected. With a planetary albedo of 30%, a 1% increase in the solar output, for example, would be equivalent to an increase of 2.3 Wm^{-2} in the mean shortwave radiative input to the earth-atmosphere system. Such a

NATO ASI Series, Vol. I 25
The Solar Engine and Its Influence
on Terrestrial Atmosphere and Climate
Edited by Elizabeth Nesme-Ribes

forcing of the climate system would by and large be equivalent in magnitude to the total increase in the greenhouse forcing since the beginning of industrialization.

The fact that the response is similar is partly an artefact of present models, which treat many aspects of the dynamics and physics of climate models in a simplified fashion. However, it is also probably related to the fact that certain fundamental dynamical aspects of the general circulation of the atmosphere are robust enough to suggest that the response of the climatic system to general perturbations in the radiative forcing may follow a chain of action as indicated below :

The dynamics of climatic warming are characterized by

(a) A decrease in the pole-to-equator temperature difference at the surface ->

(b) A decrease in the energy deficit of the winter hemisphere ->

(c) A decrease in the Hadley mass and energy fluxes ->

(d) An increase in the pole-to-equator moisture difference, due to the exponential dependancy on temperature ->

(e) An increase in the pole-to-equator temperature difference in the upper troposphere ->

(f) An increase in the upper troposphere jets ->

(g) An increase of latent heat fluxes compared to enthalpy fluxes in the energy transport, both in the tropics and in middle latitudes ->

(h) Increased precipitation in the winter hemisphere at middle and high latitudes, possible moderating the snow-albedo feedback and affecting the thermohaline circulation and oceanic northward heat flux in the Atlantic.

Alternately the opposite is likely to occur in the case of climatic cooling.

While tropospheric circulation appears to be rather insensitive to the forcing, the stratospheric response is different, as can be seen in the paper by Royer *et al.*, this volume.

While the response to an increase in the greenhouse gas concentration is a general cooling through the whole depth of the stratosphere, an increase in the solar output in an equivalent amount restricts the cooling to the lower part of the stratosphere, essentially reflecting a higher tropopause related to the warmer troposphere. However, such studies will require models with a proper stratospheric resolution, otherwise the difference cannot be properly detected.

It is apparent that the effect of solar output variations should be investigated out with more realistic general circulation models. Comparisons among models are also important, in order to identify common aspects of the response. Of particular importance is a satisfactory representation of the stratosphere including a realistic handling of the UV radiation, which is not yet properly considered in most numerical models.

Summary comments

E.N. Parker
University of Chicago, Enrico Fermi Institute and Depts. of Physics and Astronomy
Chicago, Illinois 60637 USA

I. Historical introduction

The human awareness of weather and climate goes as far back in time as the human race, although the deliberate recording of climate is a recent development. The recognition of solar activity has been a thing of only the last four centuries. It is safe to say that in neither case of climate or solar activity, do we possess an understanding of the whole phenomenon, and the effect of activity on the climate is a thing that has yet to be quantified.

The present concept of varying solar activity and a correlated climatic variation had a first glimmer in the century old recognition by Spörer (1887, 1889) and then Maunder (1890, 1894, 1922) that there were very few sunspots during the second half of the 17th century, together with the extraordinary fact that those few were essentially all in the southern hemisphere of the Sun. Clerke (1894) pointed out the near absence of a visible corona during total eclipses over the same period. When Douglas established his tree ring dating scheme in the 1920's there was an 11-year periodicity apparent in the annual growth rate of trees in some locations in southwestern United States. Maunder suggested, then, a connection between climate (rainfall in particular) and solar activity. But the data from different locations were not consistent with each other and no one paid much attention to the idea. It lay neglected except for a vague skepticism that the alleged absence of sunspots during the "Maunder" Minimum, 1645-1715 AD was nothing more than an absence of competent observational effort during that time. The ease with which sunspots can be seen with even the most meager telescope belies any such explanation, of course, but ideas that defy immediate explanation have always been subject to intellectual exile on the flimsiest of pretexts.

NATO ASI Series, Vol. I 25
The Solar Engine and Its Influence
on Terrestrial Atmosphere and Climate
Edited by Elizabeth Nesme-Ribes

Nothing much happened, then, for almost 50 years, until Jack Eddy became interested in the question of the reality of the Maunder Minimum. He began from a skeptical view and set about to decide through a thorough study of the historical records what the abundance of sunspots might have been during that period. There had been aurorae, of course, so the Sun was not entirely dead. Eddy went directly to the log books of the telescopic observatories of Western Europe, kept by such notable astronomers as Cassini. His investigation soon showed that the astronomers of the time were seriously interested in sunspots, baffled by their scarcity compared to early times (1610-1645 AD), and casually or actively engaged in their detection. In short, the Maunder Minimum was real (Eddy, 1976, 1977 a-c, 1980, 1983a, b) and more recent researchers (e.g. Ribes *et al.*, 1987, Ribes and Nesme-Ribes, 1993) have confirmed this fact, as well as filling in more scientific detail. The reader is referred to the volume edited by Sonett, Giampapa, and Matthews (1991) for a comprehensive review.

Eddy then turned to the newly constructed record of ^{14}C production. The age of samples of wood had by then been established back for 7000 years, through tree ring dating methods. Hence it was possible to determine the proportion $^{14}C/^{12}C$) of radioactive carbon in the wood at the time of growth. The ^{14}C is a consequence of an (n, p) nuclear reaction of neutrons (produced by the impact of cosmic ray protons on the atmosphere) with atmospheric ^{14}N. Therefore, the ^{14}C abundance in the sample is a direct measure of the mean cosmic ray intensity over the 40 or so years prior to the year of growth of the wood. The essential point is that the outward sweep of the solar wind pushes back the galactic cosmic rays to some degree, so that the more active the Sun, the lower the cosmic ray intensity here in the inner solar system. Therefore, the ^{14}C production rate is an inverse measure of the level of solar activity. The ^{14}C record showed a peak in production, some 10-15 percent higher than average, during the period designated as the Maunder Minimum. The record went on to show a similar peak in the 15th century, now called the Spörer Minimum. There was a localized dip of 10 percent or so in the 12th century, now called the Medieval Maximum and indicating a period of greatly enhanced solar activity. One can only speculate as to what was to be seen on the Sun during that time. We can assume that it was a quantitative change in the same sunspot, flare, coronal mass ejection, solar wind activity that we see today. Eddy found that over the 70 centuries of ^{14}C production data, 10 centuries showed enhanced production indicating a Maunder Minimum type period of greatly reduced activity. He also found 8 centuries of enhanced activity, like the Medieval Maximum in which the ^{14}C production was substantially suppressed. In summary, about one out of every 4 centuries is graced with an extended period of abnormal solar activity. Hyperactivity and hypoactivity are about equally likely, and there is no evident pattern of occurrence. (For an up to date analysis of the solar and climatological implications of the ^{14}C and ^{10}Be data see the contributions by J. Beer, *et al.*, and by M. Stuiver in these proceedings). Eddy, Gilman and Trotter (1976) studied old records of sunspots and concluded that the rotation of the Sun was substantially more strongly peaked toward the equator during the Maunder Minimum, although other studies (cf. Abarbanell and Wöhl, 1981, and Nesme-Ribes *et al.*, this volume). These puzzling facts together give some idea of the complexity of the dynamical gyrations of the convective zone of the Sun, from which it is believed all activity is driven.

With the fact of the random switching of the level of solar activity between three distinct levels (high, low, and normal) Eddy noticed that the climate in the northern temperature zone showed a remarkable correlation (cf. Eddy, 1977c). The 15th through the 17th century was notorious as the Little Ice Age, when the usually mild oceanic climates of the Netherlands, France, Germany and the British Isles, gave way to cold winters. The canals froze over and became thoroughfares for both pedestrians and freight wagons. The canals had not frozen significantly before, and have not since. Killing frosts in midsummer were noted by the occasional European traveler in the northern prairies of what is now the United States.

The advance and retreat of alpine glaciers indicated a similar direct relation between the temperatures in the northern temperate zone, and the 50 or 100 year mean level of solar activity, although the reliability of such proxy indicators of climate becomes increasingly fuzzy the farther back in time.

A direct relation between temperature in the troposphere and solar activity was hard to understand. It was well known, of course, that the solar UV varies by several percent, with the fractional change increasing rapidly with increasing frequency, so that the solar X-ray irradiance varies by factors of 20 or more with the level of activity on the Sun. The consequences can be seen in the terrestrial ionosphere and also in the upper stratosphere, where the temperature tracks the solar activity. We are more aware nowadays, with the problem of anthropogenic pollutants in the ionosphere, of the importance of the UV and X-ray part of the solar variation. But there was, and still is, no evident means by which the UV and X-rays influence the mean annual temperature in the northern temperate zone at the surface of Earth.

The polar ionosphere is often bombarded by energetic ($\sim 10^5$ - 10^7 eV) protons from solar flares and the intensity of the cosmic rays varies oppositely to the solar activity, as already noted. But the cosmic rays are about as intense as starlight, which by itself can hardly be expected to modify the climate. There is, however, one large effect of cosmic rays and that is to introduce a slight ionization of the otherwise wholly un-ionized and therefore nonconducting air in the troposphere. Thus, the electrical conductivity of the air is not identically zero, and the potential difference between the solid surface of the planet and the ionosphere is limited. For the fact is that the tropical and summertime thunderstorms charge the stratosphere positive relative to the ground, and the slight conductivity of the atmosphere provides a compensating drainage of electrical charge so that a quasi-steady state is the result. The potential difference varies inversely with the cosmic ray intensity and in a direct way with the level of solar activity. It is well known that the dynamics of thunderstorms, i.e. the structure of the vigorous thunderheads, involves both the condensation or freezing of water vapor onto ions and electrically charged microscopic particles which are influenced to some degree by the ambient electric field. So there is a serious question as to what degree the climate is affected through the solar modulation of the cosmic ray intensity (cf. Markson, 1983).

We all know the next step. The absolute radiometer was developed and placed in orbit on the NIMBUS 7 spacecraft in 1978, and on ACRIM in 1980. Subsequently these instruments showed the astonishing fact that, in addition to the daily variation of solar irradiance associated with the coming and going of sunspots and flares, there is a systematic variation in the mean level of irradiance throughout the 11-year activity cycle (Willson and Hudson, 1988). Specifically, the NIMBUS 7 instrument showed a peak irradiance in 1979 just before the conventional sunspot maximum, followed by a monotonic decline by about 2 parts in 10^3 into the activity minimum around 1985, followed by a slightly smaller rise into the maximum around 1990. The ACRIM radiometer came into operation in time to record the last half of the decline into the minimum, and to verify the accuracy and reliability of the NIMBUS 7 instrument (Hoyt, *et al.*, 1992 ; Donnelly, 1993).

Curiously this remarkable discovery provided but little excitement in the blase astronomical world of today. The cause of the variation of irradiance is not known although detailed quantitative studies by Foukal and Lean (1986) and others have shown that a large part of the variation can be attributed to the bright emission patches, called plages and faculae, that appear in the magnetic active regions. The rapid formation of a large dark sunspot may provide a decrease in the irradiance of several parts in 10^3. However the decrease is generally short lived and is more than compensated in the mean by the much larger magnetic areas ($B \gtrsim 50$ gauss) covered with plages. Unfortunately, for lack of the essential high resolution observations of plages, the cause of the plage, not to mention the cause of the sunspot, is not understood. The plage, like the X-ray corona and the expanding regions of the corona that become the solar wind, is blessed with a number of not implausible explanations, without observational substantiation.

The fact of the variation of the solar irradiance by 2 parts in 10^3 over a single activity cycle suggests an explanation for the climate variations discovered by Eddy so many years ago. The climatic variations ΔT were of the order of $\pm$ 1C in the mean annual temperature ($\sim$ 273° K) so that $\Delta T/T \sim 4 \times 10^{-3}$. The irradiance variation ΔI over the past activity cycle was half as much, $\Delta I/I \sim 2 \times 10^{-3}$ and one can only believe that ΔI may be somewhat larger between normal times and the extraordinary inactivity of the Maunder Minimum when the internal circulation in the convective zone was peculiar. Hence, it looks as though the irradiance variation of the Sun, associated with the deep and prolonged minimum in the activity, may explain the correlation with the mean annual temperature in the northern temperate zone.

Note, then, that the southern hemisphere of Earth is mostly deep ocean, as the northern hemisphere is mostly land, so that the mean temperature in the southern temperate zone is strongly controlled by the enormous thermal capacity of the oceans. Hence there should be relatively little, if any, effect of a century long change in solar irradiance and indeed none appears in the climatic records.

II. The terrestrial atmosphere

This brings us to consider the remarkably complex dynamical climate of the terrestrial oceans and atmosphere. The atmosphere is warmed from the top down by the solar X-rays and UV, which vary widely, and by the solar infrared red, some of which penetrates the ground and the ocean surface and finally the visible radiation that penetrates best of all. The atmosphere cools by infra red radiation from all levels. The essential point is that the character of the absorption of solar radiation throughout the many levels of the atmosphere is strongly affected by water vapor, carbon dioxide, methane, etc, providing an infra-red blanket of spotty and time varying structure. This brings in the water and carbon dioxide feedback mechanisms, with the remarkable fact, discovered in the study of atmospheric carbon dioxide in polar ice cores, that the carbon dioxide decreased sharply with the onset of the last ice age, 10^5 years ago, thereby reducing the greenhouse effect and enhancing (or perhaps causing ?) the ice age (Anklin *et al.*, 1993 ; Dansgaard, *et al.*, 1993). Clouds of water droplets increase the albedo and foster reduced atmospheric temperature, so that cloud cover plays a powerful role in regulating the climate. Yet we do not know quantitatively how cloud cover should respond to varying mean annual temperature, or to a small change in the solar irradiance. Snow cover on land, as well as sea ice, also plays a role, and is at the whim of ocean currents and global wind patterns. The ocean currents respond to differential heating and wind patterns but in no simple way. It is clear only that the terrestrial atmosphere can respond in striking ways to small stimuli (Broecker, 1989 ; White 1993).

It must be appreciated, then, that volcanic aerosols are a wild card introduced in unpredictable amounts at random times into the atmosphere, with the effect of reducing the penetration of visible light and infra red, and generally providing a cooling effect. Then we recall that the atmospheric circulation exhibits two distinct modes of circulation, sometimes remaining in a state of zonal (east-west) circulation for many months, with relatively little north-south mixing, and again taking up a state of enhanced north-south mixing of air and water vapor over latitude. Combine this with the fact that the thermal properties of the northern and southern hemispheres are quite different, with the southern hemisphere mostly ocean and the northern hemisphere mostly land. Finally, add in the anthropogenic pollutants, from the blanketing effect of the greenhouse gases, to the chlorine and fluorine compounds that destroy the UV blocking of the ozone layer in the ionosphere. The result is a complicated quasi-unstable system, so it si no surprise that there is no general agreement as to the quantitative effects of the pollutants. The atmosphere is too complicated a system to deduce the quantitative response to anything at this time. But it is a fact of direct measurement that the atmospheric carbon dioxide, methane, etc. are increasing rapidly at the present time, and the infra red absorption of each is easily calculated. So their total blanketing effect is well determined and is increasing.

The principal effect of the greenhouse pollutants is not wholly unlike the effect of an increase in solar irradiance, both heating the troposphere. Neither heating effect can be deduced quantitatively from global climate (numerical) models (GCM) for the simple reason that no GCM can at the present time treat more than a fraction of the principal atmospheric effects (cf. Manabe and Wetherald, 1980 ; Schneider and Thompson, 1981 ; Barnett and Schlesinger, 1987 ; Schlesinger, 1989 and references therein). On the other hand, for all their present limitations the GCM's universally show that increased heating, for whatever reason, has the effect of raising the mean annual temperature in some regions while causing the mean temperature to decline in other regions as a consequence of changes in global circulation and/or cloud cover. Similarly a decrease in world wide heating provides lower temperatures in some places and higher temperatures in others. Some regions become warmer while some become cooler as a direct consequence of the warming, at the same time that all the dynamical fluctuations of the atmosphere and oceans are carrying on their own complex variation. And then, of course, the volcanic dust is injected at random times, spreading gradually around the planet to provide a transient cool period. The essential point is that not only is the development of a physical understanding of the global climate a formidable task, but the interpretation of the observed climatic variations in association with varying solar irradiance and monotonically increasing greenhouse gases becomes extremely tricky. The reader may find it interesting to consult the volume edited by B.M. McCormack (1983) which summarizes the terrestrial atmosphere so far as its response to solar variations was understood ten years ago.

III. Interpretation

The question has been posed (and answered in both extremes) whether the observed increasing mean level of solar activity has been associated with a sufficient increase in the mean irradiance to account for some part of the observed global warming of the past 100 years. In short, how much of the warming might be attributed to greenhouse gas pollution of the atmosphere and how much might be attributed to the Sun itself, and how much might be from some other cause entirely, such as a shift in oceanic circulation or diminished cloud cover, or reduced volcanic dust etc ?

We cannot separate greenhouse warming and solar irradiance changes by studying and modeling the terrestrial atmosphere alone. But if we can obtain some idea of the magnitude and form of the time variation of the irradiance, we might be able to identify a climatic component with similar form as a likely solar contribution. Indeed Schlesinger proposed just such a separation in his presentation to this NATO Workshop, based on the monotonic character of the greenhouse warming as distinct from a broad pulse in solar irradiance.

We cannot afford to wait for 100 years of accumulating data to study the correlation of solar irradiance, solar activity (defined in several ways), and terrestrial climate. Fortunately there are other ways to speed up the process. Sallie Baliunas and co-workers have begun monitoring the irradiance of 15 solar-type stars (Radick, Lockwood, and Baliunas, 1990). These are stars with approximately the same mass, composition, and age as the Sun. The Sun is middle aged, about halfway through the expected 10^{10} years on the main sequence. It is well known that earlier in its life on the main sequence a star like the Sun is very much more active, presumably as a result of a high rate of spin. What is more, the observations of Baliunas, *et al.*, have shown that the activity of a young star has a different character, with the star becoming fainter, rather than brighter, as the activity increases. Evidently the proportion of starspots relative to plages is greatly increased. An update on this important work is providede by W. Soon , S.L. Baliunas and Q. Zhang in these proceedings (1994).

The 15 solar type stars are monitored by spectroscopic analysis of the chromospheric line structure to determine their level of activity. This is an ongoing part of the continuing program, begun 30 years ago by O.C. Wilson, to study the activity cycles of stars of many different types. The luminosities are determined by intercomparison of the solar type stars with the mean of the group serving as the absolute standard. The total error in the process is estimated at about 0.5 parts in 10^3.

It must be understood that they record only that part of the irradiance that penetrates the terrestrial atmosphere, essentially the visible and near infra red part of the spectrum. What is more their viewing of each star is at an arbitrary angle with respect to the spin axis of that star. Briefly, they find that 11 of the sample of 15 show fractional variation of irradiance (in association with the activity) up to about 2 x 10^{-3} over a time of 6 or 8 years, much like the Sun. Two show no significant change in irradiance or chromospheric line structure, suggesting that they are in a deep Maunder Minimum phase. Finally, two of the sample show a monotonic increase in irradiance of about 4 parts in 10^3 in just six years (Radick, Lockwood, and Baliunas, 1990 ; Baliunas and Jastrow, 1994). Are they rising into a Medieval Maximum ? Or coming out of a Maunder Minimum ? Or experiencing wild fluctuations in the middle of a Medieval Maximum phase ? Only time will tell. The essential point is that an expanded monitoring program of solar type stars (i.e. going to fainter magnitude stars) could in a decade or two tell us a lot about the possibilities for the Sun.

Now the study of the 15 solar type stars shows that there is a very approximate linear relation between the change in irradiance and the change in activity as determined by the structure of certain chromospheric line profiles. The Sun falls in the same general scatter of observational points. It must be appreciated that the observations extend through only a fraction of an activity cycle of the 15 solar-type stars, and only a little more than one cycle for the Sun. There is no reason to assert that the Sun, or any other star, follows precisely the same track through successive cycles, although it may very well be the case that they do. Nor can it be asserted that there is no long term drift of the irradiance over many cycles, although it may very well be that there is none. For it must be appreciated that we do not understand the basic cause of the variation of 2 parts in 10^3 through a single cycle. Nor do we know or understand much of the subsurface machinations associated with the activity cycle. The presentation by Nesme-Ribes, *et al.*, (1994) in these proceedings provides an idea of how present thinking is running. In our present state of ignorance of the physics it would be folly to assert that there is no other aspect to the phenomenon.

The nature of a plage and a sunspot, along with the source of energy that heats the X-ray corona, the expanding corona, and the solar wind, are not known. It is evident that solar physics has a difficult task before it. There are plenty of ideas. But there are no high resolution observations of the motions and structure of the magnetic fibrils at the visible surface of the Sun. The action lies at scales of the order of 100 km, which can be resolved only with a telescope with a diffraction limited primary mirror with a diameter of at least 1m placed above the atmosphere with a balloon or spacecraft to avoid the blurring of atmospheric seeing, which generally limits resolution to 500 km or more. The observations should have been made 20 years ago, because so many fundamental questions require the observational facts that such a high resolution study would supply. NASA has proved adroit at avoiding the study for all these years, and the interested reader is referred to Williams (1990) for an account of their efforts.

If we understood the physics of the plage we might be able to make scientific decisions as to other contributions to the changing solar irradiance, such as the obvious possibility that a significant part is a result of a slight change $\Delta T_{\odot}$ in the mean photospheric temperature over a large area of the solar surface. It would be remarkable if the extra energy coming to the surface of the Sun during years of enhanced activity should appear only in the plage and not at all as $\Delta T_{\odot}$. The radiation from the Sun is that of a black body with a temperature of about 5600 K, so far as the visible and infra red are concerned. So a fractional change in the irradiance of 1×10^{-3} is equivalent to a $\Delta T_{\odot}$ of about 1.4 K. Indeed, one would not expect such a $\Delta T_{\odot}$ to be spread uniformly over the solar disk, so we should scan the photosphere (outside plage regions) to determine if the photospheric temperature varies with latitude at any time during the 11-year activity cycle, and then determine how such a variation changes through an activity cycle. The observations would require great care, but if they detect a $\Delta T(\theta, \varphi)$ it would tell a great deal about what may be going on inside the solar furnace. Kuhn and Libbrecht (1991) have shown some preliminary results.This brings us back to the problem described earlier, whether there is a unique linear relation between solar irradiance and solar activity, or whether there is no such relation.There are many ways of defining the level of activity at any point in time. Sunspot area and number, coronal mass ejections, flares, UV and X-ray irradiance, 10 cm radio emission, and chromospheric line profiles each provide a possible representation of activity. The most difficult observational problem is to determine long term drift in irradiance, over periods of a decade or longer. Such drifts may, or may not, exist and may, or may not, cause major climate effects. This question is of primary concern when we address the 10^2 year climatic variations at Earth that, so far, are observed to track the mean level of activity of the Sun.

Now it would be extremely convenient if all variation of irradiance were restricted to a unique relation with the activity level. That is about the only thing that can be assumed at the moment, but it must be regarded as tentative, while detailed quantitative observations of the Sun and similar stars patiently pursue the problem over the coming decades.

Whatever the final result, it must be kept in mind that the unique coincidence of unusual terrestrial climate with the centuries of abnormal solar activity precludes the assertion that the irradiance does not change enough to have important climatic effects. So we are obliged to pursue the hypothesis that the irradiance varies sufficiently, whatever that amount may ultimately prove to be. Thus another essential feature is the pursuit of the global climate model, with particular attention to the sensitivity of mean annual temperature to small changes in solar irradiance. For if the terrestrial climate could be calibrated, then the climate record would serve as a measure of solar irradiance. The pursuit of the variation of solar irradiance impacts all aspects of the active Sun, all the way down through the convective zone to the strong shear at its base. The belief is that the activity of the Sun is entirely a product of the convection in the outer 2 x 105 km of the Sun ($R_{\odot}$ = 7 x 10^5 km). The convection is a heat engine that generates, evidently, the magnetic fields that we observe at the surface. The magnetic fields are strongly and continuallydeformed, and provide the observed activity as they dissipate the magnetic free energy continually created by their deformation. The generation of magnetic field is erratic, and cyclic if not truly periodic. The magnetic fields are not particularly symmetric about the equatorial plane, nor about the spin axis of the Sun. The fields at the surface are in an intense fibril state when they first appear ($\sim$ 500 gauss, Zwaan,

1985) and may be presumed to be in a fibril state throughout the convective zone. The mean field at the bottom of the convective zone must be at least 2 x 10^3 gauss, which stifles the weak eddy diffusivity ($\sim$ 10^{11} cm/sec) that is essential for the operation of the solar dynamo. So there is very little in contemporary kinematical models of the dynamo that is quantitative and reliable in detail (see discussion in Parker, 1993). Some models consider the back reaction of the magnetic field on the cyclonic convection, and there are several detailed studies of the suppression of turbulent diffusion, but so far as we are aware, the effects have not yet been combined in a comprehensive dynamo model.

What I am saying is that the physics of the solar activity, with which irradiance changes appear to be associated, is too vague to provide much guidance at the present time. Like the global climate models of the terrestrial atmosphere, the solar models provide a framework for discussion, and for understanding the contribution of individual effects. But in neither case do the models allow assertion of what is important, what is not important, what has happened in the past and what may happen in the future. Only precise observation over the coming decades, and careful analysis of the information already in hand, can establish the full nature of the variation of solar irradiance, the terrestrial climate, and what connection there exists between them.

The most basic and primitive of the essential observations is the direct measurement of the solar irradiance, with an absolute accuracy substantially better than one part in 10^3 over several decades. For it is the 50 year changes that have yet to be determined, in addition to the study of each successive activity cycle. Such a task can be achieved only if there are at least two independent absolte radiometers functioning at all times and preferably three, so that the instrumental drifts can be cross checked and corrected (cf. Hoyt, *et al.*, 1992). If there should be a significant gap at any time in the continuous observations, the ultimate scientific value of the accumulating data would be seriously reduced, starting all over again to look for long term drifts. And the presence of long term drift of the solar irradiance is of the highest priority. We must emphasize again that the long term precision monitoring of the irradiance of a large number of solar type stars is an essential complement to the detailed study of the Sun, for such a program is the most effective means to compress the long times that have to be endured as the Sun proceeds at its own slow pace through different phases of irradiance. What is more, it may be the only way to establish the existence of long term drift of the irradiance.

There is no reason whatever why there should not be complete coverage of the solar irradiance. The radiometers are now well developed, standardized, cheap, lightweight and require only a minimum bit rate. They can be mounted on communication satellites, or any other spacecraft that is expected to be in operation for several years at a time. A failure to provide the necessary continuing coverage is scientific negligence. Given the short memory and chronic myopia of bureaucratic officialdom, the continuing importance of the measurements of solar irradiance must be repeated over and over.

In addition to the basic measurement of solar irradiance it will be important to obtain some idea of the spectral distribution of the irradiance variations. How much is infra red as distinct from visible, and precisely how much of the total arises from the large fractional changes in UV and X-rays ? For it is the distribution over wavelength that tells us how and to what extent the irradiance interacts with each level of the terrestrial atmosphere. This brings us to the terrestrial atmosphere, whose complexity is well known, and described above in all too cursory a fashion. The EOS space program can be expected, over the next decade, to advance our knowledge of the specific effects of greenhouse warming and changes in solar irradiance. The similar climatic effects of the two means that the effort to understand the former automatically advances understanding of the latter, and vice versa. One of the foremost specific questions is to establish how much change in solar irradiance was associated with the cold climate of the Maunder Minimum, and to understand the atmospheric mechanics of how that changed irradiance came to provide the recorded climatic cooling. Zhang *et al.*, (1993) have made preliminary estimates that a reduction of irradiance by 3-5 parts in 10^3 may have been involved, while Baliunas, Soon, and Zhang (1994) estimate a decline of 3 parts in 10^3. None of these estimates would appear to conflict with the variation of 2 parts in 10^3 already observed in a single cycle of solar activity, nor is it clear how such changes in irradiance provide known climatic fluctuations that appear to be coordinated with the level of solar activity. Thus the success of EOS and the related studies of global climate models may, with some luck, provide a powerful tool for the quantitative study of the irradiance variations of the Sun, allowing the direct studies of solar irradiance to determine the role of the Sun in the present global warming.

The differences in the scientific tasks posed by the terrestrial climate on the one hand and the solar irradiance variation on the other are obvious. However, it is interesting to note that both tasks involve the global circulation of an atmosphere, the one heated largely from above and the other from below, respectively. Both exhibit more than one mode of global circulation, and both lie beyond the power of any present or conceivable computers to simulate in the detail necessary to provide reliable global climates. So, as has been said before, when the computations become too difficult, it is necessary to pursue in detail the physics of the many individual contributing effects, building up at least a qualitative understanding of the whole atmosphere before asking for a more essential numerical simulation of the climate.

It is interesting to contemplate that a relatively small solar contribution to changing mean temperature means a truly worrisome greenhouse gas warming, demanding an immediate and expensive technical response to avoid human catastrophe in the coming century. On the other hand, a large solar contribution means a less disastrous greenhouse contribution, accompanied by substantial uncontrolled climatic effects of solar variability, of the nature of those experienced in the past. Neither prospect is leasing to contemplate.

IV. Concluding remarks

In closing, it may be useful to comment in general terms on the execution of the task that lies before us all. .The facts are that the major century-long climatic variations of the past track the century-long variations in solar activity, over several ups and downs. The recently discovered variation of irradiance with the changing level of activity throughout a single cycle of solar activity, and a large part of a cycle in several solar-type stars, suggests that century long terrestrial climatic variation may have been wholly, or partly, in response to changes in solar irradiance. Unfortunately we are now confronted with the fact of accumulating greenhouse gases in the terrestrial atmosphere, as well as other anthropogenic pollutants that influence conditions in the ionosphere, etc. This accumulation is simultaneous with a general slow increase in the mean level of solar activity, with which one associates a likely increase in the mean level of solar irradiance and simultaneously a modest ($\sim$ 0.5 C) net warming of the terrestrial climate over the past 10^2 years. Some of the responses of individual members of the scientific community are not unexpected. On the one hand it has been asserted that a change in solar irradiance has had nothing to do with the observed warming since warming can be "explained" by the accumulation of greenhouse gases. On the other hand it has been declared that the entire warming follows directly from the estimated increase in solar irradiance, so that there is no basis for attributing any warming to the accumulation of greenhouse gases. Neither extreme is tenable, since the two effects are similar in nature, adding heat directly to the troposphere, and *both effects are non zero*. Our task is to quantify the phenomena so that some idea of the relative contributions can be achieved. As was emphasized in this Workshop, the economic issues involved in limiting greenhouse gases (if that should prove necessary) are immense, so that reliable quantitative information within the next decade is essential if at all possible.Our best policy as scientists is to pursue the physics of the terrestrial atmosphere and of solar variability and their interrelation as best we know how. It is imperative that we accept all the facts, as they develop, with a willingness to change our opinions as the unfolding facts require. It is imperative that our stated conclusions be worded precisely, claiming neither more nor less han the facts support, and admitting outright ignorance where there are no supporting facts. We are remiss if we are overly cautious just as much as if we are too extravagant. We are remiss if we do not exhibit enthusiasm for our scientific task. We can expect from time to time there will be gross distortion of the facts in political circles (including some portions of the scientific community). The complexity of the scientific task and the enormous implication for human welfare in future centuries, as well as the general public attitude toward long range scientific and technical-economic problems, provide fertile ground for political maneuvering. We must steer our own scientific course in spite of the political winds with which we shall be buffeted. In particular, we must not be bought off the scientific track by immediate funding in return for an unrealistic promise of a "quick fix."

References

Abarbanell, C., and Wöhl, H. (1981). Solar rotation velocity as determined from sunspots drawings of J., Hevelius, in the 17th century. Solar Phys., 70: 197-203

Anklin, M., Barnola, J.M., Beer, J., Bllunier, T., Chappellaz, J., Clausen, H.B., Dahl-Jensen, D., Dansgaard, W., De Angelis, M., Delmas, R.J., Duval, P., Fratta, M., Fuchs, A., Fuhrer, K., Gundestrup, N., Hammer, C., Iversen, P., Johnsen, S., Jouzel, J., Kipfstuhl, J., Legrand, M., Lorius, C., Maggi, V., Miller, H., Moore, J.C., Oeschger, H., Orombelli, G., Peel, D.A., Raisbeck, G., Raynaud, D., Schott-Hvidberg, C., Schwander, J., Shoji, H., Souchez, R., Stauffer, B., Steffensen, J.P., Stievenard, M., Sveinbjörnsdottir, A., Thorsteinsson, T., Wolff, E.W. (1993). Climate instability during the last interglacial period recorded in the GRIP ice core. Nature, 364: 203

Baliunas, S., and Jastrow, R. (1994). Evidence on the climate impact of solar variations. Harvard-Smithsonian Center for Astrophysics Preprint Series No. 3671

Baliunas, S., Soon, W., and Zhang, Q. (1994). On the parametrization and interpretation of stellar surface magnetic cycle periods. Science, (in press)

Barnett, T.P., and Schlesinger, M.E. (1987). Detecting changes in global climate induced by greenhouse gases. J. Geophys. Res., 92: 14772

Beer, J., Joos, F., Lukasczyk, Ch., Mende, W., Rodriguez, J., Siegenthaler, U., and Stellmacher, R.. ^{10}Be as an indicator of solar variability and climate. This volume

Broecker, W.S. (1989). Hansen and the greenhouse effect. Science, 245: 451

Clerke, A.M. (1894). A prolonged sunspot minimum. Knowledge, 17: 206

Dansgaard, W., Johnsen, S.J., Clausen, H.B., Dahl-Jensen, D., Gundestrup, N.S., Hammer, C.U., Hvidberg, C.S., Steffensen, J.P., Sveinbjörnsdottir, A.E.0., Jouzel, J., and Bond, G. (1993). Evidence for general instability of past climate from a 250 kyr ice-core record. Nature, 364: 218

Donnelly, R.F. (1993). Results of the solar electromagnetic radiation study for solar cycle 22 (SOLER 22, STEP PROJECT 1.2). Proc. 1992 STEP Symp, Adv. Space Res.

Eddy, J.A. (1976). The Maunder Minimum. Science, 192: 1189

Eddy, J.A. (1977a). Climate and the changing sun. Climatic Change, 1: 173

Eddy, J.A. (1977b). The case of the missing sunspots. Scient. Amer., 236 (May): 80

Eddy, J.A. (1977c). Historical evidence for the existence of the solar cycle. In : The Solar Output and its Variation (Colorado Associated University Press, Boulder). ed. OR White, 51

Eddy, J.A. (1980). The historical record of solar activity. In : The Ancient Sun (New York, Pergamon), ed. RO Pepin, J.A. Eddy and R.B. Merrill, 119

Eddy, J.A. (1983a). The Maunder Minimum, A reappraisal. Solar Phys., 89: 195

Eddy, J.A. (1983b). An historical review of solar variability, weather, and climate. In weather and climate responses to solar variations. (Boulder, Colorado Associated University Press), ed. B.M. McCormack 1

Eddy, J.A., Gilman, P.A., and Trotter, D.E. (1976). Solar activity during the Maunder minimum. Solar Physics., 46: 3-14

Foukal, P., and Lean, J. (1986). The influence of faculae on total solar irradiance and luminosity. Astrophys. J., 302: 826

Hoyt, D.V., Kyle, H.L., Hickey, J.R., and Maschhoff, R.N. (1992). The Nimbus 7 solar total irradiance, A new algorithm and its derivation. J. Geophys. Res., 97: 51

Kuhn, J.R., and Libbrecht, K.G. (1991). Nonfacular solar luminosity variations. Astrophys. J. Letters, 381: L35

Manabe, S., and Wetherald, R.T. (1980). The effects of doubling the CO_2 concentration on the climate of a general circulation model. J. Atmos. Sci. , 32: 3

Markson, R. (1983). Solar modulation of fair-weather and thunderstorm activity and a proposed program to test an atmospheric electrical sun-weather mechanism. In weather and climate responses to solar variations (Boulder, Colorado Associated University Press), ed. B.M. McCormack, 323

Maunder, E.W. (1890). Professor Spörer's researches on sunspots. Mon. Not. Roy. Astron. Soc., 50: 251

Maunder, E.W. (1894). A prolonged sunspot minimum. Knowledge, 17: 173

Maunder, E.W. (1922). The prolonged sunspot minimum 1645-1715. J. Brit. Astr. Assoc., 32: 140

McCormac, B.M. (1983). Weather and climate response to solar variations (Boulder, Colorado Associated University Press)

Nesme-Ribes, E., Sokoloff, D., Ribes, J.C., and Kremliovsky, M. (1994). The Maunder minimum and the solar dynamo. This volume

Parker, E.N. (1993). A solar dynamo surface wave at the interface between convection and nonuniform rotation. Astrophys. J., 408: 707

Radick, R.R., Lockwood, G.W., and Baliunas, S.L. (1990). Stellar activity and brightness variations. A glimpse at the Sun's history. Science, 24: 39

Ribes, E., Ribes, J.C., and Barthalot, R. (1987). Evidence for a larger Sun with a slower rotation during the seventeenth century. Nature, 326: 52-54

Ribes, J.C., and Nesme-Ribes, E. (1993). The solar sunspot cycle in the Maunder Minimum AD 1645 to AD 1715. Astron. Astrophys., 276: 549-563

Schlesinger ME (1989). Letter, Hansen and the greenhouse effect. Science, 245: 451

Schneider, S.H., and Thompson, S.L. (1981). Atmospheric CO_2 and climate : Importance of transient response. J. Geophys. Res., 86: 3135

Sonett, C.P., Giampapa, M.S., and Matthews, M.S. (1991). The Sun in time (Tucson, University of Arizona Press)

Soon, W.H., Baliunas, S.L., and Zhang, Q.. A technique for estimating long-term variations of solar total irradiance : Preliminary estimates based on observations of the Sun and solar-type stars. This volume

Spörer, F.W.G. (1887). Astr. Ges. Vjschr. Leipzig, 22: 323

Spörer, G.W.G. (1889). Bull. Astr., 6: 60

Stuiver, M.. Atmospheric ^{14}C as a proxy of solar and climatic change. This volume

White, J.W.C. (1993). Don't touch that dial. Nature, 364: 186

Williams, D.J. (1993). Nat. Acad. Sci./Nat. Res. Council Report, Space Physics Paradox, Appendix 2

Willson, R.C., and Hudson, H.S. (1988). Solar luminosity variations in solar cycle 21. Nature, 332: 810

Zhang, Q., Soon, W.H., Baliunas, S.L., Lockwood, C.W., Skiff, B.A., and Radick, R.R. (1993). The brightness variation of the sun in past centuries. Nature, (in press)

Zwaan, C. (1985). The emergence of magnetic flux. Solar. Phys., 100: 397

Authors index

The ASI Series Books Published as a Result of Activities of the Special Programme on Global Environmental Change

This book contains the proceedings of a NATO Advanced Research Workshop held within the activities of the NATO Special Programme on Global Environmental Change, which started in 1991 under the auspices of the NATO Science Committee.

The volumes published as a result of the activities of the Special Programme are:

Vol. 1: **Global Environmental Change.** Edited by R. W. Corell and P. A. Anderson. 1991.
Vol. 2: **The Last Deglaciation: Absolute and Radiocarbon Chronologies.** Edited by E. Bard and W. S. Broecker. 1992.
Vol. 3: **Start of a Glacial.** Edited by G. J. Kukla and E. Went. 1992.
Vol. 4: **Interactions of C, N, P and S Biogeochemical Cycles and Global Change.** Edited by R. Wollast, F. T. Mackenzie and L. Chou. 1993.
Vol. 5: **Energy and Water Cycles in the Climate System.** Edited by E. Raschke and D. Jacob. 1993.
Vol. 6: **Prediction of Interannual Climate Variations.** Edited by J. Shukla. 1993.
Vol. 7: **The Tropospheric Chemistry of Ozone in the Polar Regions.** Edited by H. Niki and K. H. Becker. 1993.
Vol. 8: **The Role of the Stratosphere in Global Change.** Edited by M.-L. Chanin. 1993.
Vol. 9: **High Spectral Resolution Infrared Remote Sensing for Earth's Weather and Climate Studies.** Edited by A. Chedin, M.T. Chahine and N.A. Scott. 1993.
Vol. 10: **Towards a Model of Ocean Biogeochemical Processes.** Edited by G. T. Evans and M.J.R. Fasham. 1993.
Vol. 11: **Modelling Oceanic Climate Interactions.** Edited by J. Willebrand and D.L.T. Anderson. 1993.
Vol. 12: **Ice in the Climate System.** Edited by W. Richard Peltier. 1993.
Vol. 13: **Atmospheric Methane: Sources, Sinks, and Role in Global Change.** Edited by M. A. K. Khalil. 1993.
Vol. 14: **The Role of Regional Organizations in the Context of Climate Change.** Edited by M. H. Glantz. 1993.
Vol. 15: **The Global Carbon Cycle.** Edited by M. Heimann. 1993.
Vol. 16: **Interacting Stresses on Plants in a Changing Climate.** Edited by M. B. Jackson and C. R. Black. 1993.
Vol. 17: **Carbon Cycling in the Glacial Ocean: Constraints on the Ocean's Role in Global Change.** Edited by R. Zahn, T. F. Pedersen, M. A. Kaminski and L. Labeyrie. 1994.

Vol. 18: **Stratospheric Ozone Depletion/UV-B Radiation in the Biosphere.** Edited by R. H. Biggs and M. E. B. Joyner. 1994.

Vol. 19: **Data Assimilation: Tools for Modelling the Ocean in a Global Change Perspective.** Edited by P. O. Brasseur and J. Nihoul. 1994.

Vol. 20: **Biodiversity, Temperate Ecosystems, and Global Change.** Edited by T. J. B. Boyle and C. E. B. Boyle. 1994.

Vol. 21: **Low-Temperature Chemistry of the Atmosphere.** Edited by G. K. Moortgat, A. J. Barnes, G. Le Bras and J. R. Sodeau. 1994.

Vol. 22: **Long-Term Climatic Variations – Data and Modelling.** Edited by J.-C. Duplessy and M.-T. Spyridakis. 1994.

Vol. 23: **Soil Responses to Climate Change.** Edited by M. D. A. Rounsevell and P. J. Loveland. 1994.

Vol. 24: **Remote Sensing and Global Climate Change.** Edited by R. A. Vaughan and A. P. Cracknell. 1994.

Vol. 25: **The Solar Engine and Its Influence on Terrestrial Atmosphere and Climate.** Edited by E. Nesme-Ribes. 1994.

Vol. 26: **Global Precipitations and Climate Change.** Edited by M. Desbois and F. Désalmand. 1994.

Vol. 27: **Cenozoic Plants and Climates of the Arctic.** Edited by M. C. Boulter and H. C. Fisher. 1994.

MIX
Papier aus verantwortungsvollen Quellen
Paper from responsible sources
FSC® C105338

If you have any concerns about our products,
you can contact us on
ProductSafety@springernature.com

In case Publisher is established outside the EU,
the EU authorized representative is:
Springer Nature Customer Service Center GmbH
Europaplatz 3, 69115 Heidelberg, Germany

Printed by Libri Plureos GmbH
in Hamburg, Germany